DECONTAMINATION OF HEAVY METALS

Processes, Mechanisms, and Applications

ADVANCES IN INDUSTRIAL AND HAZARDOUS WASTES TREATMENT SERIES

Advances in Hazardous Industrial Waste Treatment (2009)
edited by Lawrence K. Wang, Nazih K. Shammas, and Yung-Tse Hung

Waste Treatment in the Metal Manufacturing, Forming, Coating, and Finishing Industries (2009)
edited by Lawrence K. Wang, Nazih K. Shammas, and Yung-Tse Hung

Heavy Metals in the Environment (2009)
edited by Lawrence K. Wang, J. Paul Chen, Nazih K. Shammas, and Yung-Tse Hung

Handbook of Advanced Industrial and Hazardous Wastes Treatment (2010)
edited by Lawrence K. Wang, Yung-Tse Hung, and Nazih K. Shammas

Decontamination of Heavy Metals: Processes, Mechanisms, and Applications (2012)
Jiaping Paul Chen

RELATED TITLES

Handbook of Industrial and Hazardous Wastes Treatment (2004)
edited by Lawrence K. Wang, Yung-Tse Hung, Howard H. Lo, and Constantine Yapijakis

Waste Treatment in the Food Processing Industry (2006)
edited by Lawrence K. Wang, Yung-Tse Hung, Howard H. Lo, and Constantine Yapijakis

Waste Treatment in the Process Industries (2006)
edited by Lawrence K. Wang, Yung-Tse Hung, Howard H. Lo, and Constantine Yapijakis

Hazardous Industrial Waste Treatment (2007)
edited by Lawrence K. Wang, Yung-Tse Hung, Howard H. Lo, and Constantine Yapijakis

DECONTAMINATION OF HEAVY METALS

Processes, Mechanisms, and Applications

JIAPING PAUL CHEN

CRC Press
Taylor & Francis Group
Boca Raton London New York

CRC Press is an imprint of the
Taylor & Francis Group, an **informa** business

CRC Press
Taylor & Francis Group
6000 Broken Sound Parkway NW, Suite 300
Boca Raton, FL 33487-2742

ISBN-13: 9781439816677 (Hbk)

Visit the Taylor & Francis Web site at
http://www.taylorandfrancis.com

and the CRC Press Web site at
http://www.crcpress.com

This book is dedicated to my father, Dr. Bing-Xie Chen,

and mother, Dr. Shao-Zhi Wu

Contents

Preface

Heavy metals such as zinc and copper could be important constituents of most living animals, plants, and microorganisms, and many nonliving substances in the environment. Some of them are essential for the growth of biological and microbiological lives. Their absence could limit the growth of small microorganisms to large plants or animals. However, the presence of any of these heavy metals in excessive quantities is harmful to human beings and can interfere with many beneficial uses of the environment due to their toxicity and mobility. In addition, heavy metals are nonbiodegradable, which makes it more difficult to decontaminate. Therefore, it is desirable to measure, understand, and control the heavy metal concentrations in the environment.

In a deliberate effort to complement other industrial waste treatment and hazardous waste management texts published by Taylor & Francis and CRC Press, *Decontamination of Heavy Metals: Processes, Mechanisms, and Applications* covers the important results in the research of metals in environment. In Chapter 1, the general overview and recent research trends of heavy metals are covered. The concepts and technologies of pollution prevention are covered in Chapter 2. Precipitation is the most commonly used technology for metal decontamination. The chemistry in the process, the calculation methods, and the treatment systems are discussed in Chapter 3. Precipitation–crystallization is an emerging technology that combines precipitation and crystallization technologies. It can effectively remove a series of heavy metal ions. Chapter 4 provides a systematic description of the technology. Chapter 5 addresses oxidation and reduction reactions in natural and treatment systems. Chapter 6 presents electrochemical technologies for heavy metal decontamination. Sorption technology has been researched intensively in the last 10 years. Chapter 7 provides a detailed description of several important types of sorbents for metal decontamination, which may provide guidelines to the readers in the selection and application of sorbents. Chapter 8 presents a series of mathematical models for metal removal and recovery. Both equilibrium and kinetic models are presented.

A successful modern heavy metal control program for a particular industry will include not only traditional water pollution control but also soil conservation, site remediation, groundwater protection, public health management, solid waste disposal, and combined industrial-municipal heavy metal waste management. In fact, it should be a total environmental control program. It is the intention that the important technical and economical information is provided in this book on the development of the most feasible total heavy metal control program that can benefit both industry and local municipalities. Frequently, the most economically feasible methodology is a combined industrial-municipal heavy metal management.

This book can be used as a textbook for graduate students as well as a reference book for environmental professionals. It features the major hazardous heavy metals in water, land, and facilities that have significant effects on the public health and the environment. Academic members, students, researchers, managers, engineers, and scientists in environmental, civil, chemical, sanitary, mechanical, and public health engineering and science will find valuable educational materials here.

This book emphasizes in-depth presentation of environmental pollution sources, waste characteristics, control technologies, management strategies, facility innovations, process alternatives, costs, case histories, effluent standards, and future trends for industries. The extensive bibliographies for each heavy metal or metal-related industrial waste treatment or practice should be invaluable to the readers who need to trace, follow, duplicate, or improve on a specific industrial hazardous waste treatment practice.

I would like to thank many people who have encouraged and supported me during this book was prepared. My father, Dr B. X. Chen, my mother, Dr S. Z. Wu, my wife, Mdm J. C. P. Lee, my sister, Dr J. Y. Chen, and my loving kids, E. E. B. Chen, O. E. Q., L. E. Y., Chen and V. E. K. Chen have continuously encouraged me in this book project. My research staff, Dr Y. M. Zheng, Mr S. W. Zou, and my students, Mr Y. Ma, Dr K.G. Nadeeshani Nanayakkara, Miss D. D. Zhao and Miss L. Yu have assisted me on the editing work. It was Professor Sotira Yiacoumi of Georgia Tech who introduced the topic of metal contamination. Without all these people, the completion of the book would be impossible.

J. Paul Chen,
Singapore

Author

Dr Jiaping Paul Chen is an associate professor of environmental engineering at the National University of Singapore. His research interests are physicochemical treatment of water and wastewater and modeling. He has published more than 100 journal papers and book chapters with citations of over 2300 and H-index of 28. He holds four patents in the areas of adsorption technologies and ballast water management systems. He has received several honors and awards, including Guest Professor of the Hua Zhong University of Science and Technology, and Shandong University of China, and Distinguished Overseas Chinese Young Scholar of National Natural Science Foundation of China. He has been recognized as an author of highly cited papers (Chemistry and Engineering) of ISI Web of Knowledge. Professor Chen received his bachelor degree in chemical engineering from Hua Chiao University of China, master degree in environmental engineering from the Tsinghua University of China, and PhD degree in environmental engineering from Georgia Institute of Technology in Atlanta, Georgia, U.S.A.

1

Occurrence and Importance of Heavy Metal Contamination

1.1 Introduction

Meteoric iron was perhaps the first metal that our forefathers encountered. Obviously, the hardness of metal became important to them for their survival. Metals have been used by human beings for more than 10,000 years. Copper products such as pins and awls, used around 7,000 to 9,000 years ago, were found in Ali Kosh of Western Iran and Cayönü Tepesi near Ergani in Anatolia. Beads of malachite and native copper were found in the Neolithic site of Cayonü Tepesi in southeastern Turkey, which dated back to 7250–6750 BC.[1] Metalware, weapons, and farm tools have been needed for more than 5,000 years. Increasing demands have driven the exploration, extraction, and production of metals. In modern society, metal has found much wider applications, ranging from the tiny cellular phone chip to large spacecraft. It plays an essential role in our life.

Heavy metal is a general collective term applied to the group of metals and metalloids with an atomic density greater than 5 g/cm^3. Another definition of heavy metal is that its molecular weight is above 40. Any metal with molecular weight less than 40 is called a *light metal.*

Although both definitions are only loosely defined terms for heavy metals, they are widely recognized and usually applied to such elements as antimony, bismuth, cadmium, cerium, chromium, cobalt, copper, gallium, gold, iron, lead, manganese, mercury, nickel, platinum, silver, tellurium, thallium, tin, uranium, vanadium, and zinc. Most heavy metals (e.g., Cd, Cr, Cu, Hg, Ni, Pb, and Zn) are commonly associated with pollution and toxicity problems, particularly when they are present in soluble form. Sodium, potassium, calcium, and magnesium are light metals.

Common characteristics of heavy metals are

1. A metal is a good conductor of electricity.
2. The electric resistance of a metal is directly proportional to the absolute temperature.

3. A metal has high thermal conductivity.
4. A metal has high density, which is above 5 g/cm^3.
5. A metal has malleability and ductility, without which it cannot be drawn into sheets and wires.
6. A metal normally has oxidation numbers or valences such as 0, +1, +2, +3, and +6.
7. A zero-valent metal is present in solid form. Mercury is the only exception as it is in the form of a liquid.
8. Metals may exist in nature as metal oxides, metal carbonate, or metal sulfate.
9. Metals may be present in nature as soluble species that are complexed with various anionic substances called ligands. An example is the formation of a series of copper hydroxides and copper chlorides when copper chloride is in a water solution ($Cu(OH)^+$, $Cu(OH)_2$, $CuCl^+$, and $CuCl_2$).

Nonmetals normally are nonconductors of electricity. Some have characteristics that are intermediate between metal and nonmetal. They are called half-metals, or metalloids. They have low electric conductivity. However, the conductivity may increase as temperature rises. This group of elements includes boron, silicon, germanium, arsenic, and tellurium. They have economic and environmental significance. Silicon plays an important role in the electronics industry. On the other hand, arsenic is one of the most toxic metals to human beings.

Note that conductive polymers and graphite are electrically conductive and solid. However, they are not considered metals.

The need for sound management of metal waste has greatly increased in the past 10 years. The main driving forces are as follows:

1. The demand for metals has increased.
2. Numerous studies have increasingly demonstrated that many serious health problems are associated with the presence of heavy metals in drinking water, food, and air.
3. The demand for a better quality of life has been increasing.

Most organic pollutants are susceptible to biological degradation for detoxification. However, heavy metals cannot be degraded into harmless end products. They tend to accumulate in living organisms.

Nonbiodegradable toxic heavy metal contamination has become one of the most important environmental problems in the world, especially in many developing countries such as China and India. These contaminants often stream from many traditional industries such as electronics, machines, metal plating, mining, painting, tannery, jewelry, agriculture (which intensively uses fertilizer and biocides), oil extraction and production, and car manufacturing. High-tech industries may also generate heavy metal waste streams.

In addition to the industrial sources, some heavy metals may occur naturally in rock-forming and ore minerals. As a result, there is a range of normal background concentrations of these elements in soils, sediments, waters, and even living organisms. Therefore, the presence of such heavy metals may constitute environmental pollution.

Many aquatic environments have received heavy metals that have concentrations exceeding water quality criteria (from agencies such as US EPA and WHO) that are designed to protect the environment, animals, and humans. Metal contamination problems are often exacerbated because the metals have the tendency to be transported with sediments, are persistent in the environment due to nonbiodegradation nature, and can bioaccumulate in the food chain. A few of the oldest cases of metal pollution in the world may be even traced back to copper, mercury, and lead mining, smelting, and utilization by ancient civilizations, such as the Roman and the Phoenician.

Better understanding of heavy metal toxicity, frequently occurring environmental accidents from anthropogenic sources as well as natural geochemical processes, and public awareness of environmental pollution are the driving forces for enactment and enforcement of stricter legislation on the discharge of the toxic heavy metal contaminants. Consequently, development of cost-effective management systems and treatment technologies becomes paramount.

1.2 Economy and Metals

The main advantages of heavy metals are their hardness, malleability and thermal ductility, and electrical conductivity. China and India have become economically important, and the demand for heavy metals has significantly increased over the past 10 years. The construction, electronics, and chemical industries are the main consumers of heavy metals. In the past five years, the prices and production of heavy metals such as nickel, copper, iron, and aluminum have tremendously increased. As the largest mine export country, Australia has enjoyed rapid economic growth and appreciation of its currency. Figure 1.1 demonstrates the rapid rise in the gold price, which is an important indicator of the prices of heavy metals.

China, has become the largest importer of metal ores and products. The main reasons behind this are high demand from the domestic construction industry and manufacturing industries that produce goods for export. For example, China imported 60,000 tons of steel rods valued at 76,722 million US dollars in May 2009. The imports in January–May 2009 reached 260,000 tons valued at 373.016 million US dollars according to www.chinamining.org. The tremendously high demand for raw materials such as metal ores can be evidenced by the size of ports in China. It is reported that the Chinese ports

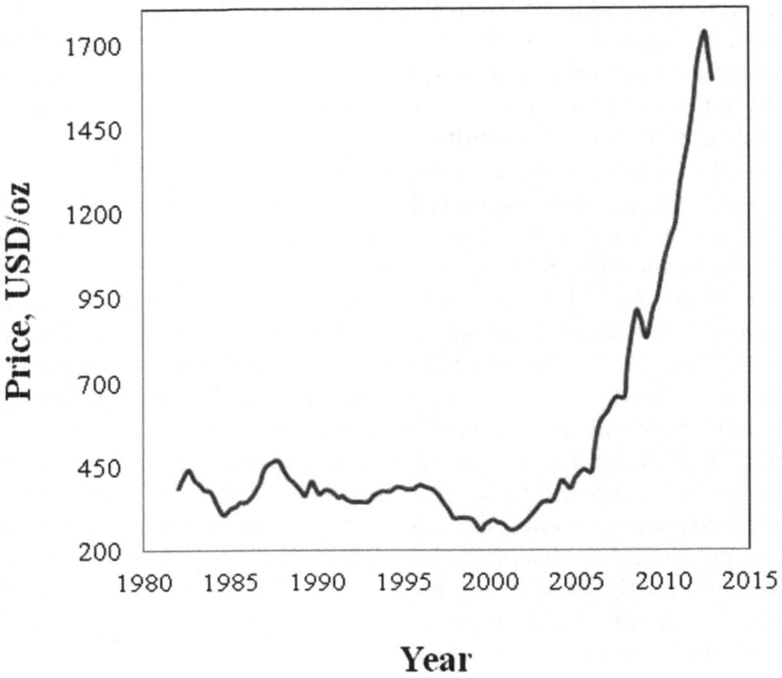

FIGURE 1.1
The price of gold from January 1974 to May 2011.

of Shanghai, Hong Kong, Shenzhen, Guangzhou, Ningbo, and Qingdao are among the top 10 largest ports in the world.

Iron is used for production of stainless steel. Steel is heavily used in the construction and automobile industries. When nickel and chromium are added during production, the hardness of steel can be greatly improved. Nickel compounds are also used for nickel plating, coloring ceramics, making batteries, and as a catalyst. China imported 5,350,000 tons of iron ore valued at 3,922.280 million US dollars and 440,000 tons of chromium ore valued at 76.143 million US dollars, respectively, in May 2009 (www.chinamining.org). The imports in January–May 2009 reached 24,189,000 tons valued at 18,853.828 million US dollars and 1,870,000 tons valued at 372.165 million US dollars, respectively, for iron and chromium ores. The world iron ore market achieved an all-time high for production in 2011 of 1.92 billion tons up, 4.7% from 2010. However, China's iron ore import started fail since early 2012.

Alumina is needed for the manufacturing industry. China imported 480,000 tons of alumina valued at 117.871 million US dollars in May 2009. The imports in January–May 2009 reached 2,150,000 tons valued at 517.462 million US dollars.

Copper demand is increasing because of its application in various industries, such as semiconductor, metal coating, and chip production.

China imported 510,000 tons of copper ore valued at 518.247 million US dollars in May 2009 (www.chinamining.org). The imports in January–May 2009 were 2,440,000 tons valued at 2,375.958 million US dollars.

Globalization is an important characteristic of economic development. There is no doubt that metal extraction, production, and application can cause contamination of water and food. The trade in various food items due to globalization may transfer food products contaminated with toxic heavy metal to other places that may not be affected by the contamination. For example, the rice grown in arsenic-contaminated regions can harm people who live in the uncontaminated regions.

1.3 Environmental Importance

Sodium, potassium, magnesium, and calcium are essential to sustain biological growth of any living organism. They are found in our bodies, as shown in Figure 1.2. Some heavy metals exist in our bodies at much lower levels; it is still not well known to us whether their presence is essential for human growth, development, and reproduction.

Metals can be divided into three groups according to their importance to human beings. The first group is essential light metals, the second group is

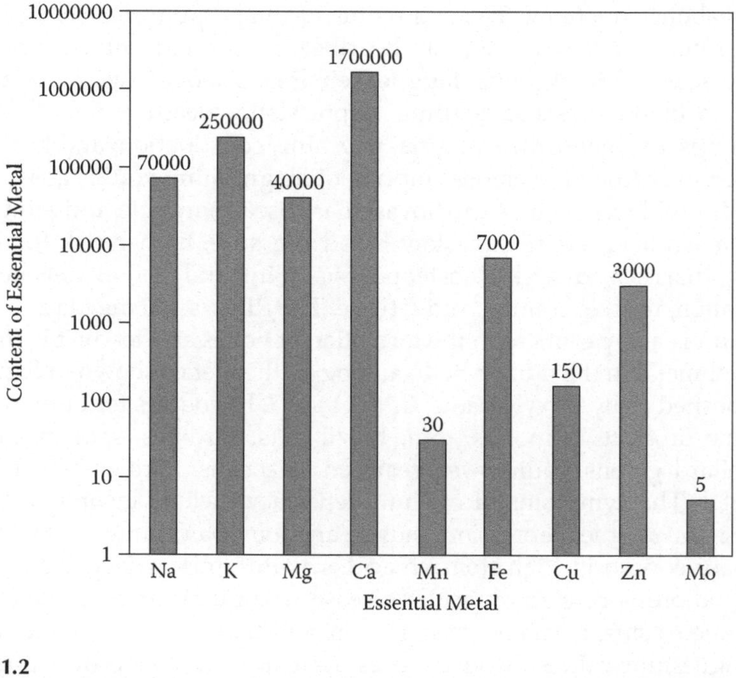

FIGURE 1.2
Content of essential metals for human beings (expressed in mg per 70 kg body weight).

essential heavy metals, and the third group is toxic heavy metals. Any metal would become hazardous to living organisms and human beings once its concentration is above a certain level.[2]

1.3.1 Essential Light Metals

Sodium, potassium, magnesium, and calcium are essential light metals that are extremely important in sustaining biological growth. The contents of these four metals are much higher than many other metals as shown in Figure 1.2.

Both sodium and potassium have high concentrations in our human body. Potassium is one of three key elements (K, N, and P) in the growth of plants. Sodium and potassium participate in metabolic processes. However, a very high concentration of sodium will cause high blood pressure, fluid retention, and swelling. The recommended daily allowance (RDA) is 2,400 mg/day (www.nutrientfacts.com). Potassium at a very high concentration will cause various problems such as nausea, diarrhea, irregular heartbeat, vomiting, muscle weakness, and fatigue. The RDA is 3,500 mg/day.

Magnesium is widely distributed throughout the human body, as shown in Figure 1.2. The sources of magnesium for human beings are almonds, nuts, wheat germ, soybeans, wheat bran, seeds, millet, dark green vegetables, legumes, fruit, hard water, seafood, and mineral water. Magnesium mainly participates in functions within the cell and is found complexed to nucleic acids. It is important for nerve impulse transmissions, muscle contractions, and metabolic functions. The symptoms of magnesium deficiency include cardiovascular disease, irregular heartbeat, insomnia, anxiety, nervousness, muscle and joint pains, fatigue, seizures, osteoporosis, high stomach acid, high blood pressure, asthma, depression, sweating, muscle spasms and cramps, dysmenorrhea, angina, migraine, constipation, and headaches. However, overdose or overconsumption of magnesium would cause a series of health problems such as cardiovascular disease, muscle and joint pains, low stomach acid, depression, low blood pressure, higher risk for several cancers, diarrhea, dry skin, osteoporosis, joint and spinal degeneration, dehydration, muscle cramps, and fatigue. The RDA is 400 mg/day.

Calcium is a key element in the formation of bones and teeth; it is the most abundant metal in the human body among all the metals shown in Figure 1.2. It is deposited as hydroxyl-apatite, $Ca_5(PO_4)_3OH$. Its sources for human beings are dairy products, almonds, tofu, Brazil nuts, sardines, salmon, broccoli, kale, collard greens, cauliflower, seaweed, soybeans, hard water, and mineral water. The symptoms of calcium deficiency include insomnia, anxiety, nervousness, fatigue, depression, muscle and joint pains, muscle spasms and cramps, osteoporosis, high stomach acid, seizures, miscarriage, birth defects, high blood pressure, cardiovascular disease, irregular heartbeat, hemorrhagic stroke, aneurysms, dysmenorrhea, rickets, and higher risk for some cancers. Like magnesium, calcium would express its toxicity when it is overconsumed. The potential symptoms are arteriosclerosis, cardiovascular disease, ischemic heart disease and stroke, high blood pressure, low stomach acid, muscle and joint pains, depression, fatigue, glaucoma, higher risk for several cancers, osteoporosis, calcification, osteoarthritis, constipation and dry skin. The RDA of calcium is 1,000 mg/day. A daily intake of 1,200 mg may be required for men older than 70. For prevention of colon cancer, 1,800 mg per day may be required.

1.3.2 Essential Heavy Metals

There are several heavy metals that may be beneficial to the biological growth of living organisms. The important heavy metal elements are manganese, iron, cobalt, copper, zinc, molybdenum, and selenium. Some important essential metals are given in Table 1.1.[3]

Manganese (Mn) is important in the activation of many enzymes in vitro and is involved in glucose utilization. It can be used in cholesterol and fat synthesis, in building bones, and in the creation of DNA and RNA. It is well distributed in our food. For example, it is present in whole grains, bran, nuts, fruits, green and leafy vegetables, and tea. The RDA of manganese is 2 mg/day.

Iron (Fe) is the most abundant transition element. It is essential in the oxygen-carrying protein molecule of blood. The symptoms of iron deficiency include fatigue and shortness of breath. Generally speaking, it is well represented in our food sources such as beef, fish, poultry, egg yolk, oysters, and enriched cereals and breads. The RDA of iron is 18 mg/day.

Cobalt (Co) is an essential enzyme activator. It is an important component in cobalamin known as vitamin B_{12}.

TABLE 1.1

Daily Dietary Allowances of Essential Metals Recommended by National Academy of Sciences, USA

	Age (Years)	Weight (kg)	Ca (mg)	Fe (mg)	Mg (mg)	Zn (mg)
Infants	0–½	6	360	10	60	3
	½–1	9	540	15	70	5
Children	1–3	13	800	15	150	10
	4–6	20	800	10	200	10
	7–10	30	800	10	250	10
Males	11–14	44	1200	18	350	15
	15–18	61	1200	18	400	15
	19+	67+	800	10	350	15
Females	11–18	44–54	1200	18	300	15
	19+	58	800	18(10)[a]	300	15
Pregnant			1200	18+[b]	450	20
Lactating			1200	15	450	25

[a] (10) for females above 50 years.

[b] Figure cannot be met by ordinary diets. Therefore, use of supplemental iron is recommended.

Copper is an important element for human beings. Its presence is necessary in the production of hemoglobin in red blood cells. Copper (I) is found in enzymes capable of carrying oxygen (doing a similar job as hemoglobin). It can help strengthen tendons and cartilage. Copper is also used in the proper functioning of several enzymes. Similar to zinc and iron, it functions as a micronutrient for humans. Copper deficiency in humans can cause anemia, bone and cardiovascular disorders, deterioration in mental and nervous systems, defective keratinization of hair, reduction in levels of neurotransmitters, dopamine, and norephedrine, and defective myelination in the brain stem and spinal cord. The RDA of copper is 2 mg/day.

Chromium (Cr) has been considered to be a carcinogen. However, most recent studies have shown that trivalent chromium (Cr(III)) is an essential element in animal and human nutrition when its concentration is low. Cr(III) is reportedly required to maintain normal glucose metabolism, which is demonstrated in rat experiments.

Zinc (Zn) is required by the human body. Zn(II) has the ability to occupy low symmetry sites in enzymes and can function as an essential constituent of enzymes. Zinc is a constituent of most cells. Several enzymes have zinc as a cofactor. The RDA of zinc is 15 mg/day.

Molybdenum (Mo) is involved in electron transfer processes. Nitrogen fixation is also associated with the element. The human body contains about 0.07 mg – Mo/kg-body weight. The daily consumption is 0.12–0.24 mg. Mo is present mainly in the liver and kidneys.

Nickel (Ni) is important in physiological processes as a cofactor in the absorption of iron. Urease needs nickel for the hydrolysis of urea.

Selenium (Se) is able to neutralize free radicals that may cause aging and some types of cancer. Its sources are beef, poultry, brown rice, and whole grains. This element is important to human beings and thus is required because of its perceived anticancer properties. The RDA of selenium is 70 mg/day. In addition, trace-level selenium is good for growth and fertility in animals and prevention of diseases.

The biological growth of any living organism would be affected if the concentration of essential heavy metal is below an optimal value, as shown in Figure 1.3. These heavy metals (e.g., zinc) are nutritionally essential in small quantities for good health. However, an essential metal would become toxic (and even extremely toxic) when the supply (intake) is excessive. It can cause acute or chronic toxicity when a large amount is consumed. Essential metal becomes toxic when its content is above 40 times what is required (i.e., when it is oversupplied).

1.3.3 Toxic Heavy Metals

Toxic heavy metal is a nonessential heavy metal. Its negative effect is less observed when its concentration is less than a certain value, as shown in

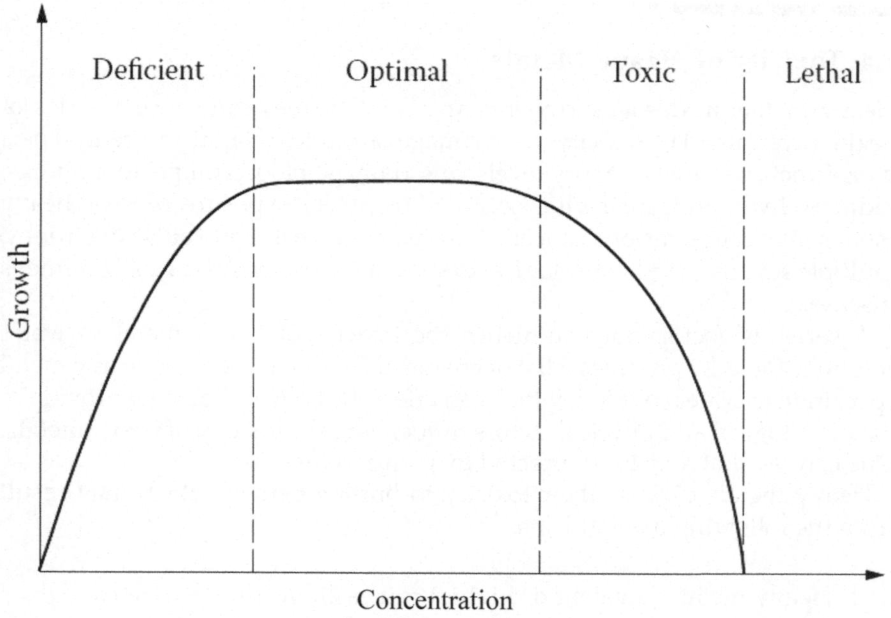

FIGURE 1.3
Effects of essential heavy metal on biological growth of living organisms.

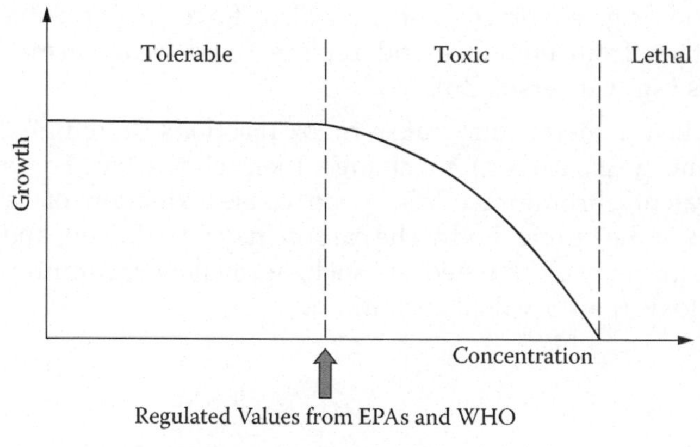

FIGURE 1.4
Effect of toxic heavy metal concentration on biological growth.

Figure 1.4. However, the heavy metal would become harmful when its concentration is above a toxic level, which is often regulated by the national environmental protection agencies (EPAs) or WHO. The metal would become a lethal toxicant when its concentration is above a lethal level. The important toxic heavy metals include arsenic, lead, mercury, and nickel.

1.4 Toxicity of Heavy Metals

Heavy metals in dosages higher than critical values can cause a series of health problems. The toxicity may damage or reduce mental and central nervous functions, lower energy levels, and damage blood composition, lungs, kidneys, liver, and other vital organs. Long-term exposure of toxic heavy metals may cause various cancers, Parkinson's disease, muscular dystrophy, multiple sclerosis, and neurological degenerative processes (e.g., Alzheimer's disease).

A series of factors may influence the toxicity of heavy metal in water solution. Table 1.2 provides a list of chemical factors influencing heavy metal speciation in water solution, which can affect the toxicity. Table 1.3 provides a list of biological and physical factors influencing the toxicity of heavy metals. One can see that toxicity is affected by many factors.

Heavy metals express their toxicity to human beings, which may result from the following mechanisms:

1. Heavy metals (divalent or higher) can easily form precipitates with anionic substances (CO_3^{2-}, SO_4^{2-}, and PO_4^{3-}). The precipitates (solids) may stay in the human body.
2. Heavy metals can be adsorbed onto organic functional groups in various tissues of the human body, which may change the biological activities (e.g., enzymes). This may also "force" the essential metals to detach from enzymes and replace them with harmful heavy metals (e.g., Cu versus Zn).
3. Some heavy metals may cause redox reactions, "forcing" the basic elements (e.g., carbon) to change their chemistry. For example, hexavalent chromium (CrO_4^{2-}) can cause oxidation of carbon in tissues of the human body. The carbon may be oxidized, and thus its chemistry may be changed. As such, hexavalent chromium is much more toxic than trivalent chromium.

1.5 Guidelines and Standards for Heavy Metals in Drinking Water

Such heavy metals as lead, copper, cadmium, zinc, mercury, and nickel are among the most common pollutants found in industrial effluents. Even at low concentration levels, they can be toxic to various living organisms, including humans. Government agencies and international organizations such as the US EPA and WHO have issued a list of rules, regulations, and guidelines for

TABLE 1.2

List of Chemical Factors Influencing Heavy Metal Speciation in Water Solution

Items	Factors	Comments
Nature of metal ions and reactions	Inorganic or organic	Metal may exist as inorganic or organic species.
	free ions or complex ions	Metal ion at lower pH (e.g., pH 1-2) exists as free metal ion. It may become complexed such as metal hydroxides.
	Particulate/solid	Metal may exist as a solid or in particulate form. Particularly, it exists as metal oxide or hydroxide when pH is higher.
	Precipitated or adsorbed	Metal ion may be precipitated in the presence of precipitation reagents such as hydroxide or sulfate. It may also adsorb onto adsorbent.
Presence of other metallic compounds	Non-competitive	Non-competitive factor would not affect metal toxicity.
	Competitive	Competitive factor would affect metal toxicity. It may enhance or reduce the toxicity.
Chemical factors	pH	pH can directly changes the metal speciation so that the metal is present in the form of free metal ion or in the form of complexed metal ion.
	Dissolved oxygen	Dissolved oxygen (DO) would affect reduction-oxidation reactions, which can change the speciation of metal ion. For example, As(III) can be oxidized to As(V).
	Salinity/TDS	Salinity or total dissolve solids (TDS) generally does not participate reaction with heavy metal ion. However, metal speciation may be influenced due to high ionic strength (salinity or TDS).
	Organics	Presence of organics may cause formation of metal-organic complexes (e.g. Cu-EDTA). It may lead to reduction or oxidation of metal ions (e.g., arsenic in groundwater).
	Acid rain	Acid rain causes low pH in water solution, which can affect metal speciation. More free metal ion will be present in the area that is affected by acid rain, which are due to the changes in the solution reaction as well as adsorption and precipitation reactions.
Physical factors	Temperature	Temperature can affect both equilibrium and kinetics of reaction. Typically, high temperature would increase the reaction kinetics.
	Light	Generally, light would not affect metal speciation. However, UV light may lead to formation of free radicals, which may oxidize metal with low valency.
	Acid rain	Acid rain causes low pH in water solution, which can affect metal speciation. More free metal ion will be present in the area that is affected by acid rain, which are due to the changes in the solution reaction as well as adsorption and precipitation reactions.

TABLE 1.3
List of Factors Influencing Toxicity of Heavy Metals

Items	Factors	Comments
Nature of living organisms	Type and size	The effect of toxicity is different on different types and sizes of organisms. For example, a large organism may tolerate a high dosage of toxicant.
	Age	The age of the organism is important for its survival. Younger organisms may be able to tolerate higher dosage of toxicant.
	Self-protection	Organisms may have their self-protection systems, such as a shell, which prevents toxicants from entering their body.
	Sex	Toxicants may behave differently for male and female organisms.
Physics	Water flow	Water flow (groundwater and surface water) would cause relocation and transportation of metals. It can lead to the dilution of metals and also spread contaminants to other locations.
	Precipitation	The precipitation referred to here is rain. It can cause the dilution of heavy metal waste. However, it can also cause relocation of the waste.
	Ground water	Ground water serves as water storage and plays an important role in water supply and balance. The extraction of groundwater may transfer contaminants such as arsenic that exist in the subsurface, together with other minerals (e.g., calcium, magnesium, and carbonate), to our water supply. On the other hand, it is a common practice that treated wastewater is injected to subsurface to maintain the water storage and geostructure. As such, heavy metals in the treated wastewater may enter the groundwater.
Biology	Organisms	Heavy metals may be immobilized onto organisms due to such mechanisms as adsorption, reduction, ion exchange, and surface precipitation.
	Nutrients/food	Food and nutrients are important for the biological growth, which may indirectly affect heavy metal immobilization.

the management of heavy metals to minimize their effects on human beings. Table 1.4 shows the important toxic metals regulated by the agencies (www.epa.gov). Table 1.5 demonstrates the maximum contaminant levels (MCLs) for air, water, and soil. Table 1.6 shows the regulatory standards for heavy metal in drinking water from the 1960s to the 1970s.[4–6] Stricter standards have been adopted in the past 20 years.

TABLE 1.4

WHO Guideline for Drinking Water and Drinking Water Standards by US EPA

Heavy Metal	WHO Guideline Value (mg/L)	US EPA Maximum Contaminant Level (mg/L)
Arsenic	0.01	0.01
Cadmium	0.003	0.005
Chromium (total)	0.05	0.1
Copper	2	1.3[a]
Lead	0.01	0.0015[a]
Nickel	0.07	0.1
Zinc	3	5[b]

[a] Copper action level 1.3 mg/L; lead action level 0.015 mg/L.
[b] Secondary drinking water regulations.

TABLE 1.5

Maximum Contaminant Levels for Heavy Metals for Air, Soil, and Water Regulated by US EPA

Heavy Metal	MCLs in Air (mg/m^3)	MCLs in Sludge (soil) (mg/kg or ppm)	MCLs in Drinking Water (mg/L or ppm)	MCLs in Water Supporting Aquatic Life (mg/L or ppm)
Cadmium	0.1–0.2	85	0.005	0.008
Lead	Not available	420	0.01	0.0058
Zinc	1 (chlorine fume) 5 (oxide fume)	7,500	5	0.0766
Mercury	Not available	<1	0.02	0.05
Silver	0.01	Not available	0	0.1
Arsenic	Not available	Not available	0.01	Not available

1.6 Sources of Heavy Metal Contamination

Four sources have contributed to heavy metal contamination: natural sources, industrial sources, domestic sources, and atmospheric sources.

1.6.1 Natural Sources

Natural sources are often referred to as geologic weathering. Mineralized areas with higher metal contents are always associated with metal contamination in both groundwater and soils. For example, the area in La Grande

TABLE 1.6
Drinking Water Quality Criteria for Heavy Metals Regulated in 1960s and 1970s

Metal	US PHS	Japan	USSR	WHO	US NAS	US EPA	FRG
Year	1962	1968	1970	1971	1972	1975	1975
As	10 ppb	50 ppb	50 ppb	50 ppb	100 ppb	50 ppb	40 ppb
Cd	10 ppb	—	10 ppb	10 ppb	10 ppb	10 ppb	6 ppb
Cr	50 ppb	50 ppb	100 ppb	—	50 ppb	50 ppb	50 ppb
Cu	1 ppm	10 ppm	100 ppb	50 ppb	10 ppm	—	—
Pb	50 ppb	100 ppb	100 ppb	100 ppb	50 ppb	50 ppb	40 ppb
Hg	—	1 ppb	5 ppb	1 ppb	2 ppb	2 ppb	4 ppb
Se	10 ppb	—	1 ppb	10 ppb	10 ppb	10 ppb	8 ppb
Ag	50 ppb	—	—	—	—	50 ppb	—
Zn	5 ppm	0.1 ppm	1 ppm	5 ppm	5 ppm	—	2 ppm

Notes: WHO: World Health Organization.
US PHS: US Public Health Service.
USSR: Union of Soviet Socialist Republics (Soviet Union).
US. NAS: USA National Academy of Sciences.
FRG: Federal Republic of Germany.

River in Canada has higher mercury content in the rocks; the people living there suffered from mercury poisoning. Weathering or change in weather (due to rain, temperature, wind, and heat) can cause metal contamination. Over many years, larger rocks that have higher heavy metal contents may be broken down to fine sands, leading to the pollution in air and water. Acid rain due to industrialization can further promote natural weathering.

Take arsenic contamination as an example. In England and Wales, the arsenic content in some of the areas affected by arsenic is 20 times higher than in areas not affected by arsenic. The high contaminant level concentration is associated with sulfide mineralization (together with copper). The arsenic content in plants in arsenic-affected areas is much higher than in soils. In Waikato River of New Zealand, the arsenic content in the submerged aquatic plants has a maximum level of 650 mg/kg-dry mass, much higher than that in plants growing in uncontaminated soils (below 12 mg/kg-dry mass).

Unavailability of water resource forces people to look for water sources other than surface water (e.g., river and lake water). Groundwater is a traditional water source. Extraction of water from the subsurface may "force" harmful metals such as arsenic to enter our water systems.

1.6.2 Industrial Sources

Metal ore processing and crude oil processing are two important industrial sources for heavy metal contamination. When these two raw materials are extracted, processed, and used, heavy metal contaminants can enter the environment as solid waste, wastewater, and waste gases.

1.6.3 Domestic Sources

Heavy metals may enter domestic wastewater due to discharge from household and industries. The waste effluent from households contains lots of organic and inorganic waste. For example, detergents contain trace-level metals such as iron, magnesium, chromium, copper, and strontium. Heavy metals can be adsorbed onto biological solids (large, small, and colloidal) due to higher affinities between metals and solids.

There is an increasing awareness that urban runoff causes a serious metal problem. Street dust (from the air), dirt, and various solids, together with various chemicals applied (e.g., pest control), have contributed significantly to the problem.

Another domestic source of heavy metal contaminants is landfill. Landfill is an engineered measure for solid waste storage. Solid waste often has heavy metals existing in solid and liquid forms. For a longer period of time, anaerobic reactions occur in the site, which leads to low pH, "pushing" the metals to the geo-environment when the lining (landfill liner) is not effective.

Waste-activated sludge may be used as a nutrient source for plant growth due to its good values of carbon, potassium, pathogen, and phosphorus. However, the metal content in the "fertilizer" (sludge) is normally quite high. When it is applied, the metals could enter the surface water and groundwater.

1.6.4 Atmospheric Sources

Industrial activities have produced increasing amounts of airborne metal particles that can be metal oxides and metal carbonates. Transportation (burning fuels) and fossil-fuel burning are two major sources. The particulates become windblown over very long distances (e.g., from China to Japan and then the United States). The anthropogenic sources contribute greatly to the metal emission in the air, as shown in Figure 1.5.

1.7 Important Heavy Metals

Heavy metals play important roles in water quality and human health. The important metals with high toxicity are discussed next.

1.7.1 Arsenic

Arsenic is considered as a half-metal as it has properties of both metals and nonmetals. Arsenic is an extremely toxic poison that can affect the human body.[7–9] It has a long-term effect on the human body. It is chemically stable. The US EPA has set the arsenic standard for drinking water at no more

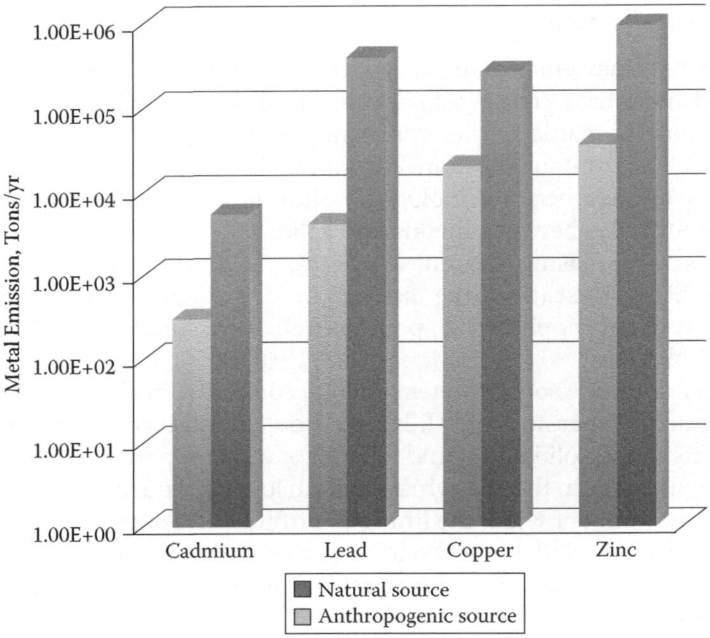

FIGURE 1.5
Comparison of metal emissions to the atmosphere contributed by anthropogenic sources and natural sources.

than 10 ppb to protect consumers served by the water treatment plants from the effects of long-term, chronic exposure to arsenic. The plants must comply with this standard by January 23, 2006, providing protection to an estimated 13 million Americans. The WHO has recommended a limit of 0.01 mg/L (10 ppb) of arsenic in drinking water. The rule has been established as a regulation for drinking water in various countries such as China and India.

Arsenic exists in some areas in the world; the groundwater in some cities and towns in India, Bangladesh, China, and the United States reportedly has arsenic with concentrations ranging from 50 to 500 ppb or even higher ppb. When groundwater is extracted from the subsurface and used as a water source, the arsenic becomes harmful to human beings. Arsenic contamination of groundwater is one of the major environmental problems in the world.

Arsenic is present in the wastewaters of industries that produce metallurgical products, glassware and ceramic products, tannery products, dyestuff, pesticides, synthetic chemicals, petroleum refinery products, semiconductor, metal coating, and electronic chip products. Because of its toxicity, it had been used as a killer of small organisms and protector for wood. For example, it has been used to kill termites, which is a legal use in some countries in

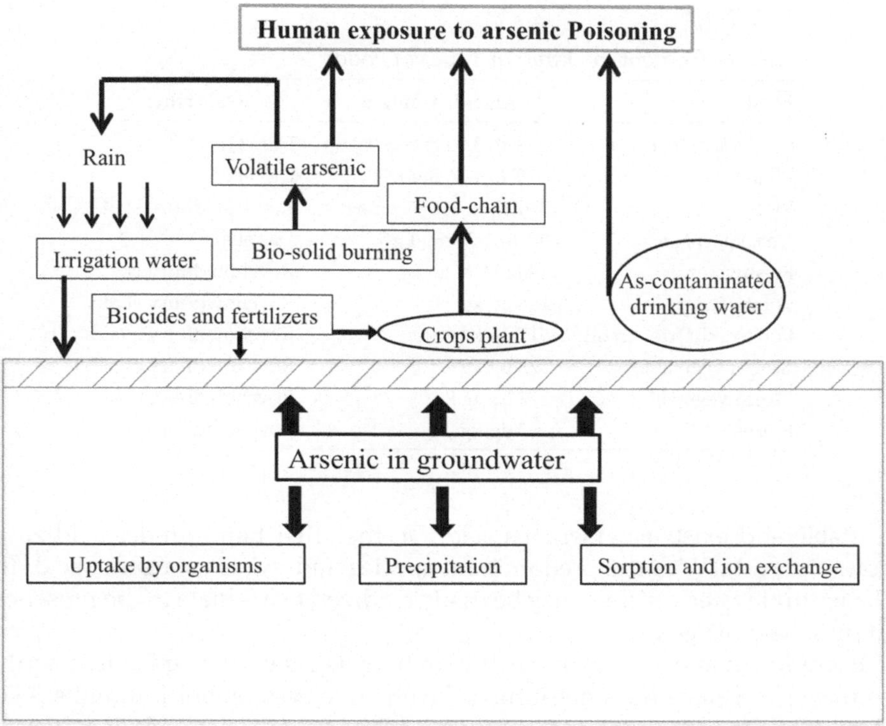

FIGURE 1.6
Possible pathways for human exposure to arsenic.

the world. It can be adsorbed into muscle tissue, the digestive tract, and the abdominal cavity.

Arsenic's effects on human beings are through direct and indirect pathways. It can directly enter the human body through drinking. It can also enter the human body through consuming food such as rice, vegetables and fishes that are produced in arsenic-contaminated water environment or irrigated by the water.

As the topmost consumer in the ecosystem, humans may be exposed to arsenic poisoning via various pathways as shown in Figure 1.6. Such pathways as plant–man and plant–animal–man may be potentially severe for human exposure to arsenic, particularly due to globalization and free trade.[9]

Arsenic mainly exists as inorganic arsenic, which are arsenate (As(V)) and arsenite (As(III)). Around 10% to 20% of arsenic species are in the form of organic arsenic. Inorganic arsenic is believed to be more toxic than organic arsenic; however, recent evidence shows that the organic arsenic is more toxic.

As(III) is more mobile in groundwater and 25–60 times more toxic than As(V). At low pH and in a mildly reducing environment (> 100 mV), As(III)

TABLE 1.7
Arsenic Content in Different Types of Foods

Food	Arsenic Content	Researchers
Colocasia antiquorum	0.09 to 3.99 mg kg^{-1}	Das et al.[8]
Potato	0.07 to 1.36 mg kg^{-1}	Das et al.[8]
Rice	374.17 to 666.57 µg kg^{-1}	Roychowdhury et al.[7]
Ipomoea reptans	0.1 to 1.53 mg kg^{-1}	Das et al.[8]
Freshwater fish	830 to 900 µg kg^{-1}	Roychowdhury et al.[7]
Potato curry	186 µg kg^{-1}	Roychowdhury et al.[7]
Potato skin fried in oil	617 µg kg^{-1}	Roychowdhury et al.[7]
Leaf of vegetables	578 µg kg^{-1}	Roychowdhury et al.[7]
Mixed vegetables	277.33 µg kg^{-1}	Roychowdhury et al.[7]
Pulses	143 µg kg^{-1}	Roychowdhury et al.[7]

is stable and exists as arsenious acid. On the other hand, under oxidizing conditions, As(V) is the predominant species and exists as arsenic acid. In the natural water, arsenite may be easily oxidized to arsenate in the presence of dissolved oxygen.

Excretion of arsenate from the human body is faster than arsenite (mainly in urine). Arsenate does not inhibit the enzyme system, but it inhibits ATP synthesis. On the other hand, arsenite inhibits enzymes and binds to tissue proteins in hair, nails, and skin. It stays in the human body for a prolonged period. Arsenic deposition in cattle bodies and in the tissues and milk of sheep has been reported. Arsenic bioaccumulation causes blackfoot disease in Taiwan.

Long-term consumption of arsenic-contaminated water from wells can cause gastrointestinal, skin, liver, and nerve tissue injuries. People who stay away from the arsenic-contaminated area may be affected through consumption of such arsenic-contaminated foods as rice and fish.

Roychowdhury et al. reported the arsenic content in individual composites of cooked items, collected from an arsenic epidemic area of West Bengal, India.[7] Table 1.7 shows the arsenic content in food. Das et al. reported that arsenic content in food from the arsenic epidemic area of Bangladesh exceeded the food safety limits as shown in Table 1.7.[8]

1.7.2 Cadmium

Cadmium has been used widely in electroplating industries, solders, batteries, television sets, ceramics, photography, insecticides, pigments, textiles, petroleum products, synthetic chemicals, electronic and metal-finishing industries, and metallurgical activities. It occurs naturally with zinc and lead in sulfide ores. The elevated concentrations of cadmium in air, water, and soil occur close to industrial emission sources, particularly those of nonferrous mining and metal processing operations. Cadmium is chemically similar to zinc. Cadmium is a highly toxic metal that can accumulate in the human body and cause irreversible damages to a number of biological systems even at an extremely low dose. It has been reported that cadmium is more effectively absorbed in the airways than in the gastrointestinal tract. As the levels of cadmium in the outdoor air are generally below 10 ng/m^3, the exposure through air is not an issue of great concern, except in heavily industrialized areas.[10]

The primary routes of cadmium intake are diet and tobacco smoking. The high accumulation levels are found in offal products of mammals, such as the kidneys and liver. High concentrations are present in some species of fish, mussels, and oysters, which are caught in the polluted coastal areas. For example, the cadmium levels in some edible crabs (*Cancer pagurus*) reportedly reached about 30–50 ppm.[11]

Consumption of cadmium-contaminated foods is the main source of intake in the nonsmoking population. Furthermore, wheat and rice, green leafy vegetables, potato, and root vegetables may also significantly contribute to the human exposure.

Tobacco smoking is an important source of exposure for smokers. It is the most significant human route of cadmium intake in noncontaminated areas. One cigarette contains approximately 1 to 2 μg Cd, which is dependent on the type, the brand, and the location of production. A person smoking a pack per day may absorb about 1 μg cadmium.

Upon entering the human body, Cd is bound tightly to metallothioneins. Above 50%, Cd in the human body accumulates in the liver and kidney because of their ability to synthesize metallothionein.[12] Cd may ultimately be eliminated via urine. However, the amount of Cd excreted daily in urine is very limited (0.005% to 0.01% of the whole cadmium content). The low excretion rates of cadmium result in high retention in the body. Cd is well known as a persistent toxin with a biological half-life of more than 20 years.

Zinc is required by the human body. Zn(II) functions as an essential constituent of enzymes. However, cadmium (and mercury) competes with zinc and replaces it.

As the main storage organ of toxicants, the kidney is always the critical target organ that displays early signs of toxicity.[13] Chronic exposure to elevated levels of cadmium can cause renal dysfunction, bone degeneration, liver damage, and blood damage. There are sufficient evidences in humans for the carcinogenicity that results from the exposure of both cadmium and cadmium compounds.

1.7.3 Chromium

Chromium is a transition metal that is extensively used in various industries. On a worldwide basis, around 80% of chromium is employed in metallurgical applications. It is added as a biocide in cooling water to prevent corrosion. It is used in the production of inks, steel, semiconductor, electronic chip, petroleum, chemicals, color, industrial dyes, paint pigments, chrome tanning and anodizing operations in the surface-finishing industry, dyeing and metal processing, metal cleaning, leather and mining industries, and other electroplating industries. For example, the steel industry uses it, as chromium can greatly enhance the hardness of steel and also prevent corrosion of steel. As chromium oxides are colorful and attractive, they have been used in various color-related products for more than 1,000 years.

Because of its powerful oxidation ability, hexavalent chromium is often used in laboratories for measurement of chemical oxygen demand (COD). The extensive use of chromium in many industries has attracted much concern because of the emission of chromium-bearing wastewater into the environment. Besides these industries, the burning of fossil fuels and waste incinerations may contribute to chromium in water.

Chromium can exist in different oxidation states, from 2^- to 6^+. Among these, Cr(III) and Cr(VI) are the two thermodynamically stable oxidation states. Trivalent chromium is the naturally occurring element, which is an essential nutrient for adults, in amounts of 50–200 µg/day. It also plays an important role in the biological functions for animals as a micronutrient. However, it should be emphasized that any heavy metal is potentially toxic if its concentration exceeds a certain threshold level. For instance, long-term exposure to trivalent chromium is known to cause allergic skin reactions and cancer. Trivalent chromium is the most prevalent form found in natural water.

All hexavalent chromium is produced because of human activities. Cr(VI) is an acute carcinogen and more mobile and toxic than Cr(III). Hence, Cr(VI) is more important than Cr(III) in water pollution control. Cr(VI) has been regarded as a Group A human Carcinogen because of its high toxicity and great mutagenicity. The chromium circulation in the polluted environment is illustrated in Figure 1.7.[14]

Cr(VI) is highly toxic to humans, in contrast to Cr(III). If Cr(VI) is ingested, it may be rapidly reduced to Cr(III) because of the action of stomach acids (and organic substances). However, if its concentration is higher than a certain level, it becomes extremely harmful to the human body. Cr(VI) is generally considered 1,000 times more toxic than Cr(III). Both acute and chronic adverse effects of chromium are mainly caused by hexavalent chromium.

The cells in the human body have active anionic transportation systems to transport sulfates and phosphates that are significant in human biological processes. These systems may take up hexavalent chromium and transport it together with those nutrients (e.g., sulfates) from the outside of the cell to its interior.[15–17]

After hexavalent chromium enters the cell, it is rapidly reduced by various intracellular reductants to trivalent chromium.[18] The intracellular levels of reducing agents determine the amount of hexavalent chromium uptake. Among the reducers, ascorbic acid plays a major role in the reduction of

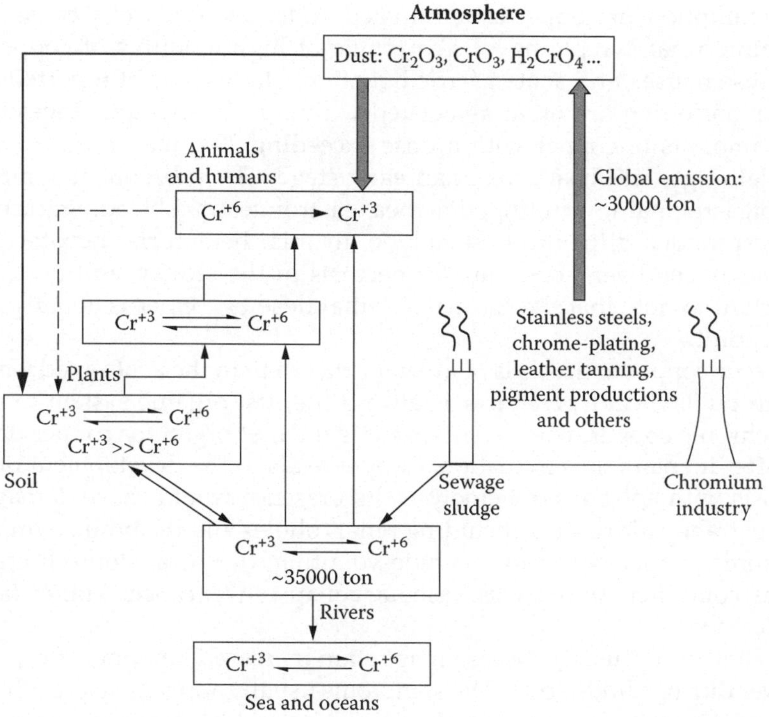

FIGURE 1.7
Chromium circulation in the polluted environment.

hexavalent chromium. A higher level of reductants would result in more chromium sorption, which consequently increases its toxicity to the cell.[17–20]

Trivalent chromium does not enter the cells via the pathway as hexavalent chromium, as it does not chemically function in the similar way to any biological nutrient.[16] This could be reason trivalent chromium is less harmful than hexavalent chromium to cells.

1.7.4 Copper

Copper is one of the commonly used heavy metals for various applications. It can be used in batteries, semiconductors, electronic chips, mobile phones, catalysts, metal processing products, and water pipes. Copper can enter the environment from the mining of copper and other metals and from factories that make or use metallic copper or copper compounds.

Copper is essential for healthy growth. However, high doses of copper salts to humans and animals can be extremely harmful. It can enter the human body through dust, food, and water. Free copper ion (Cu^{2+}) is one of the most toxic forms of copper in aquatic life.

Consumption of copper-contaminated water or foods can cause acute gastrointestinal symptoms.[21–23] An intake of high amounts of copper salt can cause nausea and acute gastric irritation. Many cases of reported acute copper poisoning are often associated with suicide attempts. Ingestion of large amounts of copper with a dose exceeding 20 g may result in dizziness, lethargy, and headache at an early stage. The subsequent symptoms are epigastric pain, vomiting, diarrhea, gastrointestinal bleeding, tachycardia, respiratory difficulties, hemolytic anemia, hematuria, hepatocellular necrosis in the liver, acute tubular necrosis in the kidney, and death. It is important to note that the estimated lethal dose of copper is about 10–20 g for an adult.

Chronic copper toxicosis is of greater concern than the acute syndrome for human public health. The liver is always the first organic system to suffer from chronic copper toxicity, because it is the first organ for copper deposition after it enters the circulation. Copper leads to the development of liver cirrhosis with episodes of hemolysis. Its presence would cause damages to some extrahepatic tissues, including renal tubules and the brain. Symptoms of chronic copper poisoning include vomiting, diarrhea, stomach cramps, nausea, coma, hepatic necrosis, vascular collapse, liver cancer, kidney failure, and death.[24]

Patients with Wilson's disease may retain increased amounts of copper in the liver during childhood.[25] The symptoms usually appear between the ages of 6 and 40 years. If untreated, the disease would become progressive and uniformly fatal by the fifth or sixth decade, typically as a consequence of liver failure.

1.7.5 Lead

The production factories of batteries, pigments, and printing products use lead. As a result, the wastewater from these factories and that from metal mining/production may contain lead ions.

Lead is the oldest metal known to be extremely toxic to human beings. Lead poisoning could be dated back to Roman times. Its sources are many, as listed in Table 1.8.

Lead has a strong affinity for thiol (-SH group) and phosphate-containing ligands, which inhibits biosynthesis and thereby affects the membrane permeability of kidney, liver, and brain cells. This would lead to the reduction of bio-functions and a breakdown of the tissues.

A series of studies have demonstrated that lead can seriously damages the important systems in the human body such as the peripheral and central nervous system, the hemopoietic system, reproductive system, and renal system, particularly in children.[26]

The uses of lead are limited and do not appear to be increasing in recent times. Indirect and direct exposure to lead through lead-based water pipes, paints, and petrol can lead to mental retardation and related dysfunctional

TABLE 1.8
Reported Lead Levels in Industrial Wastewaters

Industry	Lead (mg/L)
Battery manufacture	5–48 (particulate lead) 0.5–25 (soluble lead)
Battery manufacture	0.4–66.5 (particulate lead) 2.6–5.1 (soluble lead)
Plating	2–140
Plating pickle liquor	10
Television tube manufacture	380–400
Printed circuit board manufacture	1.65
Glass manufacture	0.43–100
Porcelain enameling	2.9
Chlor-alkali plant	1,160
Mining process water	0.02–0.098
Ammunition plant	6.5
Tetraethyl lead manufacture	126.7–144.8 (organic) 66.1–84.9 (inorganic)
Tetraethyl lead manufacture	45
Spent ink	94
Paint manufacture	1.1–10.0
Paint and ink formulation	86
Pigment manufacture	1–200
Textile dyeing	8.4
Steel manufacture, vacuum degassing process	0.47–1.39
Rubber hose manufacture, lead sheath process	63
Foundry	7.7–170
Piston ring manufacture	94.6

diseases. Lead has been found in most of the National Priorities List sites identified by the US EPA.

Lead in blood can affect pregnant women. During pregnancy, lead even at lower level is a serious threat because of the toxic effects on the fetus, including miscarriages, reduced gestational age, and low birth weights of infants.[27]

Lead poisoning can interfere with heme production and subsequent reduction of the heme body pool.[28] The deleterious damage occurs when blood lead levels are as low as 10 μg/dL. Lead inhibits the important enzymes that play vital role in the process of heme synthesis.

Symptoms of acute lead poisoning include dullness, fatigue, headache, poor attention span, irritability, and loss of memory. Children are reportedly easily affected by lead poisoning because of their rapidly developing nervous system.[29]

Lead poisoning of the kidney causes renal tubular damage and results in renal abnormalities. No effect is reportedly produced at blood lead levels below 60 μg/dl.[30–31]

1.7.6 Mercury

Mercury has wide applications in the areas of medicine, cosmetics, production of chlorine and caustic soda from the chlor-alkali industry, electronic production, herbicide and pesticide manufacturing, petrochemical production, gold and silver mining, and the military industry. It is the only metal that exists as a liquid (with a silvery color) at room temperature. Its melting point and boiling point are −38.83°C and 356.73°C, respectively.

Mercury is an extremely toxic heavy metal. In particular, mercury becomes increasingly harmful to humans because of its bioaccumulation and bioconcentration in our food chain.

It was reported that Hg concentrations in rice from Hg-contaminated areas of Wanshan, China, were as high as 569 μg/kg-weight, which was much higher than the maximum value of 20 μg/kg-weight for human consumption regulated by the Chinese government. Fish, rice, and other agricultural products from mercury and other mineral mining areas have posed a great human health risk to the humans who stay in the areas, where farming and mining coexist. Different organisms have different content of mercury.[32] For example, rice's accumulation of mercury is considerably higher than corn, rape, tobacco, and cabbage.

Mercury has a stronger affinity for sulfhydryl/thiol groups (-SH) that are involved in enzymatic reactions. Thus, it inhibits synthesis of proteins in plant leaves.

The toxic effects of mercury on plants include retardation of biological growth and premature senescence. As mercury compounds such as mercuric oxide and mercuric chloride are used in control of plant diseases, they can cause problems in plant growth. The toxic metal can enter the food chain. The US EPA classifies inorganic and methyl Hg as human carcinogens.

Mercury poisoning can result from exposure to soluble forms (such as mercuric chloride or methylmercury), inhalation of mercury vapor, or eating fish contaminated with mercury. Cinnabar (mercuric sulfide) is highly toxic through ingestion or inhalation of the dust.

Methylated organic mercury is generally more toxic and bioaccumulative than inorganic mercury because of its lipophilic nature (i.e., fat-liking nature or solubility in fat), which causes biomagnification of methyl mercury in our food chain.

Predatory aquatic wildlife species (e.g., the bald eagle), birds, and mammals are at higher risk of mercury toxicity. Some of the problems are progressive weakness in wings and legs, reproductive damage, liver and kidney damage, and neurobehavioral effects.

Elemental mercury vapor is rapidly adsorbed through human lungs but poorly through the human gastrointestinal tract. It is able to cross through both placental and blood–brain barriers. Methyl mercury can rapidly be adsorbed through the human gastrointestinal tract.

Typical symptoms of mercury diseases include impairment of peripheral vision, disturbances in sensations (e.g., numbness), poor coordination of movements, impairment of speech/hearing, and mental disturbance. The nervous system and kidney are generally damaged severely by mercury.

Pregnant women and children younger than 14 years are at increased risk from methyl mercury. Exposure leads to such symptoms as retardation and blindness.

1.7.7 Molybdenum

Molybdenum (Mo) at a trace level is an essential element for humans, plants, and animals. Generally speaking, molybdenum is not very toxic to humans. However, higher concentrations of the element are harmful to several animal species. Heart disease is due to high molybdenum and low copper content. The symptoms include bone disorders and growth retardation.

Molybdenum enters the human body via foods (e.g., milk and its products, baked food, and organ meats) and dust. When exposed to metallic molybdenum and its oxides, the blood uric acid levels of the human body increases.

1.7.8 Nickel

Nickel is one of the important heavy metals in many industrial processes. Its hardness is very high, and it is anticorrosive. The industries producing metal products, steel, chemicals, batteries, catalysts, vehicles, ceramics, and petroleum products produce huge volumes of nickel-containing wastewater.

Exposure to nickel compounds above the recommended limits can have adverse effects on human health. The most common adverse health effect of nickel in humans is an allergic reaction; large amounts of nickel can cause lung and nasal sinus cancers.

Acute toxic effects caused by an accidental inhalation of nickel carbonyl are usually classified into two types; immediate and delayed.

The symptoms of acute poisoning include headache, vertigo, nausea, vomiting, insomnia, and irritability, usually lasting a few hours. The subsequent symptoms include nonproductive cough, dyspnoea, cyanosis, palpitations, sweating, visual disturbances, vertigo, weakness, and lassitude.[33] All nickel compounds except metallic nickel have been classified as human carcinogens. The poisoning of the different nickel compounds depends on its solubility. Soluble nickel sulfate is the most toxic, while insoluble nickel oxide is the least toxic.

1.7.9 Selenium

Although trace-level selenium is essential for biological growth, it can cause several diseases when its content is high in water and foods. Higher intake will cause dullness and lack of vitality, as well as stiffness and lameness due to erosion of the joints of long bones. It may cause death because of respiratory failure. Some of its compounds may cause DNA damage and thus have carcinogenic effects on humans.

1.7.10 Silver

Some silver salts such as silver chloride have lower solubility and thus stay in the human body. Silver can also stay in tissues by forming silver–organic complexes, which are similar to adsorption of heavy metals. Silver sulfide can cause a permanent discoloration of skin and eye tissue called argyria.

1.7.11 Zinc

Zinc has been considered to be an essential micronutrient and a much less toxic metal. However, excessive amounts of zinc intake (e.g., inhaling zinc vapors and ingesting zinc-contaminated food or water) can induce system dysfunctions, which causes an impairment of growth and reproduction.[34–36] Zinc that is released into the environment mainly streams from anthropogenic activities such as mining, steel production, coal burning, and burning of waste. Unlike such heavy metals as lead and mercury, zinc contamination is generally not highly concerned. It was reported that the content of zinc was within normal range in the livers of white-fronted geese (*Anser albifrons*), eurasian wigeons (*Anas penelope*), mallards (*Anas platyrhynchos*) and spot-billed ducks (*Anas poecilorhyncha*), which was maintained by the normal homeostatic mechanism.[36]

The acute symptoms of zinc poisoning include nausea and vomiting, epigastric pain, abdominal cramps, diarrhea (frequently bloody), stomach cramps, nausea, bloody urine, icterus (yellow mucus membrane), liver failure and kidney failure.[34,35] Chronic zinc poisoning can cause anemia, pancreatic damage, and lower levels of high-density lipoprotein cholesterol (the good form of cholesterol).

References

1. De Ryck, I., Adriaens, A. and Adams, F. An overview of Mesopotamian bronze metallurgy during the 3rd millennium BC, *J. Cult. Herit.*, 6(3), 261–268, 2005.
2. Nordberg, G. F., Fowler, B. A., and Nordberg, M. *Handbook on the Toxicology of Metals (Third Edition)*. Academic Press, Burlington, MA, 2007.

3. Duruibe, J. O., Ogwuegbu, M. O. C., and Egwurugwu, J. N. Heavy metal pollution and human biotoxic effects. *IJPS*, 2(5), 112–118, 2007.
4. World Health Organization. *Guidelines for Drinking-Water Quality*. Geneva, WHO, 2008.
5. U.S. Environmental Protection Agency. *The Drinking Water Standards and Health Advisories*. Office of Water, Washington, DC, 1996.
6. U.S. Environmental Protection Agency. *2006 Edition of the Drinking Water Standards and Health Advisories, EPA 822-R-06-013*. Office of Water, Washington, DC, 2006.
7. Roychowdhury, T. U., Tokunaga H., and Ando, M. Survey of arsenic in food composites from an arsenic-affected area of West Bengal, India. *Food Chem. Toxicol.*, 40, 1611–1621, 2002.
8. Das, H. K., Mitra, A. K., Sengupta, P. K., Hossain, A., Islam, F., and Rabbani, G. H. Arsenic concentrations in rice, vegetables, and fish in Bangladesh: A preliminary study. *Environ. Int.*, 30(3), 383–387, 2004.
9. Rahman, M. A., Hasegawa, H., Rahman, M. M., Miah, M. A. M., and Tasmin, A. Arsenic accumulation in rice (*Oryza sativa* L.): Human exposure through food chain. *Ecotoxicol. Environ. Saf.*, 69 (2), 317–324, 2008.
10. Schoeters, G., DenHond, E., Zuurbier, M., Naginiene, R., Van Den Hazel, P., and Stilianakis, N. Cadmium and children: Exposure and health effects. *Acta Paediatr.*, 95, 50–54, 2006.
11. Overnell, J. Occurrence of cadmium in crabs (*Cancer pagurus*) and the isolation and properties of cadmium-metallothionein. *Environ. Health Perspect.* 65, 101–105, 1996.
12. Bernard, A. Cadmium and its adverse effects on human health. *Indian J. Med. Res.*, 128, 557–564, 2008.
13. Nordberg, G., Nogawa, K., Nordberg, M., and Friberg, L., Cadmium. In *Handbook on Toxicology of Metals*. Academic Press, New York, pp. 65–78, 2007.
14. CieslakGolonka, M. Toxic and mutagenic effects of chromium(VI). A review. *Polyhedron*, 15(21), 3667–3689, 1996.
15. Alexander, J., and Aaseth, J. Uptake of chromate in human red blood cells and isolated rat liver cells: The role of the anion carrier. *Analyst*, 120(3), 931–933, 1995.
16. Costa, M., Potential hazards of hexavalent chromate in our drinking water. *Toxicol. Appl. Pharmacol.*, 188(1), 1–5, 2003.
17. Costa, M., and Klein, C. B., Toxicity and carcinogenicity of chromium compounds in humans. *Crit. Rev. Toxicol.*, 36, 155–163, 2006.
18. Zhitkovich, A. Importance of chromium-DNA adducts in mutagenicity and toxicity of chromium (VI). *Chem. Res. Toxicol.*, 18(1), 3–11, 2005.
19. Cohen, M. D., Kargacin, B., Klein, C. B., and Costa, M. Mechanisms of chromium carcinogenicity and toxicity. *Crit. Rev. Toxicol.*, 23(3), 255–281, 1993.
20. Capellmann, M., Mikalsen, A., Hindrum, M., and Alexander, J. Influence of reducing compounds on the formation of DNA–protein cross-links in HL-60 cells induced by hexavalent chromium. *Carcinogenesis*, 16(5), 1135–1139, 1995.
21. Davanzo, F., Settimi, L., Faraoni, L., Maiozzi, P., Travaglia, A., and Marcello, I. Agricultural pesticide-related poisonings in Italy: Cases reported to the Poison Control Centre of Milan in 2000–2001. *Epidemiol. Prev.*, 28, 330–337, 2004.
22. Srivastava, A., Peshin, S. S., Kaleekal, T., and Gupta, S. K. An epidemiological study of poisoning cases reported to the National Poisons Information Centre,

All India Institute of Medical Sciences, New Delhi. *Hum. Exp. Toxicol.*, 24, 279–285, 2005.
23. World Health Organization (WHO). *IPCS Environmental Health Criteria 2000. Copper*. World Health Organization, Vammala Finland, 1998.
24. Winge, D. R., and Mehra, R. K. Host defenses against copper toxicity. *Int. Rev. Exp. Pathol.*, 31, 47–83, 1990.
25. Scheinberg, I. H., and Sternlieb, I. Wilson disease and idiopathic copper toxicosis. *Am. J. Clin. Nutr.*, 63, 842–845, 1996.
26. Papanikolaou, N. C., Hatzidaki, E. G., Belivanis, S., Tzanakakis, G. N., and Tsatsakis, A. M. Lead toxicity update: A brief review. *Med. Sci. Monit.*, 11, 329–336, 2005.
27. Nordstrom, S., Beckman, L., and Nordenson, I. Occupational and environmental risks around a smelter in northern Sweden: V. Spontaneous abortion among female employees and decreased birth weight in their offspring. *Hereditas*, 90, 291–296, 1979.
28. Piomelli, S. Childhood lead poisoning. *Pediatr. Clin. N Am.* 49, 1285–1304, 2002.
29. Lyn, P. Lead toxicity, a review of the literature. Part I: Exposure, evaluation and treatment. *Altern. Med. Rev.*, 11, 2–22, 2006.
30. Papanikolaou, N. C., Hatzidaki, E. G., Belivanis, S., Tzanakakis, G. N., and Tsatsakis, A. M. Lead toxicity update: A brief review. *Med. Sci. Monit.*, 11, 329–336, 2005.
31. Gerhardsson, L., Chettle, D. R., Englyst, V., Nordberg, G. F., Nyhlin, H., Scott, M. C., Todd, A. C., and Vesterberg, O. Kidney effects in long-term exposed lead smelter workers. *Br. J. Ind. Med.*, 49, 186–192, 1992.
32. Horvat, M., Nolde, N., Fajon, V., Jereb, V., Logar, M., Lojen, S., Jacimovic, R., Falnoga, I., Qu, L. Y., Faganeli, J., and Drobne, D. Total mercury, methylmercury and selenium in mercury polluted areas in the province Guizhou, China. *Sci. Total Environ.*, 304, 231–256, 2003.
33. Sunderman, F. W., Jr., Coulston, F., and Eichhorn, G. L. *Nickel*. National Academy of Science, Washington, DC, 97–143, 1975.
34. Samman, S. Trace elements. In Mann, J., and Truswell, S., editors. *Essentials of Human Nutrition*, 2nd ed. Oxford University Press, New York, 2002.
35. Festa, M. D., Anderson, H. L., Dowdy, R. P., and Ellersieck, M. R. Effect of zinc intake on copper excretion and retention in men. *Am. J. Clin. Nutr.*, 41(2), 285–292, 1985.
36. Kim, J. S., Oh, J. M., Metal levels in livers of waterfowl from Korea, *Ecotoxicol. Environ. Saf.*, 78, 162–169, 2012.

2

Pollution Prevention: Principles and Applications

2.1 Introduction

In the past 10 years, we have witnessed the evolution of a fully industrialized world, with global industrial production, global markets, global transportation, and global prosperity. This prospect brings with it the realization that current patterns of industrialization will not be adequate to sustain environmentally safe growth, and therefore needs dramatic improvement.[1–7] In addition, global warming and shortage in resources have curbed for our society our economic expansion. As shown in Figure 2.1, society, economy, and the environment can be related through sustainability. What is urgently needed is a total management systems approach to modern civilization's challenges by focusing on pollution prevention activities as the first step.

In the past, pollution control by media-specific control technologies improved environmental quality to a certain extent. Generally, however, not only does it fail to eliminate pollutants, but waste treatment processes produce a large amount of sludge and residue that require further treatment before disposal so that they will not create secondary pollution. Waste treatment systems require investment in design, installation, operation, and maintenance, but these systems contribute no financial benefits to industries and human beings. Pollution control technologies may also transfer pollutants from one environmental medium (air, water, or land) to another, causing potential secondary pollution problems that require further treatment and disposal. They address only short-term problems rather than eliminating pollutants. Costs of pollution control, cleanup, and liability have risen every year, as have the costs of resource inputs, energy, and raw materials. Through many years of research, the complexities of pollution management problems have been better understood.

It is believed that pollution control via such *end-of-pipe* strategies as wastewater treatment and incineration of solid waste could solve pollution problems as the treatment systems lead to the release of less harmful pollutants into the environment. Similar to pain-relieving medication, pollution control

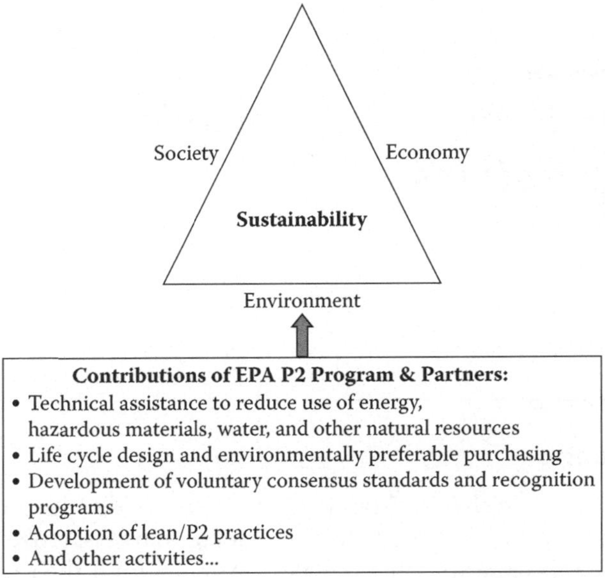

FIGURE 2.1
Illustration of the relationship between economy, society, and environment.

methods attempt, although imperfectly, to minimize the effect of releasing pollutants into the environment. Some releases and effects are curtailed, but the original toxic or environmentally harmful pollutants and products remain left behind and may be transformed into different hazardous substances to some degree in the same or different medium.

It is important to realize that *pollution prevention* (P2) applies beyond industrial sectors to a variety of economic sectors and institutional settings. Many organizations and institutions can apply pollution prevention concepts, which not only reduce generation of pollutants and wastes but also minimize use of certain environmentally harmful products and services. In practice, pollution prevention approaches can be applied to all pollution-generating activities, including in energy production and consumption, transportation, agriculture, construction, land use, city planning, government activities, and consumer behavior.

Economics plays an increasingly important role in environmental decision making. The resolution of virtually every environmental issue involves an economic component. Metal contamination is a typical example. It is well known that heavy metals, like many other commodities, have seen rapid appreciation in price, as shown in Chapter 1. On the other hand, it is becoming increasingly expensive to treat heavy metal waste and wastewater.

Traditionally, most industry and business decision makers who invested in pollution control technologies considered these nonproductive, because such added costs to production would be hard to recover. Product prices could be increased to offset these costs, but this was not an option in a competitive market. Such perspectives seemed valid in the past because decision makers did not fully understand the various benefits of P2.

Accepting the primacy of pollution management strategy and preventive technologies does not mean abandoning traditional waste control/treatment technologies or the government regulatory and legal systems designed to ensure their implementation. In fact, not all waste and pollution can be eliminated or prevented, either immediately or in the long run. The remaining waste that cannot be prevented needs to be adequately treated and disposed of. What is absolutely crucial, however, is to recognize the importance of pollution prevention in the hierarchy of environmental options.[1–3]

With their increasing shortage, appropriate management of heavy metals is important. P2 provides an essential tool for this management. In this chapter, the concept and applications of pollution prevention will be addressed. The discussions include project feasibility analyses, implementation, as well as systematic examination of industrial P2. The purpose is to provide readers with a better understanding of methodology in pollution prevention so that they can use it for better management of heavy-metal-related environmental problems.

2.2 Motivation and Concept of P2

2.2.1 Motivation

Environmental pollution results from inefficiencies in manufacturing processes, both operational practices and improperly designed and utilized equipment. Pollutants can be unused raw materials and by-products resulting from production processes.[1–5] Pollution represents a loss in profits during manufacturing. It also can be a result of human activities associated with social development. Environmental pollution from human activities is never avoidable.

End-of-pipe measures include wastewater treatment, hazardous waste incineration, landfills, and monitoring equipment. They have been used in environmental protection for many years and are an important component of P2 in environmental protection. In the past 20 years, however, many environmental accidents, complaints, and concerns have pressured industries to shift from the traditional end-of-pipe approaches to sound pollution prevention strategies.

Public concern about the environment continues to grow. Public education through various media, such as school, TV, and the Internet, have become powerful tools for spreading information about the environment and its impact on human health. Protection of the environment is increasingly becoming a social responsibility. With increasing understanding of pollutants and their long-term consequences for the environment, some pollutants that were considered less harmful have become more important. The change in arsenic regulation from 50 ppb to 10 ppb in our drinking water is a good example. Pollution is no longer a site-specific problem; it has become a global issue. For example, mercury has been detected in deep-sea animals (e.g., salmon) that are not supposed to be exposed to polluted environments. Mercury accumulation in animals is due to its transport in seawater.[3]

Pollution means loss of raw materials and production of waste (unused raw materials, by-products, and wastes). These activities definitely can cause a loss in profits. In addition, pollution created in the workplace can pose risks to the workers, and faces the public most of the time. For example, the gas from oil-refining operations has such harmful components as NO_x, SO_x, and metal oxides.

2.2.2 Principles

Socioeconomic development is necessary for meeting people's basic needs of food, clothing, transportation, and shelter and also to improve living standards, which must be sustainable. This means that development should be balanced with the environment. Environmental laws and regulations have focused on media-specific, end-of-pipe, and command-and-control regulation of pollutants and wastes. Such pollution control technologies have reduced pollution to a certain extent, but are not cost-effective and need to be upgraded to pollution prevention whenever possible. Recognizing this, Shen had addressed environmental pollution from a practical point of view by outlining the following three principles of environmental pollution, which are comparable to some of the thermodynamic laws familiar to most engineers and scientists.[1,3,5]

2.2.3 Concepts

Environmental practitioners in different organizations define P2 differently based on their own understanding and applications, resulting in different interpretations. However, the basic idea for P2 is to prevent or to reduce the sources of pollution before problems occur (2). It is in complete contrast to media-specific and end-of-pipe control approaches. The P2 concept and practices find broad applications such as waste minimization, clean production, green chemistry, green product, waste utilization, ISO 14000, and a number of other related terminologies.[3–9]

The difference between pollution prevention and pollution control can be illustrated by the following examples. Vaccines prevent illnesses, while antibiotics control illnesses; seat belts prevent injury, while casts and crutches help deal with injuries from car accidents.

According to the Pollution Prevention Act of 1990 and other related regulations, the United States defines pollution prevention as follows:[4]

1. Reduction of the amount of any hazardous substance, pollutant, or contaminant reentering any waste stream or otherwise released into the environment prior to recycling, treatment, and disposal
2. Reduction of the hazards to public health and the environment associated with the release of such substances, pollutants, or contaminants
3. Reduction or elimination of the creation of pollutants through (a) increased efficiency in the use of raw materials or (b) protection of natural resources by conservation

It has to be pointed out that P2 was defined more narrowly as waste reduction or toxic material cutback at sources, and was focused on waste releases from existing manufacturing operations. Releases of waste from production operations, including those from stacks, vents, and outfalls (called point sources) and those from leaks, open vats, paint areas, and other nonconfined sources (called fugitive emissions), are often the major sources of pollution. Certain products, while leaving the manufacturing plant for distribution through transport, storage, and consumption as well as used-product disposal, can cause serious environmental pollution problems, such as hazardous waste treatment, disposal, and remedial sites.

The definition of P2 needs to be updated as we have a better understanding of pollution problems. It should mean broadly minimizing or eliminating the sources of pollution from every place where they are created in any human activity, such as industry, agriculture, transportation, trading, and homes.[8–11]

P2 seeks not only to eliminate or reduce pollutants and wastes but also to eliminate certain harmful products and activities. P2 optimizes the total materials cycle from virgin material, to finished material, to components, to product, to obsolete product, to ultimate disposal, and to various technical and nontechnical services as well.

P2 includes practices that reduce or eliminate the creation of pollutants through increased efficiency in the use of raw materials, energy, water or other resources, or protection of natural resources by conservation. In practice, pollution prevention approaches can be applied not only to industrial sectors but also to all sectors of our society, including energy production and consumption, construction, transport, land use, city planning, government activities, and consumer behavior.[2]

2.3 P2 Laws and Regulations

Several laws have been established in the past 40 years for pollution control and prevention. Among them, the Pollution Prevention Act of 1990 is the most important. In the past five years, it has intensively addressed, together with sustainable society, conservation of resources and energy, and global environmental problems such as greenhouse gas (GHG) as illustrated in Figure 2.1.[5,7,10–17]

The US Congress enacted the Clean Water Act (CWA) of 1972 to achieve a goal of "fishable and swimmable" surface waters.[12] Most industries must meet discharge standards for the pollutants. Specific methods of control such as pollution prevention are not specified. Many facilities have used pollution prevention as a means of reducing the cost of compliance with federal regulations. State and local authorities also have responsibilities to implement the provisions of the CWA. These authorities must enforce the federal guidelines at a minimum, but may choose to enforce more stringent requirements. Some localities include pollution prevention planning requirements into discharge permits.[12]

The Emergency Planning and Community Right-to-Know Act (EPCRA, also known as SARA Title III) requires certain companies to submit an annual report of the amount of listed "toxic chemicals" entering the environments. Source reduction and waste management information must be provided for the listed toxic chemicals.

The Resource Conservation and Recovery Act (RCRA) and Hazardous and Solid Waste Amendments (HSWA) to RCRA require that the reduction or elimination of hazardous waste generation at the source should take priority over other management methods such as treatment and disposal. Hazardous waste generators are required to certify on their hazardous waste manifests that they have programs in place to reduce the volume or quantity and toxicity of hazardous waste generated to the extent economically practicable. Materials that are recycled may be exempt from RCRA regulations if certain conditions are met.

The Water Quality Act of 1987 further strengthened the Clean Water Act, and amendments to the Safe Drinking Water Act required numerous treatment facility upgrades. Although all these acts are dramatic in their protection of US citizens against waterborne diseases and the improvement of our water quality, they placed little emphasis on source reduction or elimination of the root cause of pollution.

To address this issue of the need of regulation reform, the Pollution Prevention Act of 1990 was passed.[4] It is stated as the national policy of the United States that *"pollution should be prevented or reduced at the source whenever feasible. Preventing pollution offers important economic benefits, as pollution never created avoids the need for expensive investments in waste management or cleanup."*

The rationales and the policy of the Pollution Prevention Act (PPA) are given in Table 2.2. It formalizes a national policy and commitment to waste reduction, functioning primarily to promote the consideration of pollution prevention measures at the federal government level. This act crosses media boundaries by establishing a national policy on pollution prevention, including programs that emphasize source reduction, reuse, recycling,

TABLE 2.1

Three Basic Principles of Environmental Pollution[3]

Principle	Description
Pollution from human activities is unavoidable	Pollution is created by the release of pollutants and wastes into the environment as well as by the production of certain environmentally harmful products and services as a result of careless human activities related to social and economic development.
Prevent pollution whenever possible	As a result of the first principle, pollution needs to be cost-effectively managed. Pollution can be prevented or minimized, but may not be completely eliminated. The remaining residual pollution from human activities must be properly treated and disposed of in order to protect human health and the environment.
Minimal pollution is acceptable	Ecosystems can safely handle and assimilate certain amounts of pollution. If pollution is within the environmental quality standards, its impact on human health and the environment can be acceptable. We must work within the confines of the natural laws to prevent pollution problems in an innovative, planned, and economically feasible fashion.

Note: *Human activities* cover production, distribution, transport, storage, mining, urban development, construction, consumption, and services. The word *products* can be industrial, agricultural, mineral, structural, commercial, and others. The word *services* can be conceptual, technical, and physical such as professional and nonprofessional, government and nongovernment services, including design, plan, operation, construction, transportation systems, repair, maintenance, education and training, management, and others.

and training. All these areas are important to the successful implementation of a P2 industrial wastewater management program.

According to the PPA, the US EPA should review existing and proposed programs and new regulations to determine their effect on source reduction. Source reduction activities among the US EPA programs and other federal agencies are coordinated. The US EPA provides public access to environmental data and fosters the exchange of source reduction information. It establishes pollution prevention training programs for federal and state environmental officials. The US EPA is required to develop state and tribal capacity, use federal procurement to encourage source reduction, and establish standard methods of measuring results.

The US EPA has implemented a diverse set of programs and initiatives to meet their obligations defined by the law since 1990. A series of achievements have been reported, including 33/50, Climate Wise, Green Lights, Energy Star, WAVE, the Pesticide Environmental Stewardship Program, Indoor Air, Indoor Radon, Design for the Environment, the Environmental Leadership Program, and the Common Sense Initiative.[12,13]

To achieve a sustainable society through implementation of P2 at the point of source, using greener substances and conservation of natural recourses

TABLE 2.2

Findings and Policy of Pollution Prevention Act of 1990[3]

Findings	(1) The United States of America annually produces millions of tons of pollution and spends tens of billions of dollars per year controlling this pollution. (2) There are significant opportunities for industry to reduce or prevent pollution at the source through cost-effective changes in production, operation, and raw materials use. Such changes offer industry substantial savings in reduced raw material, pollution control, and liability costs as well as help protect the environment and reduce risks to worker health and safety. (3) The opportunities for source reduction are often not realized because existing regulations, and the industrial resources they require for compliance, focus on treatment and disposal rather than on source reduction; existing regulations do not emphasize multimedia management of pollution; and businesses need information and technical assistance to overcome institutional barriers to the adoption of source reduction practices. (4) Source reduction is fundamentally different and more desirable than waste management and pollution control. The Environmental Protection Agency needs to address the historical lack of attention to source reduction. (5) As a first step in preventing pollution through source reduction, the Environmental Protection Agency must establish a source reduction program that collects and disseminates information, provides financial assistance to states, and implements the other activities provided for in this chapter.
Policy	The Congress hereby declares it to be the national policy of the United States that pollution should be prevented or reduced at the source whenever feasible; pollution that cannot be prevented should be recycled in an environmentally safe manner whenever feasible; pollution that cannot be prevented or recycled should be treated in an environmentally safe manner whenever feasible; and disposal or other release into the environment should be employed only as a last resort and should be conducted in an environmentally safe manner.

has been greatly emphasized and practiced in the past five years.[5] P2 is taken as a cornerstone of the sustainability movement in order that our next generations will have a quality of living environment as good as or even better than what we have now.

A strategic plan for 2010 to 2014 was prepared by the US EPA to guide the P2 program for the development of a sustainable world.[5] The P2 opportunities are identified and leveraged for reduction of the emission of greenhouse gases, the use of hazardous materials, and the use of natural resources by which a greener and more sustainable economy can be achieved. In addition, the plan includes an appendix that describes the P2 five-year strategy for the five sector areas: *chemicals and manufacturing industries, hospitality, electronics, building and construction, and municipalities and institutions.*

The intention of the program is to maximize the achievement of five goals for eventual reduction in the emissions of greenhouse gases and use of hazardous materials and natural resources. Table 2.3 provides the list of goals, strategies, measures, and the draft strategic target by 2014.

TABLE 2.3

List of US EPA P2 5-Year Strategic Plan (2010–2014)

Goals	Strategies	Measure	Draft Strategic Target
Goal 1: Reduce the generation of greenhouse gas (GHG) emissions to mitigate climate change	Reduce energy use through redesign Facilitate energy efficiency Promote alternative technologies Target new sources of GHG emissions for elimination	MMTCO2e reduced	115 MMTCO2e reduced by 2014
Goal 2: Reduce the manufacture and use of hazardous materials to improve human and ecological health	Encourage life cycle reduction of hazardous materials through redesign Facilitate information sharing Offer guidance on the reduction of hazardous materials	Hazardous materials reduced	20 billion pounds of hazardous materials reduced by 2014
Goal 3: Reduce the use of water and conserve other natural resources to protect ecosystems	Encourage life cycle reductions of natural resource use through redesign Facilitate conservation practices	Water reduced	190 billion gallons cumulatively reduced by 2014
Goal 4: Create business efficiencies that derive economic benefits and improve environmental performance	Demonstrate P2 effectiveness Promote sustainable business models Promote business incentives Promote technology transfer Promote cross-agency collaboration	Dollars saved	$14 billion saved by 2014
Goal 5: Institutionalize and integrate pollution prevention practices through government services, policies, and initiatives	Advance sustainability initiatives Utilize P2 to help address climate change Promote voluntary consensus standards Leverage resources to support P2 programs Offer tools to support continuous improvement within P2 programs	P2 approaches	–

Note: MMTCO2e: Million metric tons of carbon dioxide equivalent.

Three aims are: 1. Emission of greenhouse gases

2. Use of hazardous materials

3. Use of natural resources

2.4 P2 Technologies

Today's rapidly changing technologies, industrial processes, and products may generate pollutants that, if improperly managed, could threaten public health and the environment. Many pollutants, when mixed, can produce hazards through heat generation, fire, explosion, or release of toxic substances. To prevent these hazards, pollution generators must be required to describe and characterize their pollutants accurately, by including information as to the type and the nature of the pollutants, chemical compositions, hazardous properties, and special handling instructions

In practice, preventive technologies not only reduce the generation of waste materials, but also encourage environment-friendly products and services. They can be applied also to all pollution-generating activities, including energy production and consumption, transportation, agriculture, construction, land use, city planning, government activities, and consumer behavior. In the energy sector, for example, pollution management can reduce environmental damage from extraction, processing, transport, and combustion of fuels.

Pollution prevention is receiving widespread emphasis internationally within multinational organizations and within individual countries. The driving force behind the emphasis is the concept of sustainable development and the hold that this concept has over planning strategies and long-term solutions to global limits and north-south economic issues. A list of important pollution prevention technologies is given in Table 2.4.

TABLE 2.4

A List of Important Pollution Prevention Technologies

No.	Action	Descriptions
1.	Raw material substitution	Eliminating or reducing a hazardous constituent used either in the product or during its manufacture.
2.	End-product substitution	Producing a different product that accomplishes the same function with less pollution than the original product.
3.	Process modification	Changing the process design to reduce waste generation.
4.	Equipment redesign	Changing the physical design of equipment to reduce waste generation.
5.	Direct recycling	Reusing materials directly in the manufacturing process without prior treatment. These materials would otherwise become wastes.
6.	Good housekeeping	Instituting new procedures, such as preventive maintenance, to reduce waste generation.
7.	Inventory control	Minimizing the quantities of raw materials or manufactured product in stock, to eliminate surplus that could become waste when the product is changed or discontinued.

In the energy sector, for example, pollution prevention can reduce environmental damage from extraction, processing, transport, and combustion of fuels. Pollution prevention technologies include (a) increasing efficiency in energy use, (b) substituting fossil fuels by renewable energies, and (c) design changes that reduce the demand for energy. During the past few years, considerable progress and success have been made in achieving pollution prevention in various sectors of our society.

2.5 P2 Benefits

The most important benefit of P2 is that it can achieve national and international environmental goals while coinciding with the interests of human beings and industries. For the long term, businesses will have strong economic incentives to reduce the toxicity and volume of the waste they generate. The benefits of P2 practices include

- Avoiding inadvertent transfer of pollutants across media by treatment and disposal systems
- Reducting the risks from any release of pollutants and wastes into the environment
- Protecting natural resources for future generations
- Saving cost by preventing excessive use of raw material, energy, and natural resources
- Reducing costs of regulatory violations and costs for waste treatment and disposal
- Avoiding long-term potential liabilities associated with releases of wastes and disposal sites
- Enhancing production efficiency and company reputation
- Improving product quality for world trade market competition

When P2 is executed, some wastes can be reused as raw materials or even by-products. Waste reduction means increasing production efficiency and generating more profits. Reducing wastes may also lead to upstream benefits because it reduces ecological damage from raw material extraction and pollutant release during the production process as well as waste recycling, treatment, and disposal operations. For example, extraction and processing metal ores have been generating significantly high volumes of solid waste, wastewater, and waste gases that contain heavy metals, sulfate, lower pH, and many other contaminants. With sound national-level or even global P2 programs in such industries as construction and metal coating, pollution generation can be effectively minimized.

A company with effective P2 plans may well be the low-cost producer and enjoy significant benefits in a competitive world market. Costs per unit produced may drop as P2 measures reduce liability risks and operating costs. Cost savings from prevention come not only from avoiding environmental costs such as hazardous waste disposal fees but also from avoiding costs that are often more challenging to count, such as those resulting from injuries to workers and ensuing losses in productivity. In that sense, prevention is not only an environmental activity but also a tool to promote workers' health and safety.

Furthermore, P2 activities can greatly enhance the company's public image, public health, and public relations. Among all the benefits, the economic benefits of P2 have proved to be the most compelling argument for industry and business to undertake prevention projects.[7]

Many successful P2 cases are available in the literature.[8–13] The P2 program in the US EPA (www.epa.gov) provides a series of good examples. Cost savings by 100%, the reduction of BOD and SS by 75% and 45%, respectively, and the increase in the paper recovery rate by more than 100% are reported.[9,10]

The benefits from the P2 exercises based on long-term evaluation are obvious; however, their short-term advantages may not be significant. Sometimes, implementation of P2 may even have a negative impact on industrial performance.[3,11] A negative relationship between environmental efficiencies (by end-of-pipe or P2 approaches) and US corporate financial performance was reported. Therefore, a sound balance must be carefully planned and maintained.

2.6 Pollution Prevention Feasibility

The level of required analysis of feasibility depends on the complexity of the considered pollution prevention project. A simple, low-capital cost improvement method such as preventive maintenance would not need much analysis to determine whether it is technically, environmentally, and economically feasible. On the other hand, input material substitution/replacement may affect the quality and quantity of product. A major modification in process equipment could require large capital expenditures.

2.6.1 Technical Feasibility

Technical feasibility analysis requires comprehensive knowledge of relevant manufacturing processes, pollution prevention techniques, vendors, and the resources and limitations of the facility. Technical analysis should determine which technical alternative is the most appropriate for the specific P2 project. The facility-related questions to be considered in the analysis would include the following:

- Will it reduce waste?
- Is space available?
- Are utilities available or must these be installed?
- Is the new equipment or technique compatible with current operating procedures, workflow, and production rates?
- Will product quality be maintained?
- How soon can the system be installed?
- How long will production be stopped in order to install the system?
- Is special expertise required to operate or maintain the new system? Will the vendor provide acceptable service?
- Will the system create other environmental problems?
- Is the system safe?
- Are there any regulatory barriers?

All affected groups in the facility should contribute to and review the results of the technical analysis. Prior consultation and review with the affected groups (e.g., production, maintenance, and purchasing) will ensure the viability and acceptance of an option. If a change in production methods is necessary, the project's effects on the quality of the final product must be determined.

2.6.2 Environmental Feasibility

The environmental feasibility analysis weighs the advantages and disadvantages of an option on the environment. Housekeeping and direct efficiency improvements surely have obvious advantages. Some options require a thorough environmental evaluation, especially if they involve product or process changes or the substitution of raw materials. The criteria in the analysis would include

- Reduction in waste quantity and toxicity
- Risk of transfer to other media
- Reduction in waste treatment or disposal requirements
- Reduction in raw material and energy consumption
- Impact of alternate input materials and processes
- Previous successful use within the company or in other industry
- Low operating and maintenance costs

- Short implementation period and ease of implementation
- Regulatory requirements

To make a sound evaluation, as much information as possible should be gathered on the environmental aspects of the relevant product, raw material, or constituent part of the process. The information should include the environmental effects from the production phase, product life cycle, extracting and transporting the alternative raw materials, and treating any unavoidable waste. Energy consumption should be considered. The entire life cycle of both the product and the production process should be part of the evaluation.

A generalized industrial manufacturing plant is illustrated in Figure 2.2. As shown, mass and energy flow enter the manufacturing processes, which include raw materials, energy (e.g., heat and electricity), water used for manufacturing or cooling, and air. In order to enhance chemical reactions, catalysts, which may be very expensive (e.g., rhodium, gold, and silver), are added to the reactors. The inflow can include the foregoing substances and energy. During the manufacturing process, profitable products are produced, together with by-products and wastes. By-products have their own value only when they are used for adjustable applications; otherwise, they can become waste.

Waste can be categorized as harmless or harmful. The former essentially does not have an environmental impact, while the latter is important. Identification of harmful waste, design of new manufacturing processes, and retrofits of existing plants can be conducted with help of knowledge-based approaches or numerical optimization approaches. Conceptual tools have been also used in the development stages of a design.[17–24] A hierarchical decision procedure described by Douglas is a good example[17].

A knowledge-based system, sometimes called an expert system, is a system of rules based on an area of proven expert knowledge. It also can

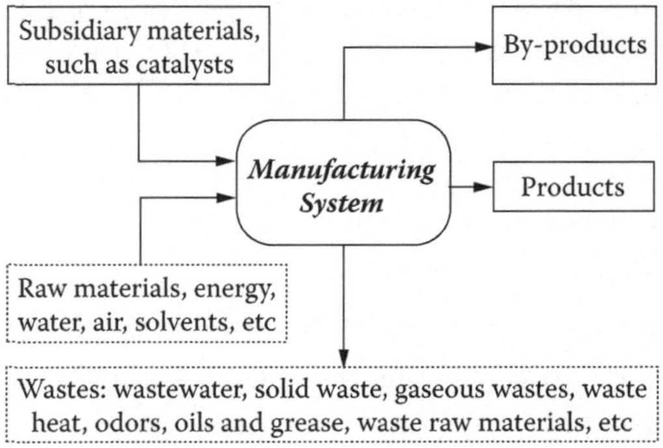

FIGURE 2.2
Illustration of manufacture and subsequent waste generation.

be used for hierarchical design and review procedures. Computer programs based on this system can simulate human thought processes and can therefore be used to design a *cleaner* manufacturing facility to produce less polluted (or *greener*) products. This system is essentially dependent on a long-term accumulation of expert knowledge. It can be used for a new plant design as well as a retrofit of an old plant. More recently, Halim and Srinivasan developed an intelligent system for qualitative waste minimization analysis.[18] A knowledge-based expert system called ENVOP Expert was used to identify practical and cost-effective P2 programs. A case study of the hydrodealkylation process was tested with satisfactory results.

The numerical optimization approaches are based on several considerations, such as energy consumption and mass transfer. Economical analysis together with consumption of both energy and mass were incorporated by some researchers.

The well-cited pinch analysis (or pinch technology), originally developed based on fundamental thermodynamics, has been used to analyze heat flows through industrial processes. It can be used for reduction of energy consumption. It also can be used to minimize wastewater in the process industries.[19] Through water reuse or the internal rearrangements in the manufacturing facility, the emission of waste to the environment can be minimized. This approach was used for the P2 in a citrus plant.[20] An initial diagnosis indicated that the maximum theoretical freshwater consumption or wastewater generation was reduced by 31%.

2.6.3 Economic Feasibility

Economic feasibility analysis is a relatively complex concept. It deals with the allocation of scarce and limited resources to various pollution prevention modifications. It will be used to determine which investments in P2 plans can contribute the most to the company.

A benefit is usually defined as anything that contributes to the objectives of the pollution prevention project; costs are defined as anything that detract from the achievement of a project's objectives. Normally, benefits and costs are evaluated from the perspective of whether they contribute to (or detract from) the maximization of a company's income. An economic cost-benefit analysis uses a number of measures of profitability, such as net present value, internal rate of return, and benefit-cost ratio.

When measuring savings, it is important to look at not only the direct savings but also the indirect savings from P2 implementation. In addition, there are intangible benefits such as a good public image, which is difficult to quantify in financial terms.

The economic feasibility analysis of pollution prevention alternatives examines the incremental costs and savings that will result from each pollution prevention option. Typically, pollution prevention measures require some investment. As with any project, the costs should be broken down as follows:

- **Capital expenditures**: For purchasing process equipment, additional equipment, materials, site preparation, designing, purchasing, installation, utility connections, training costs, start-up costs permitting costs, initial charge of catalysts and chemicals, working capital, and financing charges.
- **Operating costs**: Typically associated with costs of raw materials; water and energy; maintenance; supplies; utilities; labor; waste treatment, transportation, handling, storage, and disposal; and other fees. Revenues may partially offset operating costs from increased production or from the sale or reuse of by-products or waste.
- **Other costs**: Administrative costs; regulatory compliance costs such as permitting, record keeping and reporting, monitoring, and manifesting; insurance costs; workman's compensation; and on-site waste management and control equipment operation costs.

A pollution prevention project can obviously benefit from water, energy, and material savings as well as from waste reduction, recycling, and reuse. It may deliver substantial benefits from an improved product and company image or from improved employee health, which remain largely unexamined. Although they are often difficult to measure, they should be incorporated into the assessment whenever feasible.

2.7 P2 Implementation and Revision

After a pollution prevention project or plan of program is established and its technical, environmental, and economic feasibilities are analyzed, the project implementation begins. Funding is the most important component influencing the project. Some companies may have difficulty raising funds internally for capital investment, especially companies in developing countries. As such, they may seek external funding sources such as loans from banks, international organizations (e.g., Asian Development Bank), and government institutions.

2.7.1 Project Implementation

Implementation of a P2 project will generally follow the established procedures, including process modification, equipment change, and replacement of raw materials. Typically, the following steps are involved:

- Conducting a detailed design
- Preparing a construction bid package and equipment specifications

- Installing new equipment
- Training personnel
- Starting operation
- Operating manual and record keeping
- Monitoring and evaluating performance

2.7.2 Review and Revision of Project

The pollution prevention process does not end with implementation. After the pollution prevention plan is implemented, its performance will be recorded, which can be used in the technical and economic assessments. If the performance fails to meet the original expectations, the project will have to be revised.

2.8 Key Points in P2 Applications

2.8.1 Material Handling and Storage

Material handling and storage operations can cause two types of fugitive emissions: (a) low-level leaks from process equipment and (b) episodic fugitive emissions. Often, the methods for reducing low-level equipment leaks result in fewer episodes, and vice versa.

The methods for reducing or eliminating fugitive emissions are (a) leak detection and repair and (b) equipment modification. Such emissions can be prevented by good practices. Proper materials handling and storage ensures that raw materials reach a process without spills, leaks, or other types of losses, which could result in waste generation. Some basic guidelines for good operational practices to reduce waste are as follows:

- Space containers appropriately to facilitate inspection.
- Label all containers with material identification, health hazards, and first aid recommendations.
- Stack containers according to manufacturers' instructions to prevent cracking and tearing from improper weight distribution.
- Separate different hazardous substances to prevent cross-contamination and facilitate inventory control.
- Raise containers off the floor to inhibit corrosion from "sweating" concrete.

Spills and leaks are major sources of pollutants in industrial processes and material handling. When material arrives at a facility, it is safer to store it prior to use. Material may also be stored during stages of the production process. It is important to prevent spillage, evaporation, leakage from containers or conduits, and shelf-life expirations. Standard operating procedures to eliminate and minimize spills and leaks must take place regularly. Better technology might consist of tighter inventory practices, seal-less pumps, welded rather than flanged joints, bellows seal valves, floating roofs on storage tanks, and rolling covers versus hinged covers on openings.

The storage vessels may contain various wastes with different properties. Reducing waste from storage vessels therefore involves a variety of activities. Storage tanks for storing heavy metals are found at metal coating, semiconductor, and chemical operations. They can become disastrous sources of waste when they leak out to nearby waters such as lake and rivers.

Spills due to overfilling of storage containers are another source of emissions that occur during loading and unloading operations. These spills can be prevented through the use of appropriate overflow control equipment or overflow alarms.

2.8.2 Process Modification

Many industrial plants have prevented pollution successfully by modifying production processes. Such modifications include adopting more advanced technology through process variable controls, changing cleaning processes, chemical catalysts, and segregating and separating waste as follows.

2.8.2.1 Process Variable Controls

Temperature and pressure applications are critical variables as materials are reacted and handled in industrial processes. They can significantly alter the formation of toxins. Improvements include better control mechanisms, better sensors to measure reactions, more precise methods, and computer assistance for automation.

2.8.2.2 Replacement with Cleaning Processes

The cleaning of parts, equipment, and storage containers is a significant source of contamination. Toxic deposits are common on equipment walls. The use of solvents to remove such contamination creates two problems: disposal of the contaminants and emissions from the cleaning process itself. Some changes include the use of water-based cleansers versus toxic solvents, nonstick liners on equipment walls, nitrogen blankets to inhibit oxidation-induced corrosion, and such solvent-minimizing techniques as high-pressure nozzles for water rinse-out.

2.8.2.3 *Chemical Catalysts*

Suitable catalysts can greatly facilitate chemical reactions. Better catalysts and better ways to replenish or recycle them would induce more complete reactions and thus less waste. Substitution of feedstock materials that interact better with existing catalysts can accomplish the same objective.

2.8.2.4 *Segregation and Separation*

A drop of pollutant in a pure solution may create pollution. Segregating wastes and non-wastes reduces the quantity of waste that must be handled. Technical innovations and modifications have provided more precise and reliable separation.

Such physiochemical techniques as distillation, supercritical extraction, membranes, reverse osmosis, ultra-filtration, electrodialysis, and adsorption can efficiently separate pollutants from various industrial effluents.

2.8.3 In-Process Recycling

Materials are processed frequently in the presence of heat, pressure, and catalysts to form products. As materials are reacted, combined, shaped, painted, plated, and polished, excess materials not required for subsequent stages become waste, frequently in combination with toxic solvents used to cleanse the excess from the product. The industry disposes of these wastes either by recycling them into productive reuse or by discharging them as wastes into the air, water, or land. Often, expensive treatment is required to reduce the toxicity and pollutants in the waste discharge before final disposal. These liquid, solid, or gaseous wastes at each stage of the production process are the source of pollution problems. On-site recycling of process waste back into the production process will often allow manufacturers to reduce pollution and save costs for less waste treatment and disposal.

Recycling activities may be performed either on-site or off-site. On-site recycling activities include (a) direct use or reuse of the waste material in a process. It differs from closed-loop recycling, in that wastes are allowed to accumulate before reuse; and (b) reclamation by recovering secondary materials for a separate end use or by removing impurities so that the waste may be reused.

Advantages of on-site recycling include

- Less waste leaving the facility
- Control of reclaimed solvent's purity
- Reduced liability and cost of transporting waste off-site
- Reduced reporting (manifesting)
- Possible lower unit cost of reclaimed solvent

Disadvantages of on-site recycling must also be considered, including

- Capital outlay for recycling equipment
- Liabilities for worker health, fires, explosions, leaks, spills, and other risks as a result of improper equipment operation
- Possible need for operator training; and additional operating costs

Off-site commercial recycling services are well suited for generators of small quantities of waste since they do not have sufficient volume of waste solvent to justify on-site recycling. Commercial recycling facilities are privately owned companies that offer a variety of services, ranging from operating a waste recycling unit on the generator's property to accepting and recycling batches of solvent waste at a central facility.

2.8.4 Materials and Product Substitutions

The issues involving materials and product substitutions are complex as the economic factor, the consumer preferences, and the technological feasibilities must all be considered. Obviously, the use of less toxic materials in production can effectively prevent pollution in a decentralized society. Scientists and engineers are actively evaluating and measuring material toxicity and developing safer materials. Likewise, the life-cycle approach requires that products be designed with an understanding from the raw material stage through the final disposal stage. Examples of product life-cycle applications include substitutes for fast-food packaging, disposal diapers, plastic containers, certain drugs, and pesticides.

2.8.4.1 Materials Substitution

Industrial plants could use less hazardous materials or more efficient inputs to decrease pollution. Input substitution has been especially successful in material-coating processes, with many companies substituting water-based for solvent-based coatings. Water-based coatings decrease volatile organic compound emissions, while conserving energy. In the production of pesticides, toxic arsenic powers can be substituted by polymer-based materials, so that arsenic will not spread in the atmosphere.

2.8.4.2 Product Substitution

Manufacturers may reduce pollution by redesigning or reformulating end products to be less hazardous. For example, chemical products could be produced as pellets instead of powder, decreasing the amount of waste dust lost during packaging. Unbleached paper products could replace bleached alternatives. With uncertain consumer acceptance, redesigning products could be one of the most challenging avenues for preventing pollution in the industrial sector. Moreover, product redesign may require substantial alterations in production technology and inputs, but refined market research and consumer education strategies, such as product labeling, will encourage consumer support.

Changes in end products could involve reformulation and a rearrangement of the products' requirements to incorporate environmental considerations. For example, the end product could be made from renewable resources, have an energy-efficient manufacturing process, have a long life, and be nontoxic as well as easy to reuse or recycle. In the design of a new product, these environmental considerations could become an integral part of the program of requirements.

In both the redesign of existing products and the design of new products, additional environmental requirements will affect the methods applied and procedures followed. These new environmental criteria will be added to the list of traditional criteria. Environmental criteria for product design include

- Using renewable natural resource materials
- Using recycled materials
- Using fewer toxic solvents or replacing solvent with less toxic solvent
- Reusing scrap and excess material
- Using water-based inks instead of solvent-based ones
- Reducing packaging requirements
- Producing more replaceable component parts
- Minimizing product filter
- Producing more durable products
- Producing goods and packaging that can be reused by consumers
- Manufacturing recyclable final products

2.8.5 Materials Separation

In the chemical process industry, separation processes account for a significant portion of investments and energy consumption. For example, distillation of liquids is the dominant separation process in the chemical industry. Pollution prevention technology aims to find methods that provide a sharper separation than distillation, thus reducing the amounts of waste, improving the use of raw materials, and yielding better energy economy. The available technologies are supercritical extraction, membrane filtration, electrodialysis, electrolysis, adsorption, distillation, and supercritical extraction.

2.9 Case Studies

2.9.1 33/50 Program

A successful example is the reduction of important pollutants through the 33/50 program (13). The program targeted 17 priority chemicals (e.g., lead, cadmium, mercury, nickel, benzene, tetrachloroethylene, and toluene) and set as its goal a 33% reduction in releases and transfers of these chemicals by 1992 and a 50% reduction by 1995, measured against a 1988 baseline. Its primary purpose was to demonstrate whether voluntary partnerships could augment the US EPA's traditional command-and-control approach by bringing about targeted reductions more quickly than would regulations alone. The program sought to foster a pollution prevention ethic, encouraging companies to consider and apply pollution prevention approaches to reducing their environmental releases rather than traditional end-of-the-pipe methods for treating and disposing of chemicals in waste. The 33/50 Program achieved its goal in 1994, one year ahead of schedule, primarily through program participants' efforts. Facilities also reduced releases and transfers of the other 33/50 chemicals by 50% from 1988 to 1995.

2.9.2 Water Reduction in Pulp Mill

Single- and multi-objective optimization approaches have been used in the analysis of pollution prevention. The integration approach has been used in pollution prevention/wastewater minimization programs.[21–24] The fundamentals of the approach are optimization/minimization of capital and operating costs with minimum waste production and energy consumption. A series of case studies are described in the literature. Parthasarathy and Krishnagopalan used mass integration for the systematic reallocation of aqueous resources in a Kraft pulp mill.[25] An optimal allocation of chloride in different streams throughout the plant was achieved, which led to the built-up concentration below undesirable levels. More important, the fresh water requirement was reduced by 57%.

2.9.3 P2 Plan in LBNL

There are successful cases reported in the P2 application for metal waste and water management in organizations and companies.[26–29] A P2 project was developed for a metal-cleaning facility in Lawrence Berkeley National Laboratory (LBNL).[26] It was an effort to meet the water conservation and pollution prevention goals of LBNL.

Heavy metal precipitation methods and various chromium reduction methods were tested for the treatment of wastewater samples containing chrome by (1) sodium metabisulfite, (2) ferrous sulfate, and (3) dimethyl dithio carbamate, ferrous sulfate, and alum. It was found that (a) the first three treatment methods were all effective for chromium reduction or absorption, (b) coagulants (or coprecipitants) and polymer must be used in order to effectively precipitate heavy metals out of wastewater. By using a proper combination of chromium reduction and precipitation methods, the treated wastewater sample could meet the discharge limits.

In the water-saving and recycling study, a spray rinse system and two wastewater, closed-loop recycling systems (reverse osmosis and ion exchange) were selected and installed for this facility. With the spray rinse system, the rinse water usage was reduced by at least 40%. With the closed-loop systems, another 50% of the wastewater (7.5 million liters/year) was recycled.

References

1. Shen, T. T. *Industrial Pollution Prevention Book*, 2nd ed. Springer-Verlag, Germany, 1999.
2. US EPA. *Pollution Prevention 1997—A National Progress Report*, EPA-742-R-97-000. Washington, DC, 1997.
3. Chen, J. P., Shen, T. T., Hung, Y. T., and Wang, L. K. P. Pollution prevention. In *Handbook of Industrial Hazardous Waste Treatment*, 2nd ed., chap. 21, pp. 971–1004, L. K. P. Wang, Y. T. Hung, H. H. Lo, and C. Yapidakis, Eds. Dekker/CRC Press, New York, 2004.
4. US EPA. Pollution Prevention Act of 1990. www.epa.gov/p2/pubs/p2policy/act1990.htm, Washington, DC, 2011.
5. US EPA. 2010–2014 Pollution Prevention (P2) Program Strategic Plan. Washington, DC, 2010.
6. US EPA. *Facility Pollution Prevention Guide*, EPA/600/R-92/088. Washington, DC, 1992.
7. US EPA. *Evaluation of EPA Efforts to Integrate Pollution Prevention Policy throughout EPA and at Other Federal Agencies*, Washington, DC, 2008.
8. Overcash, M. The evolution of US pollution prevention, 1976–2001: A unique chemical engineering contribution to the environment—a review. *J. Chem. Technol. Biotechnol.*, 77, 1197, 2002.
9. Bendavid-Val, A., Overcash, M., Kramer, J., and Ganguli, S. *EP3-Environmental Pollution Prevention Project Paper.* US Agency for International Development, Project No. 936-5559, Washington, DC, p. 71, 1992.
10. Das, L. K., and Jain, A. K. Pollution prevention advances in pulp and paper processing, *Environ. Progress*, 20(2), 87, 2001.
11. Sarkis, J., and Cordeiro, J. J. An empirical evaluation of environmental efficiencies and firm performance: Pollution prevention versus end-of-pipe practice. *Eur. J. Operational Res.*, 135, 102, 2001.
12. US EPA. *Printed Wiring Board Pollution Prevention and Control Technology: Analysis of Updated Survey Results*, EPA 744-R-95-006. Washington, DC, 1995.

13. US EPA. *33/50 Program, The Final Record*, Office of Pollution Prevention and Toxics, EPA-745-R-99-004. Washington, DC, March 1999.
14. Bui, L. T. M., and Kapon, S. The impact of voluntary programs on polluting behavior: Evidence from pollution prevention programs and toxic releases. *J. Environ. Econ. Manage.*, 64, 31–44, 2012.
15. Evans, J. W., and Hamner, B. Cleaner production at the Asian Development Bank. *J. Cleaner Production*, 11, 639–649, 2003.
16. Krewer, U., Liauw, M. A., Ramakrishna, M., Babu, M. H., and Raghavan, K. V. Pollution prevention through solvent selection and waste minimization. *Ind. Eng. Chem. Res.*, 41, 4534, 2002.
17. Douglas, J. M. Process synthesis for waste minimization. *Ind. Eng. Chem. Res.*, 31(1), 238, 1992.
18. Halim, I., and Srinivasan, R. Integrated decision support system for waste minimization analysis in chemical processes. *Environ. Sci. Technol.*, 36, 1640, 2002.
19. Wang, Y. P., and Smith, R. Wastewater minimization. *Chem. Eng. Sci.*, 49(7), 981, 1994.
20. Thevendiraraj, S., Klemes, J., Paz, D., Aso, G., and Cardenas, G. J. Water and wastewater minimization study of a citrus plant. *Resour. Conserv. Recycling*, 37, 227, 2003.
21. El-Halwagi, M. M. *Pollution Prevention through Process Integration: Systematic Design Tools*. Academic Press, San Diego, 1997.
22. Alva-Argaez, A., Kokossis, A. C., and Smith, R. Wastewater minimization of industrial systems using an integrated approach. *Comput. Chem. Eng.*, 22, 741, 1998.
23. Savelski, M. J., and Bagajewicz, M. J. On the optimality conditions of water utilization systems in process plants with single contaminants. *Chem. Eng. Sci.*, 55, 5035, 2000.
24. Bagajewicz, M., Rodera, H., and Savelski, M. Energy efficient water utilization systems in process plants. *Comput. Chem. Eng.*, 26, 59, 2002.
25. Parthasarathy, G., and Krishnagopalan, G. Systematic reallocation of aqueous resources using mass integration in a typical pulp mill. *Adv. Environ. Res.*, 5, 61, 2001.
26. Chang, L. Y. Chrome reduction and heavy metals removal from wastewater—A Pollution Prevention Approach, WM'01 Conference, February 25–March 1, Tucson, AZ, 2001.
27. US EPA. *Final Report for Verification of the Metal Finishing Facility Pollution Prevention Tool (MFFPPT)*, EPA/600/R-07/067. Washington, DC, 2007.
28. Washington State Department of Ecology. Pollution Prevention Progress for 23 Washington Electroplating Facilities: An Industry Sector Report, Publication No. 96–426. Olympia, WA, 1996.
29. Cheremisinoff, N. P. Pollution and pollution prevention. In *Handbook of Pollution Prevention and Cleaner Production: Best Practices in the Agrochemical Industry*. Elsevier, Burlington, MA, 2011.

3

Precipitation Technology

3.1 Introduction

Precipitation is the process by which dissolved metals are made insoluble, usually as metal hydroxides, metal sulfide, and metal carbonate. It is the formation of a solid in a solution or inside another solid during a chemical reaction or by diffusion in a solid. When the reaction occurs in a liquid, the solid formed is called the *precipitate*, and the liquid remaining above the solid is called the *supernatant*. The solids or powders derived from precipitation are also known as *flowers*. The conventional metal removal process is often divided into three steps: (1) pretreatment, (2) precipitation, and (3) posttreatment.

Precipitation is the most conventional technology used for production of inorganic materials (e.g., hydrometallurgy) or removal/recovery of toxic (and precious) wastes (e.g., heavy metals). It has been widely used in the treatment of waste streams from industries such as metal mining, metal production, and semiconductor industry.

Over 75% of metal treatment operations employ this technology. There are several precipitation treatment reagents available for metal treatment. Among them, hydroxide precipitation is the standard method of removing heavy metals from wastewater. This may be achieved by increasing the pH of metal-containing wastewater with an alkaline reagent to reduce the concentration of the dissolved metals and removing the resultant metal hydroxide precipitates by sedimentation and dewatering operations.

Precipitation is suitable for the treatment of various heavy metal ions from wastewater. However, it creates a large amount of metal precipitates, which require further treatment and disposal. This would result in both high capital cost and operation cost. An alternative measure called *precipitation–crystallization* has been developed. It can overcome the major problems of the traditional precipitation technology. The technology will be presented in Chapter 4 of this book.

In this chapter, the basic theory of precipitation technology is first presented. The calculation of precipitation is illustrated, and a few examples are provided. The pre- and posttreatment technologies of conventional precipitation technology are presented with detailed explanations of mechanisms.

Finally, several important case studies are provided to illustrate the systems for the treatment of various metal wastewaters.

3.2 Theory

Conventional chemical precipitation is the most frequently applied technology in the removal and recovery of heavy metals when their concentrations in the waste streams are above a few hundred parts per million. The precipitation chemicals used include lime, soda ash, and phosphate, which are called *ligands*. The efficiency of precipitation is highly dependent on the water chemistry, which includes the solution pH, the initial concentrations of metals and ligands, and the presence of competing factors (e.g., complexing agent and other metal ions).[1–13]

3.2.1 Calculation of Precipitation Reaction

Precipitation occurs as follows:

$$A_mB_n(s) = mA^{+n}\,(aq) + nB^{-m}\,(aq) \tag{3.1}$$

where A and B represent metal ion and anionic treatment reagents with the valencies of +n and −m, respectively. At equilibrium, we can define a constant called *solubility product constant* (K_{s0} or K_{sp}) as:

$$K_{s0} = [A^{+n}\,(aq)]^m \times [B^{-m}\,(aq)]^n \tag{3.2}$$

where $[A^{+n}\,(aq)]$ and $[B^{-m}\,(aq)]$ are concentrations of A^{+n} and B^{-m} (aq) at equilibrium, respectively. The solubility product constants of important metal precipitates are given in Table 3.1.[4] Equation 3.2 states the equilibrium relationship of concentrations A and B, which are related by the solubility product constant. In reality, a solution may have different situations of undersaturation, saturation, and oversaturation. As such, a parameter called the *actual ion activity product* (IAP and also termed as Q) is defined as follows:

$$IAP = Q = [A^{+n}\,(aq)]_{act}{}^m \times [B^{-m}\,(aq)]_{act}{}^n \tag{3.3}$$

where $[A^{+n}\,(aq)]_{act}$ and $[B^{-m}(aq)]_{act}$ are concentrations of A^{+n} and B^{-m} at actual state or initial state, respectively. If IAP is above K_{sp}, the solution is oversaturated; if IAP equals K_{sp}, the solution is at equilibrium; and if IAP is less than K_{sp}, the solution is undersaturated.

TABLE 3.1
Solubility Constants for Important Metal Precipitates

Compound	Formula	K_{sp}
Aluminum phosphate	$AlPO_4$	9.84×10^{-21}
Barium bromate	$Ba(BrO_3)_2$	2.43×10^{-4}
Barium carbonate	$BaCO_3$	2.58×10^{-9}
Barium chromate	$BaCrO_4$	1.17×10^{-10}
Barium fluoride	BaF_2	1.84×10^{-7}
Barium hydroxide octahydrate	$Ba(OH)_2 \cdot 8H_2O$	2.55×10^{-4}
Barium iodate	$Ba(IO_3)_2$	4.01×10^{-9}
Barium iodate monohydrate	$Ba(IO_3)_2 \cdot H_2O$	1.67×10^{-9}
Barium molybdate	$BaMoO_4$	3.54×10^{-8}
Barium selenate	$BaSeO_4$	3.40×10^{-8}
Barium sulfate	$BaSO_4$	1.08×10^{-10}
Barium sulfite	$BaSO_3$	5.0×10^{-10}
Beryllium hydroxide	$Be(OH)_2$	6.92×10^{-22}
Bismuth arsenate	$BiAsO_4$	4.43×10^{-10}
Bismuth iodide	BiI_3	7.71×10^{-19}
Cadmium arsenate	$Cd_3(AsO_4)_2$	2.2×10^{-33}
Cadmium carbonate	$CdCO_3$	1.0×10^{-12}
Cadmium fluoride	CdF_2	6.44×10^{-3}
Cadmium hydroxide	$Cd(OH)_2$	7.2×10^{-15}
Cadmium iodate	$Cd(IO_3)_2$	2.5×10^{-8}
Cadmium oxalate trihydrate	$CdC_2O_4 \cdot 3H_2O$	1.42×10^{-8}
Cadmium phosphate	$Cd_3(PO_4)_2$	2.53×10^{-33}
Calcium carbonate (calcite)	$CaCO_3$	3.36×10^{-9}
Calcium fluoride	CaF_2	3.45×10^{-11}
Calcium hydroxide	$Ca(OH)_2$	5.02×10^{-6}
Calcium iodate	$Ca(IO_3)_2$	6.47×10^{-6}
Calcium iodate hexahydrate	$Ca(IO_3)_2 \cdot 6H_2O$	7.10×10^{-7}
Calcium molybdate	$CaMoO_4$	1.46×10^{-8}
Calcium oxalate monohydrate	$CaC_2O_4 \cdot H_2O$	2.32×10^{-9}
Calcium phosphate	$Ca_3(PO_4)_2$	2.07×10^{-33}
Calcium sulfate	$CaSO_4$	4.93×10^{-5}
Calcium sulfate dihydrate	$CaSO_4 \cdot 2H_2O$	3.14×10^{-5}
Calcium sulfite hemihydrate	$CaSO_3 \cdot 0.5H_2O$	3.1×10^{-7}
Cesium perchlorate	$CaClO_4$	3.95×10^{-3}
Cesium periodate	$CsIO_4$	5.16×10^{-6}
Cobalt(II) arsenate	$Co_3(AsO_4)_2$	6.80×10^{-29}
Cobalt(II) hydroxide (blue)	$Co(OH)_2$	5.92×10^{-15}
Cobalt(II) iodate dihydrate	$Co(IO_3)_2 \cdot 2H_2O$	1.21×10^{-2}
Cobalt(II) phosphate	$Co_3(PO_4)_2$	2.05×10^{-35}
Copper(I) bromide	$CuBr$	6.27×10^{-9}

(*Continued*)

TABLE 3.1 (CONTINUED)

Solubility Constants for Important Metal Precipitates

Compound	Formula	K_{sp}
Copper(I) chloride	$CuCl$	1.72×10^{-7}
Copper(I) cyanide	$CuCN$	3.47×10^{-20}
Copper(I) iodide	CuI	1.27×10^{-12}
Copper(I) thiocyanate	$CuSCN$	1.77×10^{-13}
Copper(II) arsenate	$Cu_3(AsO_4)_2$	7.95×10^{-36}
Copper(II) iodate monohydrate	$Cu(IO_3)_2 \cdot H_2O$	6.94×10^{-8}
Copper(II) oxalate	CuC_2O_4	4.43×10^{-10}
Copper(II) phosphate	$Cu_3(PO_4)_2$	1.40×10^{-37}
Europium(III) hydroxide	$Eu(OH)_3$	9.38×10^{-27}
Gallium(III) hydroxide	$Ga(OH)_3$	7.28×10^{-36}
Iron(II) carbonate	$FeCO_3$	3.13×10^{-11}
Iron(II) fluoride	FeF_2	2.36×10^{-6}
Iron(II) hydroxide	$Fe(OH)_2$	4.87×10^{-17}
Iron(III) hydroxide	$Fe(OH)_3$	2.79×10^{-39}
Iron(III) phosphate dihydrate	$FePO_4 \cdot 2H_2O$	9.91×10^{-16}
Lanthanum iodate	$La(IO_3)_3$	7.50×10^{-12}
Lead(II) bromide	$PbBr_2$	6.60×10^{-6}
Lead(II) carbonate	$PbCO_3$	7.40×10^{-14}
Lead(II) chloride	$PbCl_2$	1.70×10^{-5}
Lead(II) fluoride	PbF_2	3.3×10^{-8}
Lead(II) hydroxide	$Pb(OH)_2$	1.43×10^{-20}
Lead(II) iodate	$Pb(IO_3)_2$	3.69×10^{-13}
Lead(II) iodide	PbI_2	9.8×10^{-9}
Lead(II) selenate	$PbSeO_4$	1.37×10^{-7}
Lead(II) sulfate	$PbSO_4$	2.53×10^{-8}
Lithium carbonate	Li_2CO_3	8.15×10^{-4}
Lithium fluoride	LiF	1.84×10^{-3}
Lithium phosphate	Li_3PO_4	2.37×10^{-11}
Magnesium carbonate	$MgCO_3$	6.82×10^{-6}
Magnesium carbonate trihydrate	$MgCO_3 \cdot 3H_2O$	2.38×10^{-6}
Magnesium carbonate pentahydrate	$MgCO_3 \cdot 5H_2O$	3.79×10^{-6}
Magnesium fluoride	MgF_2	5.16×10^{-11}
Magnesium hydroxide	$Mg(OH)_2$	5.61×10^{-12}
Magnesium oxalate dihydrate	$MgC_2O_4 \cdot 2H_2O$	4.83×10^{-6}
Magnesium phosphate	$Mg_3(PO_4)_2$	1.04×10^{-24}
Manganese(II) carbonate	$MnCO_3$	2.24×10^{-11}
Manganese(II) iodate	$Mn(IO_3)_2$	4.37×10^{-7}
Manganese(II) oxalate dihydrate	$MnC_2O_4 \cdot 2H_2O$	1.70×10^{-7}
Mercury(I) bromide	Hg_2Br_2	6.40×10^{-23}
Mercury(I) carbonate	Hg_2CO_3	3.6×10^{-17}

TABLE 3.1 (CONTINUED)
Solubility Constants for Important Metal Precipitates

Compound	Formula	K_{sp}
Mercury(I) chloride	Hg_2Cl_2	1.43×10^{-18}
Mercury(I) fluoride	Hg_2F_2	3.10×10^{-6}
Mercury(I) iodide	Hg_2I_2	5.2×10^{-29}
Mercury(I) oxalate	$Hg_2C_2O_4$	1.75×10^{-13}
Mercury(I) sulfate	Hg_2SO_4	6.5×10^{-7}
Mercury(I) thiocyanate	$Hg_2(SCN)_2$	3.2×10^{-20}
Mercury(II) bromide	$HgBr_2$	6.2×10^{-20}
Mercury(II) iodide	HgI_2	2.9×10^{-29}
Neodymium carbonate	$Nd_2(CO_3)_3$	1.08×10^{-33}
Nickel(II) carbonate	$NiCO_3$	1.42×10^{-7}
Nickel(II) hydroxide	$Ni(OH)_2$	5.48×10^{-16}
Nickel(II) iodate	$Ni(IO_3)_2$	4.71×10^{-5}
Nickel(II) phosphate	$Ni_3(PO_4)_2$	4.74×10^{-32}
Palladium(II) thiocyanate	$Pd(SCN)_2$	4.39×10^{-23}
Potassium hexachloroplatinate	K_2PtCl_6	7.48×10^{-6}
Potassium perchlorate	$KClO_4$	1.05×10^{-2}
Potassium periodate	KIO_4	3.71×10^{-4}
Praseodymium hydroxide	$Pr(OH)_3$	3.39×10^{-24}
Radium iodate	$Ra(IO_3)_2$	1.16×10^{-9}
Radium sulfate	$RaSO_4$	3.66×10^{-11}
Rubidium perchlorate	$RbClO_4$	3.00×10^{-3}
Scandium fluoride	ScF_3	5.81×10^{-24}
Scandium hydroxide	$Sc(OH)_3$	2.22×10^{-31}
Silver(I) acetate	$AgCH_3COO$	1.94×10^{-3}
Silver(I) arsenate	Ag_3AsO_4	1.03×10^{-22}
Silver(I) bromate	$AgBrO_3$	5.38×10^{-5}
Silver(I) bromide	$AgBr$	5.35×10^{-13}
Silver(I) carbonate	Ag_2CO_3	8.46×10^{-12}
Silver(I) chloride	$AgCl$	1.77×10^{-10}
Silver(I) chromate	Ag_2CrO_4	1.12×10^{-12}
Silver(I) cyanide	$AgCN$	5.97×10^{-17}
Silver(I) iodate	$AgIO_3$	3.17×10^{-8}
Silver(I) iodide	AgI	8.52×10^{-17}
Silver(I) oxalate	$Ag_2C_2O_4$	5.40×10^{-12}
Silver(I) phosphate	Ag_3PO_4	8.89×10^{-17}
Silver(I) sulfate	Ag_2SO_4	1.20×10^{-5}
Silver(I) sulfite	Ag_2SO_3	1.50×10^{-14}
Silver(I) thiocyanate	$AgSCN$	1.03×10^{-12}
Strontium arsenate	$Sr_3(AsO_4)_2$	4.29×10^{-19}
Strontium carbonate	$SrCO_3$	5.60×10^{-10}
Strontium fluoride	SrF_2	4.33×10^{-9}

(Continued)

TABLE 3.1 (CONTINUED)
Solubility Constants for Important Metal Precipitates

Compound	Formula	K_{sp}
Strontium iodate	$Sr(IO_3)_2$	1.14×10^{-7}
Strontium iodate monohydrate	$Sr(IO_3)_2 \cdot H_2O$	3.77×10^{-7}
Strontium iodate hexahydrate	$Sr(IO_3)_2 \cdot 6H_2O$	4.55×10^{-7}
Strontium sulfate	$SrSO_4$	3.44×10^{-7}
Thallium(I) bromate	$TlBrO_3$	1.10×10^{-4}
Thallium(I) bromide	$TlBr$	3.71×10^{-6}
Thallium(I) chloride	$TlCl$	1.86×10^{-4}
Thallium(I) chromate	Tl_2CrO_4	8.67×10^{-13}
Thallium(I) iodate	$TlIO_3$	3.12×10^{-6}
Thallium(I) iodide	TlI	5.54×10^{-8}
Thallium(I) thiocyanate	$TlSCN$	1.57×10^{-4}
Thallium(III) hydroxide	$Tl(OH)_3$	1.68×10^{-44}
Tin(II) hydroxide	$Sn(OH)_2$	5.45×10^{-27}
Yttrium carbonate	$Y_2(CO_3)_3$	1.03×10^{-31}
Yttrium fluoride	YF_3	8.62×10^{-21}
Yttrium hydroxide	$Y(OH)_3$	1.00×10^{-22}
Yttrium iodate	$Y(IO_3)_3$	1.12×10^{-10}
Zinc arsenate	$Zn_3(AsO_4)_2$	2.8×10^{-28}
Zinc carbonate	$ZnCO_3$	1.46×10^{-10}
Zinc carbonate monohydrate	$ZnCO_3 \cdot H_2O$	5.42×10^{-11}
Zinc fluoride	ZnF_2	3.04×10^{-2}
Zinc hydroxide	$Zn(OH)_2$	3×10^{-17}
Zinc iodate dihydrate	$Zn(IO_3)_2 \cdot 2H_2O$	4.1×10^{-6}
Zinc oxalate dihydrate	$ZnC_2O_4 \cdot 2H_2O$	1.38×10^{-9}
Zinc selenide	$ZnSe$	3.6×10^{-26}
Zinc selenite monohydrate	$ZnSeO_3 \cdot H_2O$	1.59×10^{-7}

Sulfides

Compound	Formula	K_{spa}
Cadmium sulfide	CdS	8×10^{-7}
Copper(II) sulfide	CuS	6×10^{-16}
Iron(II) sulfide	FeS	6×10^{2}
Lead(II) sulfide	PbS	3×10^{-7}
Manganese(II) sulfide (green)	MnS	3×10^{7}
Mercury(II) sulfide (red)	HgS	4×10^{-33}
Mercury(II) sulfide (black)	HgS	2×10^{-32}
Silver(I) sulfide	Ag_2S	6×10^{-30}
Tin(II) sulfide	SnS	1×10^{-5}
Zinc sulfide(sphalerite)	ZnS	2×10^{-4}
Zinc sulfide (wurtzite)	ZnS	3×10^{-2}

Take note that some of metal precipitates are amorphous. The term amorphous *refers to a solid that does not have a well-organized crystal structure and is almost always more soluble than the corresponding crystalline solid.*[10,11]

EXAMPLE 3.1

Assume an MF_2 solid is placed in a closed container that has deionized water. Identify the regions of oversaturation, equilibrium, and undersaturation when the concentration of metal (M^{2+}) is fixed.

SOLUTION

We have the following precipitation reaction:

$$MF_2 = M^{2+} + 2F^-$$

At equilibrium, we have:

$$K_{sp} = [M^{2+}]_e \times [F^-]_e^2$$

where $[M^{2+}]_e$ and $[F^-]_e$ are concentrations of M^{2+} and F^- at equilibrium, respectively.

$$\log K_{sp} = \log[M^{2+}]_e + 2\log[F^-]_e$$

$$\log[M^{2+}]_e = -pK_{sp} + 2\,p[F^-]_e$$

$\log[M^{2+}]_e$ can be plotted as a function of $p[F^-]_e$, which is the straight line in Figure 3.1.

For any situation, we have:

$$Q = [M^{2+}] \times [F^-]^2$$

We can fix $[M^{2+}]$ and vary $[F^-]$ as shown in Figure 3.1. When the solution is at Region I, it is at oversaturation as the Q value is greater than K_{sp}. As such, the precipitation occurs leading to consumption of metal ion M^{2+}. When the fluoride concentration further moves to the right, and would reach Region II, which is at equilibrium. On the other hand, if the precipitate is placed in a clean solution, it is in Region III. The precipitate would be dissolved, leading to an increase in the fluoride concentration in the solution. The fluoride concentration, however, cannot keep increasing. There would be a point of equilibrium (Region II) where the rate of precipitation equals that of dissolution. At equilibrium (Region II), we have: $[M^{2+}] = [M^{2+}]_e$ and $[F^-] = [F^-]_e$.

If more reactions such as precipitation reactions and solution reactions (metal–ligand complex formations) take place, the (total) metal concentration (solubility) will consist of its free metal concentration as well as the concentrations of metal complexes, determined by the following equation.

$$[Me]_T = [Me^{+m}] + \text{sum of all soluble metal complexes} \qquad (3.4)$$

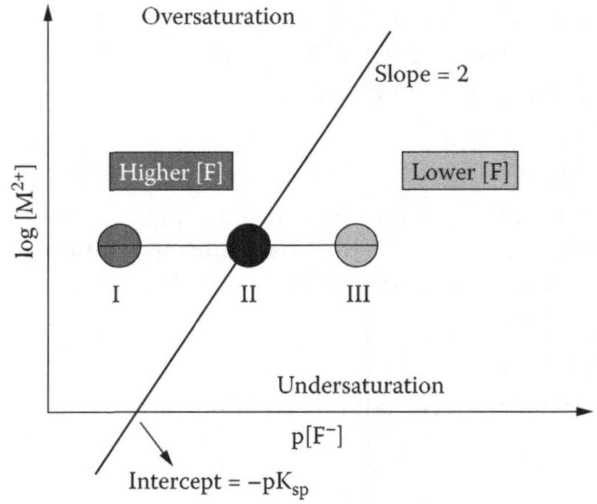

FIGURE 3.1
Illustration of precipitation, equilibrium, and dissolution of metal ions.

where $[Me]_T$ is the (total) metal concentrations (solubility), and $[Me^{+m}]$ is the free metal concentration. For example, the copper solubility in the copper hydroxide containing solution consists of free copper ions (Cu^{2+}) and the important copper complexes of $CuOH^+$, $Cu(OH)_2$, and $Cu_2(OH)_2^{2+}$ as follows:

$$[Cu]_T = [Cu^{2+}] + [CuOH^+] + [Cu(OH)_2] + 2[Cu_2(OH)_2^{2-}]$$

Equation 3.3 can be used to find out the potential of precipitation or dissolution. The determination of metal ion and the anionic reagent concentrations can be done based on the stoichiometry and the solubility product constant. These concentrations can accurrately be determined via computational tools such as MINEQL+.[7] The following two examples are given to illustrate the calculation of metal precipitation. Note that only one precipitation reaction is considered, and solution reactions are assumed to be less important.

EXAMPLE 3.2

Two solutions with silver ion (1 L and 2 mM) and chloride ion (1 L and 2 mM), respectively, are added to a container. The $\log K_{sp}$ is −9.750. Calculate the final concentrations of both silver and chloride. What happens if $[Ag^+]_0$ and $[Cl^-]_0$ are 0.1 M and 10^{-3} M?

SOLUTION

In this case, we have silver ion (A) and chloride ion (B). Let the initial and final concentrations of A and B as $[A]_0$, $[B]_0$, $[A]_f$, and $[B]_f$. When A and B are added together, we will have the following reaction:

$$A + B = AB, K_{sp}$$

$$K_{sp} = [A]_f \times [B]_f$$

According to the stoichiometry of the above precipitation reaction, we can have

$$[A]_0 - [A]_f = [B]_0 - [B]_f$$

$$[A]_f \times [B]_f = K_{sp}$$

$$[B]_f = K_{sp} / [A]_f$$

$$[A]_0 - [A]_f = [B]_0 - K_{sp}/[A]_f$$

$$[A]_f = \frac{-([B]_0 - [A]_0) + \sqrt{([B]_0 - [A]_0)^2 + 4Ksp}}{2} \tag{3.5a}$$

$$[B] = \frac{2Ksp}{-([B]_0 - [A]_0) + \sqrt{([B]_0 - [A]_0)^2 + 4Ksp}} \tag{3.5b}$$

Now, we have $[A]_0 = [Ag^+]_0 = 10^{-3}$ M, $[B]_0 = [Cl^-]_0 = 10^{-3}$ M, and $K_{sp} = 10^{-9.750}$. Thus, we have:

$$[Ag^+]_f = [Cl^-]_f = 1.33 \times 10^{-5} \text{ M}$$

If $[Ag^+]_0$ and $[Cl^-]_0$ are 0.1 M and 10^{-3} M, we have

$$[Ag^+]_f = 9.9 \times 10^{-2} \text{ M and } [Cl^-]_f = 1.8 \times 10^{-9} \text{ M}$$

EXAMPLE 3.3

We have a precipitation reaction as follows:

$$Ca^{2+} + 2F^- = CaF_2 \text{ (s)}$$

Derive the expressions for the final concentrations of Ca^{2+} and F^-. The initial concentrations of Ca^{2+} and F^-, and solubility product constant (K_{sp}) are known. Assume that calcium and fluoride have the initial concentrations of 10^{-3} M, calculate the final concentrations of calcium and fluoride. The solubility product constant is $10^{-10.50}$.

SOLUTION

The precipitation reaction can be expressed as:

$$Ca^{2+} + 2F^- = CaF_2 \text{ (s)}$$

A relationship between the initial concentrations and final concentrations of reactants at equilibrium can be derived as follows according to the stoichiometry in the above reaction:

$$\frac{1}{[Ca^{2+}]_0 - [Ca^{2+}]} = \frac{2}{[F^-]_0 - [F^-]}$$

$$2([Ca^{2+}]_0 - [Ca^{2+}]) = [F^-]_0 - [F^-]$$

Moreover, we can have the equation for solubility product constant as:

$$K_{sp} = [Ca^{2+}]\cdot[F^-]^2$$

The final concentration of calcium ion at equilibrium can be calculated by:

$$[Ca^{2+}] = \frac{K_{sp}}{[F^-]^2}$$

A re-arrangement of the above equations will lead to

$$2\left([Ca^{2+}]_0 - \frac{K_{sp}}{[F^-]^2}\right) = [F^-]_0 - [F^-]$$

$$[F^-]^3 + (2[Ca^{2+}]_0 - [F^-]_0)[F^-]^2 - 2K_{sp} = 0 \tag{3.6}$$

Since the initial concentration of reactants $[Ca^{2+}]_0$ and $[F^-]_0$, and the solubility product constant are known, the final concentration of fluoride can be calculated, which can then be used to determine the final concentration of calcium at equilibrium.

We have $[Ca^{2+}]_0 = 10^{-3}$ M, $[F^-]_0 = 10^{-3}$ M and $K_{sp} = 10^{-10.50}$. The above equation can be solved by numerical technique such as Newton-Raphson Method as follows:

$$x_{n+1} = x_n - \frac{f(x)}{f'(x)}$$

Let $x = [F^-]$ and we can have:

$$f(x) = x^3 + (2[Ca^{2+}]_0 - [F^-]_0)x - 2K_{sp}$$

$$f'(x) = 3x^2 + (2[Ca^{2+}]_0 - [F^-]_0)$$

Noted that $f'(x)$ denotes the derivative of the function $f(x)$.

We have:

$$K_{sp} = 10^{-10.5}$$

$$[Ca^{2+}]_0 = 10^{-3} M$$

$$[F^-]_0 = 10^{-3} M$$

Assume an initial value of 2×10^{-4} M for x, we can solve the above equation for $[F^-]$ through several iterations based on the Newton-Raphson Method. Finally, we can obtain the final concentrations of fluoride as 2.27×10^{-4} M. The final concentration of calcium is 6.14×10^{-4} M.

3.2.2 Typical Treatment Reagents

There are several anionic treatment reagents (ligands) for the precipitation of heavy metal ions. These are hydroxide, carbonate, phosphate, and sulfide ions. The chloride ion, although seldom used, can be employed for the precipitation of silver ion. The important selection criteria are cost, efficacy of treatment, and ease in posttreatment (postseparation). It is obviously easier to understand the importance of cost and efficacy of treatment. However, the last criterion is sometime ignored. As wastewater typically contains other metal ions that may not be toxic, they may be precipitated out together with the heavy metal ions. For example, sodium carbonate can be used to remove copper ions from wastewater. It can also react with calcium and magnesium ions, which leads to additional consumption of carbonate (more costly) and more precipitates (sludge) to treat.

3.2.2.1 Hydroxide

Among treatment reagents, alkali reagent (hydroxide, OH^-) is the most common; the two alkalis commonly used are sodium hydroxide (NaOH) (referred to as caustic soda or simply caustic) and lime (namely, calcium hydroxide or hydrated lime, $Ca(OH)_2$). The addition of alkalis causes an increase in the wastewater pH to the point where the lower.

Sodium hydroxide is more expensive than lime. However, lime has the advantage of being and acting as a coagulant during the precipitation or settling process, whereas sodium hydroxide does not.

Lime must be produced on site (via slaking) from CaO (quicklime), and the resulting slurry continuously stirred to prevent cake formation. Lime may react with sulfates in the wastewater, producing a hard precipitate and resulting in formation of scaling. Thus, it is not advisable to use lime for the pH adjustment, coagulation, and precipitation if the concentration of sulfate is too high.

In the metals precipitation processes by alkalis, most operations use caustic only (over 80% of operations) according to the National Metal Finishing Resource Center (NMFRC) in the United States;[12] lime is less used, and a combination of caustic and lime is also used in around 10% of operations. Magnesium hydroxide is seldom used.

Hydrated lime ($Ca(OH)_2$, traditionally called *slaked lime*, is cheaper than caustic soda. The metal hydroxide precipitants produced by the use of lime have much faster settling rates because of the formation of calcium solids (coprecipitation of calcium with hydroxide). As the settled precipitate sludge from the lime precipitation treatment has a higher solid content, it is easier to dewater the sludge.

On the other hand, the precipitation rate of caustic soda is faster than that of lime and thus a less complicated feed system is needed. The precipitation tank is smaller when caustic soda is used. The sludge from it is much less than that from lime.

In addition to the above added chemicals, hydroxide can also be generated through biological approaches. Remoudaki et al. reported the microbial strains that could generate hydroxide.[13] The strains were *Alcaligenes eutrophus* CH34 (from the sediment of a zinc decontamination tank, Belgium), *A. eutrophus* ER121 (from the zinc contaminated soil, Belgium), *Pseudomonas mendocina* AS302 (from a copper mine in Shituru, Zair), and *P. stutzeri* EM77 (from Rhodium Madeleine Mines in Canada). They could grow by using an easily available and inexpensive carbon source such as acetate. The growth was observed for all the strains in the metal-bearing wastewaters. The solution pH increased from neutral to alkaline values during growth. Metal precipitation thus occurred due to the metabolically generated alkalinity in two pilot moving bed sand filters.

3.2.2.2 Carbonate

When carbonate is present, metals such as lead, cadmium, and nickel can form insoluble metal carbonates that can achieve the metal removal. Some wastewaters may already contain enough carbonates to cause precipitation to occur when heavy metals are introduced. Normally, inorganic carbonates (e.g., Na_2CO_3) can be added for the metal removal/precipitation.

Carbonate precipitation takes place only if carbonate ions (CO_3^{-2}) are present. The modeling by MINEQL+ shows that free carbonate ions are present only if the pH is high as shown in Figure 3.2. A caustic is often added to raise the pH. High pH can also promote the precipitation of the metals as hydroxides. Hence, carbonate precipitation is often a coprecipitation that involves the formation of metal carbonates and metal hydroxides. The metal carbonate precipitates can settle and be dewatered more easily than the metal hydroxide precipitates. However, when pH in solution is above a certain level (e.g., 10 in some cases), the formation of metal hydroxyl complexes is enhanced, which can increase the metal solubility and reduce the precipitation effectiveness.

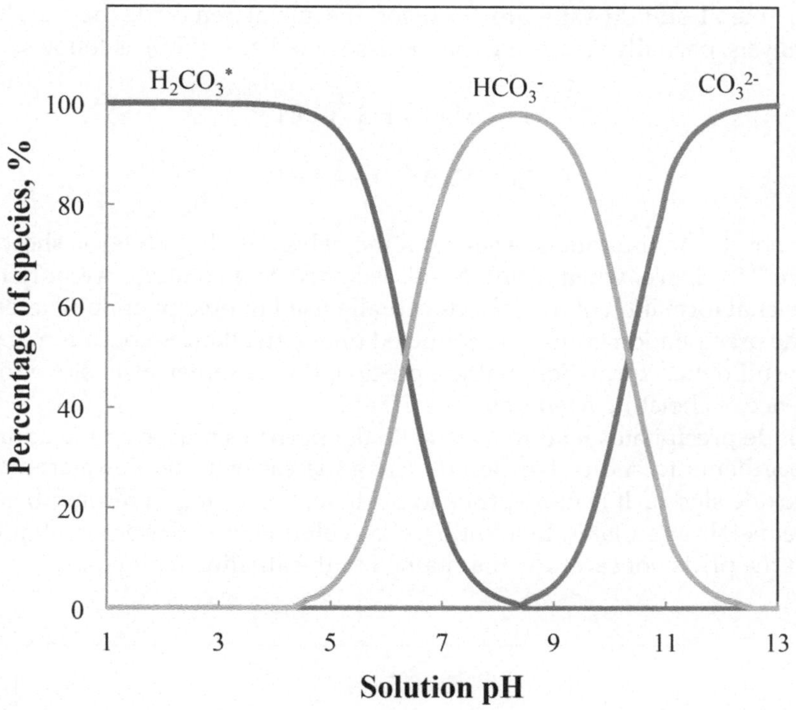

FIGURE 3.2
pC-pH diagram of carbonate.

3.2.2.3 Sulfide

The metal sulfide precipitation can lead to extremely low metal concentrations due to the very low solubility as shown in Table 3.1. This process is particularly powerful for the treatment of metal ions in the chelated waste streams as the residual metal concentrations in the treated wastewater are much lower than those by other precipitation reagents. Sodium sulfide and calcium sulfide can be used as the precipitation reagent. The metal sulfide precipitates are less acceptable for the off-site recovery than metal hydroxide precipitates. In addition, the volume of metal sulfide sludge generated is much larger than that of metal hydroxide sludge, which is normally due to the such metal sulfides as Fes and Cas. When sulfide is used, we would have hydrogen sulfide (H_2S), HS^-, and S^{-2} in the water. There could be some odor if the pH of the solution is too low because of the presence of hydrogen sulfide.

Hydrogen sulfide is a weak acid and dissociates in two steps according to:

$$H_2S = H^+ + HS^-$$

$$HS^- = H^+ + S^{-2}$$

When metal sulfide salts are used for the metal removal, they undergo hydrolysis, partially reforming the undissociated acid (H_2S) as follows:

$$S^{-2} + H_2O = HS^- + OH^-$$

$$HS^- + H_2O = H_2S + OH^-$$

S^{-2} is not the dominating species until the pH is very high (~14) as shown in Figure 3.3. Hence, when a sulfide salt is added to a solution, we only have the partial formation of the S^{-2} ions actually used in precipitation. Therefore, sulfide precipitation is always conducted under an alkaline condition to promote sulfide ion formation. Sulfide precipitation is rather insensitive to the presence of chelating agents such as EDTA.

Sulfide precipitates tend to form colloidal particles requiring the addition of coagulants to settle. The sulfide sludge is easier to be dewatered than hydroxide sludge. It is more prone to oxidation, resulting in resolubilization of the metals as sulfates. In addition, the evolution of H_2S is a potential hazard if the pH is not carefully maintained in the alkaline region.

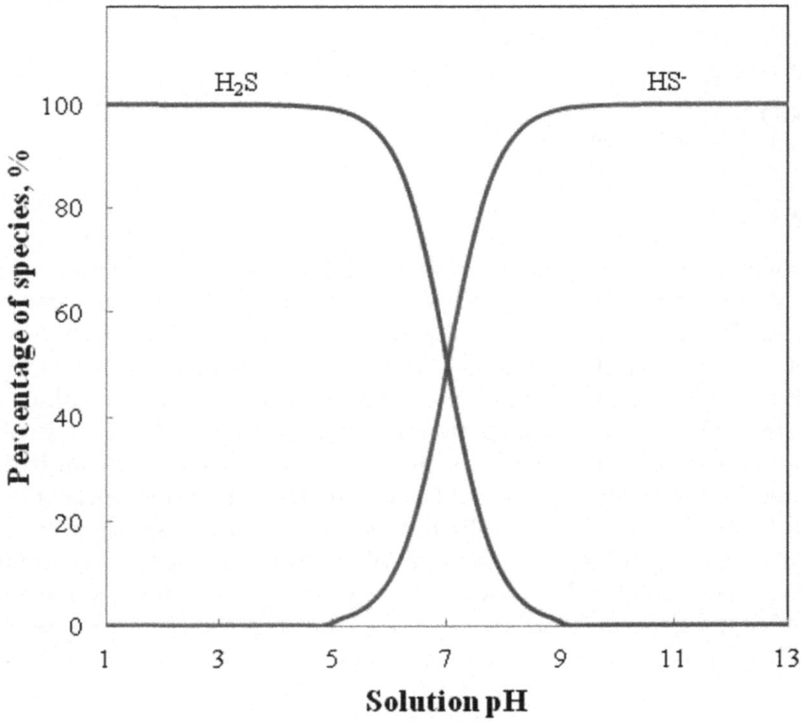

FIGURE 3.3
pC-pH diagram of sulfide species.

In some cases, metal ions can be removed through coprecipitation that involves the formation of more than one metal precipitate. Adsorption may be involved in the removal. For example, the toxic arsenic ion from industrial mineral-processing effluents can be removed and immobilized by typically the addition of lime and coprecipitation of arsenate with ferric iron. It has been widely practiced and has been proved to be effective in the decontamination of the arsenic.

3.2.3 Important Operational Parameters

It is known that pH plays a critical role in the metal precipitation reactions. Each metal precipitate has a characteristic solubility that depends on pH. The pH is usually adjusted to between 6.0 and 12.0, with 9 being the most frequent target. Figures 3.4–3.6 provide the solubility of metals of important precipitates as a function of pH, which is determined by MINEQL.[7] Note that

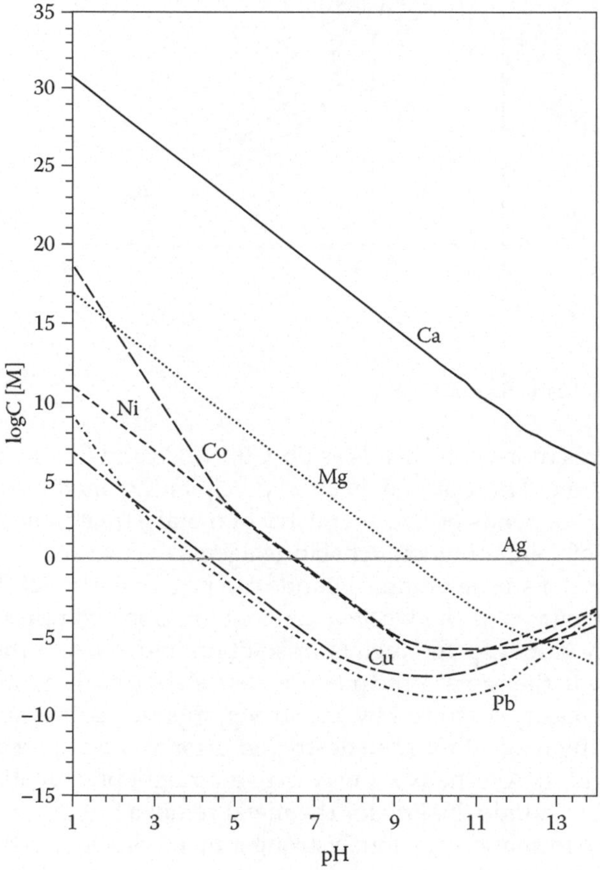

FIGURE 3.4
Solubility of metal hydroxide ($M(OH)_2$).

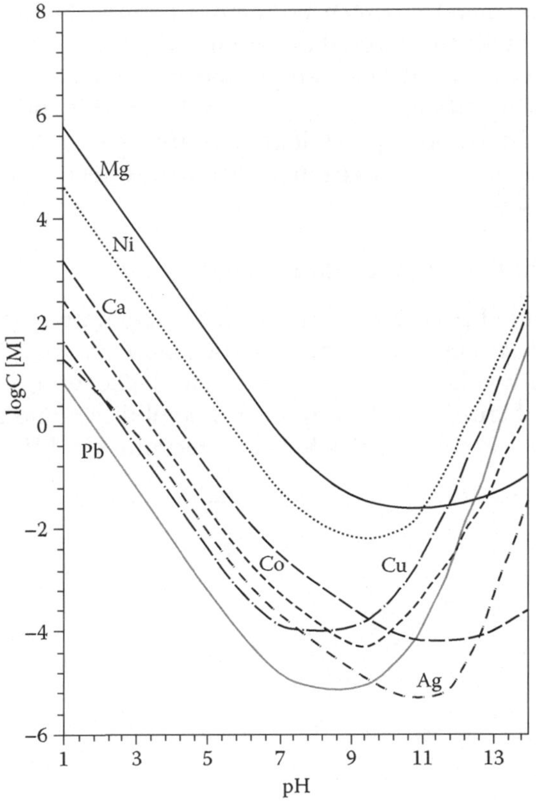

FIGURE 3.5
Solubility of metal carbonate (MCO_3).

M in the figures represents metals as Pb, Cu, Ni, Co and Ca; the unit is mol/L of concentration. The optimal pH for precipitating metals from a waste stream greatly depends on the chemistry of metals, treatment reagents, and the presence of competing ions/chelating agents.

The presence of strong organic ligands can negatively affect the formation of metal precipitates. If the wastewater contains such complexing ligands/reagents, they can keep the metals in solution and prevent them from the precipitation. If the metal precipitate is less stable, such as metal hydroxides, it can be easily dissolved by the strong organic ligands such as EDTA. The complexing reagent must be destroyed prior to metal precipitation (e.g., metal hydroxides). Alternatively, we can use stronger precipitation treatment reagent such as sulfide (NaHS) for the metal removal; in this case, a reagent can lead to the formation of a much stronger metal precipitate that cannot be dissolved by the complexing reagent.

As a strong complexing reagent, the EDTA can greatly reverse the precipitation reaction. The pH-dependent EDTA speciation is given in Figure 3.7.

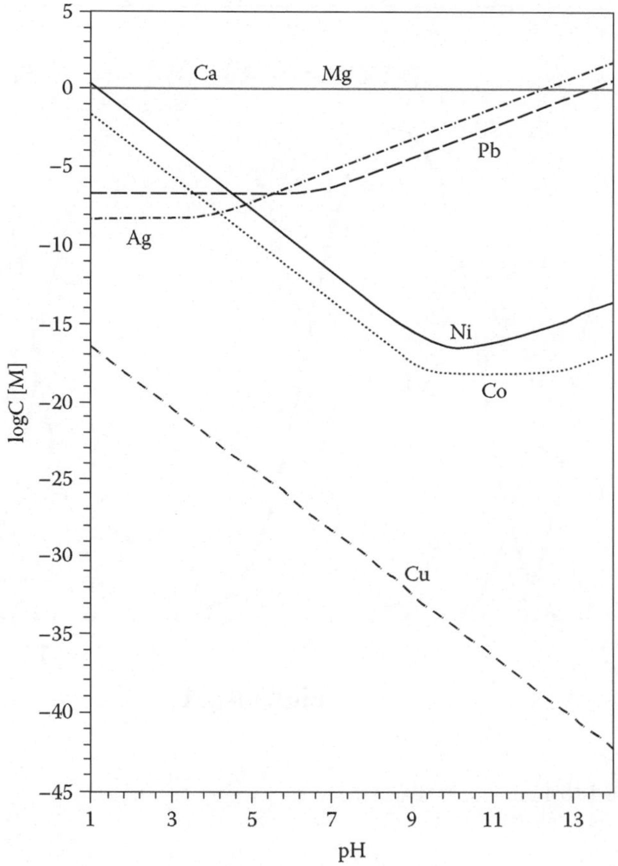

FIGURE 3.6
Solubility of metal sulfur (MS).

Table 3.2 shows that the EDTA can significantly affect the formation of copper precipitate. When 2×10^{-2} M sodium hydroxide is added to 1×10^{-2} M copper, a complete precipitation of copper occurs, and there is no soluble copper ion in the solution. However, the precipitate is totally dissolved once 2×10^{-2} M EDTA is added into the solution. The soluble copper ions are in the form of $CuOHEDTA^{3-}$ (5.85×10^{-3} M) and $CuEDTA^{2-}$ (4.15×10^{-3} M), respectively.

Cyanide is important because it can form complexes with many heavy metals. Note that cyanide is commonly used in metal coating and electronic chip productions. Cyanide is typically removed via oxidation with chlorine under alkaline conditions. When operated, the solution pH is extremely important. The cyanide becomes extremely toxic when it exists as HCN. As shown in Figure 3.8, the cyanide exists as HCN when the pH is less than 6. Thus, the operations of chemical reduction or precipitation must be conducted at pH above 12, where the HCN does not exist.

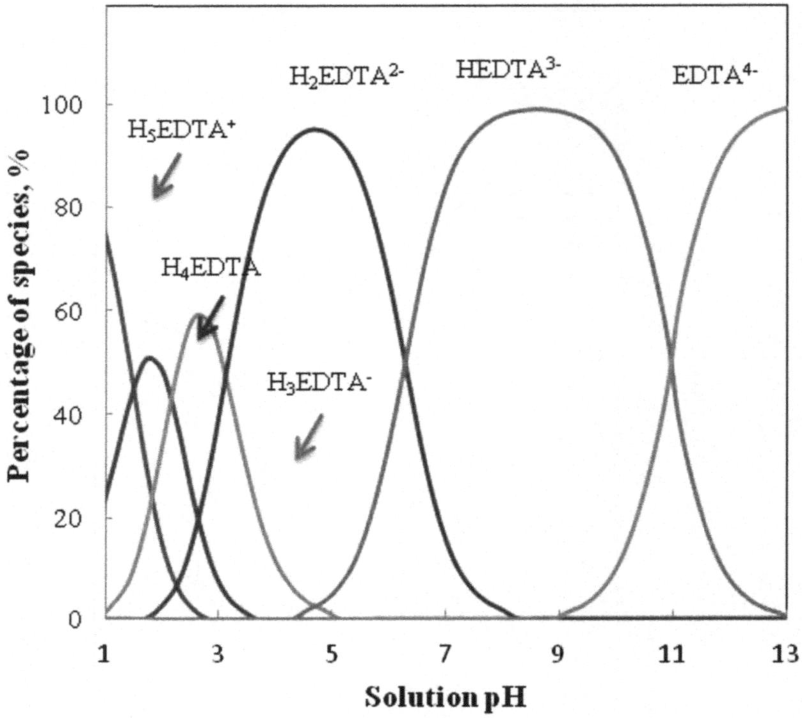

FIGURE 3.7
pH effect on the distribution of EDTA species.

TABLE 3.2
Effect of EDTA on Copper Precipitation Predicted by MINEQL+

$[CuCl_2]$	NaOH	Na_4EDTA	pH	Soluble Copper Ion	Precipitates
10^{-2} M	2×10^{-2} M	0	7.39	0	10^{-2} M (Tenorite)
10^{-2} M	2×10^{-2} M	0.5×10^{-2} M	7.57	5×10^{-3} M (in Cu_2EDTA)	5×10^{-3} M (Tenorite)
10^{-2} M	2×10^{-2} M	10^{-2} M	11.79	9.99×10^{-3} M	4.36×10^{-6} (Tenorite)
10^{-2} M	2×10^{-2} M	2×10^{-2} M	12.15	10^{-2} M*	0

*: $[CuOHEDTA^{3-}] = 5.85 \times 10^{-3}$ M; $[CuEDTA^{2-}] = 4.15 \times 10^{-3}$ M.

3.2.4 Treatability of Individual Metals

3.2.4.1 Arsenic

Arsenite (As(III)) can be removed by precipitation as arsenic sulfide through the addition of sodium sulfide according to the following equation that cause a very low solubility precipitate (As_2S_3), leading to the effluent concentration as low as 0.05 ppm.

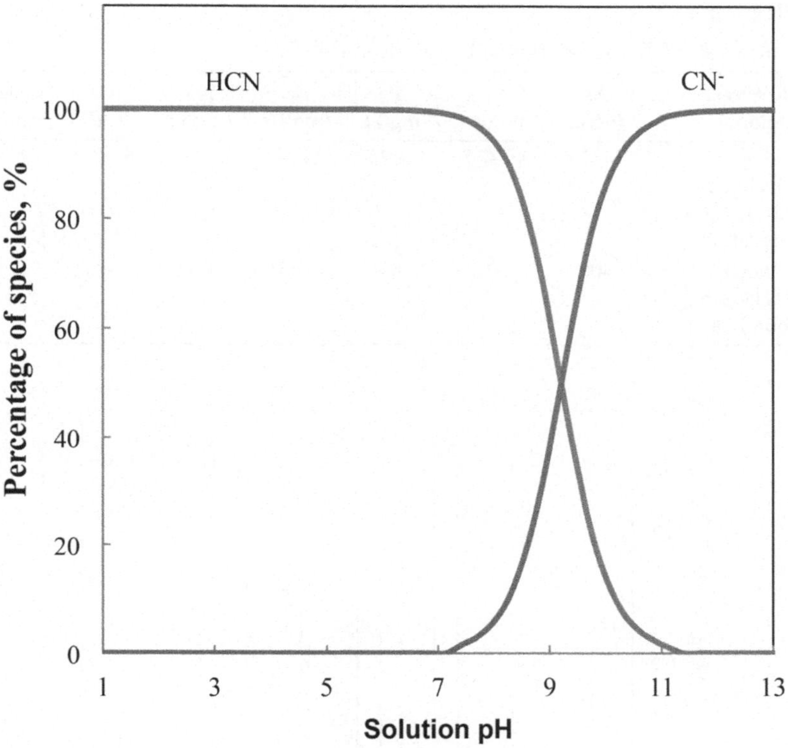

FIGURE 3.8
pH effect on the distribution of cyanide species.

$$2AsO_3^{3-} + 3HS^- + 9H^+ = 6H_2O + As_2S_3$$

Arsenic can be removed by coprecipitation with an addition of $FeCl_3$ forming a $Fe(OH)_3$ floc. The effluent concentration can be as low as 0.005 ppm.

Meng et al. conducted laboratory and field tests to evaluate the effectiveness of a household filtration process and investigate the effects of phosphate and silicate on the removal of arsenic from Bangladesh groundwater by ferric hydroxides. It was reported that the Fe/As ratios of greater than 40 (mg/mg) were required to reduce arsenic to less than 50 μg/L in the Bangladesh well water and the simulated groundwater in the presence of phosphate, silicate, and other impurities as shown in Table 3.3. The household filtration process included coprecipitation of arsenic by adding a packet (approximately 2 g) of ferric and hypochlorite salts to 20 L of well water and subsequent filtration of the water through a bucket sand filter.[14] A field demonstration study was performed to test the treatment system in seven households in Bangladesh in early 2000. The field data obtained from the participating families demonstrated that the household treatment system was able to remove arsenic from

TABLE 3.3

Groundwater Characteristics in Bangladesh [14]

Well Water Location	As (μg/L)	Fe (mg/L)	P (mg/L)	Si (mg/L)	Na (mg/L)	Ca (mg/L)	Mg (mg/L)
Kishoreganj, Munshiganj, and Chandpur, Bangladesh	280–600	4.7–7.7	1.6–2.7	14–20	15–78	65–151	14–42
Simulated Bangladesh groundwater	400	0	0–1.9	0–18	50	100	20

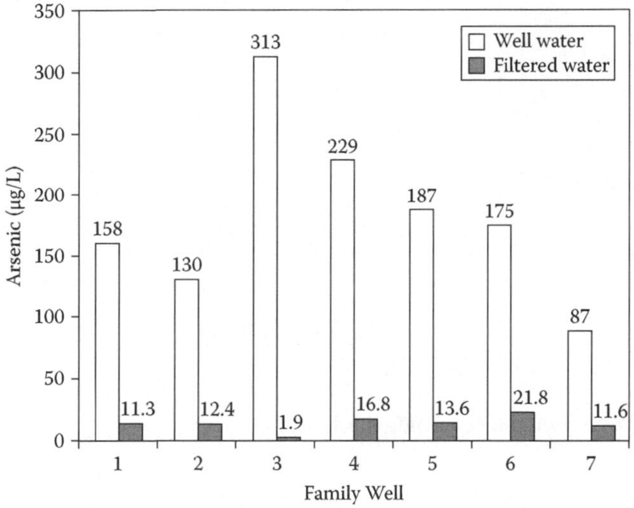

FIGURE 3.9
Treatment efficacies of arsenic-contaminated groundwater by ferric hydroxides.[14]

approximately 300 μg/L in the well water to less than 50 μg/L as demonstrated in Figure 3.9. The participating families were willing to accept the simple and affordable process and use it to prepare clean water for drinking and cooking.

3.2.4.2 Cadmium

The removal of cadmium can be achieved by precipitation as hydroxide at pH ranging from 8 to 11, which leads to less than 1 ppm of cadmium in the treated solution. It can be removed by coprecipitation/coagulation at pH 6 to 7 with $FeCl_3$. Sulfide can be used for the removal of cadmium. Carbonate can also be used to precipitate cadmium in wastewater under pH controlled at 7.5 to 8.5. The effluent concentration is less than 1 ppm, which is comparable to the precipitation by hydroxide.

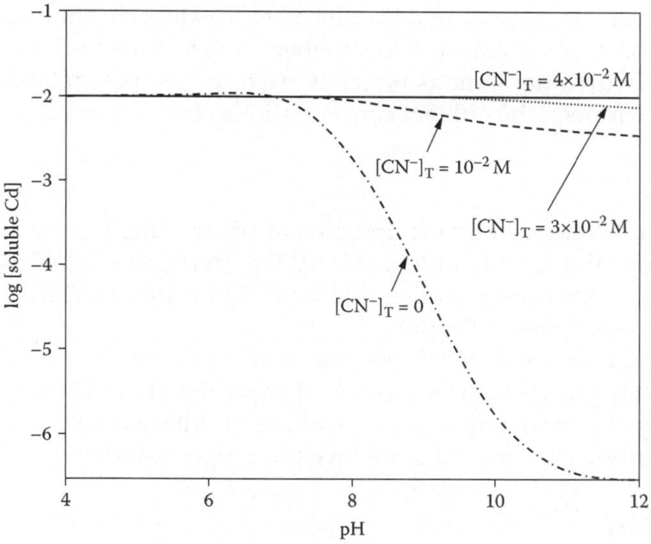

FIGURE 3.10
Effect of cyanide on precipitation of cadmium precipitation by hydroxide, assuming that the initial cadmium concentration is 10^{-2} M.

Cyanides interfere with any of these precipitation treatment reagents. Figure 3.10 shows the effect of cyanide on the formation of cadmium hydroxide, which is determined by MINEQL+. At the initial cadmium concentration of 0.01 M, great removal by precipitation at pH 12 can be obtained. However, the precipitation cannot occur when the solution has cyanide with a concentration of 4×10^{-2} M. As such, the complexing reagent must be removed prior to cadmium precipitation. Chemical oxidation at an alkaline condition as a pretreatment step can effectively remove the cyanides.

3.2.4.3 Chromium

Trivalent chromium is typically precipitated through the addition of hydroxide, which forms chromium hydroxide ($Cr(OH)_3$). Other precipitation treatment reagents such as carbonate and sulfide can also be used.

A two-step process is needed when a wastewater contains hexavalent chromium. First, the hexavalent chromium can be reduced to trivalent chromium by reducing reagents such as $FeSO_4$, $Na_2S_2O_5$ (sodium bisulfite), or SO_2 (sulfur dioxide) at pH < 3. The reduction reactions are given as follows:

$$Cr_2O_7^{-2} + SO_2 \rightarrow Cr^{+3} + SO_4^{-2}$$

$$Cr_2O_7^{-2} + Fe^{+2} + H^{+} \rightarrow Cr^{+3} + Fe^{+3} + H_2O$$

Once the hexavalent chromium is reduced to trivalent chromium, hydroxide can be added or pH is adjusted to a higher level, leading to the formation of chromium hydroxide. Lime is typically used for the precipitation reaction. After the treatment, the effluent concentration is below 1 ppm.

3.2.4.4 Copper

Copper can be removed by precipitation at pH ranging from 9 to 11, resulting in the metal solubility of less than 0.05 ppm (Figure 3.4). Copper can be precipitated in the presence of sulfide at pH 8 to 9, and the resulting effluent concentration is below 0.02 ppm.

However, the presence of complexing reagents such as cyanide, EDTA, and ammonia may greatly interfere with copper precipitation. The reagent can be removed by chemical methods (e.g., oxidation). Alternatively, adsorption by activated carbon can be used to remove the copper complexes.

3.2.4.5 Nickel

Nickel can be removed by adding hydroxide, which increases pH to 9 or 11.5 in the wastewater. Under such a condition, the nickel solubility (metal concentration) is neglible (Figure 3.4). In addition, carbonate and sulfide can also cause effective removal of the metal. Complexing reagents such as EDTA and cyanide can interfere with the precipitation. They must be removed to ensure the precipitation treatment efficiencies.

3.2.4.6 Mercury

Mercury can be removed by precipitation through the addition of sodium sulfide to the wastewater, leading to the effluent concentration below 0.01 ppm. Alum, $FeCl_3$, and other coagulants may be used for the removal; the effluent concentration may be less than 0.01 ppm.

3.2.4.7 Lead

Lead can be removed by sodium hydroxide or lime. Under pH 9-11, the lead ions can be greatly removed, leading to Sulfide and carbonate can also be used for the removal of lead. Typically, the pH is controlled between 7–9, which results in residual concentrations comparable to that by hydroxide.

3.3 Pretreatment

In addition to heavy metal ions, there are several types of impurities in raw heavy metal water stream. Typically, they are waste acids, organic compounds, oils and greases (O&G), suspended solids (SS), and microbial contaminants. The waste acids can lead to a low pH such as less than 2, which presents an unfavorable condition for precipitation. The organic compounds may be chelating agents and form metal complexes that may retard the efficiencies of precipitation. For example, the precipitation cannot work for metal removal if a strong ligand EDTA is in the solution, as illustrated in Table 3.2. Oils and greases, which are also organic, could cause interference in the precipitation process (as well as adsorption/ion exchange). SS can be organic, inorganic, and microbial. They may affect the precipitation. In addition, SS must be reduced to a certain level in the treated stream in order to meet regularity requirement.

Prior to the key unit, pretreatment processes are required to remove the aforementioned substances that would interfere on the subsequent precipitation treatment. Coagulation/flocculation, sedimentation, sand filtration, membrane filtration, and dissolved air flotation (DAF) are the available technologies used for the pretreatment.

3.4 Posttreatment

After the precipitation and the separation of precipitates from the treated stream, a pH adjustment may be required if it does not meet the regulatory requirement. The pH neutralizer requires a different retention time that depends on the type of pH neutralizer used. Both sodium hydroxide and lime can be used; however, the retention time is different, even if a good mixing is applied. The retention time may be longer if the solution is not well stirred, which is always the case in the operation. The retention time is above 15 min when sodium hydroxide is used; it would take a minimum of 30 min when lime is used.

3.5 Key Devices in Pre- and Posttreatment Steps

3.5.1 Coagulation and Flocculation

Coagulation and flocculation are two processes associated with each other, which is used in water and wastewater treatment. Coagulation is a process that destabilizes the fine and individual SS (and organics) with a great stability. Flocculation is a process that grows the destabilized SS to a series of large floc. These two processes are commonly termed *coagulation*.

Most of the metals removal systems use a chemical called *coagulant* (and other aids) to foster the growth of SS (and organics) before the metal waste stream enters a clarifier. In operation, the coagulant is added to the water, which may be alum, ferrous sulfate, and ferric chloride. In addition, a highly diverse range of organic polyelectrolytes suitable for different wastewaters may be used. They are normally superior to inorganic coagulants. The long length of the polymer allows the SS and larger organic compounds to attach each other.

A rapid mixing must be applied in order that the coagulant can be mixed well in the water. The stable SS may become destabilized due to the presence and the effects of the coagulant. The mechanisms include neutralization of surface charge of the SS, adsorption of the SS onto the metal precipitate surfaces (e.g., $Al(OH)_3$ or $Fe(OH)_3$), and attachment of SS onto the long-chained macromolecules (e.g., $Al_n(OH)_m^{3n-m}$). As a result, the SS, large organic compounds, and O&G can be agglomerated into, fast-settling particles (a series of much large floc) that may settle rapidly and then be easily separated by the clarifier or float to the surface of water that can be removed by the dissolved air flotation (DAF). Normally, the appropriate coagulant is selected by testing a range of concentrations, pH, and observing settling behavior after mixing, which is called as *jar testing*.

Commonly used coagulants are alum (aluminum sulfate), ferrous sulfate ($FeSO_4$), sodium hydrosulfite, soda ash, and sodium dithiocarbamate. Ferrous sulfate can cause both reduction (due to the divalent iron) and precipitation (due to the $Fe(OH)_2$ and/or $Fe(OH)_3$) in the removal of metallic wastewater. While being an effective precipitation reagent, the process by ferrous sulfate generates approximately four times as much sludge as sodium hydroxide alone.[12] Calcium chloride and alum are typically used in the pretreatment for the removal of fluoride, phosphate, silicates, and emulsified oil. A better efficacy may be achieved when both calcium chloride and alum are simultaneously used.

3.5.2 Sedimentation

Sedimentation is a process that separates solids from water by the difference of gravity between the solids and the water. It is also called as *clarification*. It is the most common technology for the removal of SS from the incoming metallic wastewater in the pretreatment and for the removal of metal precipitates in the posttreatment prior to its final discharge to the sewage works.

The two most successful types of sedimentation process are sludge blanket clarifier and plate settler. In the sludge blanket clarifier, the wastewater first passes through a sludge blanket that has agglomerated particles as shown in Figures 3.11 and 3.12. The mixing can promote the particle growth (i.e., flocculation) and reduce the concentration of SS. The plate settler has a series of inclined plates, through which the wastewater flows in an upward direction. The particles settle a few inches before impinging on the plate surface. The particles subsequently slide down the plate surface to the base of the separator as shown in Figure 3.13.

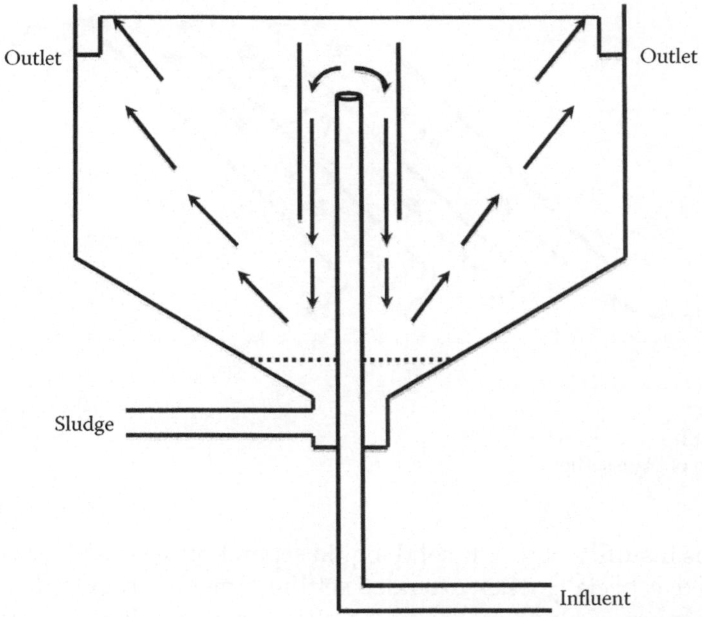

FIGURE 3.11
Illustration of circular basin.

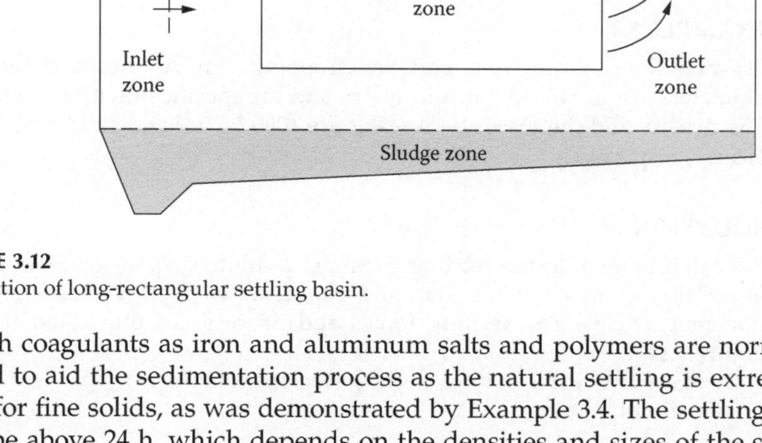

FIGURE 3.12
Illustration of long-rectangular settling basin.

Such coagulants as iron and aluminum salts and polymers are normally added to aid the sedimentation process as the natural settling is extremely slow for fine solids, as was demonstrated by Example 3.4. The settling time may be above 24 h, which depends on the densities and sizes of the solids. Thus, coagulant must be added to shorten the settling time. If one single

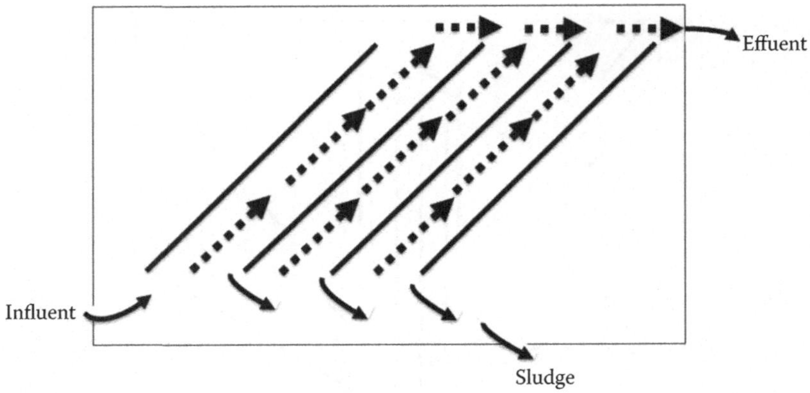

FIGURE 3.13
Illustration of plate settler.

clarifier is insufficient for the solid–liquid separation, a secondary clarifier is often used, especially when extended settling times are required.

Performance of clarifiers varies significantly, depending on the type of wastewater, the design of upstream components, the design of the clarifier, and the dosage of coagulants. A properly operated clarifier should have SS less than 0.30 mg/L in the effluent. A well-operated clarifier would have 5 to 10 mg/L of SS. If the SS is higher (e.g., 30 mg/L), a filter system can be used to remove remaining suspended solids that are not effectively removed by the clarification. A sand bed or mixed media filter can be used. Another alternative technology is the membrane filtration operated at microfiltration (MF) or ultrafiltration (UF) mode, which could either supplement or even replace the sedimentation.[15–18]

EXAMPLE 3.4

Estimate the time for three particles dropping 1 m. Assume that the diameters are 10^{-4} m, 10^{-6} m, and 10^{-9} m and the specific gravity is 2.65. The density and the viscosity of water are 1000 kg/m^3 and 1.518×10^{-3} N·s/m^2, respectively.

SOLUTION

We can assume that the settling occurs in a dilute suspension. There is no change in particles' size and shape, as well as no aggregation of particles. The settling forces and process are illustrated in Figure 3.14.

There are two forces that act on the particle as follows:

The force of gravity *is*

$$f_g = \rho_p g V_p$$

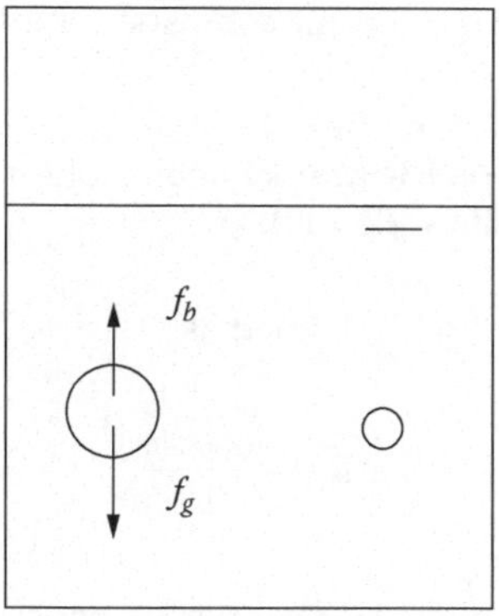

FIGURE 3.14
Settling of particle.

The buoyant force *is*

$$f_b = \rho_w g\, V_p$$

Thus, the net force exerted is

$$f_{net} = (\rho_p - \rho_w)\, g\, V_p$$

A third force (drag force) is created due to viscous friction as:

$$f_d = C_D A_p \rho_p v^2/2$$

where ρ_p and ρ_w are densities of the particle and water, respectively V_p is volume of the particle, A_p is the cross-sectional area, and v is particle settling velocity.

As the net force exerted is balanced with the drag force, we have

$$f_{net} = f_d$$

$$(\rho_p - \rho_w)\, g\, V_p = C_D A_p \rho_w v^2/2$$

$$\frac{V_p}{A_p} = \frac{\frac{4}{3}\pi\left(\frac{d}{2}\right)^3}{\pi\left(\frac{d}{2}\right)^2} = \frac{2d}{3}$$

$$v^2 = \frac{4}{3}\frac{(\rho_p - \rho_w)\, gd}{C_D \rho_w} \tag{3.7}$$

where d is diameter of particle and C_D is the coefficient of drag.

The parameter C_D varies with characteristics of different flow regimes.

$$C_D = 24/\text{Re (laminar flow, Re} < 1)$$
$$= 24/\text{Re} + 3/\text{Re}^{1/2} + 0.34 \text{ (transitional, } 1 < \text{Re} < 10^4)$$
$$= 0.4 \text{ (turbulent flow, Re} > 10^4)$$

The Reynolds number (Re) is defined as:

$$Re = \frac{\phi v \rho_w d}{\mu}$$

ϕ is shape factor and equals 1.0 for a perfect sphere.

First, we assume that the Re number is less than 1. Thus, we can use the following equation for the C_D calculation:

$$C_D = 24/\text{Re}$$

This equation can be combined with Equation 3.7, yielding the below stokes equation:

$$v = \frac{g(\rho_p - \rho_w)d^2}{18\mu} \tag{3.8}$$

Assume $\phi = 1$. The calculation results are given in Table 3.4. It is shown that it takes a very long time for smaller particles to settle down and separate. It further indicates the importance of using coagulation for enhanced particle–water separation. Alternatively, membrane technology may be engaged for the separation.

EXAMPLE 3.5

$FeCl_3$ of 50 mg/L is added to a wastewater from a metal finishing operation, which has suspended solids of 100 mg/L and flow rate of 1,000 m3/day. The addition of the salt forms $Fe(OH)_3$, which can greatly improve the settling of SS and remove heavy metals (in the form of metal precipitates). An experiment shows that $FeCl_3$ can remove the metal waste. It is known that 1 mg/L of $FeCl_3$ can produce 0.25 mg/L of

TABLE 3.4

Comparison of Settling of Different Particles

Diameter (m)	Particle Settling Velocity (m/s)	Time for a Settling of 1 m
10^{-4}	5.91×10^{-3}	2.82 minutes
10^{-6}	5.91×10^{-7}	19.6 days
10^{-9}	5.91×10^{-13}	53586 years

metal sludge, and pH is controlled around 7. The $FeCl_3$ has the following reaction:

$$FeCl_3 + 3H_2O = Fe(OH)_3(s) + 3HCl$$

How many kilograms of sludge is produced every day? Assume that the removal efficiency of the sedimentation tank is 85% and the specific gravity of the sludge is 1.1, calculate how much sludge (m^3) is produced every day.

SOLUTION

The sludge from the process includes the SS from the raw water, $Fe(OH)_3$, and metal sludge.

SS = 100 mg/L.

Metal sludge = 50 × 0.25 = 12.5 mg/L.

According to the chemical reaction given, we have

$FeCl_3 \sim Fe(OH)_3$

$$\frac{162.2}{50} = \frac{107}{\mu}$$

x = 32.98 mg/L.

Total sludge concentration = 100 + 12.5 + 32.98 = 145.8 (mg/L) = 145.8 (g/m^3).

Q = 1000 m^3/day.

Total weight (Wt) = Q × Total sludge concentrations × 1000 × 145.8 = 145.8 (kg/day).

Weight = volume × density.

Wt = V × ρ.

Total sludge volume (V) = Wt/ρ = 145.8/(1000 × 1.1) = 0.132 (m^3/day).

Therefore the volume of sludge to be handled daily is 0.132 m^3/day.

3.5.3 Filtration

Conventional filtration by granular media is commonly used for the removal of SS that ranges from 5 to 50 mg/L in the influent to less than 1 NTU in the effluent. Approximately 20% of operations in the United States use the filtration for polishing the treated water from the precipitation unit. Such filter media as garnet, ground anthracite, silica sand, and activated carbon can be used. In the operation, the water flows downward through the filter media (medium), leading to the removal of SS. After the filter is operated for a period of time, the solids form a mat (cake) on the surface of filter media. As the time goes, more SS with different sizes accumulate inside of the media/medium, leading to breakthrough, a process that SS starts to bypass or penetrate through the filter. The filter bed is then cleaned by an upward washing of the bed with water or with water and air.

Multimedia filters employ two or more filter media with different grain size and densities.[12] In the selection, the smaller particles are the most dense (e.g., garnet with a specific gravity of 4.5 and particle size of 0.2 to 0.4 mm);

the medium-sized particles have an average density (e.g., sand grains with a specific gravity of 2.65 and a grain size of 0.5 mm); and the largest particles are the least dense (e.g., anthracite grains with a specific gravity of 1.6 and a grain size of 1.0 mm). The larger and less dense particles are always on the upper layer in the filter (top). The smaller particles locate at the bottom. During the filtration operation, larger (coarser) SS are removed in the upper layer, while the smaller SS are removed by the lower layer/bottom. Backwashing is performed in the opposite direction.

3.5.4 Dissolved Air Flotation

Dissolved air flotation can be used to either supplement or even replace the sedimentation/filtration. A DAF unit generates fine air bubbles that attach to the hydroxide particles and cause them to rise to the surface of a tank, where they are removed by skimming. It is commonly used by industries with oily waste streams (e.g., petroleum refining and industrial laundries).

3.5.5 Sludge Thickening and Dewatering

After the precipitation, the metal ions are converted to metal precipitate solids. The solution with numerous fine solids or particles is called *sludge* or *slurry*. A sludge dewatering process is then required to separate the solids from the water. It is also used for the removal of solids generated in the pretreatment. The dewatering is usually performed in a series of steps utilizing two or more separation devices, with each subsequent step for increasing a certain percentage of solids from the sludge.

The solid content from either precipitation tank or clarifier is in a range of 0.5% to 3%. A slightly higher percentage of 2% to 5% is first accomplished by a sludge-thickening device. The objective of sludge thickening is to reduce both capital and operational costs of subsequent sludge dewatering steps by substantially reducing the volume of the sludge. It can be accomplished by gravity thickening, flotation thickening, or centrifugal thickening with the gravity thickening being the most commonly used technology. The design variables to be considered in the selection of a thickening process are solids concentrations of both influent and effluent, volumetric flow rate of the feed, coagulant requirement, operation and maintenance costs, reliability of process, ease in operation, training of operator, and safety.

After the thickening, mechanical dewatering and thermal dehydration devices are used to further dewater the sludge that has less water than that before the thickening. Filter press, vacuum filter, centrifuges, and belt presses are the available mechanical tools, with the filter press being the most commonly used.[12]

Through mechanical dewatering, the sludge becomes "drier" with the solid content of approximately 10% to 60%. A sludge drying/dehydration device (e.g., heat driven) can then be further used to remove the water content which leads to the solid content of above 90%. As such, the costs for transportation and disposal of the sludge are significantly due to lower volume of solid waste (sludge).

Sludge disposal cost primarily depends on sludge volume. It has been proved that mechanical dewatering devices are cost-effective in the dewatering of the watering sludge. Figure 3.15 shows the reduction in the volume of sludge by the dewatering as a function of solid concentration. Take sludge with 1,000 gallons and 1% of solid as an example. The sludge can be dewatered to that with 30% of solids and sludge volume of 30 gal, which achieve 97% reduction in the sludge volume. The mechanical dewatering devices (pressure, vacuum, and compression filters, as well as centrifuges) have proved effective in the dewatering of metal sludge.

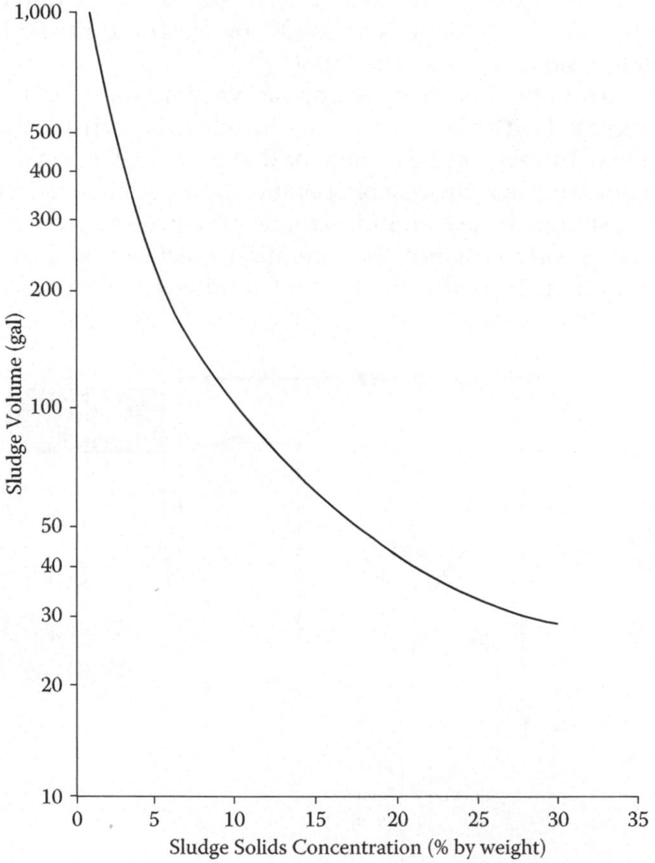

FIGURE 3.15
Sludge volume versus sludge solids concentration.

3.5.5.1 Pressure Filter

As the most popular mechanical sludge dewatering device, the pressure filter dewaters the sludge by pressurizing it and forcing the water to penetrate through a membrane. It is simple in construction and operation that has wide variation in SS as illustrated in Figure 3.16. It can effectively remove water in the sludge. Normally, a filter operated at such a low pressure as 80 to 150 psig can dewater the sludge to a SS percentage of 20%–40%. On the other hand, a filter operated at a high pressure, ranging from 150 to 250 psig, can achieve the SS of the dewatered sludge to 30%–60%. The treatment time and magnitude of applied pressure are two key factors affecting the SS concentration in the dewatered sludge.

3.5.5.2 Vacuum Filter

The vacuum filter removes the water from the sludge by the vacuum force. The water is drawn through the membrane by the force. Vacuum filters work well when the SS of the sludge is above 3% by weight. If the SS is less than that, the sludge should be thickened first.

The rotary-drum filter is the most popular vacuum filter in the industries. It is often precoated with diatomaceous earth, which is particularly important for the sludge with very low SS content or that is difficult to treat. The added materials would increase the cost of operation and also increase the quantity of dewatered sludge. However, industrial experience shows that the added materials can greatly enhance the operation efficiency and achieve high solid concentration. Typically, the SS content after the vacuum filtration is

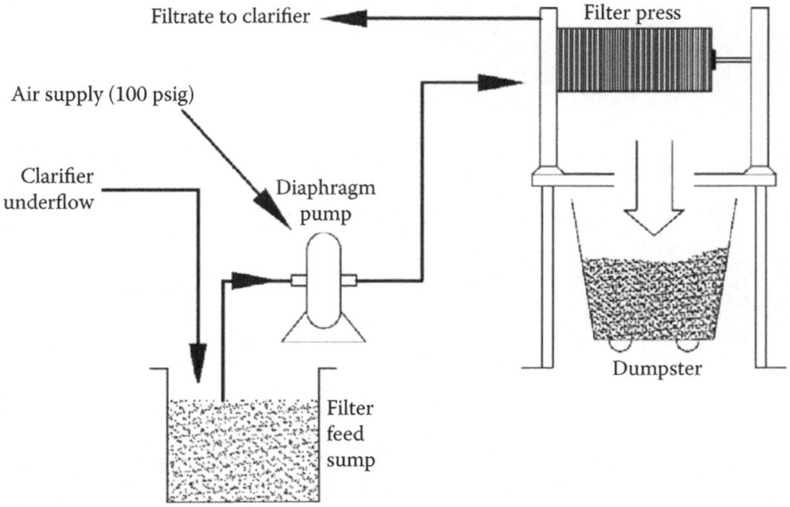

FIGURE 3.16
Illustration of filter compress for dewatering of metal sludge.

20%–40%, which depends on the depth of sludge and rotational speed in the drum.

3.5.5.3 Compression Filter

The compression filter dewaters the sludge by squeezing it between water permeable membranes. A series of belts and rollers are operated to gradually increase the compressive force that is applied to the sludge. The compression filter is effective primarily in the removal of highly compressible sludge that consists of large and soft particle floc associated with polyelectrolyte conditioning. The energy requirement of compression filter is less than that for the vacuum filter and centrifuge. However, the capital cost of the technology becomes relatively higher if it is operated at a smaller scale.

The filter press is available for operation in a wide range of capacities of 0.6 to 200 ft^3. A typical operating cycle ranges from 4 to 8 h, which depends on the physical characteristics of the sludge. It can be operated for the sludge with a solid content ranging from 20% to 70%, resulting from the types and concentrations of precipitation treatment reagents and those of the coagulants. On the other hand, the basket centrifuge and vacuum filter are able to cater for sludge with solid contents of 10% to 25 % and 15% to 40 %, respectively, which underperform the filter press. The disadvantage of the filter press is that it is operated in batch mode.

3.5.5.4 Centrifuge Device

The centrifuge device dewaters the sludge by the centrifugal force. The stream with SS is preconditioned with such a coagulant as polyelectrolyte in order that the settling ability is enhanced. After the operation of separation by centrifuge, the SS content ranges from 10% to 25%. As the coagulant is needed and the SS content is not as high as other technologies, centrifuge is less favorable in the industries.

3.6 Case Studies

3.6.1 Treatment of Heavy Metals in Wastewater from Electroplating Operation

In the electroplating industry, we have different types for different processing units. The electroplating unit produces such toxic heavy metals as cadmium, copper, chromium, nickel, zinc, and tin. It also contains less harmful metal ions such as aluminum and iron. In addition, such precious heavy metals as gold, silver, indium, platinum, and rhodium are used in the electroplating process. In the coating process, chromate and phosphate can be used. Once the waste streams leave different processing units, they must be treated prior to being discharged to the sewage network. There are various technologies available for the selection of treatment systems. Proper selection of key treatment technology and pre- and posttreatment technologies is essential. The selection, design, and operation play key roles to ensure that the national and location requirements can be met.

A treatment system consisting of the different processes is shown in Figure 3.17.[19] There are two wastewater streams from the chrome plating and the cyanide baths. They are isolated and directly channeled to two different subsystems, which are treated by different approaches. The two treated effluents are then combined in an averaging tank, which is mixed with other wastewater streams such as from acid and alkali baths and rinses, other plating baths and rinses, and chemical dumps. The mixture is subsequently sent to the neutralization tank.

In the chromium reduction (lower part of the diagram), the hexavalent chromium is reduced to the trivalent chromium by SO_2. The trivalent chromium can be removed by alkali neutralization. Note that the hexavalent chromium may also be reduced through electrochemical reduction.

The toxic cyanide-bearing waste stream is treated or oxidized through chlorination or ozonation (upon part of the diagram). This leads to less toxic compounds typically with lower molecular weights.

These two streams are mixed and flow to the neutralization and precipitation tank. The combined stream is treated with either acid or alkali for

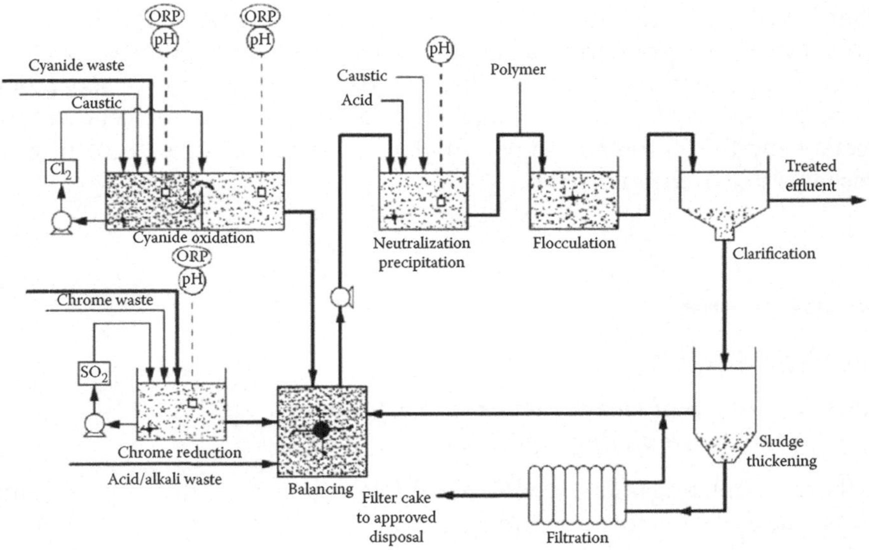

FIGURE 3.17
Treatment system of heavy metals in wastewater from electroplating operation.

the pH adjustment, and the heavy metals are precipitated to form metal hydroxide. The waste stream after the precipitation contains solids. Which is treated by coagulant for further enhancement in the separation of metal precipitates as well as settling of metal hydroxide sludge. The solids and the water are separated by the clarifier. The sludge (mainly metal hydroxide) is dewatered through the gravity thickening and mechanical dewatering. The resulting dried sludge can then be sent for final disposal (e.g., landfill).

3.6.2 Metal Removal by Insoluble Sulfide Precipitation

Wastewater from Holly Carburetor (Colt Industries in Tennessee) consisted of iron, zinc, and hexavalent or trivalent chromium ions with varying concentration. They were treated by sulfide precipitation instead of hydroxide precipitation.[19] The technology marketed as Sulfex™ was selected, as the metal sulfides were considerably less soluble than the metal hydroxides. As shown in Figure 3.18, the treatment system included pH adjustment by lime, addition of FeS and polymers for precipitation and coagulation/flocculation, sedimentation, and sand filtration. The effluent quality of the system was able to meet the levels required by the US EPA. Most devices of the system were the same as those of the hydroxide precipitation system. Consequently, the conventional hydroxide precipitation system may be slightly modified by replacement of hydroxide with sulfide to achieve the improved metal removal capability and efficacy with lower cost for the modification and perhaps operation. The main precipitation treatment reagent was ferrous sulfide (FeS), an insoluble sulfide precipitation (ISP) reagent. It was selected because it was relatively insoluble. Hence, the concentration of SS in the treated effluent was kept at a lower level. A main advantage of ISP reagent over the soluble sulfide precipitation (SSP) reagent is that there is no toxic odor due to the usage of hydrogen sulfide in the treatment system and the effluent.

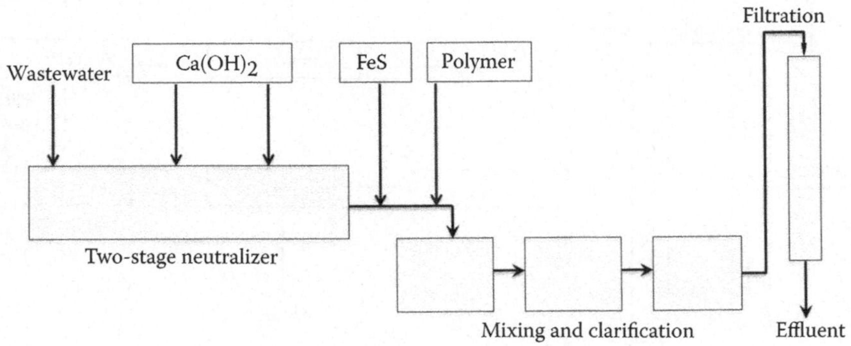

FIGURE 3.18
Metal removal by insoluble sulfide precipitation.

Note that the SSP reagents are the water-soluble compounds such as sodium hydrosulfide (NaHS) and sodium sulfide (Na_2S).

3.6.3 Hybrid System for Metal Removal

A hybrid system consisting of key units such as pH adjustment, precipitation, and ion exchange was used to treat the heavy metal wastewater, as shown in Figure 3.19. Four waste streams were first collected.[19] The hexavalent chromium was reduced by the bisulfate under an acidic condition. The water pH was adjusted by lime in order to facilitate the maximum metal removal for each waste processed. A polymer as coagulant was then added to the water for the removal of the SS (consisting of SS from the influent and the metal precipitates) in the device labeled "flocculation." A pressure filtration system with diatomaceous earth precoat was used to separate the SS from the water. The SS (sludge) was sent to the final disposal.

The water from the filtration was sent to the ion-exchange polishing unit, by which the remaining heavy metal ions were further removed to the required levels by the US EPA. A two-stage ion-exchange (IX) unit was used with the first stage using a hydrogen ion resin and the second using sodium ion resin. As the IX proved more selective for heavy metals in the presence of calcium ions, the low-cost lime was used for neutralization instead of using high-cost caustic soda.

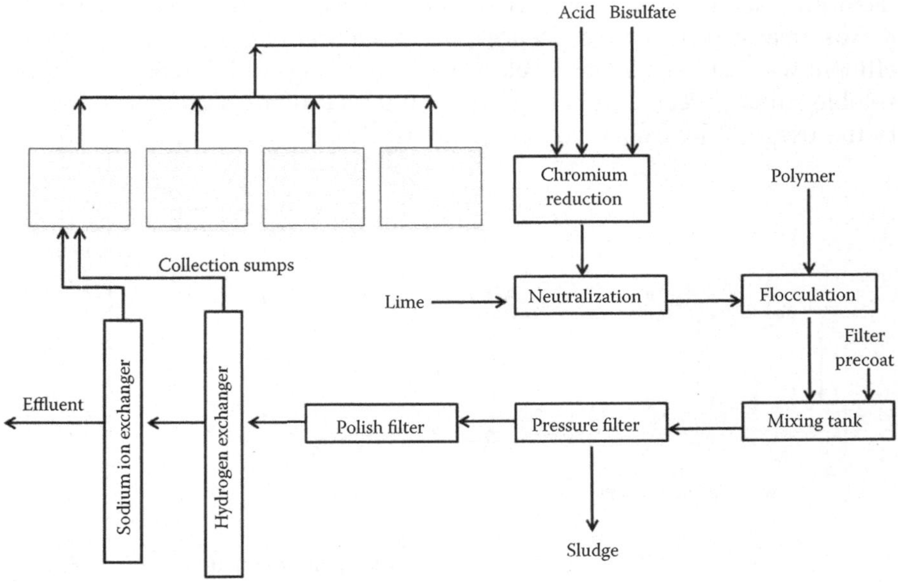

FIGURE 3.19
Hybrid system for metal removal.

Note that if organic compounds are present in the wastewater, we may add an oxidation process before the water enters the IX process, by which the COD can be greatly removed. Chorine or ozone can be used. Alternatively, an activated carbon adsorption column can be used instead of oxidation, by which the organic compounds can be adsorbed onto the activated carbon. The treated effluent can be good enough for reuse by the production processes, leading to a saving in the overall water consumption. From the operational standpoint, it would be beneficial as the fouling of IX resin would be reduced due to the removal of organic compounds.

3.6.4 Segregated Treatment of Difficult-To-Treat Metal

In some situations, we may see that the concentration of one metal cannot meet the discharge permission even though others can. We will have to segregate the waste stream of that metal from the streams containing other metals.

An operation of metal plating by using copper, nickel, and chromium on the plastic components in Medfrod Plastics Company in Wisconsin (USA), produced a wastewater that was particularly difficult to treat.[19] It was treated by the treatment system demonstrated in Figure 3.20.

A preliminary study demonstrated that the concentration of nickel did not be reduced by a conventional treatment system to the level required by the discharge permit. Hence, the electroless nickel plating rinse flow was segregated from the rest of the wastewater, which was based on the principle of pollution prevention. As shown in Figure 3.20, the nickel wastewater was precipitated at a high pH and then clarified. The wastewater of copper and

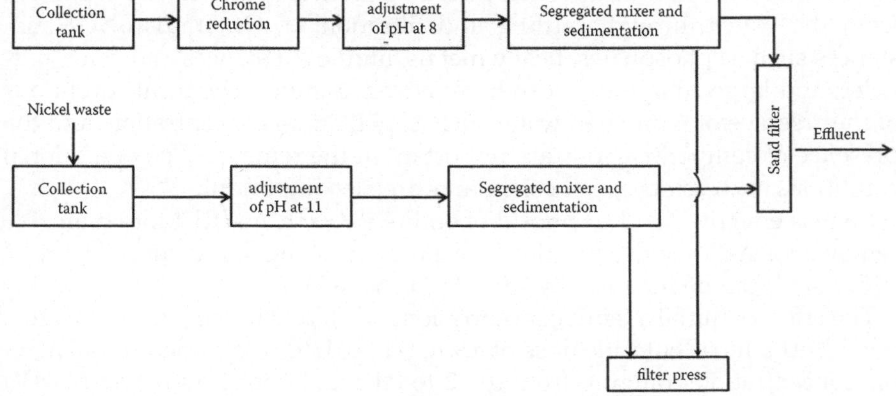

FIGURE 3.20
Illustration of segregated treatment of difficult-to-treat metal.

chromium was treated by chromium reduction, pH adjustment for precipitation, and sedimentation. The stream from the nickel treatment and that from the copper or chromium treatment were combined in the sand filtration and finally discharged. The testing results from the preliminary studies showed that the nickel concentration (and copper or chromium concentrations) in the total effluent met the permit specifications.

However, the concentration of nickel in the total effluent could not meet the discharge permit when the full-scale treatment system was operated. As such, a modification was conducted. The plant reduced the flow to the nickel treatment system by 50% by further enhancement of rinse water recycle. This led to addition dilution in the combined stream as the flow rate of effluent from the copper or chromium was remained unchanged. Consequently, the nickel concentration in the total effluent finally met the discharge permit.[19]

3.6.5 Treatment of Arsenic by Precipitation–Coagulation

Arsenate (As(V)) and arsenite (As(III)) are primary forms of arsenic in soils and natural waters. Because of increasingly reported evidences of toxic effects from the arsenic, more strict regulations have been made to limit its discharge to sewers and water bodies.

Calcium, iron, or aluminum salts such as lime, alum $Al_2(SO_4)_3 \cdot 14.3H_2O$ and ferric chloride can be added into the arsenic-contaminated water for the removal.[20] These metal ions may form a series of metal precipitates such as calcium, iron, or aluminum hydroxides. The arsenic species and other soluble or insoluble substances may attach onto the insoluble aluminum and ferric hydroxides through electronic attraction or chemical adsorption. The metal hydroxides may act as adsorption sites for the arsenic ions. In addition, the arsenic ions can form precipitates together with aluminum or ferric ions. The arsenic species attached on the metal hydroxides can be separated from the water through settling and filtration.[21,22] Other dissolved substances such as phosphates, heavy metals, humic substances and color, particles, and large compounds can be removed. Solution chemical conditions of the arsenic-contaminated water such as pH, initial concentration, and the presence of competitive ions are important in the removal. The operational conditions such as dosage of coagulants are also important.

Lime is effective for the removal of both As(V) and As(III). More than 90% removal for As(V) with an initial concentration 0.4 mg/L is achieved at pH > 10.5. As(III) can be removed by 75% at pH above 11.0.

The effect of initial arsenic concentration is different in the removal of arsenic.[23–25] At a ferric chloride dose of 4.9 mg/L, pH 7.0, and varied initial arsenic concentrations ranging from 0.002 to 0.1 mg/L, both As(III) and As(V) removal efficiencies were independent upon the initial concentrations.[24] For initial As(V) concentrations between 0.1 and 1.0 mg/L, a dosage of 30 mg/L of either alum or ferric sulfate in the optimum pH range removed over 95%

of As(V). Above an initial concentration of 1.0 mg/L, the removal decreased with an increasing concentration. For the initial concentrations of As(III) greater than 0.1 mg/L, neither alum nor ferric sulfate dosed at 30 mg/L could remove As(III) to concentrations below 50 ppb.

Edwards reported more than 90% removal of As(V) at a dosage of ferric chloride above 20 mg/L or alum above 40 mg/L.[26] At a lower dosage, the removal did not follow a trend because of poor particle removal, high initial As(V) concentrations, and possible interference from other anions in the different waste streams. The optimum removal efficiencies of As(V) by ferric salts and alum were reported at pH 5–7.2 and pH 5–7.0, respectively.[24] The solids due to the operation could be directly removed by filtration. The final concentration of below 10 ppb was achieved. The removal percentage of As(III) was 40% to 50%, which was worse than that of As(V). Thus, a pre-oxidation of As(III) to As(V) should be conducted for the enhancement in removal.

Several competitive inorganic substances commonly exist in natural waters. It was demonstrated that the presence of sulfate hindered the As(III) removal at pH below 7.0.[25] Only a slight decrease in the removal of As(V) was observed. On the other hand, the presence of calcium ion increases the removal of As(V) at higher pH.

Since disinfection is always involved in drinking water treatment. Preozonation may also be used as one of first steps, leading to the oxidation of As(III) to As(V). A survey of several drinking water treatment shows that soluble As(V) is converted to particulate As(V) by adsorption during rapid mixing, and is removed along with naturally occurring particulate arsenic predominantly by sedimentation.[27] As(III) may be converted to As(V) during the chlorination. Thus, the difficulty of treatment due to the presence of As(III) may not be an issue of concern if pre-ozonation and/or pre-chlorination is used.

3.7 Limitations and Solutions

The precipitation technology has a major advantage of low cost. However, it has some disadvantages discussed as follows.

3.7.1 Presence of Chelating Agents

The precipitation process cannot reduce metals to an acceptable level in the presence of chelating compounds unless additional treatment reagents are employed.[19] The use of additional treatment reagents may significantly increase the volume of sludge generated from the operation, which posts additional costs in the overall treatment.

Some industries use chelating agents in the processes, which react with heavy metals. They would interfere with the metal precipitation when hydroxides are used. Such chelating agents as ammonia, phosphates, tartrates, and EDTA are commonly used in the plating operations. As a result, they find their way into the metal waste streams. The chelating agents react with the dissolved metal ions to form a series of metal complexes that are quite soluble in neutral and alkaline solutions.[19]

We may have two solutions to overcome the solubilizing effects of the chelating agents:

1. The heavy metal ions in the wastewater are precipitated by a precipitation reagent other than hydroxides, the approach that is relatively resistant to chelating effects.
2. The wastewater is first treated by such a chemical approach as pH adjustment or oxidation, which frees the metal ions from the chelating agents.

In the first solution, we can use such methods as sulfide precipitation and water-insoluble starch xanthate (ISX) precipitation.[28] In addition, adsorption and ion exchange can be used, which is not discussed in Chapter 7.

The sulfide precipitation process is used to precipitate heavy metal ions with sulfide. The metal precipitates are in the forms of metal sulfide instead of metal hydroxides. The metal sulfides have been found capable of achieving low levels of metal solubility in highly chelated waste streams. This method has been proven as an alternative measure to the conventional hydroxide precipitation process. It can also serve as a process to further reduce soluble metal ion concentrations in the solution treated by a hydroxide precipitation system.

The second solution would have to be used if the first approach does not work. In the operation, the wastewater is segregated and pretreated by either raising the pH to a highly alkaline level through such additions of chemicals as lime or lowering the pH to a very low level (acidic level). Under both extreme pH levels, the metal complexes can be fully dissociated, freeing the free metal ions. A low-cost and suitable cationic substance such as calcium ion is then used to tie up the chelating agent, which prevents it from the recombination with the heavy metal ions when the waster is neutralized. This approach uses a large amount of reagents. However, it has proved to be an effective means of treatment of metal waste stream that contains chelating agents.

3.7.2 Production of Solids

The precipitation operation generates a large amount of precipitates (solid waste). The sedimentation operation is generally not able to remove suspended solids below approximately 5 to 10 mg/L SS. The effluent from the sedimentation tank typically has SS up to 50 mg/L. The SS can be greatly reduced if a polishing filtration process is employed. As the SS remaining in the effluent could contain metals, such an effluent may not be suitable for reuse and discharge.

The metal hydroxide precipitate sludge resulting from treatment of wastewater from metal-coating operation has been classified as a hazardous waste. As such, it must be managed, handled, transported, and recovered or disposed of under the national and local environmental rules and regulations.

3.7.3 Importance of Process Control

Control is extremely important to achieve better treatment efficacy in precipitation. Among various factors in the control, pH is the most important. If it were not well controlled, the whole process would fail. Complete metal precipitation only occurs at a specified pH and metal concentrations may increase if pH is either higher or lower than the specific pH, which is clearly illustrated in Figures 3.4–3.6. In addition, pH plays an important role on the surface properties of precipitates (e.g., surface charges), by which the settling may be affected.[28]

References

1. Stumm, W. *Chemistry of the Solid-Water Interface: Processes at the Mineral-Water and Particle-Water Interface in Natural Systems*. John Wiley & Sons, New York, 1992.
2. Stumm, W., and Morgan, J. J. *Aquatic Chemistry: Chemical Equilibria and Rates in Natural Waters*, 3rd ed. John Wiley & Sons, New York, 1996.
3. Pepper, I. L., Gerba, C. P., and Brusseau, M. L. *Pollution Science*. Academic Press, New York, 1996.
4. Haynes, W. M., eds. *CRC Handbook of Chemistry and Physics*, 92nd Edition, Taylor and Francis, Boca Raton, FL, 2012.
5. vanLoon, G. W., and Duffy, S. J. *Environmental Chemistry: A Global Perspective*, 2nd ed. Oxford University Press, Oxford, 2001.
6. Sharma, H. D., and Reddy, K. R. *Geoenvironmental Engineering: Site Remediation, Waste Containment, and Emerging Waste Management Technologies*. John Wiley & Sons, Hoboken, NJ, 2004.
7. Schecher, W. D., and McAvoy, D. C. *MINEQL+ Chemical Equilibrium Modeling System*. Version 4.5 for Windows. Environmental Research Software. Hallowell, ME, 2001.
8. Wang, L. K. P., Chen, J. P., Hung, Y. T., and Shammas, N. K., eds. *Heavy Metals in the Environment*. CRC Press/Taylor and Francis Group, Boca Raton, FL, 2009.
9. Wang, L. K. P., Chen, J. P., Hung, Y. T., and Shammas, N. K., eds. *Membrane and Desalination Technologies*. Totowa, NJ, Springer, 2011.

10. Snoeyink, V. L., and Jenkins, D. *Water Chemistry*. John Wiley & Sons, New York, 1980.
11. Sohnel, O. and Garside, J., Precipitation: Basic Principles and Industrial Applications, Butterworth, Boston, 1992.
12. National Metal Finishing Resource Center, http://www.nmfrc.org.
13. Remoudaki, E., Hatzikioseyian, A., Kousi, P., and Tsezos, M. The mechanism of metals precipitation by biologically generated alkalinity in biofilm reactors. *Water Res.*, 37, 3843–3854, 2003.
14. Meng, X. G., Korfiatis, G. P., Christodoulatos, C., and Bang, S. Treatment of arsenic in Bangladesh well water using a household co-precipitation and filtration system. *Water Res.*, 35, 2805–2810, 2001.
15. Kim, S. L., Chen, J. P., and Ting, Y. P. Study on feed pretreatment for membrane filtration of secondary effluent. *Sep. Sci. Technol.*, 29, 171–179, 2002.
16. Chen, J. P., Kim, S. L., and Ting, Y. P. Optimization of feed pretreatment for membrane filtration of secondary effluent. *J. Membrane Sci.*, 219, 27–45, 2003.
17. Chen, J. P., Mou, H. H., Wang, L. K. P., and Matsuura, T. Membrane filtration. In *Advanced Physicochemical Treatment Processes, Handbook of Environmental Engineering Series* (L. K. P. Wang, Y. T. Hung, and N. K. Shammas, eds.), 4, 199–256. Springer/Humana Press, Totowa, NJ, 2006.
18. Zheng, Y. M., Zou, S. W., Nanayakkara , K. G. N., Matsuura, T., and Chen, J. P. Adsorptive removal of arsenic from aqueous solution by a PVDF/Zirconia blend flat-sheet membrane. *J. Membrane Sci.*, 374, 1–11, 2011.
19. U.S. Environmental Protection Agency. The Electroplating Industry: Environmental Regulations and Technology. EPA/625/10-85/001, Washington, DC, 1985.
20. U.S. Environmental Protection Agency. Arsenic Rule Benefits Analysis: An SAB Review. EPA-SAB-EC-01-008. Washington, DC. 2001.
21. Kumar, P. R., Chaudhari, S., Khilar, K. C., and Mahajan, S. P. Removal of arsenic from water by electrocoagulation. *Chemosphere*, 55, 1245–1252, 2004.
22. Chen, J. P., Chang, S. Y., and Hung, Y.T. Electrolysis. In *Physicochemical Treatment Processes*, Handbook of Environmental Engineering Series (L. K. P. Wang, Y. T. Hung, and N. K. Shammas, eds.), 3, 359–378, Humana Press, Totowa, NJ, 2004.
23. Logsdon, G. S., Sorg, T. J., and Symons, J. M., Removal of heavy metals by conventional treatment, Proc. 16th Water Quality Conference—Trace metals in Water Occurrence, Significance, and Control. University Bulletin. U. of Illinois., 71, 1974.
24. Sorg, T. J., and Logsdon, G. S., Treatment technology to meet the interim primary drinking water regulations for inorganics. Part 2. *J. Am. Water Works Assoc.*, 70, 379–393, 1978.
25. Hering, J. G., Chen, P., Wilkie, J. A., Elimelech, M., and Liang, S. Arsenic removal by ferric chloride. *J. Am. Water Works Ass.*, 88, 155–167, 1996.
26. Edwards, M. A. Chemistry of arsenic removal during coagulation and Fe-Mn oxidation. *J. Am. Water Works Ass.*, 86, 64–77, 1994.
27. Gregor, J. Arsenic removal during conventional aluminium-based drinking-water treatment. *Water Res.*, 35, 1659–1664, 2001.
28. Mokone, T. P., Lewis, A. E., and van Hille, R. P. Effect of post-precipitation conditions on surface properties of colloidal metal sulphide precipitates. *Hydrometallurgy*, 119–120, 55–66, 2012.

4

Precipitation–Crystallization Technology

4.1 Introduction

Precipitation–crystallization is an innovative technology that combines precipitation and crystallization. Typically, it is operated in a fluidized bed reactor (FBR) and thus is called *fluidized bed crystallization* (FBC) *technology*. The FBC was originally developed for the process of removing phosphates and softening drinking water.[1–8] A series of researches have been conducted on the removal and recovery of various precious metals such as lead, zinc, copper, cobalt, nickel, arsenic, and silver from heavy metal–bearing wastewater streams by using this approach. It can also be used for treatment of waste streams containing anionic contaminants such as fluoride and phosphate ions.

During the operation of the precipitation–crystallization technology, seeding medium (granular solids or sand pellets) such as sand grains are first added into a FBR. The heavy metal waste stream and the precipitation reagent are channeled to the reactor. Numerous fine metal precipitates or fine particles are formed. They are subsequently coated on the surfaces of added solids, instead of staying in a suspended form that can be seen in the conventional precipitation operation. The effluent is free of heavy metal ions or has very low concentrations of heavy metal ions. The post-treatment for the precipitates (sludge) becomes unnecessary, leading to cost saving. Removal and recovery of heavy metals from wastewater by the precipitation/crystallization in a FBR is cost-effective according to the literature.[1–8]

In this chapter, the theory for precipitation–crystallization technology will be illustrated first. The important control measure that leads to the success in the technology will be discussed, and a detailed study of the technology for lead removal will be presented. Finally, several important case studies will be provided to further illustrate the technology.

4.2 Description of Technology

As discussed in Chapter 3, conventional precipitation technology for treating soluble heavy metal ions (e.g., lead ion) can be accomplished by chemical precipitation reactions in a mixing tank, followed by sedimentation and dewatering of the precipitated lead sludge. For any continuous operation, it requires a large space for installing chemical mixing tanks, settling tanks, and mechanical sludge dewatering facilities. Furthermore, in order to recover the heavy metals from sludge, dewatering and solidification of sludge must be applied, which makes this approach quite costly. Thus, conventional precipitation is not practical, especially for many small plants, even though it has been used for many years.

An alternative is to use precipitation–crystallization technology, in which a fluidized bed crystallizer is served as "metal stripper" to remove metal from solution. The FBC for metal removal is based on the principle/mechanism that the fine metal precipitates are first formed by adding precipitate reagents via such mechanisms as surface precipitation, and then are deposited/grow on the seeding medium (e.g., sand) in an FBR. The development of an FBC was based on experience with the softening of drinking water and the removal of phosphate from wastewater.[3,4] The operation of precipitation–crystallization process in an FBR has been widely used in the purification of sparingly soluble substances. Since 1985, DHV Water in the Netherlands and K. U. Leuven in Belgium have used this process for recovering and removing heavy metals from wastewater.[2]

Compared to traditional treatment technologies, FBR and surface crystallization are two major factors for the successful operation of heavy metal removal. The fluidized bed provides a surface area of 3,000–10,000 m^2/m^3, much larger than a fixed-bed reactor. Such a surface area allows the occurrence of surface precipitation reactions on the sand surface with a large reaction area. As a result, the precipitating ions and precipitated particles have more chance to contact granular sand (pellets) in the bed. As the system can be operated at high rates and with high efficiency, the cementing of the sand pellets can be prevented.[5] The system is compact, requires a relatively low investment, and has low operational costs. As it contains no moving parts other than pumps, its operation is much simpler.

Precipitation–crystallization technology makes it possible to produce pure and compact metal precipitate pellets. The pellets can be easily recovered. The sand and recovered metal solution can be reused in industrial processes. On the other hand, the metal sludge from conventional precipitation process has very high water content; the impurities are often incorporated into the sludge. Large volumes of metal sludge are produced that cannot be directly reused.

Moreover, this system is very flexible in use. It can be very small, suitable for small plants with low metal treatment capacities. It can also be used to handle a large volume of wastewater. For example, the plant for the municipal water works of Amsterdam consists of 10 FBRs with a total capacity of 8,500 m^3/h. The operating cost comes from chemicals and personnel; the electricity consumption is relatively low.

4.3 Theoretical Background

4.3.1 Surface Precipitation

Precipitation is generally achieved by mixing solutions of reacting components (precipitation reagents). The initial supersaturation of these reacting components is essential to accomplish the precipitation process. There is no clear dividing line between the two phenomena "precipitation" and "crystallization." Precipitation can be treated as embodying fast crystallization.[6] Chemical precipitation reactions from aqueous solutions can often be described by Equation 4.1 as follows:

$$mA^{n+} + nB^{m-} \rightarrow A_mB_n(s) \tag{4.1}$$

With respect to the solid phase formed, the supersaturation S is defined as follows:

$$S = \left(\frac{[A^{n+}]^m [B^{m-}]^n}{K_{sp}} \right)^{\frac{1}{m+n}} \tag{4.2}$$

$[A^{n+}]$ and $[B^{m-}]$ are the concentrations of the heavy metal and precipitation reagent in the solution. K_{sp} is the solubility product of metal precipitate defined in Chapter 3 and is available in Table 3.1.

For example, Equation 4.2 for the precipitation of lead ion by using carbonate can be revised as

$$Pb^{2+} + CO_3^{2-} \rightarrow PbCO_3\ (s)$$

$$S = \sqrt{\frac{[Pb^{2+}][CO_3^{2-}]}{K_{sp}(PbCO_3)}}$$

where K_{sp} $(PbCO_3)$ is the solubility product of lead carbonate.

The fluidized bed provides a large surface area to the metal precipitation reactions. As the system is operated at a pH above 7 for most precipitation operation, the surfaces of sand grains are negatively charged. Metal ions are able to be adsorbed to the surfaces of sand grains due to various effects, such as electrical attraction. The metal ions on the sand surface then react with the precipitation reagent (e.g., carbonate ions) in the bulk solution, which causes the metal ion to be precipitated on the sand surface. This process is the "surface precipitation," an interaction between sand grains and metal precipitate particles. As time proceeds, more and more metal ions are stripped from solution, and the metal precipitate particles stick on the sand surface.

The surface precipitation plays a key role in the operation of the treatment system, which is illustrated in the precipitation–crystallization for lead ion removal schematically given in Figure 4.1. The lead carbonate crystals deposited on sand grains can be further visualized by the microscopic photographs given in Figure 4.2.[1] The crystals of lead carbonate are virtually distributed evenly on the sand surface.

As the particle cohesive sites on the sand surface for the deposition of metal precipitate particles is not unlimited, there is an optimal mass loading of metal precipitates at any time to achieve the best disposition, which is related to supersaturation. Once the supersaturation is higher than the limit, a large amount of fine metal precipitate particles will be formed in the bulk solution, other than on the sand surface. Thus, it causes a decrease in metal removal efficiency.

4.3.2 Crystallization Kinetics

Crystallization is usually the result of three processes:

1. Crystal nucleation
2. Crystal growth
3. Secondary changes in the resulting crystal suspension by such mechanisms as agglomeration, aging, and recrystallization[7]

Depending on the reaction conditions, these processes can proceed either in series or in parallel throughout the operation.

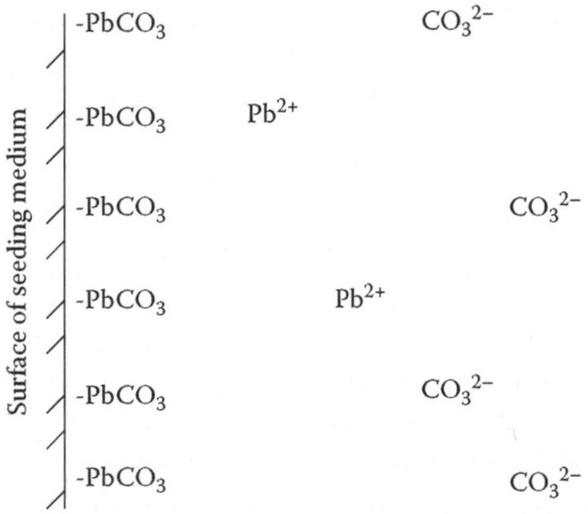

FIGURE 4.1
Schematic illustration for precipitation-crystallization for removal of lead ions.

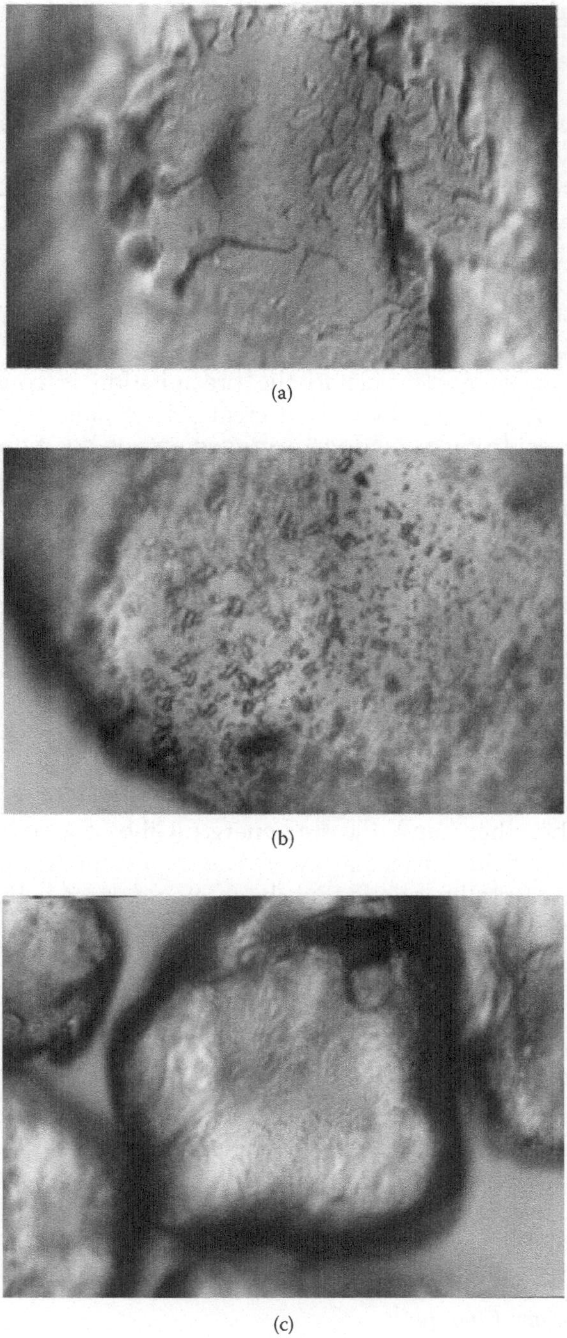

FIGURE 4.2
Formation of lead carbonate on the surface of (a) sand grains before crystallization; (b) sand grains after crystallization; (c) sand grains after being coated by lead carbonate and then washed by HCl.

4.3.2.1 Crystal Nucleation

Crystal nucleation is the first step in the whole process of crystallization, which results in the formation of numerous nuclei. The rapid precipitation is a consequence of high supersaturation. The nucleation plays a major role in the precipitation–crystallization process.[6] Two types of primary nucleation can be identified, homogeneous and heterogeneous nucleation. For homogeneous nucleation, the formation of precipitate phase is not brought by the presence of other particles. On the other hand, for heterogeneous nucleation, the formation of new precipitate is catalyzed by a foreign solid phase. The heterogeneous nucleation often occurs on the sand grains surface because of the surface precipitation reactions.

The formation of nuclei due to the precipitation is considered as an energy-consuming process because it creates an organized structure with defined surfaces from a random arrangement of solution constituents. Because of the energy requirement, it is necessary for the solution to be supersaturated, or have a concentration greater than that predicted by equilibrium with the precipitate.[8] It is the driving force for precipitation and crystallization. The catalytic effect of a foreign solid phase on nucleation is explained by the resulting decrease in the energy barrier to nucleation if the nucleus forms on the surface of the foreign phase. The Volmer nucleation model for a liquid is often used to describe the heterogeneous nucleation of a solid.[6]

The foreign particles then make the nucleation easier and serve as a nucleation catalyst. As a result, the degree of supersaturation could be lower as more crystallizing ions become available on the surface of the foreign particle. On the other hand, the free energy (Gibbs energy) for forming a crystallite on a similar surface is lower. The important requirement for the heterogeneous crystallization is that the foreign surface behaves similar to the precipitate in lattice structure.

4.3.2.2 Crystal Growth

The second step in the whole process of crystallization is the crystal growth, which leads to the multilayer deposition of the precipitates on the nuclei. The equilibrium normally cannot be reached in the whole process of crystallization in the metal treatment by precipitation processes. The supersaturation can be found as a critical factor influencing the crystal growth rate. However, the supersaturation at a much higher level is not beneficial to the crystal growth.

4.3.2.3 Secondary Changes

The third process in crystallization involves the secondary changes, including aging, aggregation, and recrystallization. Under high supersaturation, this process plays a role in removing heavy metal ions in the precipitation.

4.3.3 Degree of Crystal Dispersion

The degree of crystal dispersion can be used to indicate the magnitude of crystallization, which depends on the supersaturation in the solution. It can be defined by Von Weimarn's empirical equation as follows:

$$\text{The degree of crystal dispersion} = \frac{K(Q-s)}{S} \tag{4.3}$$

where K is a constant, Q is concentration of metal ion before reacting with the precipitation reagent, s is the solubility of precipitate, $Q-s$ is the instantaneous supersaturation, $(Q-s)/S$ and represents the relative supersaturation.

The degree of crystal dispersion is higher when the instantaneous supersaturation of metal ion is higher because the solubility is a constant for a given system at a certain temperature. With a high degree of dispersion, the number of nuclei in the solution is large, which indicates the occurrence of homogeneous nucleation leading to the production of small crystals. On the other hand, a low degree of dispersion indicates the formation of large crystals.

The growth of precipitated particles strongly depends on the surface precipitation. The supersaturation in the solution is a critical factor influencing the precipitation reaction.

At a low supersaturation, heterogeneous nucleation dominates, and the particle growth relies on the surface reaction. Typically, large crystals are produced on the sand surface. However, at a high supersaturation, the rate of crystal growth can become very high and the number of precipitated particles is much more than the adsorptive sites on the sand grains. Homogeneous nucleation inevitably happens, even with foreign solids in the system. The crystals formed in the solution are small and stay in the treated solution. These fine precipitate crystals can eventually be eluted from the FBR. To maximize the metal precipitation efficiency, the supersaturation inside the reactor cannot be very high. In other words, the heavy metal concentrations before the precipitation reactions cannot be very high, which can be achieved by using dilution. As illustrated in Figure 4.3, a stream from the effluent can be used to dilute the concentrations of both metal and precipitation reagent, leading to better coating of metal precipitates on the surface of sand.

4.4 Important Control Factors

Several factors play important roles in the treatment efficiency of the precipitation–crystallization process. Their effects are illustrated by the batch and continuous experiments studies conducted by Chen and Yu.[1] The findings are given as follows.

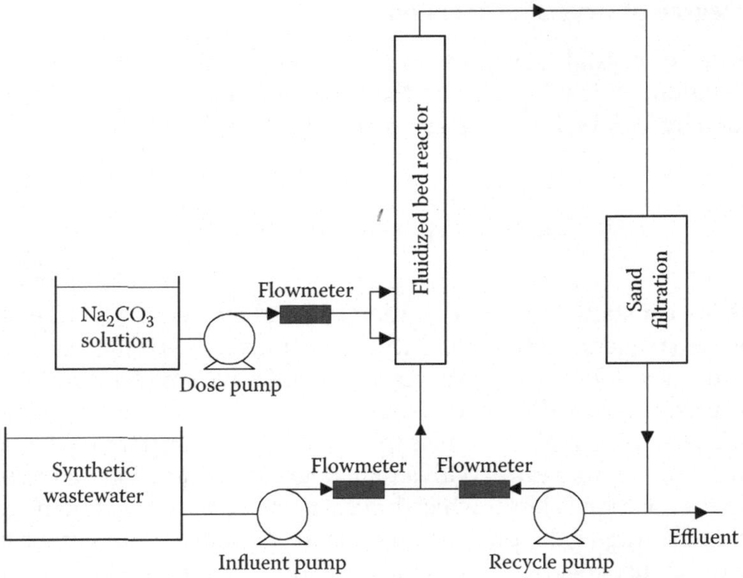

FIGURE 4.3
Illustration of precipitation–crystallization process for lead treatment.

TABLE 4.1
General Process Conditions in the Fluidized Bed Reactor for Lead Removal

Item	Value	Unit
Influent Pb concentration	20, 40	mg/L
Feeding C_T/Pb ratio	3.0	mol/mol
Operation pH	7.0–8.0	
Influent flow rate	6	L/h
Recycle ratio	0.67	
Sand grains type	SiO_2, diameter = 0.2–0.3 mm, density = 1466 kg/m^3	
Fixed bed height	18	cm

The continuously operated process used in the study is illustrated in Figure 4.3, and the operation conditions are given in Table 4.1. The precipitation and formation of lead carbonate occurred in a clear acrylic FBR. A polyester net of 80 mesh was installed at the bottom of the column in order to distribute the influent flow smoothly. A dual-media sand filter was used to polish the effluent quality as the solution may contain the disperse crystals (precipitates), which could cause a high turbidity. During the experiments, the sand filter was back-washed periodically to ensure the good filtration efficiency.

The synthetic lead-bearing wastewater driven by a pump was fed constantly from the bottom of the FBR at certain flow rate in order to fluidize the inside medium. The precipitation reagent (Na_2CO_3) was pumped into the column through two nozzles at height of 5.5 cm and 16 cm from the bottom, respectively. Finally, a portion of effluent was pumped back to the FBR in order to facilitate the fluidization, as well as to dilute the metal concentration that can reduce the supersaturation.

4.4.1 Total Carbon Concentration versus Metal Concentration

The total carbonate concentration (C_T) plays an important role in the metal precipitation and the formation of particles with different sizes and concentrations. It is defined as:

$$C_T = [H_2CO_3^*] + [HCO_3^-] + [CO_3^{2-}] \tag{4.4}$$

Batch experiments for the determination of $C_T/[Pb^{2+}]$ ratio were first carried out. The final pH, turbidity, and lead concentration versus the ratio are shown in Figure 4.4. It is seen that the final pH increases from 8 to 9.8 as the ratio is enhanced from 1 to 2 mol/mol. However, it remains unchanged from the ratio of 2 to 5 mol/mol. As the ratio is increased, sodium carbonate concentration (used for increase of C_T) is increased, which results in an increase in the solution pH. Both final lead concentration and turbidity reach their lowest values when the $C_T/[Pb^{2+}]$ ratio is ranged between 2.5 and 3 mol/mol.

As discussed, supersaturation plays a key role in the formation of precipitates. As the ratio becomes high, the fraction of carbonate concentration increases, which causes serious supersaturation in solution and thus faster reaction kinetics. This is favorable to the spontaneous nucleation, which leads to the production of numerous fine precipitate solids, while the crystal growth becomes unfavorable. The amount of fine lead carbonate particles in the solution could be significantly increased. Consequently, the final turbidity and lead concentration are found to be higher as the $C_T/[Pb^{2+}]$ ratio is enhanced. It can be determined that the optimum ratio for the FBR is 3 mol/mol.

4.4.2 Start-Up of the System

When the operation is started with bare sand grains, the system may not immediately reach its best performance due to the low crystal lattice compatibility between silica dioxide (on sand particles) and lead carbonate. The start-up effect was studied through a series of experiments with influent concentration of 20 mg/L. The removal efficiency, lead concentration in the effluent, and the turbidity versus time are demonstrated in Figure 4.5.

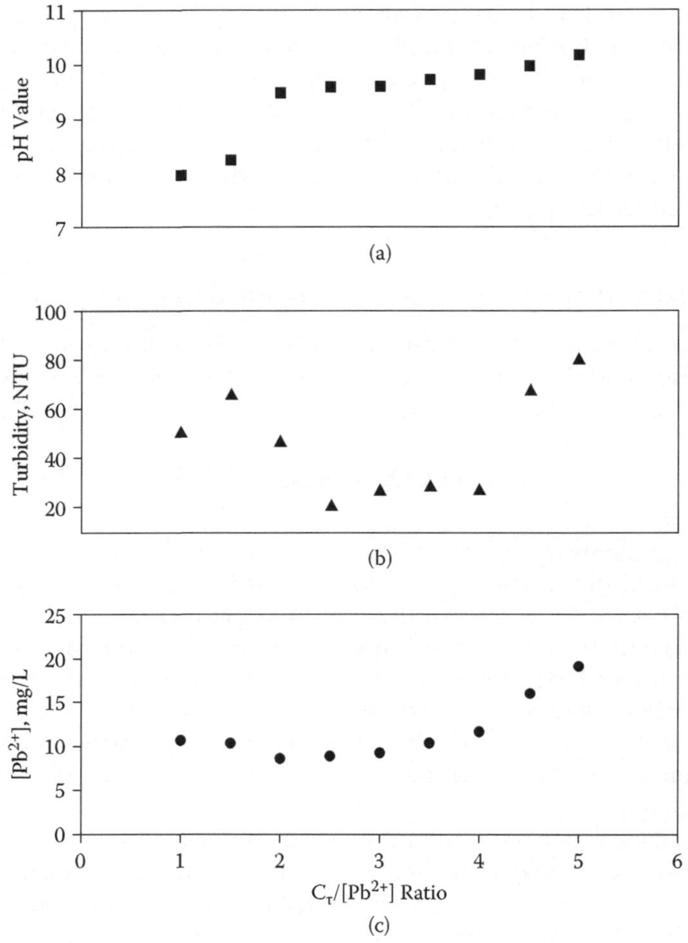

FIGURE 4.4
Determination of $C_T/[Pb^{2+}]$ ratio. (a) pH value; (b) Turbidity; (c) lead concentration.

The removal and coating efficiency are defined as follows:

$$\text{Removal efficiency (\%)} = \left(1 - \frac{[M^{2+}]_{total} \text{ after sand filtration}}{[M^{2+}]_{influent}}\right) \times 100\% \tag{4.5}$$

$$\text{Coating efficiency (\%)} = \left(1 - \frac{[Pb^{2+}]_{total} \text{ before sand filtration}}{[Pb^{2+}]_{influent}}\right) \times 100\% \tag{4.6}$$

The lead removal efficiency depends on the precipitation of amorphous lead carbonate on the surface of bare sand grains. As time proceeds, lead carbonate deposition is developed on the sand grains. As shown, the stable removal

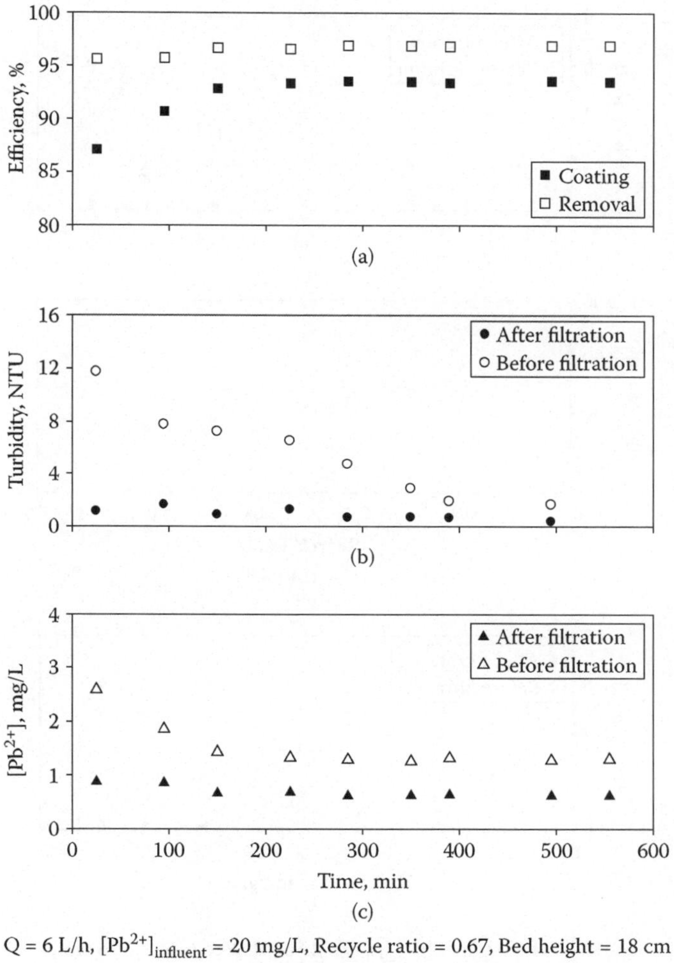

FIGURE 4.5
Start-up of the system.

of lead and lower turbidity are obtained after the system is operated for 380 min. Interestingly, when the system is operated for 150 min, the turbidity is still quite high (Figure 4.5b), while the concentrations both before and after the filtration reach low levels (Figure 4.5c). The "discrete" precipitates could be recoated on surface of the sands after a period of operation, causing the subsequent turbidity removal.

4.4.3 Recycle Ratio and Hydraulic Loading

As shown in Figure 4.6a, the lead concentrations both before and after filtration significantly decrease, with the recycle ratio increasing from 0 to 0.25.

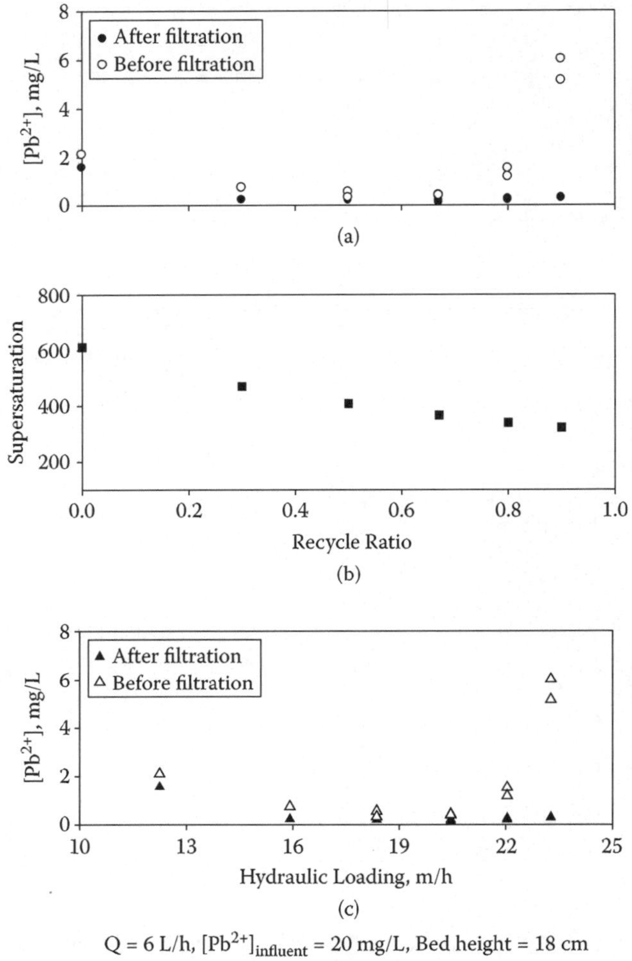

FIGURE 4.6
Recycle ratio and hydraulic loading on system performance. (a) $[Pb^{2+}]$ vs. recycle ratio; (b) supersaturation vs. recycle ratio; (c) $[Pb^{2+}]$ vs. hydraulic loading.

Minimum lead concentrations both before and after filtration are observed when the ratio ranges from 0.25 to 0.67.

When the recycle ratio is higher than 0.67, however, the lead concentrations before the sand filtration increase remarkably, which indicates poor crystallization in the FBR. At the recycle ratio of 0.9, the lead concentration before the filtration is as high as 6 mg/L. The concentration after the filtration, however, remains very low due to the filtration. The optimum recycle ratio of 0.67 is therefore recommended.

As illustrated in Figure 4.6b, the supersaturation in the FBR decreases gradually from 611 to 321 when the recycle ratio is increased from 0 to 0.9.

Slightly low supersaturation could prevent the formation of numerous nuclei, which could improve the crystallization of lead carbonate on the sand surface.

Several problems, however, could emerge if the supersaturation is too low. First, when the supersaturation in the FBR is lower, fewer metal ions may be adsorbed on the surface of the sands because of the "diluted solution," which causes the formation of fewer precipitates and "slows down" precipitation reactions. Second, if a higher recycle ratio is applied to reduce the supersaturation, it could "wash out" the precipitates originally coated on the surface of the grains due to the increase in the turbulence/mixing strength in the FBR. This could be the major reason that caused higher concentration in the effluent.

The relationship between effluent lead concentration and hydraulic loading is illustrated in Figure 4.6c. When the hydraulic loading is increased from 12.2 to 20.0 m/h, the effluent lead concentrations of both before and after filtration decrease greatly and reach their values at 20.0 m/h. However, the concentration before the sand filtration increases to 6.0 mg/L when the hydraulic loading is increased to 23.5 m/h.

This could be due to the occurrence of sand abrasion (the wash-out effect discussed previously) in the FBR. The high hydraulic loading causes an increase in turbulence, which induces particle breakage. Subsequently, the erosion of lead carbonate loosely deposited on the sand surface takes place, and the lead concentration before sand filtration may be inevitably augmented. To avoid the erosion, the hydraulic load, therefore, should not exceed 22 m/h.

4.4.4 pH Effect

The effect of pH on the removal and coating efficiencies before and after sand filtration is illustrated in Figure 4.7 together with the lead concentrations. The lead concentration in the effluent decreases when pH is increased from 6.2 to 7.3. The effluent lead concentrations reach the low levels in the pH ranging from 7.3 to 8.8. The concentrations increase slightly as pH is increased from 8.8 to 9.5. Theoretical calculation by the MINEQL+ shows in Figure 3.5 of Chapter 3 that the minimum lead concentration in lead carbonate precipitation reaches the lowest value at a pH ranging from 7 to 10, which is consistent with other researches reported in the literature.[9] The findings in Figure 4.7 coincide with these values. It can be concluded that the pH should be controlled around 8 to 9.

4.4.5 Lead Loading and Supersaturation

The lead loading may be defined as the mass of lead proceeding through the unit surface area in unit time (g/M^2h). Figure 4.8a shows the relationship between the lead loading and effluent concentrations. When the lead loading remains lower than 500 g/M^2h, the lead concentrations before or after filtration are below 1.0 mg/L. Such lower concentrations can meet the limit as 5 mg/L for discharging into the sewer systems issued by government authorities such

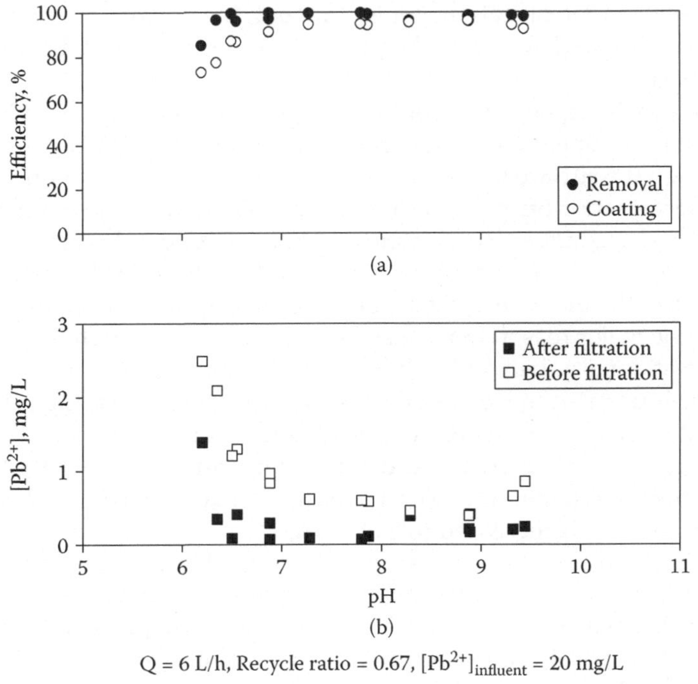

FIGURE 4.7
pH effect on system performance.

as the Ministry of Environment in Singapore. If the lead loading is increased above 500 g/m^2h, the effluent quality before the sand filtration becomes significantly poor, while the concentration after the filtration still remains less than 1 mg/L, which is attributed to the filtration in this system. It could be concluded that the lead loading should be maintained at a reasonable value in order to have better crystallization and to avoid the wash-out effect in the FBR. The lead loading should be controlled below 500 g/m^2h if the filtration is not used.

A similar trend can be found in the relationship between the supersaturation and the lead concentrations shown in Figure 4.8b. When the supersaturation in the FBR is higher than 800, the effluent quality before sand filtration becomes dramatically deteriorated. With the sand filtration, it is possible to keep the effluent lead concentration below 1.0 mg/L. However, if the filter is not employed, the supersaturation in the FBR should be below 800.

4.4.6 Bed Height

The total surface area of the sand is proportional to the sand bed height in the reactor. The larger the reaction area, the higher the applicable metal loading. As a consequence, the treatment capacity of the FBR would be higher.

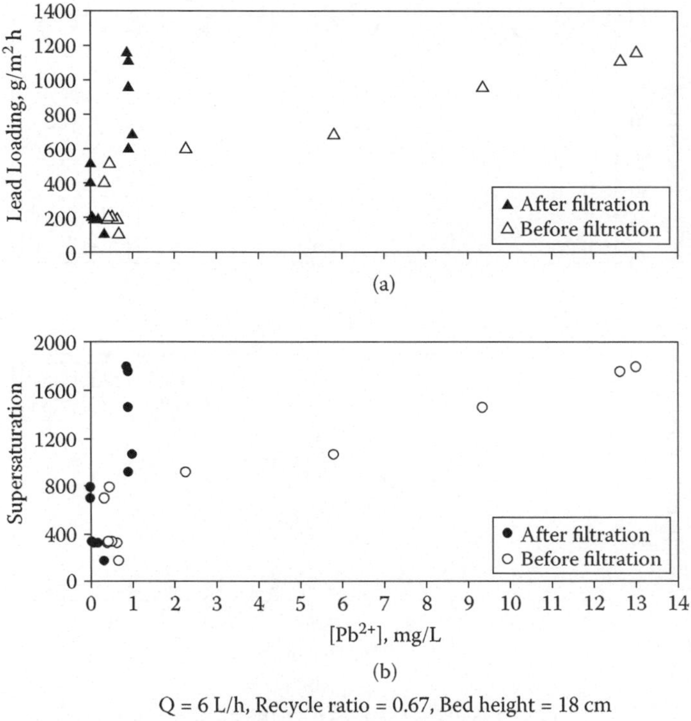

FIGURE 4.8
Lead loading and supersaturation on system performance.

During the study, the reactor was first filled with bare sand grains. The influent concentration was 40 mg/L. After a start-up period of 16 h, a mineral layer was deposited on the sand surface. During the experiments, sand grains were then removed from the FBR so as to reduce the bed height.

Figure 4.9 illustrates the variation of the efficiencies and turbidity at different sand bed heights. The coating efficiency increases dramatically from 35% to 95% as the bed height is increased from 5 to 14 cm. It remains 95% as the height ranges from 14 to 28 cm. The lead removal efficiency is above 95% as the height ranges from 5 to 28 cm. Because of the filter, the efficiency is always very high, even if the height is not very high. Similarly, the effluent turbidity before and after filtration is very low when the height is from 14 to 28 cm. As the height is increased, more sands are in the reactor, which provides large surface areas for crystallization of the lead precipitates. As a result, the effluent turbidity is significantly decreased. It has to be pointed out that the lead removal here is still high, even though the influent concentration is 40 mg/L compared to 20 mg/L used in the previous cases (e.g., Figures 4.5 to 4.7).

From the point of economic view, it is always better to have less sand in the FBR as long as the effluent quality can meet the regulations. The optimum

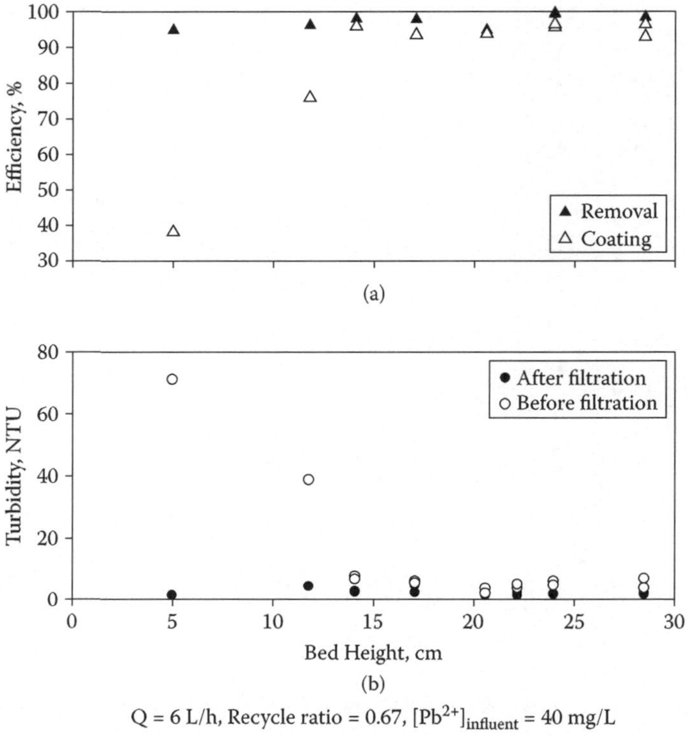

FIGURE 4.9
Bed height on system performance.

TABLE 4.2
Composition of the Crystal Coated on Sand Surface

Metals	Pb	K	Ca	Fe	Al	Trace
% (Wt)	98.56	0.09	0.52	0.11	0.12	0.6

bed height of 15 cm is therefore recommended. The corresponding ratio of bed height to total height of the FBR is 0.23. In practice, the range from 0.25 to 0.3 should be reasonable and applicable.

4.4.7 Properties of Sand Grains and Suspended Solids

To examine the deposition of lead on the sand grains (seeding material), the samples of the lead crystal–coated sand grains after a 100 h continuous operation were collected from the FBR. The samples were easily dissolved by 1 M hydrochloride acid solution. The concentrations of various metal ions in the solution were determined by the ICP-ES with results given in Table 4.2.

4.4.7.1 Metal Contents on Sand Grains

It can be found that the crystals deposited on sand surface consist of almost 99% of lead, and the rest are potassium, calcium, aluminum, iron, and traces of sodium and magnesium. These results illustrate that the effective removal of lead prior to sand filtration is mainly fulfilled by the crystallization. Moreover, the density of the crystal-coated sand grains is 1595 kg/m^3, which represents an increase of 8.8% (weight percentage) from the bare sand grains (1466 kg/m^3). As the resulting solution contains higher concentration of lead ions, the solution may be used to recover valuable metals for the industrial applications.

4.4.7.2 Microscopic Examination of Lead-Coated Sand Grains

The surfaces of bare and lead-coated sand grains after a 100-h operation are illustrated in Figure 4.2a,b. Figure 4.2c shows the surface of metal-precipitated coated sand after being washed by HCl.

The surface of the bare sand grains is smooth. At the higher magnification (400 times), numerous little rectangular lead carbonate crystals are visible on the metal-coated sand surface. This further visualizes that the lead removal by the FBR is attributed to the crystallization on the surface of sand grains.

4.4.7.3 Suspended Solids in FBR

In addition to the lead carbonate crystals deposited on the sand surface, there are some fine suspended particles formed in the FBR. Size distribution of the suspended particles is demonstrated in Figure 4.10. It is shown that particles with diameters ranging from 10 to 50 μm are dominant. Average diameter is approximately 22.15 μm. The suspended particles with such dimensions can easily be removed by a sand filter.

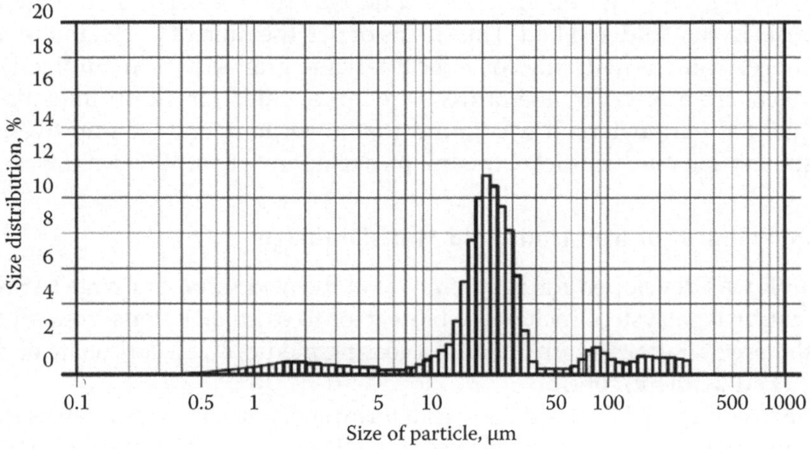

FIGURE 4.10
Size distribution of fine suspended particles.

4.5 Case Studies

4.5.1 Recovery of Silver

Wilms et al. investigated the optimal reaction conditions for crystallizing silver carbonate in a laboratory scale FBR.[2] The seeding media used in this operation was the pure quartz sand with a diameter between 0.2 and 0.3 mm. The important factors were studied on the effects on the recovery efficiency and the silver concentration in the effluent, which included the pH, the feed ratio of C_T/Ag, silver load, the hydraulic load, and the total sand grains mass in the FBR. The C_T is defined in Equation 4.4.

It was found that the optimal pH and the feed ratio of C_T/Ag in the silver carbonate crystallization were around 10.2 and 3 mol/mol, respectively. To gain a lower effluent concentration of Ag, the silver load should not be higher than 2 kg Ag/m^2h. The hydraulic load was less important on the removal efficiency of silver.

4.5.2 Recovery of Ni-Bearing Electroplating Wastewater

Lee et al. used the FBR for the treatment and recovery of nickel from the electroplating wastewater.[10] Eight different seeding materials were tested: silica sand, high-quality silica sand, beach sand, calcite, brick powder, granule-activated carbon, glass beads, and plastic beads. From the batch experiments, the amount of nickel crystal aggregated on the surface of solid grains followed a descending order: calcite > high-quality silica sand > silica sand > brick powder > beach sand > granule activated carbon > glass beads > plastic beads.

As it was impossible to recover the nickel from calcite, the calcite was not selected. The high-quality silica sand was thus selected as the seeding material in the fluidized bed. The effects of pH, the ratio of C_T/Ni in the feed, the nickel load, the hydraulic load, the total sand grains mass in fluidized bed, as well as the effect of temperature, were investigated for the optimization of operational parameters. It was found that lower temperature improved the crystallization rate. For nickel treatment, a temperature of 10°C was favorable.

4.5.3 Removal of Iron from Acid Mine Drainage

An FBR was developed for iron removal in the acid mine drainage (AMD).[11] The multistage system included a bioreactor to oxidize ferrous iron, an FBR for the precipitation of ferric iron as a coating media, and a carbonate bed for the pH adjustment/control.

The optimum pH of about 3.5 was determined. An acidic system was used, in contrast to the base system used for other metals. Ninety-nine percent iron removal efficiency was achieved, with the effluent iron concentration remaining <3 mg/L and pH > 6.

4.5.4 Removal of Multispecies Heavy Metals

Zhou and co-workers developed a precipitation–crystallization process for removing copper, nickel, and zinc from industrial wastewater.[12] The process diagram is given in Figure 4.11. It was demonstrated that the solution pH had a great effect on the heavy metal removal. The optimum pH, ranging from 9.0 to 9.1, was found at which heavy metal removal efficiency could be achieved above 92% and 95% when the influent heavy metal concentrations were 10 and 20 mg/L, respectively. Above 92.4% of the precipitated solid was composed of metal hydroxide, slightly lower than that reported by Chen and Yu.[1] The hydraulic retention time had almost no effect on the heavy metal removal when it was greater than 7.1 min. The scanning electron microscope study indicated that carbonate injection with a drip-wise manner was more efficient than that with a slug dosing manner mechanisms.

The researchers differentiated the precipitation into two types of reactions, namely, "nucleated precipitation" and "discrete precipitation" as shown in Figure 4.12. The nucleated precipitation allowed the metal precipitation to become coated on the sand surface, possibly due to adsorption or charge attraction, while the discrete precipitation is referred to the formation of fine precipitates that occur in and stay in the bulk solution instead of on the sand surface.

When the ratio of carbonate to metal was very high (e.g., 6:1 and 3:1 with the influent metal concentrations of 10 and 20 mg/L, respectively), many metal ions were precipitated out in the form of discrete precipitation before they were coated on the sand surface to form the nucleated precipitation. If that happens, a filtration device becomes essential. This is the reason that an internal dilution shown in Figure 4.11 plays an important role.

In addition, the start-up by the addition of calcium is very important in the process performance. At the beginning of the study, only sodium carbonate

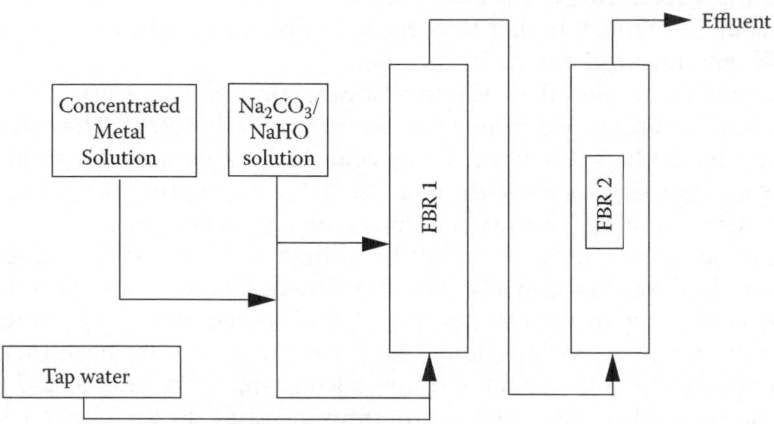

FIGURE 4.11
Process diagram of precipitation–crystallization for treatment of wastewater containing copper, lead, nickel, and zinc.[12]

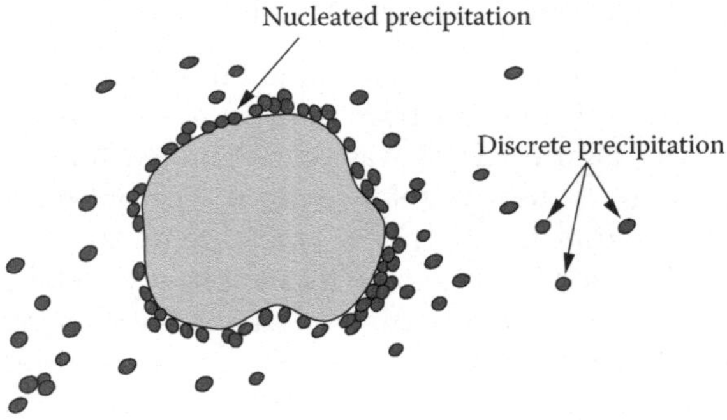

FIGURE 4.12
Illustration of formation mechanisms for the metal precipitates.

was injected into the FBR that was filled with virgin sands and had heavy metal concentrations of 10 mg/L each for Cu, Ni, and Zn in the influent. The molar ratio of [carbonate]/[metals] was 6:1. It was observed that the discrete precipitation prevailed in the reactor. A very little buildup in the metal coating on the sand occurred. The effluent contained a higher level of suspended solids. The heavy metal removal efficiency was 30% to 40%. To improve the efficacy, a calcium chloride solution was injected in to the reactor, leading to the direct nucleation of calcium carbonate on the sand surface to induce a coplating effect. As shown in Figure 4.13, the treatment efficiency was greatly improved after 4 days when the calcium was added. The process became stable. No more injection was needed after the injection of calcium. The sand essentially becomes a metal precipitate aggregate, as shown in Figure 4.14. The metal contents were quite high, as demonstrated in Figure 4.15, which is suitable for metal extraction and reuse.[12] Note that Reactor 1 and Reactor 2 in the figure are referring to the FBR and filter, respectively. The metal contents in Reactor 1 are much higher than those in Reactor 2, indicating success in the FBC and no requirement for the filter.

Huang et al. studied the treatment of removing multispecies heavy metals by fluidized sand bed from a metal-bearing wastewater.[13] The carbonate was present in the wastewater. Three columns in a treatment system were employed in order to achieve the better metal removal efficiency, which was higher than the other treatment systems with only one column.

The removal efficiency could still be as high as nearly 95% even though the lime (in a form of $Ca(OH)_2$) was not dosed. However, the coated layer on the sand grain was not as dense as that observed with the introduction of calcium (through an injection). The coated layer did not have the same high physical strength as that with the calcium injection. This coated metal carbonate could become "rubbed off" from the sand surface due to particle collision and abrasion, producing numerous finely divided metal particles that would eventually be flushed out of the column system. Meanwhile, the injection of calcium also helped to overcome the slowness and instability

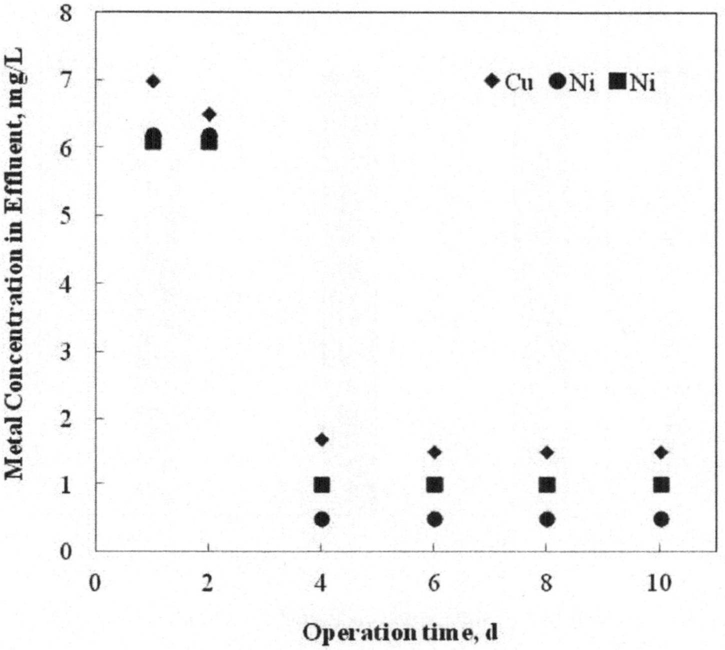

FIGURE 4.13
Improvement of process operation for the metal removal due to the addition of calcium ions.[12]

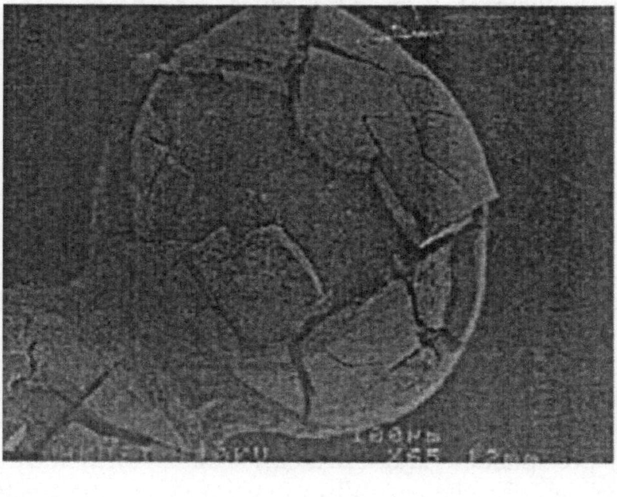

FIGURE 4.14
SEM photograph of a metal-coated sand.[12]

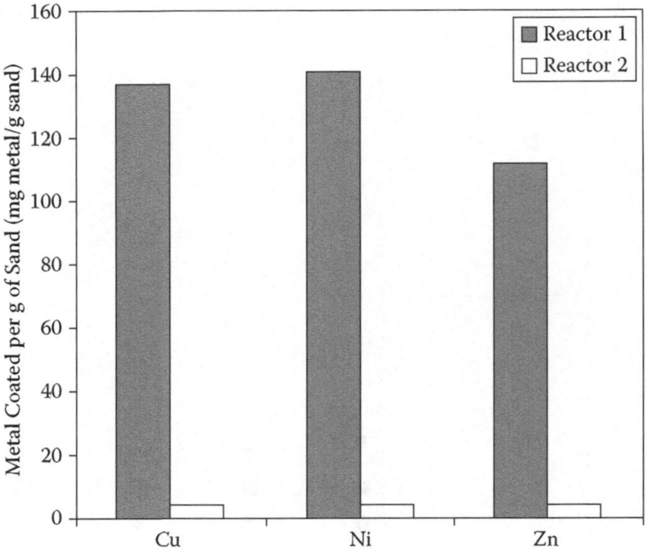

FIGURE 4.15
Contents of heavy metals in the sand particles collected from fluidized-bed reactor (Reactor 1) and filter (Reactor 2).

in the start-up. With the presence of calcium ion, the direct nucleation of calcium carbonate on sand surface would induce a coplating effect, which improved the metal stripping from the solution in the initial stage.

Based on the studies, the injection of some calcium appears to be necessary in the removal of heavy metals in the FBC.[10,12] It has to be pointed out that adding too much calcium ion is not economically sound as the purity of heavy (and precious) metals could be reduced significantly, and there is too much calcium carbonate in the solids.

4.5.5 Removal of Phosphate

Several research studies have been conducted for the removal of phosphate by fluidized bed crystallization technology.[3,4,14] The phosphate was essentially removed through the formation of calcium phosphate. By selecting sand grains of small sizes (0.1–0.3 mm) and a low superficial velocity (7×10^{-3} m/s), and by spreading the addition of NaOH solution (reactant) over two dosage points, the optimal phosphate removal with the efficiency of 80% was achieved.

In addition, the researchers developed a mathematical model for the calcium phosphate precipitation–crystallization system. The particle growth rate in a fluidized bed was described as two competing mechanisms: orthokinetic precipitation and breakage. This mathematical model can be applied to describe the crystallization process in the treatment of the heavy metal-bearing wastewater. However, it cannot be used in the design and operation of FBR as it is highly complicated and has many unknown parameters in the calculation.

Berg et al. developed a lab-scale, seed-induced crystallization process for phosphorus removal and recovery from wastewater.[15] An FBR and an expanded-bed reactor were used with calcite and tobermorite-rich crushed gas concrete (CGC) as the seeds. The results showed that the CGC seed was much more efficient in phosphorus removal from organic-enriched wastewater than the calcite seed.

The phosphorus removal efficiency of 80%–100% could be achieved over 4 months. The solid from the operation contained 10% phosphorous (weight percentage). Moreover, the FT-IR and XRD studies confirmed that the CGC was transformed into calcite during the process. The hydroxyl apatite-like compounds were formed on the seed material.

4.5.6 Copper Removal and Recovery

A two-stage water treatment system with an integration of a sulfate-reducing bioreactor with a FBC reactor demonstrated in Figure 4.16 was developed for the removal and recovery of copper.[16] During the testing, a simulated semiconductor manufacturing wastewater was employed that contained high levels of copper, sulfate, and a mixture of citrate, isopropanol, and polyethylene glycol. The fine sand was used as a seeding material for the metal crystallization.

It was demonstrated that this two-stage reactor was very efficient for copper removal; over 99% removal efficiency for copper was achieved, and the copper concentration in the effluent was only 89 μg/L. The copper crystals deposited on

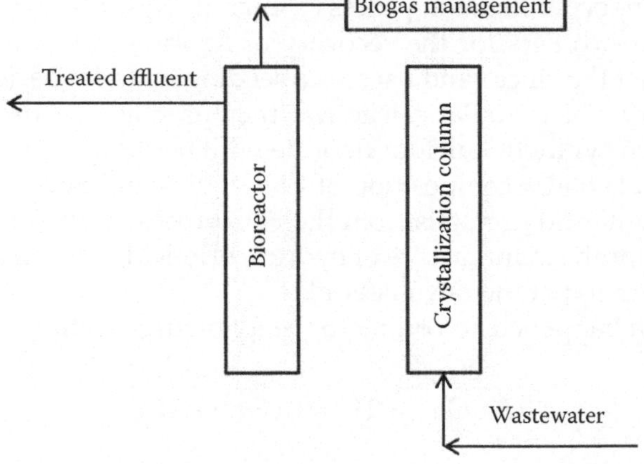

FIGURE 4.16
Process diagram of sulfate-reducing bioreactor/crystallization reactor system. Noted that 1 M NaOH was used for absorption of biogas in the "Biogas management" system.

sand grains were identified as covellite (CuS) with a high purity. Because of the incomplete degradation of isopropanol and its breakdown product (acetone), the removal of organic substances (in term of chemical oxygen demand) was less than 70%. The system showed a great potential for the simultaneous removal of heavy metal and organic contaminants.

Copper removal from oil-field brine by the precipitation–crystallization process with $CaCO_3$ was investigated.[17] The experiments were carried out in a bench-scale crystallizer equipped with a draft tube at room temperature. The coprecipitation induced by the reaction of Na_2CO_3 and $CaCl_2$ can decrease the copper concentration from 0.27 ppm in the synthetic brine to around 0.06 ppm. The copper removal efficiency of 78% was demonstrated.

The mixing intensity mainly affected the particle size distribution during the development of $CaCO_3$ particles precipitated through the reaction of soda ash with $CaCl_2$. However, the size distribution cannot be improved by using initial seeds. They were not effective in the enlargement in the sizes of the solids possibly because of the prevalence of primary nucleation in the precipitation of $CaCO_3$. The arrangement of reactant addition to the brine was found to be effective in the enlargement in the sizes of solids as well as the copper removal from the brine. The experimental results suggested that 5% of copper removal was due to the incorporation into the crystal lattice, while the rest happened through adsorption onto the crystals.

4.5.7 Fluoride Removal and Recovery

Fluoride removal by conventional precipitation technology generates huge amounts of water-rich sludge. Calcium fluoride due to the precipitation cannot be recovered. Hence, it cannot be recycled due to the high water content and the low quality of the sludge.

Precipitation–crystallization process shown in Figure 4.17 was studied by Aldaco and co-workers for the recovery of fluoride from water.[18] Initially, they found that the silica sand used as a seeding material performed poorly in the recovery. The granular calcite was then used as a seeding material in order to obtain synthetic calcium fluoride with better quality. It was shown that the crystals had a composition of $CaF_2 > 97\%$ and $SiO_2 < 1\%$. The synthetic calcium fluoride crystals from the FBC process could be recycled as a raw material for the manufacture of hydrofluoric acid, leading to a reduction in the consumption of the raw material.

The reaction happened according to the following reaction:

$$CaCO_2(s) + 2F^- = CaF_2(s) + CO_3^{2-}$$

Figure 4.18 shows photographs of the product from the crystallization process (B) when granular calcite (A) is used as a seeding material. The obtained products are homogeneous white spherical particles (pellets).

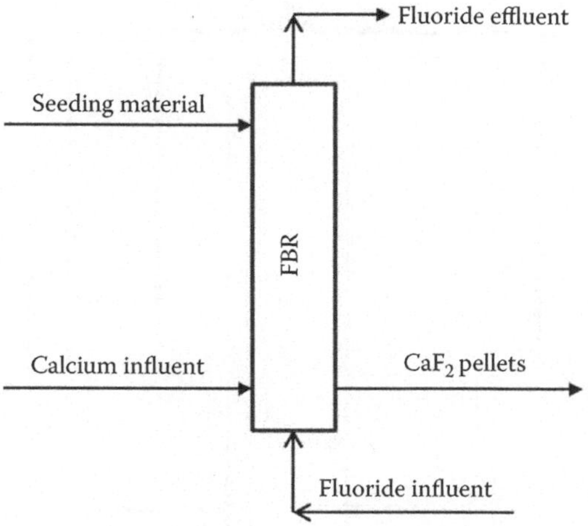

FIGURE 4.17
Illustration of fluidized-bed reactor for fluoride removal.

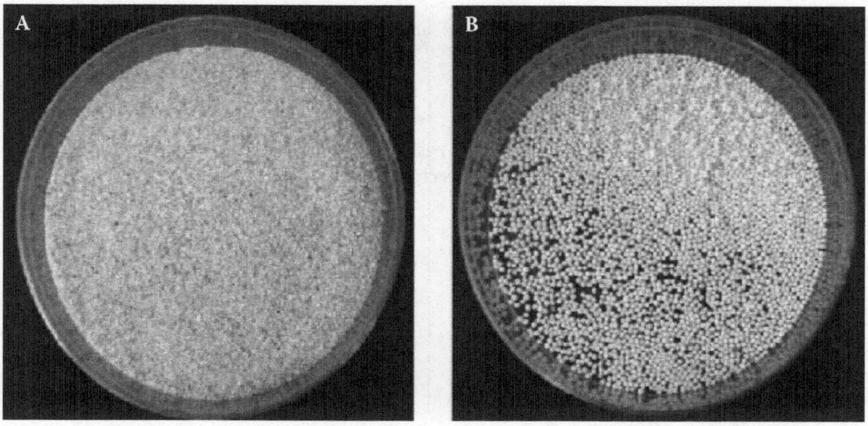

FIGURE 4.18
Comparison of seed material (a) and (b) CaF_2 pellets.[19]

Removal and recovery of fluoride in wastewaters in a pilot plant of crystallization process in a FBR were further studied with the process diagram given in Figure 4.19.[18] The fluoride was recovered as calcium fluoride (CaF_2) in the form of pellets. The process achieved up to 97% fluoride removal. The fluoride recovery of 70%–80% was obtained, leading to synthetic CaF_2 as a raw material for the production of hydrofluoric acid. The working mechanism of the precipitation–crystallization is given in Figure 4.20. This crystallization process in an FBR contributes to an environmentally friendly production due to the simultaneous reduction of waste production and recovery of materials.

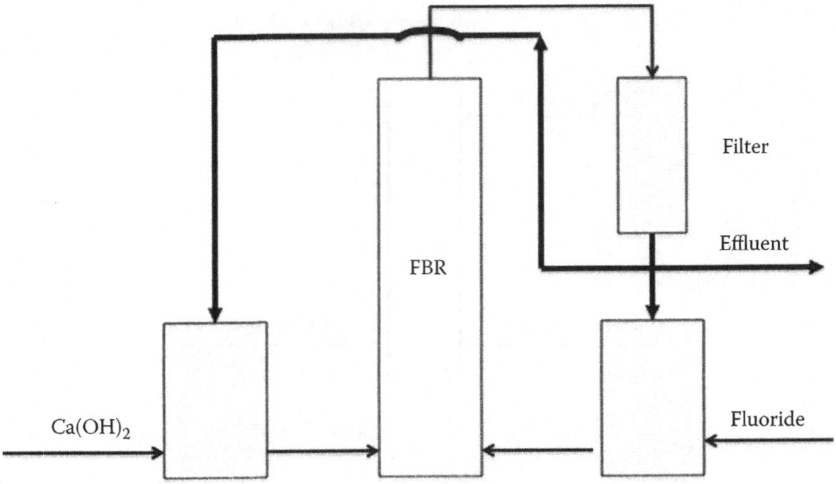

FIGURE 4.19
Illustration of treatment system for removal and recovery of fluoride.

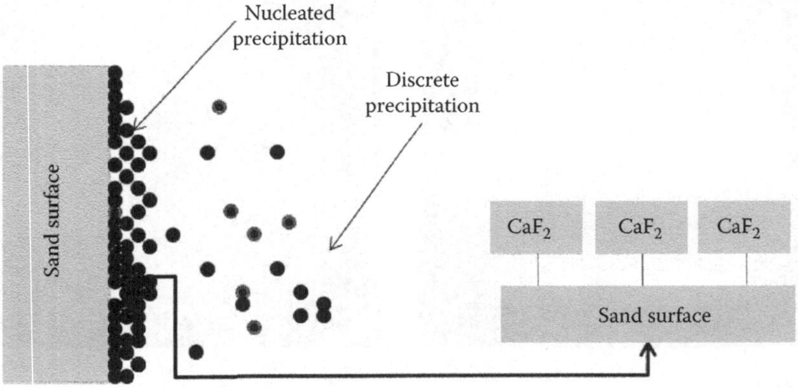

FIGURE 4.20
Precipitation–crystallization mechanism for the formation of CaF_2 pellets.

4.5.8 Arsenic Removal

FBC was employed to remove arsenate anions from aqueous solution.[19–21] The Fe/As ratio was maintained at a constant level, whereas the solution pH and iron (III) concentration were changed in order to control the degree of supersaturation of ferrihydrite. The relationship between supersaturation and arsenic removal was found to follow an exponential curve. A higher arsenic removal occurred at higher supersaturation for each tested pH value. Moreover, higher supersaturation was important to achieve a certain level of arsenic removal at pH 7 than that for the same level of removal at pH 3.5. It was found that the overall efficiency of arsenic removal depended on a combination of both pH effects and supersaturation, which controlled the affinity of arsenic for the ferrihydrite surface as well as the surface area and physical properties of the ferrihydrite.

A precipitation–crystallization study was performed at 22°C by 10 min neutralization of industrially relevant concentrated arsenate–iron(III) (Fe/As = 2 and 4) acidic sulfate solutions to pH 4, 6, and 8 in batch reactors, and the concentration of arsenic was monitored up to 1 year.[19] It was found that arsenic immobilization was due to the formation of a coprecipitate (Ca(II)–Fe(III)–As(V)) as illustrated in Figure 4.21. No sulfate incorporation was found.

Huang et al. used a precipitation–crystallization process to treat wastewater containing high-strength arsenic with the reactor and process diagram shown in Figures 4.22 and 4.23.[21] Jar tests were initially carried out to evaluate the precipitation of arsenic sulfide. A lab-scale FBR was then applied to screen critical operational parameters. Two most significant parameters were sulfide dosage and operational pH. The process was optimized to meet the wastewater discharge standard. It was found that the technology exhibited a high capability of arsenic removal and the arsenic concentration in the effluent was at the ppb level. Under the optimum conditions (pH = 2 and S/As = 2.2), the arsenic level could meet the regularity value in Taiwan.

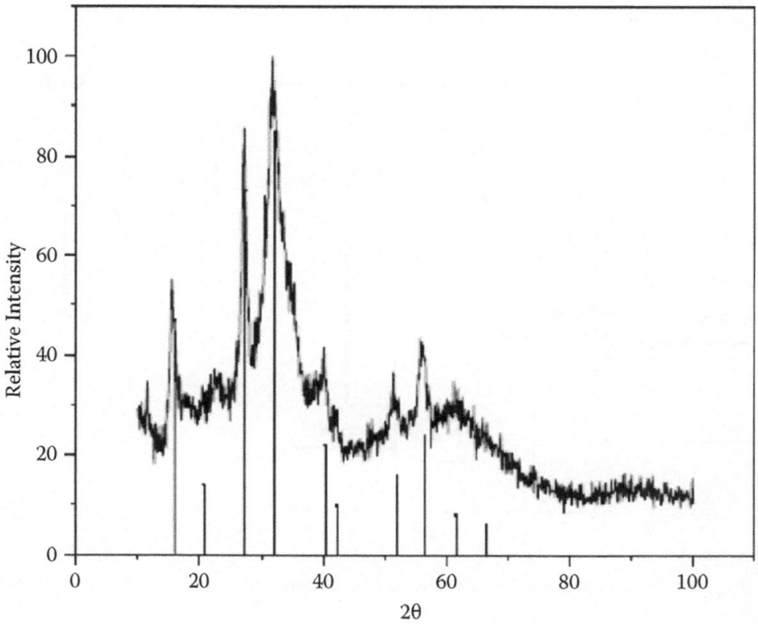

FIGURE 4.21
XRD patterns of the solid obtained in the arsenic removal (Fe/As = 2, pH 8).

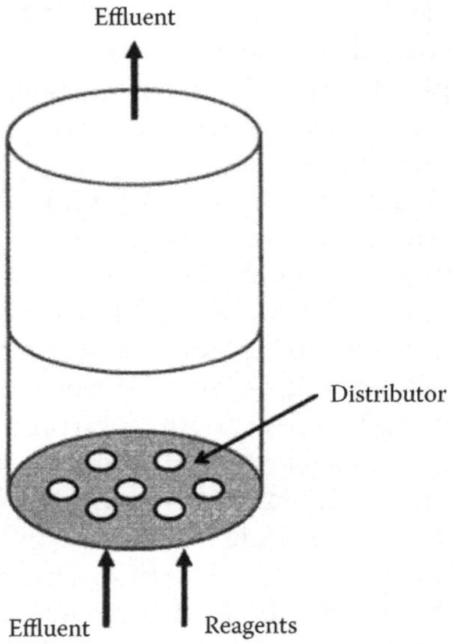

FIGURE 4.22
Illustration of fluidized-bed reactor for arsenic removal.

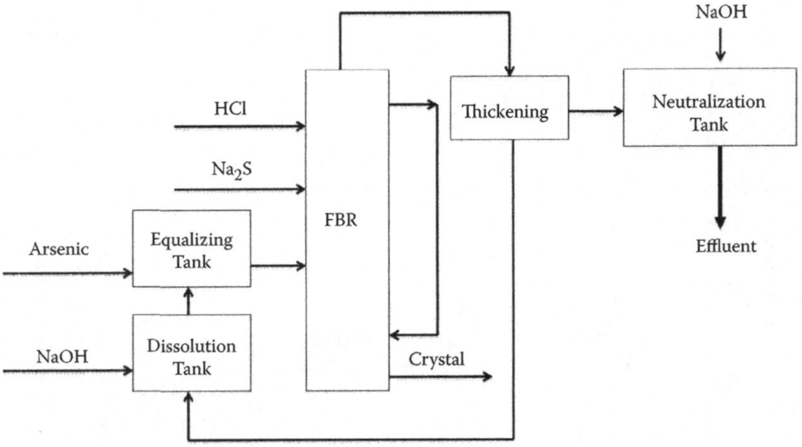

FIGURE 4.23
Process diagram of FBC process for wastewater treatment containing high-strength arsenic.

References

1. Chen, J. P., and Yu, H. Lead removal from synthetic wastewater by crystallization in a fluidized bed reactor. *J. Environ. Sci. Health,* 35(6), 817–835, 2000.
2. Wilms, D., Vercaemst, K., and Vandijk, J. C. Recovery of silver by crystallization of silver carbonate in a fluidized-bed reactor. *Water Res.,* 26(2), 235–239, 1992.
3. Seckler, M. M., Bruinsma, O. S. L., and VanRosmalen, G. M. Phosphate removal in a fluidized bed. 1. Identification of physical processes. *Water Res.,* 30(7), 1585–1588, 1996.
4. Seckler, M. M., vanLeeuwen, M. L. J., Bruinsma, O. S. L., and vanRosmalen, G. M. Phosphate removal in a fluidized bed. 2. Process optimization. *Water Res.,* 30(7), 1589–1596, 1996.
5. Scholler, M., Dijk, J. C., and Wilms, D. Fluidized bed pellet reactor to recover metals or anions, *Metal Finishing,* 89(11), 46–50, 1991.
6. Sohnel, O., and Garside, J. *Precipitation: Basic Principles and Industrial Applications.* Butterworth-Heinemann, Oxford, 1992.
7. Demopoulos, G. R. Aqueous precipitation and crystallization for the production of particulate solids with desired properties. *Hydrometallurgy,* 96(3), 199–214, 2009.
8. Snoeyink, V. L., and Jenkins, D. *Water Chemistry.* Wiley-Interscience, New York, 1980.
9. Patterson, J. W., Allen, H. E., and Scala, J. J. Carbonate precipitation for heavy metals pollutants. *J. Water Pollut. Control Fed.,* 30, 2397–2410, 1977.
10. Lee, M., You, H., and Huang, J. Treatment and recovery technology for nickel containing electroplating wastewater. *International Conference for Water Supply and Wastewater Treatment.* ICEWW'94, Beijing, 488–494, 1994.
11. Diz, H. R., and Novak, J. T. Fluidized bed for removing iron and acidity from acid mine drainage. *J. Environ. Eng-ASCE,* 124(8), 701–708, 1998.
12. Zhou, P., Huang, J. C., Li, A. W. F., and Wei S. Heavy metal removal from wastewater in fluidized bed reactor. *Water Res.,* 33(8), 1918–1924, 1999.
13. Huang, J., Wei, K. M., Li, W. F., and Zhou, P. Removal of heavy metals by nucleated precipitation in fluidized sand bed. *WEFTEC,* Asia'98, Singapore, 606–613, 1998.
14. Seckler, M. M., Leeuwen, M. L. J., Bruinsma, O. S. L., and Rosmalen, G. M. Phosphate removal in a fluidized bed. II. Process optimization. *Water Res.,* 30(7), 1589–1596, 1996.
15. Berg, U., Donnert, D., Ehbrecht, A., Bumiller, W., Kusche, I., Weidler, P. G., and Nuesch, R. "Active filtration" for the elimination and recovery of phosphorus from waste water. *Colloids Surf. A,* 265(1–3), 141–148, 2005.
16. Sierra-Alvarez, R., Hollingsworth, J., and Zhou, M. S. Removal of copper in an integrated sulfate reducing bioreactor-crystallization reactor system. *Environ. Sci. Technol.,* 41(4), 1426–1431, 2007.
17. Khosravi, J., and Alamdari, A. Copper removal from oil-field brine by coprecipitation. *J. Hazard. Mater.,* 166(2–3), 695–700, 2009.
18. Aldaco, R., Garea, A., and Irabien, A., Calcium fluoride recovery from fluoride wastewater in a fluidized bed reactor. *Water Res.,* 41(4), 810–818, 2007.

19. Jia, Y. F., and Demopoulos, G. P. Coprecipitation of arsenate with iron(III) in aqueous sulfate media: Effect of time, lime as base and co-ions on arsenic retention. *Water Res.*, 42(3), 661–668, 2008.
20. Richmond, W. R., Loan, M., Morton, J., and Parkinson, G. M. Arsenic removal from aqueous solution via ferrihydrite crystallization control. *Environ. Sci. Technol.*, 38(8), 2368–2372, 2004.
21. Huang, C. P., Pan, J. R., Lee, M. S., and Yen, S. Treatment of high-level arsenic-containing wastewater by fluidized bed crystallization process. *J. Chem. Technol. Biotechnol.*, 82(3), 289–294, 2007.

5

Reduction-Oxidation Processes

5.1 Introduction

Reduction-oxidation (redox) processes may be defined as the processes that result from redox reactions, and can be used for treatment of metallic waste streams or understanding of natural/engineered processes that can change the nature of heavy metals.[1–12] Depending on the objective, there is always one half-reaction (reduction or oxidation) playing a major role in the process. Table 5.1 shows the important half-reactions.[13]

When reduction is used, heavy metal ions (mono-, di-, or trivalent, M^+, M^{2+}, or M^{3+}) can be converted to metal ions with lower valency or zero-valent metal (M^0). For example, hexavalent chromium (Cr(VI)) can be reduced to less toxic trivalent chromium in the biosorption process, and monovalent silver ion can be reduced to zero-valent silver by using hydrazine or electroplating.

When oxidation is used, heavy metal (with valency of zero or higher) can be oxidized. For example, zero-valent iron as a sacrificial metal can be oxidized, which produces coagulant for the heavy metal removal, a process called electrocoagulation (EC), which is addressed in Chapter 6. Trivalent arsenic can be oxidized to pentavalent arsenic, which is less toxic and more easily removed from groundwater.

Redox reactions can be further classified into two groups based on the nature of the oxidizing agent (oxidant) or reducing agent (reductant or reducer): chemically induced redox reactions and electrochemically induced redox reactions. The first type reaction occurs when a chemical is added, which leads to redox reactions. The chemical is often in the form of a liquid such as hydrazine. The second type reaction occurs when direct current is applied, which is discussed in Chapter 6.

Redox reactions play important roles in natural and engineered processes for metal waste management. It is one of the major technologies for metal wastewater treatment. It is particularly suitable for decontamination of metallic waste stream when the concentration is above 1 mM. Redox distinguishes itself from other technologies as it can convert metal ions to the zero-valent form, which can be further processed and used in industries (Table 5.1).

TABLE 5.1
List of Half-Reactions and Standard Potentials

Reaction	$E°/V$
$Ac^{3+} + 3\,e \rightleftharpoons Ac$	−2.20
$Ag^{+} + e \rightleftharpoons Ag$	0.7996
$Ag^{2+} + e \rightleftharpoons Ag^{+}$	1.980
$Ag(ac) + e \rightleftharpoons Ag + (ac)^{-}$	0.643
$AgBr + e \rightleftharpoons Ag + Br^{-}$	0.07133
$AgBrO_3 + e \rightleftharpoons Ag + BrO_3^{-}$	0.546
$Ag_2C_2O_4 + 2e \rightleftharpoons 2Ag + C_2O_4^{2-}$	0.4647
$AgCl + e \rightleftharpoons Ag + Cl^{-}$	0.2223
$AgCN + e \rightleftharpoons Ag + CN^{-}$	−0.017
$Ag_2CO_3 + 2e \rightleftharpoons 2Ag + CO_3^{2-}$	0.47
$Ag_2CrO_4 + 2e \rightleftharpoons 2Ag + CrO_4^{2-}$	0.4470
$AgF + e \rightleftharpoons Ag + F^{-}$	0.779
$Ag_4[Fe(CN)_6] + 4e \rightleftharpoons 4Ag + [Fe(CN)_6]^{4-}$	0.1478
$AgI + e \rightleftharpoons Ag + I^{-}$	−0.1522
$AgIO_3 + e \rightleftharpoons Ag + IO_3^{-}$	0.354
$Ag_2MoO_4 + 2e \rightleftharpoons 2Ag + MoO_4^{2-}$	0.4573
$AgNO_2 + e \rightleftharpoons Ag + NO_2^{-}$	0.564
$Ag_2O + H_2O + 2e \rightleftharpoons 2Ag + 2OH^{-}$	0.342
$Ag_2O_3 + H_2O + 2e \rightleftharpoons 2AgO + 2OH^{-}$	0.739
$Ag^{3+} + 2e \rightleftharpoons Ag^{+}$	1.9
$Ag^{3+} + e \rightleftharpoons Ag^{2+}$	1.8
$Ag_2O_2 + 4H^{+} + 4e \rightleftharpoons 2Ag + 2H_2O$	1.802
$2AgO + H_2O + 2e \rightleftharpoons Ag_2O + 2OH^{-}$	0.607
$AgOCN + e \rightleftharpoons Ag + OCN^{-}$	0.41
$Ag_2S + 2e \rightleftharpoons 2Ag + S^{2-}$	−0.691
$Ag_2S + 2H^{+} + 2e \rightleftharpoons 2Ag + H_2S$	−0.0366
$AgSCN + e \rightleftharpoons Ag + SCN^{-}$	0.08951
$Ag_2SO_4 + 2e \rightleftharpoons 2Ag + SO_4^{2-}$	0.654
$Ag_2WO_4 + 2e \rightleftharpoons 2Ag + WO_4^{2-}$	0.4660
$Al^{3+} + 3e \rightleftharpoons Al$	−1.662
$Al(OH)_3 + 3e \rightleftharpoons Al + 3OH^{-}$	−2.31
$Al(OH)_4^{-} + 3e \rightleftharpoons Al + 4OH^{-}$	−2.328
$H_2AlO_3^{-} + H_2O + 3e \rightleftharpoons Al + 4OH^{-}$	−2.33
$AlF_6^{3-} + 3e \rightleftharpoons Al + 6F^{-}$	−2.069
$Am^{4+} + e \rightleftharpoons Am^{3+}$	2.60
$Am^{2+} + 2e \rightleftharpoons Am$	−1.9
$Am^{3+} + 3e \rightleftharpoons Am$	−2.048
$Am^{3+} + e \rightleftharpoons Am^{2+}$	−2.3
$As + 3H^{+} + 3e \rightleftharpoons AsH_3$	−0.608
$As_2O_3 + 6H^{+} + 6e \rightleftharpoons 2As + 3H_2O$	0.234
$HAsO_2 + 3H^{+} + 3e \rightleftharpoons As + 2H_2O$	0.248

TABLE 5.1 (CONTINUED)

List of Half-Reactions and Standard Potentials

Reaction	$E°/V$
$AsO_2^- + 2H_2O + 3e \rightleftharpoons As + 4OH^-$	−0.68
$H_3AsO_4 + 2H^+ + 2e \rightleftharpoons HAsO_2 + 2H_2O$	0.560
$AsO_4^{3-} + 2H_2O + 2e \rightleftharpoons AsO_2^- + 4OH^-$	−0.71
$At_2 + 2e \rightleftharpoons 2At^-$	0.3
$Au^+ + e \rightleftharpoons Au$	1.692
$Au^{3+} + 2e \rightleftharpoons Au^+$	1.401
$Au^{3+} + 3e \rightleftharpoons Au$	1.498
$Au^{2+} + e \rightleftharpoons Au^+$	1.8
$AuOH^{2+} + H^+ + 2e \rightleftharpoons Au^+ + H_2O$	1.32
$AuBr_2^- + e \rightleftharpoons Au + 2Br^-$	0.959
$AuBr_4^- + 3e \rightleftharpoons Au + 4Br^-$	0.854
$AuCl_4^- + 3e \rightleftharpoons Au + 4Cl^-$	1.002
$Au(OH)_3 + 3H^+ + 3e \rightleftharpoons Au + 3H_2O$	1.45
$H_2BO_3^- + 5H_2O + 8e \rightleftharpoons BH_4^- + 8OH^-$	−1.24
$H_2BO_3^- + H_2O + 3e \rightleftharpoons B + 4OH^-$	−1.79
$H_3BO_3 + 3H^+ + 3e \rightleftharpoons B + 3H_2O$	−0.8698
$B(OH)_3 + 7H^+ + 8e \rightleftharpoons BH_4^- + 3H_2O$	−0.481
$Ba^{2+} + 2e \rightleftharpoons Ba$	−2.912
$Ba(OH)_2 + 2e \rightleftharpoons Ba + 2OH^-$	−2.99
$Be^{2+} + 2e \rightleftharpoons Be$	−1.847
$Be_2O_3^{2-} + 3H_2O + 4e \rightleftharpoons 2Be + 6OH^-$	−2.63
p-benzoquinone + $2H^+ + 2e$ = hydroquinone	0.6992
$Bi^+ + e \rightleftharpoons Bi$	0.5
$Bi^{3+} + 3e \rightleftharpoons Bi$	0.308
$Bi^{3+} + 2e \rightleftharpoons Bi^+$	0.2
$Bi + 3H^+ + 3e \rightleftharpoons BiH_3$	−0.8
$BiCl_4^- + 3e \rightleftharpoons Bi + 4Cl^-$	0.16
$Bi_2O_3 + 3H_2O + 6e \rightleftharpoons 2Bi + 6OH^-$	−0.46
$Bi_2O_4 + 4H^+ + 2e \rightleftharpoons 2BiO^+ + 2H_2O$	1.593
$BiO^+ + 2H^+ + 3e \rightleftharpoons Bi + H_2O$	0.320
$BiOCl + 2H^+ + 3e \rightleftharpoons Bi + Cl^- + H_2O$	0.1583
$Bk^{4+} + e \rightleftharpoons Bk^{3+}$	1.67
$Bk^{2+} + 2e \rightleftharpoons Bk$	−1.6
$Bk^{3+} + e \rightleftharpoons Bk^{2+}$	−2.8
$Br_2(aq) + 2e \rightleftharpoons 2Br^-$	1.0873
$HBrO + H^+ + 2e \rightleftharpoons Br^- + H_2O$	1.331
$HBrO + H^+ + e \rightleftharpoons 1/2Br_2(aq) + H_2O$	1.574
$HBrO + H^+ + e \rightleftharpoons 1/2Br_2(l) + H_2O$	1.596
$BrO^- + H_2O + 2e \rightleftharpoons Br^- + 2OH^-$	0.761
$BrO_3^- + 6H^+ + 5e \rightleftharpoons 1/2Br_2 + 3H_2O$	1.482
$BrO_3^- + 6H^+ + 6e \rightleftharpoons Br^- + 3H_2O$	1.423
$BrO_3^- + 3H_2O + 6e \rightleftharpoons Br^- + 6OH^-$	0.61

(Continued)

TABLE 5.1 (CONTINUED)
List of Half-Reactions and Standard Potentials

Reaction	E°/V
$(CN)_2 + 2H^+ + 2e \rightleftharpoons 2HCN$	0.373
$2HCNO + 2H^+ + 2e \rightleftharpoons (CN)_2 + 2H_2O$	0.330
$(CNS)_2 + 2e \rightleftharpoons 2CNS^-$	0.77
$CO_2 + 2H^+ + 2e \rightleftharpoons HCOOH$	−0.199
$CO_2 + HCO_3^- + 8H^+ + 8e \rightleftharpoons CH_3COO^- + 3H_2O$	−0.28
$2CO_2 + HCO_3^- + NH_4^+ + 11H^+ + 12e \rightleftharpoons CH_3CHNH_2COO^- + 5H_2O$	−0.33
$6CO_2 + HCO_3^- + 30H^+ + 30e \rightleftharpoons C_6H_5COO^- + 13H_2O$	−0.28
$3CO_2 + 3HCO_3^- + 18H^+ + 18e \rightleftharpoons (COO^-)CH_2COH(COO^-)CH_2COO^- + 8H_2O$	−0.34
$2CO_2 + 12H^+ + 12e \rightleftharpoons CH_3CH_2OH + 3H_2O$	−0.32
$HCO_3^- + 2H^+ + 2e \rightleftharpoons HCOO^- + H_2O$	−0.41
$6CO_2 + 24H^+ + 24e \rightleftharpoons C_6H_{12}O_6 + 6H_2O$	−0.43
$3CO_2 + 2HCO_3^- + NH_4^+ + 18H^+ + 18e \rightleftharpoons COOHCH_2CH_2CHNH_2COO^- + 8H_2O$	−0.32
$3CO_2 + 14H^+ + 14e \rightleftharpoons CH_2OHCHOHCH_2OH + 3H_2O$	−0.40
$CO_2 + HCO_3^- + NH_4^+ + 6H^+ + 6e \rightleftharpoons CH_2NH_2COOH + 3H_2O$	−0.41
$2CO_2 + HCO_3^- + 12H^+ + 12e \rightleftharpoons CH_3CHOHCOO^- + 4H_2O$	−0.33
$CO_2 + 8H^+ + 8e \rightleftharpoons CH_4 + 2H_2O$	−0.24
$CO_2 + 6H^+ + 6e \rightleftharpoons CH_3OH + H_2O$	−0.38
$15CO_2 + HCO_3^- + 92H^+ + 92e \rightleftharpoons CH_3(CH_2)_{14}COO^- + 31H_2O$	−0.29
$2CO_2 + HCO_3^- + 14H^+ + 14e \rightleftharpoons CH_3CH_2COO^- + 5H_2O$	−0.28
$2CO_2 + HCO_3^- + 10H^+ + 10e \rightleftharpoons CH_3COCOO^- + 4H_2O$	−0.36
$2CO_2 + 2HCO_3^- + 14H^+ + 14e \rightleftharpoons (CH_2)_2(COO^-)_2 + 6H_2O$	−0.30
$2CO_3^{2-} + 7H^+ + 8e \rightleftharpoons CH_3COO^- + 2H_2O$	0.075
$Ca^+ + e \rightleftharpoons Ca$	−3.80
$Ca^{2+} + 2e \rightleftharpoons Ca$	−2.868
$Ca(OH)_2 + 2e \rightleftharpoons Ca + 2OH^-$	−3.02
$Cd^{2+} + 2e \rightleftharpoons Cd$	−0.4030
$CdSO_4 + 2e \rightleftharpoons Cd + SO_4^{2-}$	−0.246
$Cd(OH)_4^{2-} + 2e \rightleftharpoons Cd + 4OH^-$	−0.658
$CdO + H_2O + 2e \rightleftharpoons Cd + 2OH^-$	−0.783
$Ce^{3+} + 3e \rightleftharpoons Ce$	−2.336
$Ce^{4+} + e \rightleftharpoons Ce^{3+}$	1.72
$CeOH^{3+} + H^+ + e \rightleftharpoons Ce^{3+} + H_2O$	1.715
$Cf^{4+} + e \rightleftharpoons Cf^{3+}$	3.3
$Cf^{3+} + e \rightleftharpoons Cf^{2+}$	−1.6
$Cf^{3+} + 3e \rightleftharpoons Cf$	−1.94
$Cf^{2+} + 2e \rightleftharpoons Cf$	−2.12
$Cl_2(g) + 2e \rightleftharpoons 2Cl^-$	1.3583
$HClO + H^+ + e \rightleftharpoons 1/2Cl_2 + H_2O$	1.611
$HClO + H^+ + 2e \rightleftharpoons Cl^- + H_2O$	1.482
$ClO^- + H_2O + 2e \rightleftharpoons Cl^- + 2OH^-$	0.81

TABLE 5.1 (CONTINUED)
List of Half-Reactions and Standard Potentials

Reaction	$E°/V$
$ClO_2 + H^+ + e \rightleftharpoons HClO_2$	1.277
$HClO_2 + 2H^+ + 2e \rightleftharpoons HClO + H_2O$	1.645
$HClO_2 + 3H^+ + 3e \rightleftharpoons 1/2Cl_2 + 2H_2O$	1.628
$HClO_2 + 3H^+ + 4e \rightleftharpoons Cl^- + 2H_2O$	1.570
$ClO_2^- + H_2O + 2e \rightleftharpoons ClO^- + 2OH^-$	0.66
$ClO_2^- + 2H_2O + 4e \rightleftharpoons Cl^- + 4OH^-$	0.76
$ClO_2(aq) + e \rightleftharpoons ClO_2^-$	0.954
$ClO_3^- + 2H^+ + e \rightleftharpoons ClO_2 + H_2O$	1.152
$ClO_3^- + 3H^+ + 2e \rightleftharpoons HClO_2 + H_2O$	1.214
$ClO_3^- + 6H^+ + 5e \rightleftharpoons 1/2Cl_2 + 3H_2O$	1.47
$ClO_3^- + 6H^+ + 6e \rightleftharpoons Cl^- + 3H_2O$	1.451
$ClO_3^- + H_2O + 2e \rightleftharpoons ClO_2^- + 2OH^-$	0.33
$ClO_3^- + 3H_2O + 6e \rightleftharpoons Cl^- + 6OH^-$	0.62
$ClO_4^- + 2H^+ + 2e \rightleftharpoons ClO_3^- + H_2O$	1.189
$ClO_4^- + 8H^+ + 7e \rightleftharpoons 1/2Cl_2 + 4H_2O$	1.39
$ClO_4^- + 8H^+ + 8e \rightleftharpoons Cl^- + 4H_2O$	1.389
$ClO_4^- + H_2O + 2e \rightleftharpoons ClO_3^- + 2OH^-$	0.36
$Cm^{4+} + e \rightleftharpoons Cm^{3+}$	3.0
$Cm^{3+} + 3e \rightleftharpoons Cm$	−2.04
$Co^{2+} + 2e \rightleftharpoons Co$	−0.28
$Co^{3+} + e \rightleftharpoons Co^{2+}$	1.92
$[Co(NH_3)_6]^{3+} + e \rightleftharpoons [Co(NH_3)_6]^{2+}$	0.108
$Co(OH)_2 + 2e \rightleftharpoons Co + 2OH^-$	−0.73
$Co(OH)_3 + e \rightleftharpoons Co(OH)_2 + OH^-$	0.17
$Cr^{2+} + 2e \rightleftharpoons Cr$	−0.913
$Cr^{3+} + e \rightleftharpoons Cr^{2+}$	−0.407
$Cr^{3+} + 3e \rightleftharpoons Cr$	−0.744
$Cr_2O_7^{2-} + 14H^+ + 6e \rightleftharpoons 2Cr^{3+} + 7H_2O$	1.36
$CrO_2^- + 2H_2O + 3e \rightleftharpoons Cr + 4OH^-$	−1.2
$HCrO_4^- + 7H^+ + 3e \rightleftharpoons Cr^{3+} + 4H_2O$	1.350
$CrO_2 + 4H^+ + e \rightleftharpoons Cr^{3+} + 2H_2O$	1.48
$Cr(V) + e \rightleftharpoons Cr(IV)$	1.34
$CrO_4^{2-} + 4H_2O + 3e \rightleftharpoons Cr(OH)_3 + 5OH^-$	−0.13
$Cr(OH)_3 + 3e \rightleftharpoons Cr + 3OH^-$	−1.48
$Cs^+ + e \rightleftharpoons Cs$	−3.026
$Cu^+ + e \rightleftharpoons Cu$	0.521
$Cu^{2+} + e \rightleftharpoons Cu^+$	0.153
$Cu^{2+} + 2e \rightleftharpoons Cu$	0.3419
$Cu^{3+} + e \rightleftharpoons Cu^{2+}$	2.4

(Continued)

TABLE 5.1 (CONTINUED)

List of Half-Reactions and Standard Potentials

Reaction	E°/V
$Cu_2O_3 + 6H^+ + 2e \rightleftharpoons 2Cu^{2+} + 3H_2O$	2.0
$Cu^{2+} + 2CN^- + e \rightleftharpoons [Cu(CN)_2]^-$	1.103
$CuI_2^- + e \rightleftharpoons Cu + 2I^-$	0.00
$Cu_2O + H_2O + 2e \rightleftharpoons 2Cu + 2OH^-$	−0.360
$Cu(OH)_2 + 2e \rightleftharpoons Cu + 2OH^-$	−0.222
$2Cu(OH)_2 + 2e \rightleftharpoons Cu_2O + 2OH^- + H_2O$	−0.080
$2D^+ + 2e \rightleftharpoons D_2$	−0.013
$Dy^{2+} + 2e \rightleftharpoons Dy$	2.2
$Dy^{3+} + 3e \rightleftharpoons Dy$	−2.295
$Dy^{3+} + e \rightleftharpoons Dy^{2+}$	−2.6
$Er^{2+} + 2e \rightleftharpoons Er$	−2.0
$Er^{3+} + 3e \rightleftharpoons Er$	−2.331
$Er^{3+} + e \rightleftharpoons Er^{2+}$	−3.0
$Es^{3+} + e \rightleftharpoons Es^{2+}$	−1.3
$Es^{3+} + 3e \rightleftharpoons Es$	−1.91
$Es^{2+} + 2e \rightleftharpoons Es$	−2.23
$Eu^{2+} +2e \rightleftharpoons Eu$	−2.812
$Eu^{3+} + 3e \rightleftharpoons Eu$	−1.991
$Eu^{3+} + e \rightleftharpoons Eu^{2+}$	−0.36
$F_2 + 2H^+ + 2e \rightleftharpoons 2HF$	3.053
$F_2 + 2e \rightleftharpoons 2F^-$	2.866
$F_2O + 2H^+ + 4e \rightleftharpoons H_2O + 2F^-$	2.153
$Fe^{2+} + 2e \rightleftharpoons Fe$	−0.447
$Fe^{3+} + 3e \rightleftharpoons Fe$	−0.037
$Fe^{3+} + e \rightleftharpoons Fe^{2+}$	0.771
$2HFeO_4^- + 8H^+ + 6e \rightleftharpoons Fe_2O_3 + 5H_2O$	2.09
$HFeO_4^- + 4H^+ + 3e \rightleftharpoons FeOOH + 2H_2O$	2.08
$HFeO_4^- + 7H^+ + 3e \rightleftharpoons Fe^{3+} + 4H_2O$	2.07
$Fe_2O_3 + 4H^+ + 2e \rightleftharpoons 2FeOH^+ + H_2O$	0.16
$[Fe(CN)_6]^{3-} + e \rightleftharpoons [Fe(CN)_6]^{4-}$	0.358
$FeO_4^{2-} + 8H^+ + 3e \rightleftharpoons Fe^{3+} + 4H_2O$	2.20
$[Fe(bipy)_2]^{3+} + e \rightleftharpoons [Fe(bipy)_2]^{2+}$	0.78
$[Fe(bipy)_3]^{3+} + e \rightleftharpoons [Fe(bipy)_3]^{2+}$	1.03
$Fe(OH)_3 + e \rightleftharpoons Fe(OH)_2 + OH^-$	−0.56
$[Fe(phen)_2]^{3+} + e \rightleftharpoons [Fe(phen)_2]^{2+}$	1.147
$[Fe(phen)_3]^{3+} + e \rightleftharpoons [Fe(phen)_3]^{2+}$ {1 molar H_2SO_4}	1.06
$[Ferricinium]^+ + e \rightleftharpoons$ ferrocene	0.400
$Fm^{3+} + e \rightleftharpoons Fm^{2+}$	−1.1
$Fm^{3+} + 3e \rightleftharpoons Fm$	−1.89
$Fm^{2+} + 2e \rightleftharpoons Fm$	−2.30

TABLE 5.1 (CONTINUED)

List of Half-Reactions and Standard Potentials

Reaction	E^o/V
$Fr^+ + e \rightleftharpoons Fr$	–2.9
$Ga^{3+} + 3e \rightleftharpoons Ga$	–0.549
$Ga^+ + e \rightleftharpoons Ga$	–0.2
$GaOH^{2+} + H^+ + 3e \rightleftharpoons Ga + H_2O$	–0.498
$H_2GaO_3^- + H_2O + 3e \rightleftharpoons Ga + 4OH^-$	–1.219
$Gd^{3+} + 3e \rightleftharpoons Gd$	–2.279
$Ge^{2+} + 2e \rightleftharpoons Ge$	0.24
$Ge^{4+} + 4e \rightleftharpoons Ge$	0.124
$Ge^{4+} + 2e \rightleftharpoons Ge^{2+}$	0.00
$GeO_2 + 2H^+ + 2e \rightleftharpoons GeO + H_2O$	–0.118
$H_2GeO_3 + 4H^+ + 4e \rightleftharpoons Ge + 3H_2O$	–0.182
$2H^+ + 2e \rightleftharpoons H_2$	0.0000
$H_2 + 2e \rightleftharpoons 2H^-$	–2.23
$HO_2 + H^+ + e \rightleftharpoons H_2O_2$	1.495
$2H_2O + 2e \rightleftharpoons H_2 + 2OH^-$	–0.8277
$H_2O_2 + 2H^+ + 2e \rightleftharpoons 2H_2O$	1.776
$Hf^{4+} + 4e \rightleftharpoons Hf$	–1.55
$HfO^{2+} + 2H^+ + 4e \rightleftharpoons Hf + H_2O$	–1.724
$HfO_2 + 4H^+ + 4e \rightleftharpoons Hf + 2H_2O$	–1.505
$HfO(OH)_2 + H_2O + 4e \rightleftharpoons Hf + 4OH^-$	–2.50
$Hg^{2+} + 2e \rightleftharpoons Hg$	0.851
$2Hg^{2+} + 2e \rightleftharpoons Hg_2^{2+}$	0.920
$Hg_2^{2+} + 2e \rightleftharpoons 2Hg$	0.7973
$Hg_2(ac)_2 + 2e \rightleftharpoons 2Hg + 2(ac)^-$	0.5116
$Hg_2Br_2 + 2e \rightleftharpoons 2Hg + 2Br^-$	0.1392
$Hg_2Cl_2 + 2e \rightleftharpoons 2Hg + 2Cl^-$	0.2681
$Hg_2HPO_4 + 2e \rightleftharpoons 2Hg + HPO_4^{2-}$	0.6359
$Hg_2I_2 + 2e \rightleftharpoons 2Hg + 2I^-$	–0.0405
$Hg_2O + H_2O + 2e \rightleftharpoons 2Hg + 2OH^-$	0.123
$HgO + H_2O + 2e \rightleftharpoons Hg + 2OH^-$	0.0977
$Hg(OH)_2 + 2H^+ + 2e \rightleftharpoons Hg + 2H_2O$	1.034
$Hg_2SO_4 + 2e \rightleftharpoons 2Hg + SO_4^{2-}$	0.6125
$Ho^{2+} + 2e \rightleftharpoons Ho$	–2.1
$Ho^{3+} + 3e \rightleftharpoons Ho$	–2.33
$Ho^{3+} + e \rightleftharpoons Ho^{2+}$	–2.8
$I_2 + 2e \rightleftharpoons 2I^-$	0.5355
$I_3^- + 2e \rightleftharpoons 3I^-$	0.536
$H_3IO_6^{2-} + 2e \rightleftharpoons IO_3^- + 3OH^-$	0.7
$H_3IO_6 + H^+ + 2e \rightleftharpoons IO^-_3 + 3H_2O$	1.601

(Continued)

TABLE 5.1 (CONTINUED)

List of Half-Reactions and Standard Potentials

Reaction	E°/V
$2HIO + 2H^+ + 2e \rightleftharpoons I_2 + 2H_2O$	1.439
$HIO + H^+ + 2e \rightleftharpoons I^- + H_2O$	0.987
$IO^- + H_2O + 2e \rightleftharpoons I^- + 2OH^-$	0.485
$2IO_3^- + 12H^+ + 10e \rightleftharpoons I_2 + 6H_2O$	1.195
$IO_3^- + 6H^+ + 6e \rightleftharpoons I^- + 3H_2O$	1.085
$IO_3^- + 2H_2O + 4e \rightleftharpoons IO^- + 4OH^-$	0.15
$IO_3^- + 3H_2O + 6e \rightleftharpoons IO^- + 6OH^-$	0.26
$In^+ + e \rightleftharpoons In$	−0.14
$In^{2+} + e \rightleftharpoons In^+$	−0.40
$In^{3+} + e \rightleftharpoons In^{2+}$	−0.49
$In^{3+} + 2e \rightleftharpoons In^+$	−0.443
$In^{3+} + 3e \rightleftharpoons In$	−0.3382
$In(OH)_3 + 3e \rightleftharpoons In + 3OH^-$	−0.99
$In(OH)_4^- + 3e \rightleftharpoons In + 4OH^-$	−1.007
$In_2O_3 + 3H_2O + 6e \rightleftharpoons 2In + 6OH^-$	−1.034
$Ir^{3+} + 3e \rightleftharpoons Ir$	1.156
$[IrCl_6]^{2-} + e \rightleftharpoons [IrCl_6]^{3-}$	0.8665
$[IrCl_6]^{3-} + 3e \rightleftharpoons Ir + 6Cl^-$	0.77
$Ir_2O_3 + 3H_2O + 6e \rightleftharpoons 2Ir + 6OH^-$	0.098
$K^+ + e \rightleftharpoons K$	−2.931
$La^{3+} + 3e \rightleftharpoons La$	−2.379
$La(OH)_3 + 3e \rightleftharpoons La + 3OH^-$	−2.90
$Li^+ + e \rightleftharpoons Li$	−3.0401
$Lr^{3+} + 3e \rightleftharpoons Lr$	−1.96
$Lu^{3+} + 3e \rightleftharpoons Lu$	−2.28
$Md^{3+} + e \rightleftharpoons Md^{2+}$	−0.1
$Md^{3+} + 3e \rightleftharpoons Md$	−1.65
$Md^{2+} + 2e \rightleftharpoons Md$	−2.40
$Mg^+ + e \rightleftharpoons Mg$	−2.70
$Mg^{2+} + 2e \rightleftharpoons Mg$	−2.372
$Mg(OH)_2 + 2e \rightleftharpoons Mg + 2OH^-$	−2.690
$Mn^{2+} + 2e \rightleftharpoons Mn$	−1.185
$Mn^{3+} + e \rightleftharpoons Mn^{2+}$	1.5415
$MnO_2 + 4H^+ + 2e \rightleftharpoons Mn^{2+} + 2H_2O$	1.224
$MnO_4^- + e \rightleftharpoons MnO_4^{2-}$	0.558
$MnO_4^- + 4H^+ + 3e \rightleftharpoons MnO_2 + 2H_2O$	1.679
$MnO_4^- + 8H^+ + 5e \rightleftharpoons Mn^{2+} + 4H_2O$	1.507
$MnO_4^- + 2H_2O + 3e \rightleftharpoons MnO_2 + 4OH^-$	0.595
$MnO_4^{2-} + 2H_2O + 2e \rightleftharpoons MnO_2 + 4OH^-$	0.60
$Mn(OH)_2 + 2e \rightleftharpoons Mn + 2OH^-$	−1.56

TABLE 5.1 (CONTINUED)
List of Half-Reactions and Standard Potentials

Reaction	E^o/V
$Mn(OH)_3 + e \rightleftharpoons Mn(OH)_2 + OH^-$	0.15
$Mn_2O_3 + 6H^+ + 2e \rightleftharpoons 2Mn^{2+} + 3H_2O$	1.485
$Mo^{3+} + 3e \rightleftharpoons Mo$	−0.200
$MoO_2 + 4H^+ + 4e \rightleftharpoons Mo + 2H_2O$	−0.152
$H_3Mo_7O_{24}^{3-} + 45H^+ + 42e \rightleftharpoons 7Mo + 24H_2O$	0.082
$MoO_3 + 6H^+ + 6e \rightleftharpoons Mo + 3H_2O$	0.075
$N_2 + 2H_2O + 6H^+ + 6e \rightleftharpoons 2NH_4OH$	0.092
$3N_2 + 2H^+ + 2e \rightleftharpoons 2HN_3$	−3.09
$N_2 + 4H_2O + 4e \rightleftharpoons 4OH^- + N_2H_4$	−1.17
$N_2 + 5H^+ + 4e \rightleftharpoons N_2H_5^+$	−0.23
$N_2H_5^+ + 3H^+ + 2e \rightleftharpoons 2NH_4^+$	1.27
$N_2O + 2H^+ + 2e \rightleftharpoons N_2 + H_2O$	1.766
$H_2N_2O_2 + 2H^+ + 2e \rightleftharpoons N_2 + 2H_2O$	2.65
$N_2O_4 + 2e \rightleftharpoons 2NO_2^-$	0.867
$N_2O_4 + 2H^+ + 2e \rightleftharpoons 2NHO_2$	1.065
$N_2O_4 + 4H^+ + 4e \rightleftharpoons 2NO + 2H_2O$	1.035
$2NH_3OH^+ + H^+ + 2e \rightleftharpoons N_2H_5^+ + 2H_2O$	1.42
$2NO + 2H^+ + 2e \rightleftharpoons N_2O + H_2O$	1.591
$2NO + H_2O + 2e \rightleftharpoons N_2O + 2OH^-$	0.76
$HNO_2 + H^+ + e \rightleftharpoons NO + H_2O$	0.983
$2HNO_2 + 4H^+ + 4e \rightleftharpoons H_2N_2O_2 + 2H_2O$	0.86
$2HNO_2 + 4H^+ + 4e \rightleftharpoons N_2O + 3H_2O$	1.297
$NO_2^- + H_2O + e \rightleftharpoons NO + 2OH^-$	−0.46
$2NO_2^- + 2H_2O + 4e \rightleftharpoons N_2O_2^{2-} + 4OH^-$	−0.18
$2NO_2^- + 3H_2O + 4e \rightleftharpoons N_2O + 6OH^-$	0.15
$NO_3^- + 3H^+ + 2e \rightleftharpoons HNO_2 + H_2O$	0.934
$NO_3^- + 4H^+ + 3e \rightleftharpoons NO + 2H_2O$	0.957
$2NO_3^- + 4H^+ + 2e \rightleftharpoons N_2O_4 + 2H_2O$	0.803
$NO_3^- + H_2O + 2e \rightleftharpoons NO_2^- + 2OH^-$	0.01
$2NO_3^- + 2H_2O + 2e \rightleftharpoons N_2O_4 + 4OH^-$	−0.85
$Na^+ + e \rightleftharpoons Na$	−2.71
$Nb^{3+} + 3e \rightleftharpoons Nb$	−1.099
$NbO_2 + 2H^+ + 2e \rightleftharpoons NbO + H_2O$	−0.646
$NbO_2 + 4H^+ + 4e \rightleftharpoons Nb + 2H_2O$	−0.690
$NbO + 2H^+ + 2e \rightleftharpoons Nb + H_2O$	−0.733
$Nb_2O_5 + 10H^+ + 10e \rightleftharpoons 2Nb + 5H_2O$	−0.644
$Nd^{3+} + 3e \rightleftharpoons Nd$	−2.323
$Nd^{2+} + 2e \rightleftharpoons Nd$	−2.1
$Nd^{3+} + e \rightleftharpoons Nd^{2+}$	−2.7

(Continued)

TABLE 5.1 (CONTINUED)

List of Half-Reactions and Standard Potentials

Reaction	E°/V
$Ni^{2+} + 2e \rightleftharpoons Ni$	−0.257
$Ni(OH)_2 + 2e \rightleftharpoons Ni + 2OH^-$	−0.72
$NiO_3 + 4H^+ + 2e \rightleftharpoons Ni^{2+} + 2H_2O$	1.678
$NiO_2 + 2H_2O + 2e \rightleftharpoons Ni(OH)_2 + 2OH^-$	−0.490
$No^{3+} + e \rightleftharpoons No^{2+}$	1.4
$No^{3+} + 3e \rightleftharpoons No$	−1.20
$No^{2+} + 2e \rightleftharpoons No$	−2.50
$Np^{3+} + 3e \rightleftharpoons Np$	−1.856
$Np^{4+} + e \rightleftharpoons Np^{3+}$	0.147
$NpO_2 + H_2O + H^+ + e \rightleftharpoons Np(OH)_2$	−0.962
$O_2 + 2H^+ + 2e \rightleftharpoons H_2O_2$	0.695
$O_2 + 4H^+ + 4e \rightleftharpoons 2H_2O$	1.229
$O_2 + H_2O + 2e \rightleftharpoons HO_3^- + OH^-$	−0.076
$O_2 + 2H_2O + 2e \rightleftharpoons H_2O_2 + 2OH^-$	−0.146
$O_2 + 2H_2O + 4e \rightleftharpoons 4OH^-$	0.401
$O_2 + 2H^+ + 2e \rightleftharpoons O_2 + H_2O$	2.076
$O_2 + H_2O + 2e \rightleftharpoons O_2 + 2OH^-$	1.24
$O(g) + 2H^+ + 2e \rightleftharpoons H_2O$	2.421
$OH + e \rightleftharpoons OH^-$	2.02
$HO_2^- + H_2O + 2e \rightleftharpoons 3OH^-$	0.878
$OsO_4 + 8H^+ + 8e \rightleftharpoons Os + 4H_2O$	0.838
$OsO_4 + 4H^+ + 4e \rightleftharpoons OsO_2 + 2H_2O$	1.02
$[Os(bipy)_2]^{3+} + e \rightleftharpoons [Os(bipy)_2]^{2+}$	0.81
$[Os(bipy)_3]^{3+} + e \rightleftharpoons [Os(bipy)_3]^{2+}$	0.80
$P(red) + 3H^+ + 3e \rightleftharpoons PH_3(g)$	−0.0111
$P(white) + 3H^+ + 3e \rightleftharpoons PH_3(g)$	−0.063
$P + 3H_2O + 3e \rightleftharpoons PH_3(g) + 3OH^-$	−0.87
$H_2P_2^- + e \rightleftharpoons P + 2OH^-$	−0.82
$H_3PO_2 + H^+ + e \rightleftharpoons P + 2H_2O$	−0.508
$H_3PO_2 + 2H^+ + 2e \rightleftharpoons H_3PO_2 + H_2O$	−0.499
$H_3PO_2 + 3H^+ + 3e \rightleftharpoons P + 3H_2O$	−0.454
$HPO_3^{2-} + 2H_2O + 2e \rightleftharpoons H_2PO_2^- + 3OH^-$	−1.65
$HPO_3^{2-} + 2H_2O + 3e \rightleftharpoons P + 5OH^-$	−1.71
$H_3PO_4 + 2H^+ + 2e \rightleftharpoons H_3PO_3 + H_2O$	−0.276
$PO_4^{2-} + 2H_2O + 2e \rightleftharpoons HPO_3^{2-} + 3OH^-$	−1.05
$Pa^{3+} + 3e \rightleftharpoons Pa$	−1.34
$Pa^{4+} + 4e \rightleftharpoons Pa$	−1.49
$Pa^{4+} + e \rightleftharpoons Pa^{3+}$	−1.9
$Pb^{2+} + 2e \rightleftharpoons Pb$	−1.1262
$Pb^{2+} + 2e \rightleftharpoons Pb(Hg)$	−1.1205

TABLE 5.1 (CONTINUED)

List of Half-Reactions and Standard Potentials

Reaction	$E°/V$
$PbBr_2 + 2e \rightleftharpoons Pb + 2Br^-$	−0.284
$PbCl_2 + 2e \rightleftharpoons Pb + 2Cl^-$	−0.2675
$PbF_2 + 2e \rightleftharpoons Pb + 2F^-$	−0.3444
$PbHPO_4 + 2e \rightleftharpoons Pb + HPO_4^{2-}$	−0.465
$PbI_2 + 2e \rightleftharpoons Pb + 2I^-$	−0.365
$PbO + H_2O + 2e \rightleftharpoons Pb + 2OH^-$	−0.580
$PbO_2 + 4H^+ + 2e \rightleftharpoons Pb^{2+} + 2H_2O$	1.455
$HPbO_2^- + H_2O + 2e \rightleftharpoons Pb + 3OH^-$	−0.537
$PbO_2 + H_2O + 2e \rightleftharpoons PbO + 2OH^-$	0.247
$PbO_2 + SO_4^{2-} + 4H^+ + 2e \rightleftharpoons PbSO_4 + 2H_2O$	1.6913
$PbSO_4 + 2e \rightleftharpoons Pb + SO_4^{2-}$	−0.3588
$PbSO_4 + 2e \rightleftharpoons Pb(Hg) + SO_4^{2-}$	−0.3505
$Pd^{2+} + 2e \rightleftharpoons Pd$	0.951
$[PdCl_4]^{2-} + 2e \rightleftharpoons Pd + 4Cl^-$	0.591
$[PdCl_6]^{2-} + 2e \rightleftharpoons [PdCl_4]^{2-} + 2Cl^-$	1.288
$Pd(OH)_2 + 2e \rightleftharpoons Pd + 2OH^-$	0.07
$Pm^{2+} + 2e \rightleftharpoons Pm$	−2.2
$Pm^{3+} + 3e \rightleftharpoons Pm$	−2.30
$Pm^{3+} + e \rightleftharpoons Pm^{2+}$	−2.6
$Po^{4+} + 2e \rightleftharpoons Po^{2+}$	0.9
$Po^{4+} + 4e \rightleftharpoons Po$	0.76
$Pr^{4+} + e \rightleftharpoons Pr^{3+}$	3.2
$Pr^{2+} + 2e \rightleftharpoons Pr$	−2.0
$Pr^{3+} + 3e \rightleftharpoons Pr$	−2.353
$Pr^{3+} + e \rightleftharpoons Pr^{2+}$	−3.1
$Pt^{2+} + 2e \rightleftharpoons Pt$	1.18
$[PtCl_4]^{2-} + 2e \rightleftharpoons Pt + 4Cl^-$	0.755
$[PtCl_6]^{3-} + 2e \rightleftharpoons [PtCl_4]^{2-} + 2Cl^-$	0.68
$Pt(OH)_2 + 2e \rightleftharpoons Pt + 2OH^-$	0.14
$PtO_3 + 2H^+ + 2e \rightleftharpoons PtO_2 + H_2O$	1.7
$PtO_3 + 4H^+ + 2e \rightleftharpoons Pt(OH)_2^{2+} + H_2O$	1.5
$PtOH^+ + H^+ + 2e \rightleftharpoons Pt + H_2O$	1.2
$PtO_2 + 2H^+ + 2e \rightleftharpoons PtO + H_2O$	1.01
$PtO_3 + 4H^+ + 4e \rightleftharpoons Pt + 2H_2O$	1.00
$Pu^{3+} + 3e \rightleftharpoons Pu$	−2.031
$Pu^{4+} + e \rightleftharpoons Pu^{3+}$	1.006
$Pu^{5+} + e \rightleftharpoons Pu^{4+}$	1.099
$PuO_2(OH)_2 + 2H^+ + 2e \rightleftharpoons Pu(OH)_4$	1.325
$PuO_3(OH)_2 + H^+ + e \rightleftharpoons PuO_2OH + H_2O$	1.062

(Continued)

TABLE 5.1 (CONTINUED)

List of Half-Reactions and Standard Potentials

Reaction	E°/V
$Ra^{2+} + 2e \rightleftharpoons Ra$	–2.8
$Rb^{+} + e \rightleftharpoons Rb$	–2.98
$Re^{3+} + 3e \rightleftharpoons Re$	0.300
$ReO_4^{-} + 4H^{+} + 3e \rightleftharpoons ReO_2 + 2H_2O$	0.510
$ReO_2 + 4H^{+} + 4e \rightleftharpoons Re + 2H_2O$	0.2513
$ReO_4^{-} + 2H^{+} + e \rightleftharpoons ReO_2 + H_2O$	0.768
$ReO_4^{-} + 4H_2O + 7e \rightleftharpoons Re + 8OH^{-}$	–0.584
$ReO_4^{-} + 8H^{+} + 7e \rightleftharpoons Re + 4H_2O$	0.368
$Rh^{+} + e \rightleftharpoons Rh$	0.600
$Rh^{3+} + 3e \rightleftharpoons Rh$	0.758
$[RhCl_6]^{3-} + 3e \rightleftharpoons Rh + 6Cl^{-}$	0.431
$RhOH^{2+} + H^{+} + 3e \rightleftharpoons Rh + H_2O$	0.83
$Ru^{2+} + 2e \rightleftharpoons Ru$	0.455
$Ru^{3+} + e \rightleftharpoons Ru^{2+}$	0.2487
$RuO_2 + 4H^{+} + 2e \rightleftharpoons Ru^{2+} + 2H_2O$	1.120
$RuO_4^{-} + e \rightleftharpoons RuO_4^{2-}$	0.59
$RuO_4 + e \rightleftharpoons RuO_4^{-}$	1.00
$RuO_4 + 6H^{+} + 4e \rightleftharpoons Ru(OH)_2^{2+} + 2H_2O$	1.40
$RuO_4 + 8H^{+} + 8e \rightleftharpoons Ru + 4H_2O$	1.038
$[Ru(bipy)_2]^{3+} + e \rightleftharpoons [Ru(bipy)_2]^{2+}$	1.24
$[Ru(H_2O)_6]^{3+} + e \rightleftharpoons [Ru(H_2O)_6]^{2+}$	0.23
$[Ru(NH_3)_6]^{3+} + e \rightleftharpoons [Ru(NH_3)_6]^{2+}$	0.10
$[Ru(en)_3]^{3+} + e \rightleftharpoons [Ru(en)_3]^{2+}$	0.210
$[Ru(CN)_6]^{3-} + e \rightleftharpoons [Ru(CN)_6]^{4-}$	0.86
$S + 2e \rightleftharpoons S^{2-}$	–0.4763
$S + 2H^{+} + 2e \rightleftharpoons H_2S(aq)$	0.142
$S + H_2O + 2e \rightleftharpoons SH^{-} + OH^{-}$	–0.478
$2S + 2e \rightleftharpoons S_2^{2-}$	–0.4284
$S_2O_6^{2-} + 4H^{+} + 2e \rightleftharpoons 2H_2SO_3$	0.564
$S_2O_4^{2-} + 2e \rightleftharpoons 2SO_4^{2-}$	2.010
$S_2O_4^{2-} + 2H^{+} + 2e \rightleftharpoons 2HSO_4^{-}$	2.123
$S_4O_6^{2-} + 2e \rightleftharpoons 2S_2O_3^{2-}$	0.08
$2H_2SO_3 + H^{+} + 2e \rightleftharpoons HS_2O_4^{-} + 2H_2O$	–0.056
$H_2SO_3 + 4H^{+} + 4e \rightleftharpoons S + 3H_2O$	0.449
$2SO_3^{2-} + 2H_2O + 2e \rightleftharpoons S_2O_4^{2-} + 4OH^{-}$	–1.12
$2SO_3^{2-} + 3H_2O + 4e \rightleftharpoons S_2O_3^{2-} + 6OH^{-}$	–0.571
$SO_4^{2-} + 10H^{+} + 8e \rightleftharpoons H_2S + 4H_2O$	0.303
$SO_4^{2-} + 4H^{+} + 2e \rightleftharpoons H_2SO_3 + H_2O$	0.172
$2SO_4^{2-} + 4H^{+} + 2e \rightleftharpoons S_2O_6^{2-} + H_2O$	–0.22
$SO_4^{2-} + H_2O + 2e \rightleftharpoons SO_3^{2-} + 2OH^{-}$	–0.93

TABLE 5.1 (CONTINUED)
List of Half-Reactions and Standard Potentials

Reaction	$E°/V$
$Sb + 3H^+ + 3e \rightleftharpoons SbH_3$	−0.510
$Sb_2O_3 + 6H^+ + 6e \rightleftharpoons 2Sb + 3H_2O$	0.152
Sb_2O_5 (senarmontite) $+ 4H^+ + 4e \rightleftharpoons Sb_2O_3 + 2H_2O$	0.671
Sb_2O_5 (valentinite) $+ 4H^+ + 4e \rightleftharpoons Sb_2O_3 + 2H_2O$	0.649
$Sb_2O_3 + 6H^+ + 4e \rightleftharpoons 2SbO^+ + 3H_2O$	0.581
$SbO^+ + 2H^+ + 3e \rightleftharpoons Sb + H_2O$	0.212
$SbO_2^- + 2H_2O + 3e \rightleftharpoons Sb + 4OH^-$	−0.66
$SbO_3^- + H_2O + 2e \rightleftharpoons SbO_2^- + 2OH^-$	−0.59
$Sc^{3+} + 3e \rightleftharpoons Sc$	−2.077
$Se + 2e \rightleftharpoons Se^{2-}$	−0.924
$Se + 2H^+ + 2e \rightleftharpoons H_2Se(aq)$	−0.399
$H_2SeO_3 + 4H^+ + 4e \rightleftharpoons Se + 3H_2O$	0.74
$Se + 2H^+ + 2e \rightleftharpoons H_2Se$	−0.082
$SeO_3^{2-} + 3H_2O + 4e \rightleftharpoons Se + 6OH^-$	−0.366
$SeO_4^{2-} + 4H^+ + 2e \rightleftharpoons H_2SeO_3 + H_2O$	1.151
$SeO_4^{2-} + H_2O + 2e \rightleftharpoons SeO_3^{2-} + 2OH^-$	0.05
$SiF_6^{2-} + 4e \rightleftharpoons Si + 6F^-$	−1.24
$SiO + 2H^+ + 2e \rightleftharpoons Si + H_2O$	−0.8
SiO_2 (quartz) $+ 4H^+ + 4e \rightleftharpoons Si + 2H_2O$	0.857
$SiO_3^{2-} + 3H_2O + 4e \rightleftharpoons Si + 6OH^-$	−1.697
$Sm^{3+} + e \rightleftharpoons Sm^{2+}$	−1.55
$Sm^{3+} + 3e \rightleftharpoons Sm$	−2.304
$Sm^{2+} + 2e \rightleftharpoons Sm$	−2.68
$Sn^{2+} + 2e \rightleftharpoons Sn$	−0.1375
$Sn^{4+} + 2e \rightleftharpoons Sn^{2+}$	0.151
$Sn(OH)_3^+ + 3H^+ + 2e \rightleftharpoons Sn^{2+} + 3H_2O$	0.142
$SnO_2 + 4H^+ + 2e \rightleftharpoons Sn^{2+} + 2H_2O$	−0.094
$SnO_2 + 4H^+ + 4e \rightleftharpoons Sn + 2H_2O$	−0.117
$SnO_2 + 3H^+ + 2e \rightleftharpoons SnOH^+ + H_2O$	−0.194
$SnO_2 + 2H_2O + 4e \rightleftharpoons Sn + 4OH^-$	−0.945
$HSnO_2^- + H_2O + 2e \rightleftharpoons Sn + 3OH^-$	−0.909
$Sn(OH)_6^{2-} + 2e \rightleftharpoons HSnO_2^- + 3OH^- + H_2O$	−0.93
$Sr^+ + e \rightleftharpoons Sr$	−4.10
$Sr^{2+} + 2e \rightleftharpoons Sr$	−2.899
$Sr^{2+} + 2e \rightleftharpoons Sr(Hg)$	−1.793
$Sr(OH)_2 + 2e \rightleftharpoons Sr + 2OH^-$	−2.88
$Ta_2O_5 + 10H^+ + 10e \rightleftharpoons 2Ta + 5H_2O$	−0.750
$Ta^{3+} + 3e \rightleftharpoons Ta$	−0.6
$Tc^{2+} + 2e \rightleftharpoons Tc$	0.400

(Continued)

TABLE 5.1 (CONTINUED)

List of Half-Reactions and Standard Potentials

Reaction	E°/V
$TcO_4^- + 4H^+ + 3e \rightleftharpoons TcO_2 + 2H_2O$	0.782
$Tc^{3+} + e \rightleftharpoons Tc^{2+}$	0.3
$TcO_4^- + 8H^+ + 7e \rightleftharpoons Tc + 4H_2O$	0.472
$Tb^{4+} + e \rightleftharpoons Tb^{3+}$	3.1
$Tb^{3+} + 3e \rightleftharpoons Tb$	−2.28
$Te + 2e \rightleftharpoons Te^{2-}$	−1.143
$Te + 2H^+ + 2e \rightleftharpoons H_2Te$	−0.793
$Tb^{4+} + 4e \rightleftharpoons Te$	0.568
$TeO_2 + 4H^+ + 4e \rightleftharpoons Te + 2H_2O$	0.593
$TeO_3^{2-} + 3H_2O + 4e \rightleftharpoons Te + 6OH^-$	−0.57
$TeO_4^- + 8H^+ + 7e \rightleftharpoons Te + 4H_2O$	0.472
$H_6TeO_6 + 2H^+ + 2e \rightleftharpoons TeO_2 + 4H_2O$	1.02
$Th^{4+} + 4e \rightleftharpoons Th$	−1.899
$ThO_2 + 4H^+ + 4e \rightleftharpoons Th + 2H_2O$	−1.789
$Th(OH)_4 + 4e \rightleftharpoons Th + 4OH^-$	−2.48
$Ti^{2+} + 2e \rightleftharpoons Ti$	−1.630
$Ti^{3+} + e \rightleftharpoons Ti^{2+}$	−0.9
$TiO_2 + 4H^+ + 2e \rightleftharpoons Ti^{2+} + 2H_2O$	−0.502
$Ti^{3+} + 3e \rightleftharpoons Ti$	−1.37
$TiOH^{3+} + H^+ + e \rightleftharpoons Ti^{3+} + H_2O$	−0.055
$Tl^+ + e \rightleftharpoons Tl$	−0.336
$Tl^+ + e \rightleftharpoons Tl(Hg)$	−0.3338
$Tl^{3+} + 2e \rightleftharpoons Tl^+$	1.252
$Tl^{3+} + 3e \rightleftharpoons Tl$	0.741
$TlBr + e \rightleftharpoons Tl + Br^-$	−0.658
$TlCl + e \rightleftharpoons Tl + Cl^-$	−0.5568
$TlI + e \rightleftharpoons Tl + I^-$	−0.752
$Tl_2O_3 + 3H_2O + 4e \rightleftharpoons 2Tl^+ + 6OH^-$	0.02
$TlOH + e \rightleftharpoons Tl + OH^-$	−0.34
$Tl(OH)_2 + 2e \rightleftharpoons TlOH + 2OH^-$	−0.05
$Tl_2SO_4 + 2e \rightleftharpoons Tl + SO_4^{2-}$	−0.4360
$Tm^{3+} + e \rightleftharpoons Tm^{2+}$	−2.2
$Tm^{3+} + 3e \rightleftharpoons Tm$	−2.319
$Tm^{2+} + 2e \rightleftharpoons Tm$	−2.4
$U^{3+} + 3e \rightleftharpoons U$	−1.798
$U^{4+} + e \rightleftharpoons U^{3+}$	−0.607
$UO_2^+ + 4H^+ + e \rightleftharpoons U^{4+} + 2H_2O$	0.612
$UO_2^{2+} + e \rightleftharpoons UO_2^+$	0.062
$UO_2^{2+} + 4H^+ + 2e \rightleftharpoons U^{4+} + 2H_2O$	0.327
$UO_2^{2+} + 4H^+ + 6e \rightleftharpoons U + 2H_2O$	−1.444
$V^{2+} + 2e \rightleftharpoons V$	−1.175

TABLE 5.1 (CONTINUED)

List of Half-Reactions and Standard Potentials

Reaction	E^o/V
$V^{3+} + e \rightleftharpoons V^{2+}$	−0.255
$VO^{2+} + 2H^+ + e \rightleftharpoons V^{3+} + H_2O$	0.337
$VO_2^+ + 2H^+ + e \rightleftharpoons VO^{2+} + H_2O$	0.991
$V_2O_3 + 6H^+ + 2e \rightleftharpoons 2VO^{2+} + 3H_2O$	0.957
$V_2O_5 + 10H^+ + 10e \rightleftharpoons 2V + 5H_2O$	−0.242
$V(OH)_4^+ + 2H^+ + e \rightleftharpoons VO^{2+} + 3H_2O$	1.00
$V(OH)_4^+ + 4H^+ + 5e \rightleftharpoons V + 4H_2O$	−0.254
$[V(phen)_3]^{3+} + e \rightleftharpoons [V(phen)_3]^{2+}$	0.14
$W^{3+} + 3e \rightleftharpoons W$	0.1
$W_2O_5 + 2H^+ + 2e \rightleftharpoons 2WO_2 + H_2O$	−0.031
$WO_2 + 4H^+ + 4e \rightleftharpoons W + 2H_2O$	−0.119
$WO_3 + 6H^+ + 6e \rightleftharpoons W + 3H_2O$	−0.090
$WO_3 + 2H^+ + 2e \rightleftharpoons WO_2 + H_2O$	0.036
$2WO_3 + 2H^+ + 2e \rightleftharpoons W_2O_5 + H_2O$	−0.029
$H_4XeO_6 + 2H^+ + 2e \rightleftharpoons XeO_3 + 3H_2O$	2.42
$XeO_3 + 6H^+ + 6e \rightleftharpoons Xe + 3H_2O$	2.10
$XeF + e \rightleftharpoons Xe + F^-$	3.4
$Y^{3+} + 3e \rightleftharpoons Y$	−2.372
$Yb^{3+} + e \rightleftharpoons Yb^{2+}$	−1.05
$Yb^{3+} + 3e \rightleftharpoons Yb$	−2.19
$Yb^{2+} + 2e \rightleftharpoons Yb$	−2.76
$Zn^{2+} + 2e \rightleftharpoons Zn$	−0.7618
$Zn^{2+} + 2e \rightleftharpoons Zn(Hg)$	−0.7628
$ZnO_2^{2-} + 2H_2O + 2e \rightleftharpoons Zn + 4OH^-$	−1.215
$ZnSO_4 \cdot 7H_2O + 2e \rightleftharpoons Zn(Hg) + SO_4^{2-} + 7H_2O$ (Saturated $ZnSO_4$)	−0.7993
$ZnOH^+ + H^+ + 2e \rightleftharpoons Zn + H_2O$	−0.497
$Zn(OH)_4^{2-} + 2e \rightleftharpoons Zn + 4OH^-$	−1.199
$Zn(OH)_2 + 2e \rightleftharpoons Zn + 2OH^-$	−1.249
$ZnO + H_2O + 2e \rightleftharpoons Zn + 2OH^-$	−1.260
$ZrO_2 + 4H^+ + 4e \rightleftharpoons Zr + 2H_2O$	−1.553
$ZrO(OH)_2 + H_2O + 4e \rightleftharpoons Zr + 4OH^-$	−2.36
$Zr^{4+} + 4e \rightleftharpoons Zr$	−1.45

In this chapter, the redox processes will be presented according to the natures of reactants. Reduction, oxidation, and biological reduction are discussed in detail. A series of examples are provided through the chapter to provide better illustrations to ensure readers have a better understanding of the redox processes.

5.2 Chemical-Induced Reduction Processes

The chemical-induced reduction uses chemical agents to reduce/convert metal ion to its elemental form. It is particularly suitable for metal ion that has higher concentration. The metal ion should not be strongly complexed with ligands such as EDTA. This technology is able to recover metal waste stream due to chemical precipitation (e.g., metal hydroxide and carbonate) and desorption of used adsorbent.

Because of the dwindling resources of valuable metals, it is becoming more economical to recover rather than remove heavy metals from wastewater, to allow reusability of these metals or to generate a by-product of substantial economical value. Hence, the chemical reduction has increasingly become more attractive.

To reduce metal ions to their elemental form, a reducing agent will be used. The commonly used reducing agents are sodium borohydride, hydrazine, and formaldehyde. This technology is relatively simple. It is similar to chemical precipitation technology, except that the soluble metal cations are recovered as the useful elemental metal instead of metal hydroxide or carbonate sludge.

5.2.1 Sodium Borohydride

Metal reduction by sodium borohydride generally proceeds according to the following reactions[1,2]

$$M^{m+} + BH_4^- + H_2O \rightarrow M^\circ \downarrow + B(OH)_3 + H^+ \tag{5.1a}$$

$$M^{m+} + BH_4^- + H_2O \rightarrow M^\circ \downarrow + B(OH)_3 + H_2 \tag{5.1b}$$

where M represents metal (e.g., copper) with a valency of m ranging from 1 to 3.

Take reduction of Pb^{2+} as an example.

If we consider Equation 5.1a, we have:

$$4Pb^{2+} + BH_4^- + 3H_2O = 4Pb^0 + B(OH)_4 + 6H^+$$

If we consider both Equations 5.1a and 5.1b, we have:

$$2Pb^{2+} + BH_4^- + 3H_2O = 2Pb^0 + B(OH)_3 + 2H_2 + 3H^+$$

Prior addition of sodium dithionite is required to avoid reoxidation problems arising from dissolved oxygen. Because of immediate formation of ultrafine copper metal particles, flocculation-sedimentation and sand filtration have to be used for sludge separation. Meanwhile, a new impurity ($B(OH)_3$) is produced in the effluent according to Equation 5.1, which might be toxic, thus defeating the purpose of removing toxic heavy metals.

TABLE 5.2

Level of Sodium Borohydride Theoretically Required for Reduction of Metal Cations

Metal	Oxidation State	Sodium Borohydride[a] (g $NaBH_4$/kg metal)
Cobalt	2+	167
Copper	2+	143
Gold	3+	72
Iridium	4+	100
Lead	2+	46
Mercury	2+	48
Nickel	2+	167
Palladium	2+	91
Platinum	4+	100
Rhodium	3+	143
Silver	1+	43

[a] The values are based on Equation 5.1a. Up to 10 times the theoretical value may be required for operation.

The high cost of sodium borohydride retards its application. Because of these disadvantages, this reductant seems less favorable in the metal recovery.

Sodium borohydride was reportedly utilized to treat copper and cobalt ions.[1] A 1000 L/h continuous pilot plant operation with the influent metal concentrations in the range of 25 to 40 mg/L was tested. Residual concentrations lower than 0.1 mg/L were achieved. The requirement of sodium borohydride theoretically for the reduction of various metal cations is given in Table 5.2.

5.2.2 Hydrazine

Hydrazine (N_2H_4) is one of the powerful reductants widely used in various chemical operations. It is used as a reducing agent for the production of finely divided metal particles, metal-on-glass films, and metallic hydrosols, and electroless plating.[3]

Degen and Maček used hydrazine as a reducing agent in nonaqueous solutions of nickel salts to prepare nickel powders that had submicrometer sizes. Paraffin oil, di-, and triethanolamine, and ethylene glycol were used as the reaction media.[4] The rate and yield of the reaction were both enhanced at higher reaction temperatures but were limited by the relatively low boiling point of water. Nickel powders with mean particle sizes ranging from 0.1 mm to several mm and with up to 99.8% purity were obtained by this method.

Ducampsanguesa et al. reported that the monodisperse spherical palladium particles were produced by the hydrazine reduction of $[Pd(NH_3)_4]^{2+}$ in ethylene glycol in the temperature range from 20°C to −9°C.[5] Nickel et al.

studied the production of a silver colloid by reduction with hydrazine as a support for highly sensitive surface-enhanced Raman spectroscopy.[6] The reduction of aqueous silver nitrate by hydrazine dihydrochloride in weakly alkaline solution resulted in a polydisperse colloid that was stable for many months without the addition of stabilizing compounds. The average size (diameter) of the predominantly spherical particles was dependent upon the initial concentration of silver ions and ranged from 40 to 70 nm.

An important half-reaction involving hydrazine is

$$4OH^- + N_2H_4 = N_2 + 4H_2O + 4e^- \quad (5.2)$$

It can effectively reduce metal cations (M^{m+}) to the elemental state (M°) according to the following reaction:

$$M^{m+} + N_2H_4 \rightarrow M^{\circ}\downarrow + N_2 + H^+ \quad (5.3a)$$

Metal ions may also be reduced according to the following reaction:[7]

$$M^{m+} + N_2H_4 + OH^- \rightarrow M^{\circ}\downarrow + N_2 + NH_3 + H_2O \quad (5.3b)$$

Hydrazine can react with dissolved oxygen (DO) in water according to the following equation:[8]

$$N_2H_4 + O_2 \rightarrow N_2 + H_2O \quad (5.4)$$

It can undergo self-oxidation and reduction in both alkaline and acidic solutions:

$$N_2H_4 \rightarrow N_2 + NH_3 \quad (5.5)$$

As metal cations are immediately reduced to the metallic state, there are very limited concentrations of metal ions present in the solution. Complexation between metal and ammonia (due to Equations 5.3b and 5.5) therefore is unlikely. By using air stripping, ammonia is easily removed from the solution. In addition, the unutilized hydrazine can be removed by aeration (Equation 5.4).

Any metal waste stream has its own characteristics such as pH, ionic strength, and presence of other competitive and noncompetitive factors. The reduction caused by the hydrazine would be affected by these factors. Hydrazine is chemically similar to ammonia as it can be regarded as an ammonia derivative, in which one of the hydrogen atoms in NH_3 is replaced by the more negative NH_2 group. The effects of important operational parameters (e.g. pH) were studied by chen and Lim[9] and the findings are given as follows.

5.2.2.1 Effect of pH

As solution pH can change the solution chemistry of metal wastewater, it may significantly affect the treatment results of reduction. When pH is high, metal precipitates are formed prior to the addition of hydrazine. However, the particles are dissolved immediately when hydrazine is added.[9]

The hydrazine has some buffering effect due to the following reaction:[8]

$$N_2H_4 + H_2O = N_2H_5^+ + OH^-, \quad K = 8.5\times10^{-7} \quad (\text{at } 25°\text{c}) \tag{5.6a}$$

$$N_2H_4 + H^+ = N_2H_5^+, \quad K^* = 8.5\times10^{-7} \quad (\text{at } 25°\text{c}) \tag{5.6b}$$

The pH effect on metal removal was experimentally studied.[9] Figure 5.1a shows that the final copper concentration significantly decreases as the initial pH is increased from 2 to 6. At initial pH > 6.0, the copper concentrations ranged from 1.42×10^{-4} to 2.39×10^{-4} M; in other words, a recovery of 97%–98% is obtained. When the initial pH ranges from 6 to 9, the final pH is always 9. The buffering capacity is limited; the final pH is the same as the initial pH when it is 2 to 5 and 9 to 13.

The dramatic increase in the recovery shown in Figure 5.1a is due to the change in the solution chemistry. The speciation of hydrazine versus pH based on Equation 5.6 indicates that percentage of N_2H_4 reaches 100% at pH > 11. The half-reaction of hydrazonium ($N_2H_5^+$) is

$$N_2H_5^+ + 3H^+ + 2e = 2NH_4^+ \tag{5.7}$$

One can see that hydrazonium is a strong oxidizing regent. As such, it cannot reduce and recover metal ions from the aqueous solution. It is therefore recommended that the solution pH above 11 be controlled in order that the percentage of N_2H_4 is 100 % (Figure 5.1b).

The pH-dependent reduction reaction can be applied for recovery of mono- and divalent metal ions. Copper and silver ions reduction reactions are as follows:

$$2Cu^{2+} + N_2H_4 = 2Cu\downarrow + N_2 + 4H^+ \tag{5.8a}$$

$$Cu^{2+} + 2N_2H_4 + 2OH^- = Cu\downarrow + N_2 + 2NH_3 + 2H_2O \tag{5.8b}$$

$$4Ag^+ + N_2H_4 = 4Ag\downarrow + N_2 + 4H^+ \tag{5.9a}$$

$$2Ag^+ + 2N_2H_4 + 2OH^- = 2\,Ag\downarrow + N_2 + 2NH_3 + 2H_2O \tag{5.9b}$$

In addition to the reduction that results from the presence of hydrazine, metal precipitation may also play a role. Take the removal of copper ions

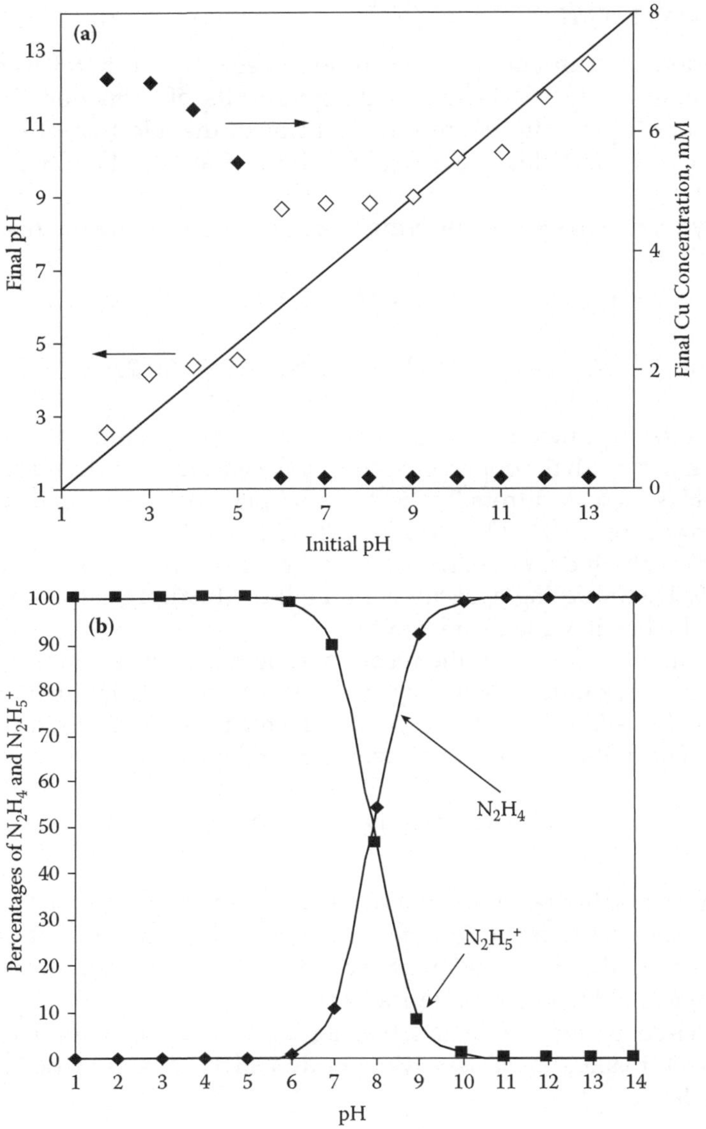

FIGURE 5.1
Effect of pH on metal reduction: (a) final pH and final Cu copper concentration as a function of pH; (b) pH vs. N_2H_4% or $N_2H_5^+$%.

illustrated in Figure 5.1a as an example. We can see that the final concentration begins to achieve its minimum value at initial pH of 6 (the corresponding final pH being 8.7). As shown in the pH-dependent hydrazine distribution in Figure 5.1b, the hydrazine-hydrazonium speciation diagram shows that the former is 80%. In other words, there is not enough hydrazine for complete copper reduction. The only reason for the complete removal is the copper precipitation reactions as follows:[10]

$$Cu^{2+} + H_2O = CuO + 2H^+ \tag{5.10a}$$

$$Cu^{2+} + 2OH^- = Cu(OH)_2 \tag{5.10b}$$

It should be pointed out that Cu^{2+} may also be reduced to Cu^+, which instantly forms metal crystal. This phenomenon has widely been reported in the literatures on electroless plating. For example, Lim et al. found from X-ray photoelectron spectroscopic study that during the electroless plating process, the Cu^{2+} ions were first reduced to Cu^+ by formaldehyde on the polypyrrole film surface and subsequently reduced to its elemental form.[11]

The importance of pH for the metal removal can be further illustrated by the experiments of copper sulfate and silver nitrate.[9] As shown in Figure 5.2a, the final copper concentration for the "no fixed pH" case (pH ranging from 4.1 to 7.8 for copper sulfate solution) is much higher than in the case where the pH is fixed at 11. In other words, the poor removal is obtained when the pH is ranged from 4.1 to 7.8. The removal is mainly due to the precipitation reactions rather than the reduction reactions given in the foregoing equations.

When the constant pH 11 is maintained, the final copper concentration of 0 to 1.8×10^{-4} M is observed for the initial concentration of $9.0 \times 10^{-4} \sim 8.0 \times 10^{-3}$ M. Based on the stoichiometry (Equation 5.8a), the final copper concentration can be calculated. As illustrated in the figure, the calculated final concentrations (termed theoretical values) match very well with the measured ones when the initial concentrations are below 6.25×10^{-3} M. When the initial copper concentration is above 6.25×10^{-3} M, the theoretical values are, however, higher than the measured values because none of the precipitation reactions or additional reductions ($Cu^{2+} \rightarrow Cu^+$) are accounted for in the calculation. For lower initial copper concentrations, the copper solids formed were very fine and tended to be suspended in the solution, taking an extremely long time to settle down.[9]

Similar findings can be obtained for the removal monovalent metal ions such as silver nitrate.[9] As shown in Figure 5.2b, the final silver concentration at pH controlled at 11 is lower than 2.3×10^{-5} M for the concentration ranging from 5.0×10^{-4} to 4.7×10^{-3} M. The final concentration for the "no fixed pH" case is also lower than 2.3×10^{-5} M for the initial concentration $< 3.7 \times 10^{-3}$ M; however, a sharp jump is observed at the initial concentration $> 3.7 \times 10^{-3}$ M, due to the incomplete reduction of silver at the lower pH.

Final silver concentrations can be calculated based on the stoichiometry, as shown in Equation 5.9a. Generally speaking, the theoretical values matched the observed ones well. There are, however, slight deviations due to the simplification in the calculations due to the possible reactions such as complexation reactions that form $Agcl_2^-$ etc.

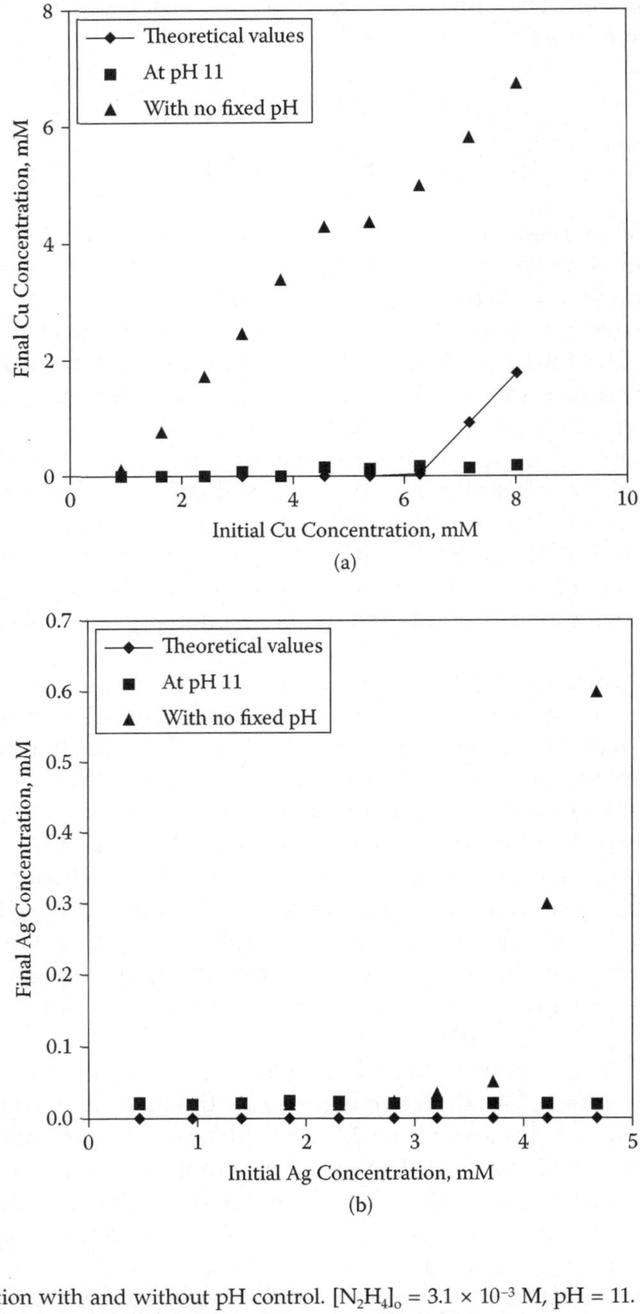

FIGURE 5.2
Metal reduction with and without pH control. $[N_2H_4]_o = 3.1 \times 10^{-3}$ M, pH = 11. (a) $[CuSO_4]_o = 9.0 \times 10^{-4} \sim 8.0 \times 10^{-3}$ M; (b) $[AgNO_3]_o = 4.7 \times 10^{-4} \sim 4.7 \times 10^{-3}$ M.

5.2.2.2 Effect of Humic Acid

Organic matters are always present in wastewater. Humic acid/substance can be used to represent organic matters. It is observed in Figure 5.3a that humic acid does not always hinder the recovery of copper, but rather, the recovery is much higher than when there is no presence of humic acid.[9] This could be due to the immediate formation of Cu^{+} and the subsequent

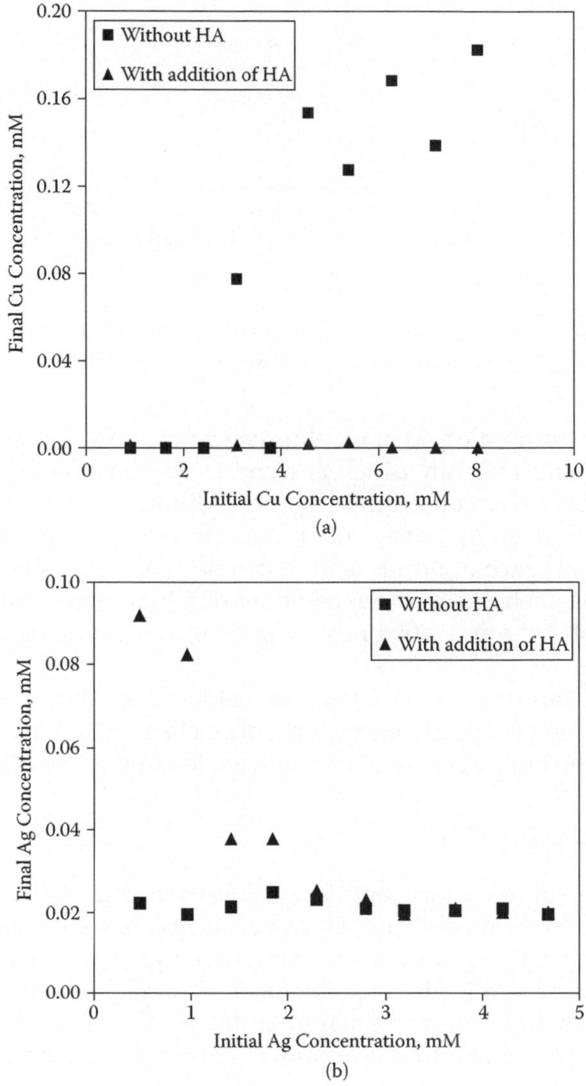

FIGURE 5.3
Effect of HA on metal reduction, $[N_2H_4]_o = 3.1 \times 10^{-3}$ M, [HA] = 0.5 ppm, pH = 11. (a) Copper, $[Cu^{2+}]_o = 9.0 \times 10^{-4} \sim 8.0 \times 10^{-3}$ M; (b) Silver, $[Ag^{+}]_o = 4.7 \times 10^{-4} \sim 4.7 \times 10^{-3}$ M.

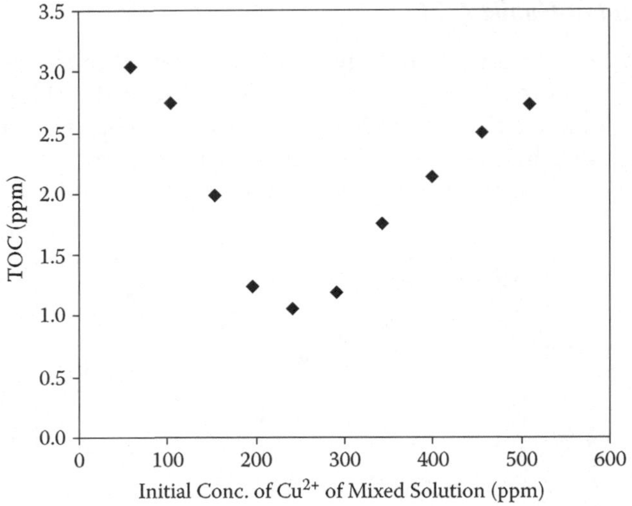

FIGURE 5.4
Removal of humic acid during reduction of copper ions. Conditions: 58–510 ppm Cu^{2+}, 0.5 ppm HA, 100 ppm N_2H_4, pH at 11 adjusted with 1 M NaOH and 0.1 M HNO_3.

production of Cu-humic acid or Cu^+-humic acid crystals. The silver recovery, however, exhibits a slightly different trend as demonstrated in Figure 5.3b. At lower initial silver concentrations, the addition of humic acid hindered the recovery of silver. At higher initial concentrations, the recovery is similar to that in the absence of humic acid. It can be speculated that at low silver concentrations, there is a sufficient amount of humic acid to bind a substantial amount of Ag^+ ions, thus preventing them from undergoing reduction by hydrazine.

Removal of humic acid (TOC) can be achieved as shown in Figure 5.4. As the reduction of copper ions may produce chemically active monovalent copper ions, the humic acid may be oxidized, leading to good removal.

5.2.2.3 *Effect of DO*

We have seen success in the application of hydrazine in the removal and reduction of heavy metal ions. However, it may be an issue of concern that hydrazine may be harmful to human beings when its concentration reaches a certain level. The physiological effects of an aqueous solution of hydrazine include corrosive action on the eyes, skin, and mucus membranes. It is a probable human carcinogen. Furthermore, hydrazine fumes with air, and its vapor attacks the nose and throat through inhalation and can also irritate the eyes, causing temporary blindness. Hence, it is important to control the concentration of hydrazine used for the metal reduction.

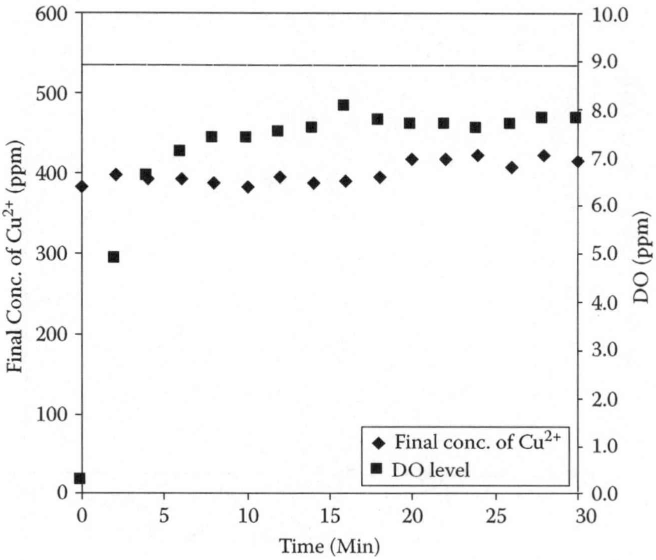

FIGURE 5.5
Effect of DO in hydrazine on the final concentration of copper. Conditions: 510 ppm Cu^{2+}, 125.9 ppm N_2H_4, temperature of 21°C. The straight line represents the saturated oxygen concentration.

Hydrazine may go through decomposition in the presence of oxygen according to Equation 5.4. Figure 5.5 shows that a higher level of DO in hydrazine solution has a very significant effect on the removal of copper from aqueous solution.[9] In addition, the dissolved oxygen does not change during the experiment.

It may be speculated that the reduction kinetics is faster than the decomposition. A study showed that an alkaline solution of 0.1 M hydrazine was found to have undergone decomposition to the extent of 2% during a 10-min exposure to the atmosphere.[8] A 0.05 M hydrazine solution was found to have decomposed to the extent of 20% after 16 h exposure to the atmosphere.

5.2.2.4 Competition in Metal Reduction

When a solution has more than one metal ion, the reduction of one metal may behave differently from others. From Figure 5.6, it can be observed that the final copper concentration is lower when competing silver ion is present than when silver ion is absent.[9] Similarly, the final silver concentrations were found to be very low in the Ag-Cu solution.

The enhanced metal recovery is probably due to codeposition. The standard electrode potentials (in volts) of silver and copper at 25°C are 0.8 V and 0.34 V, respectively.[12] The recovery of silver is much easier and faster than

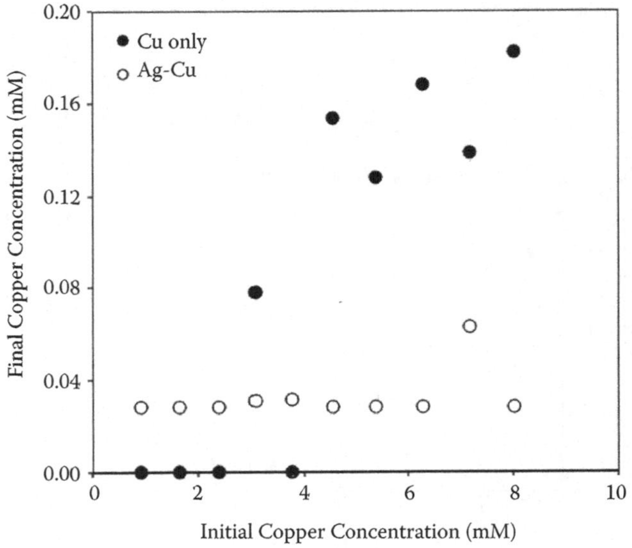

FIGURE 5.6
Effect of competitive metal ions on copper reduction, $[Cu^{2+}]_o = 9.0 \times 10^{-4} \sim 8.0 \times 10^{-3}$ M, $[Ag^+]_o = 5.0 \times 10^{-4} \sim 4.7 \times 10^{-3}$ M, $[N_2H_4]_o = 3.1 \times 10^{-3}$ M, pH = 11.

that of copper. Therefore, the silver ions can first be reduced to elemental silver (as fine particles). The fine silver particles subsequently act as nuclei (or crystal seeds) on which the elemental copper can then be deposited due to electrostatic abstraction forces. The process is similar to the novel crystallization process for lead removal discussed in Chapter 4.[13]

5.2.2.5 Effect of Seeding and Aging Process

The recovery process may include a metal reduction reaction by hydrazine, nucleation, and crystal growth of elemental metals. If the amount of elemental metal is too low, nucleation and sequent crystal growth cannot occur. Faster reaction kinetics, which leads to a huge amount of the elemental metal rapidly in the solution, however, may cause the formation of numerous nuclei in the solution, which is also not desirable.

Figure 5.7 shows that when the mixed solution is placed in the oscillating water bath for 39 h at 60°C, the mean particle size of the solid obtained is 30 μm, much larger than the solids aged for 2 h (averaged particle size of 7 μm).[9] Figure 5.8 shows the copper particles after reduction and ageing for 39 hours. It was observed that the copper particles aged for 39 h turned from reddish brown to black in color, suggesting that partial oxidation of copper might have taken place, thus resulting in the formation of CuO.[9]

Additionally, one can observe that the higher initial concentration does not lead to the formation of larger particles. When the initial concentration

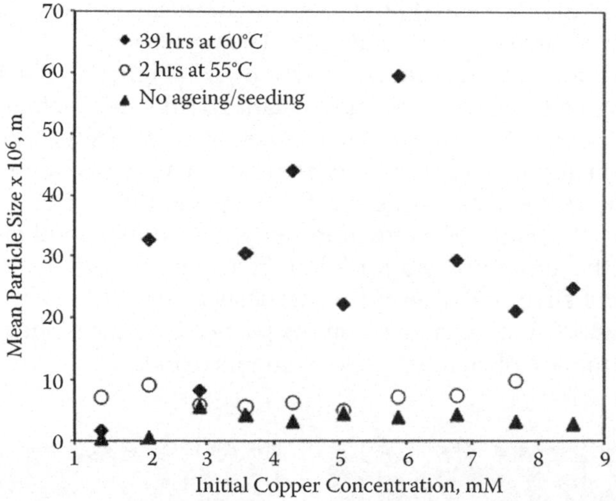

FIGURE 5.7
Effect of the seeding/aging process on mean particle size of copper obtained, $[Cu^{2+}]_o = 9.0 \times 10^{-4} \sim 8.0 \times 10^{-3}$ M, $[N_2H_4]_o = 3.1 \times 10^{-3}$ M, pH = 11.

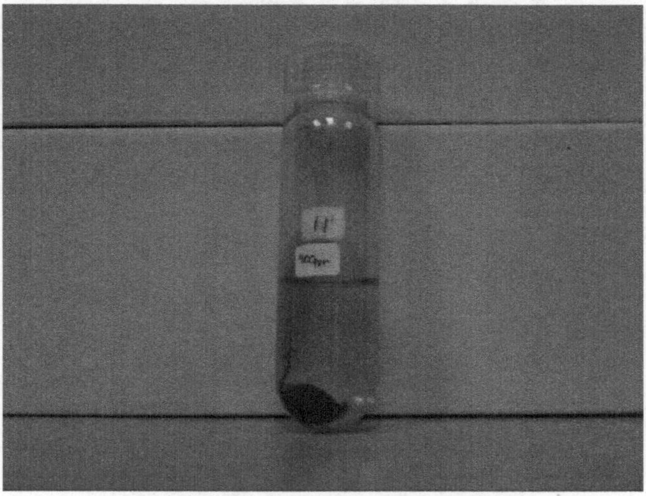

FIGURE 5.8
Copper particles after aging for 6 h. pH 11 and initial copper concentration of 900 ppm.

is higher, a rapid chemical reduction reaction occurs, which produces more fine particles. As a result, the largest particles are not observed at the highest initial concentration.

Practically speaking, it would not make sense to keep the fine metallic particles in the solution for a longer time so as to achieve large particles through aging, as this would lead to additional cost in the metal recovery. Thus, an aging time of a few hours should be sufficient. The particle size fluctuates between 3.3 and 7.0 μm for an aging time of 2 h as shown in Figure 5.7. Centrifugation is therefore needed to separate the supernatant and the fine metallic particles as they take a longer time to settle down. Alternatively, coagulation followed by sedimentation/filtration or membrane filtration may be used for the separation. Figure 5.8 shows the metal particles after aging and then being centrifuged.

Figure 5.9a shows the SEM micrograph of the copper solid obtained after conducting the aging process for 2 h at 55°C. Figure 5.9b demonstrates the micrograph of silver particles. One can observe that there are significantly huge amounts of smaller crystals on the particle's surface. They could result from the formation of elemental metal and its oxides.

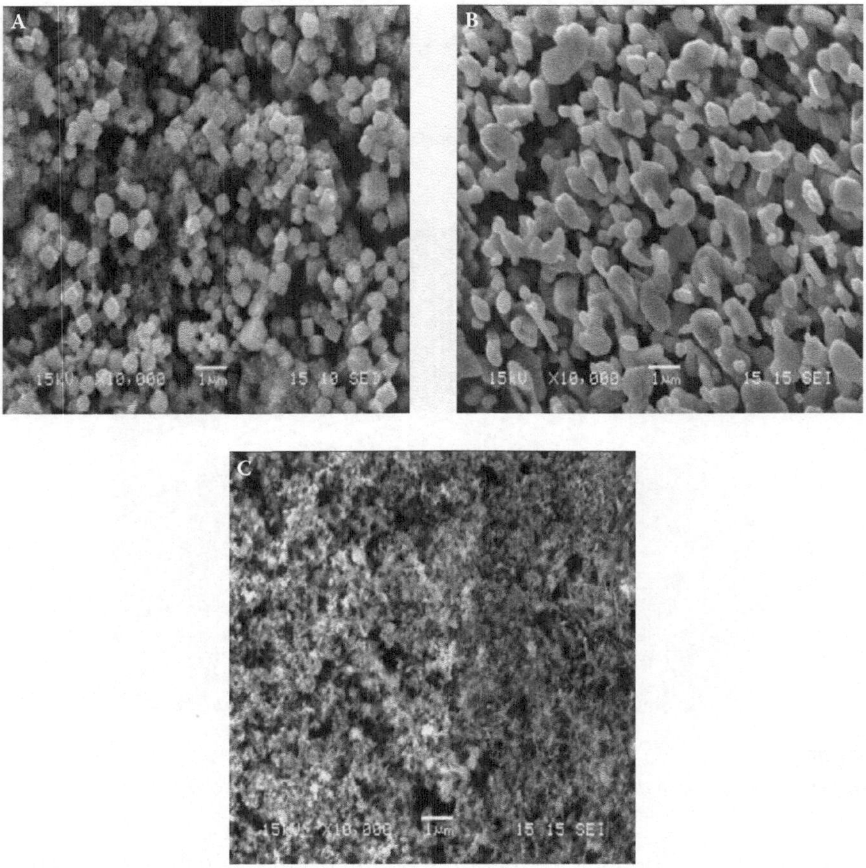

FIGURE 5.9
SEM micrograph of elemental metals: (a) copper particles after the aging process at 55°C for 2 h, (b) silver particles, (c) SEM micrograph of elemental metals in competitive recovery of copper and silver ions.

Interestingly, less defined shapes in the competitive recovery are observed in Figure 5.9c. They could be due to the slightly slower reduction reaction and the aforementioned codeposition in the competitive recovery.

5.2.3 HCHO

Formaldehyde has a wider range of industrial applications. It can be found in toothpaste, cosmetics, detergents, perfumes, deodorants, disinfectants, clothing, paper towels, inks, dyes, paints, veneers, and shampoos. It is an excellent preservative and a powerful germ killer. It has been used by farmers, dentists, biologists, bookbinders, car builders, ink makers, and wood manufacturers. The industrial output of formaldehyde-related goods in Canada alone was above $2 billion per year.[14]

Formaldehyde can be used to recover precious metals by reduction of the dissolved metal ions to the elemental metal. The reaction is important in the metal-coating process and called electroless plating. The overall electroless metal-plating reaction (e.g. copper) is theoretically given as

$$Cu^{2+} + HCHO + 3OH^- = Cu^0 + H_2 + H_2O + HCO_3^- \tag{5.11}$$

This equation employs formaldehyde as the reducing agent. Theoretically speaking, it requires 3 moles of hydroxyl ions and 1 mole of formaldehyde in order to produce 1 mole of elemental copper. In reality, however, other side reactions usually occur, such as the Cannizzaro reaction, by which formaldehyde is disproportionated as follows:

$$2HCHO + OH^- = CH_3OH + HCOO^- \tag{5.12}$$

This Cannizzaro reaction consumes additional formaldehyde and hydroxide. In addition, formaldehyde may reduce the cupric ions to form cuprous oxide, which is an unwanted product:

$$2Cu^{2+} + HCHO + 5OH^- = Cu_2O + HCOO^- + 3H_2O \tag{5.13}$$

The overall treatment efficacy depends on complexing agents, buffer, stabilizer, accelerators, and surfactants in the solution.

5.2.4 Iron

Zero- and divalent iron (Fe^0 and Fe^{2+}) can serve as electron donors for reduction of heavy metal ions. Scrap with high content of zero- and divalent iron (e.g., Fe^0 and FeO) is an industrial waste or by-product. The cost is much lower than that of many other commercial reductants. Reduction of heavy metal ions with the scrap by iron species is normally thermodynamically favorable, which can be confirmed by the calculation of change in the redox potentials based on Table 5.1.

Take reduction of hexavalent chromium by zero-valent iron as an example. We can have the following half redox reactions for iron and chromium that are given in Table 5.1.

$$1/2Fe^{2+} + e = 1/2Fe \qquad \Delta E_0 = -0.447\ V \tag{5.14a}$$

$$Fe^{3+} + e = Fe^{2+} \qquad \Delta E_0 = 0.770\ V \tag{5.14b}$$

$$1/3Fe^{3+} + e = 1/3Fe \qquad \Delta E_0 = -0.037\ V \tag{5.14c}$$

$$CrO_4^{2-} + 8H^+ + 3e = Cr^{3+} + 4H_2O \qquad \Delta E_0 = 1.48\ V \tag{5.15}$$

The zero-valent iron reduces the hexavalent chromium as follows:

$$\frac{1}{3}CrO_4^{2-} + \frac{8}{3}H^+ + \frac{1}{2}Fe = \frac{1}{3}Cr^{3+} + \frac{1}{2}Fe^{2+} + \frac{4}{3}H_2O,\ \Delta E_0 = 1.927\ V \tag{5.16a}$$

$$\frac{1}{3}CrO_4^{2-} + \frac{8}{3}H^+ + \frac{1}{3}Fe = \frac{1}{3}Cr^{3+} + \frac{1}{3}Fe^{3+} + \frac{4}{3}H_2O,\ \Delta E_0 = 1.517\ V \tag{5.16b}$$

The divalent iron can also reduce the hexavalent chromium as:

$$\frac{1}{3}CrO_4^{2-} + \frac{8}{3}H^+ + Fe^{2+} = \frac{1}{3}Cr^{3+} + Fe^{3+} + \frac{4}{3}H_2O,\ \Delta E_0 = 0.71\ V \tag{5.17}$$

El-Shazly et al. developed a fixed-bed treatment system for the removal of potassium dichromate by scrap-bearing iron spheres as shown in Figure 5.10.[15] The initial concentration of potassium dichromate was 0.05 M. The height of the fixed bed was 4 cm, and the diameter of the iron sphere was 10 mm. At a solution velocity of 12.16 cm/s, a removal efficiency of 63.2% was achieved after a 20-min reaction. The divalent iron is ineffective in alkaline conditions.

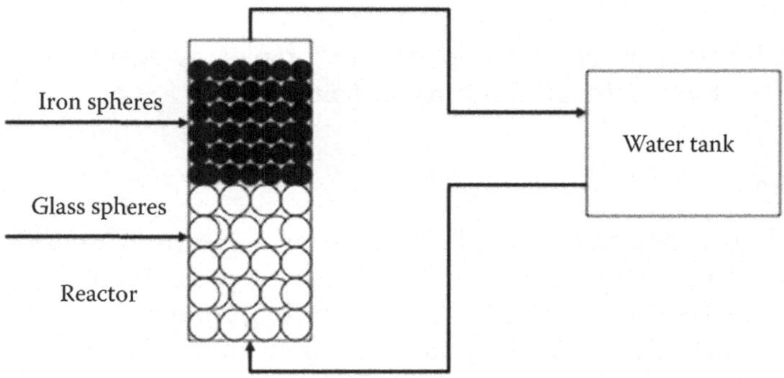

FIGURE 5.10
Reduction of hexavalent chromium by iron spheres in fixed bed reactor.

5.2.5 Other Reducing Reagents

In addition to the aforementioned reagents, strong reductants such as H_2S, SO_2, vitamin C, and glycerol are reportedly used. H_2S and SO_2 can effectively reduce metal ions with higher valencies (e.g., hexavalent chromium). However, both species have toxicities and lead to additional environmental problems. Organic reductants may become more favorable in the decontamination of heavy metals with higher valencies due to less toxicity to the environment.

Yurkow et al. reported a study on the photochemical reduction of hexavalent chromium to trivalent chromium in aqueous solutions containing glycerol.[16] The reaction occurred rapidly during irradiation with either unfiltered sunlight or a UVA-emitting light source. The photochemical reduction of Cr(VI) was pH dependent and did not occur in alkaline solution. The reduction occurred at elevated rates in acidified solutions at lower concentrations of glycerol. It was found that the reduction reaction was dependent on the unsubstituted alcohol groups of glycerol because alpha-phosphoglycerol and beta-phosphoglycerol did not support the photochemical reduction of Cr(VI). The findings from the study suggest that glycerol or related polyols can be used for the remediation of hexavalent chromium-contaminated sites.

Xu and co-workers used ascorbic acid (vitamin C) to chemically reduce hexavalent chromium ($K_2Cr_2O_7$).[17] Vitamin C is environmentally friendly and serves as an important biological reductant ubiquitous in humans and animals. It is water soluble and can easily permeate through various types of soils, which potentially can serve as a remediation reagent for cleanup of chromium-contaminated soils. It was reported that the concentration of vitamin C had played a key role the reduction of Cr(VI). The molar ratio of vitamin C to chromium was 3:1 in order that the complete reduction of Cr(VI) could be achieved. The study suggested the following stoichiometry that could be used to describe the reaction in acidic solution or in alkaline condition, respectively:

$$\begin{aligned} &Cr_2O_7^{2-} + 3\,\text{Ascorbic Acid} + 8H^+ \\ &\quad = 2Cr^{3+} + 3\,\text{Dehydroascorbic Acid} + 7H_2O \quad pH < 7 \end{aligned} \tag{5.18a}$$

$$\begin{aligned} &Cr_2O_7^{2-} + 3\,\text{Ascorbic Acid} + H_2O \\ &\quad = 2Cr(OH)_3 + 3\,\text{Dehydroascorbic Acid} + 2OH \quad pH > 7 \end{aligned} \tag{5.18b}$$

Great reduction occurred at pH values below 7. However, less reduction in Cr(VI) was observed in weakly alkaline solution (pH 8–10). The reduction could be completed within 120 min. The reductive reaction could occur in the dark. The reaction efficiency was further improved to some extent with either UV or unfiltered sunlight. Finally, it was concluded that vitamin C could be used in effective remediation of Cr(VI)-contaminated soils and groundwater in a wide range of pH, with or without sunlight.

5.3 Biological Reduction of Metal Sulfate

Reduction of metallic ions that are normally multivalent by biological means is referred to as biological reduction. The reduction agent (reductant) can be organic compounds from living or dead biomass natural organic matter (NOM), and industrial organic waste. Bacteria may also act as a catalyst or medium for the metal reduction. The reduction here refers to reduction of metal ions from higher valency to lower valency, such as reduction of As(V) to As(III) and that of Cr(VI) to Cr(III), or removal of soluble metallic ions to insoluble metals (e.g., $CuSO_4$ to CuS). The process may or may not reduce the toxicity of metals. For example, As(V) can reportedly be reduced to As(III), which leads to an increase in the toxicity. However, biosorbents or biopolymers can reduce hexavalent chromium (Cr(VI)) to less toxic trivalent chromium (Cr(III)).

Metal sulfate (M^{n+} and SO_4^{2-}, $n = 2$) may exist in contaminated groundwater and industrial waste. As both are toxic, they must be treated in order to minimize the risks to human beings and organisms. In the technology of biological reduction, the sulfate ion is first reduced to sulfide. As metal sulfide species have very low solubilities (K_{sp}), the sulfide can react with the metal ions salts (M^{n+}, $n = 2$) to form insoluble metal precipitates (MS). As a result, both metal and sulfate can be removed from aqueous solution. This process/technology is called biological reduction of metal sulfate.

5.3.1 Importance of Sulfate Removal

Sulfur exists in our natural environment in different oxidation states and in various chemical forms. This multivalent element is abundant in the forms of water in both solid and gaseous phases. Typically, what we can have are sulfate (sulfate) ions (SO_4^{2-}), sulfuric acid (H_2SO_4), hydrogen sulfide (H_2S), and elemental sulfur (S). Being one of essential elements for life, it is found in two essential amino acids, namely, cysteine and methionine. Its applications in industries are primarily as fertilizers; it is also widely used in black gunpowder, matches, insecticides, and fungicides.

Many industrial processes generate sulfur-containing wastewater. Typically, they are in the production of fertilizers and biocides, tannery operations, food production, coal power generation, metal mining and processing, and pulp and paper processing. The sulfur species in industrial wastewater may exist as sulfate, sulfide, sulfite, thiosulfate, and dithionite, which have different concentrations.

Sulfuric acid and sulfite can be used for pH adjustment and as a bleaching agent. They can be found in the wastewaters of many manufacturing processes. Sulfate (SO_4^{2-}, S: +6) is the most oxidized form of sulfur. One of the sources is via the natural generation through oxidation of sulfide ores in acid mine drainage.

Sulfate pollution is a great issue of concern as it has caused severe contamination in both surface water and groundwater. It was reported that diarrhea was associated with the ingestion of water containing high levels of sulfate. Excessive quantities of sulfate in drinking water can pose a health threat to many living organisms and human beings, according to a series of studies reported by the US EPA (www.epa.gov). Health concerns regarding sulfate in drinking water have been raised. People may be at greater risk if there is an abrupt change in drinking water from low sulfate concentration to high concentration. Sulfate in drinking water currently has a secondary maximum contaminant level (SMCL) set by US EPA as 250 mg/L based on aesthetic effects (namely, taste and odor). This regulation is not a federally enforceable standard in the United States; however, it is provided as a guideline for states and public water systems. The US EPA estimated that about 3% of the public drinking water systems in the country may have sulfate levels of 250 mg/L or greater.

In the absence of dissolved oxygen and nitrate in water, sulfate may become a major electron acceptor, by which may be converted to sulfide (H_2S). The sulfide causes odor and corrosion problems. In addition, H_2S as a fatally toxic gas can cause death of humans within a short period of time when the concentration is high. As the US EPA recommended that the upper concentration limit of sulfide in water intended for human drinking/consumption be 250 mg/L, the sulfate-rich wastewater must be treated prior to its being released into the environment. Anaerobic biological treatment technology has been recognized as an efficient method for removing sulfate from wastewater. It is noted that the anaerobic biological process for the sulfate-containing wastewater treatment is always associated with the removal of organic matter (BOD and COD) and nitrogen in the wastewater. Typically, BOD and COD removal efficiencies of above 90% can be expected.

Under aerobic conditions, sulfate is thermodynamically stable in the form of sulfur. On the other hand, hydrogen sulfide is stable under anaerobic conditions. A sulfate reduction reaction may occur at appropriate reduction potentials. Sulfate is a much less favorable electron acceptor than oxygen and nitrate. As the standard potential of SO_4^{2-}/H_2S is 0.333 V, the reduction potential of the reductant must be negative in order to maximize the sulfate reduction in wastewater.

5.3.2 Mechanisms and Controlling Factors

The sulfate reduction process can be biologically mediated by a group of microorganisms known as sulfate-reducing bacteria (SRB). SRB produces hydrogen sulfide and sulfide ions in the presence of a suitable electron donor. If the electron donor is hydrogen, a carbon source must be present, which can be carbon dioxide. Tables 5.3 to 5.6 show some key information on the biological reduction process for treatment of metal sulfate in wastewater.

TABLE 5.3
Summary of Sulfate Reduction Rate with Different Electron Donors for Various Reactor Types

Reactor Type	Electron Donor	Sulfate Reduction Rate ($mmol\ L^{-1}\ d^{-1}$)	Researchers
Fluidized bed	Ethanol	44.8	Kaksonen et al.[18]
Fluidized bed	Ethanol/lactate	100–125	Nevatalo et al.[19]
Membrane bioreactor	Formic acid	143	Bijmans et al.[20]
Membrane bioreactor	Ethanol/acetate	69	Vallero et al.[21]
Gas-lift	Hydrogen	288	Van Houten et al.[22]
Gas-lift	Hydrogen	51	Bijmans et al.[20]
Gas-lift	Synthesis	156	Van Houten et al.[23]
Expanded-granular-sludge blanket	Ethanol	73	de Smul et al.[24]
Column bioreactor	Biodiesel manufacturing waste	4.7	Zamsow et al.[25]
Column bioreactor	Ethanol	0.283 ($g\ L^{-1}\ d^{-1}$)	Costa et al.[26]
Column bioreactor	Wine wastes	0.174–0.263 ($g\ L^{-1}\ d^{-1}$)	Costa et al.[26]
Upflow anaerobic sludge blanket (UASB)	Molasses	4.3 ($g\ L^{-1}\ d^{-1}$)	Annachhatre and Suktrakoolvait[27,28]
Continuous stirred tank reactor (CSTR)	Molasses	0.84 ($g\ L^{-1}\ d^{-1}$)	Maree and Hill[29]
Anaerobic RBC	Molasses	0.35 ($g\ L^{-1}\ d^{-1}$)	Lo et al.[30]
Packed bed	Molasses	6.5 ($g\ L^{-1}\ d^{-1}$)	Maree and Strydom[31]
Packed bed	Molasses + mine water	1.36 ($g\ L^{-1}\ d^{-1}$)	Maree et al.[32]
Packed bed	Hydrogen/carbon monoxide	1.2 ($g\ L^{-1}\ d^{-1}$)	du Preez and Maree[33]
Packed bed	Carbon monoxide	2.4 ($g\ L^{-1}\ d^{-1}$)	du Preez and Maree[33]
Packed bed	Acetate	15–20 ($g\ L^{-1}\ d^{-1}$)	Stucki et al.[34]
Baffled reactor	Mixture of volatile fatty acids (VFA)	30 ($g\ L^{-1}\ d^{-1}$)	Vallero et al.[35]
Plug flow	Lactate	0.41 ($g\ L^{-1}\ d^{-1}$)	Hammack et al.[36]
Anaerobic digester	Glucose/acetate	1.92 ($g\ L^{-1}\ d^{-1}$)	Polpresert and Haas[37]
Anaerobic filter	High strength leachate	0.02 ($g\ L^{-1}\ d^{-1}$)	Henry and Prasad[38]

TABLE 5.4

Comparison of the Sulfate Reduction Rates Reported for Sulfate-Reducing Bioreactors Utilizing Low-Cost Substrates as Electron Donor for SRB

Reactor Type	Substrate	pH	*T* (°C)	SRR ($g\ L^{-1}\ d^{-1}$)	Researchers
Upflow packed bed reactor	Molasses	8.1	25	2.2	Maree and Strydom[39]
Stirred tank	Primary sewage sludge	7.2	20–22	0.8	Whiteley et al.[40]
Anaerobic filter	Ethanol and spent manure	6.5	5	1.1–1.4	Tsukamoto et al.[41]
Anaerobic filter	Methanol and spent manure	6.5	5	1.0–1.3	Tsukamoto et al.[41]
Batch	Mixture of wood chips, leaf compost, and poultry manure	7.5	22	0.01	Zagury et al.[42]
Column	Mixture of cheese whey, sawdust, and cow manure	6.5–7.5	14–24	11.5–24	Drury[43]
Column	Biodiesel manufacturing waste (glycerol)	6–8	NR	0.5	Zamsow et al.[25]
Fluidized bed bioreactor	Soluble reed canary grass hydrolysate	7	35	2.2–3.3	Lakaniemi et al.[44]
	Cellulose	NR	NR	1.18	Wakemana et al.[45]
	Leaf mulch	NR	NR	1.52–2.15	Wakemana et al.[45]
	Solage			1.33	Wakemana et al.[45]

Note: SRR = sulfate reduction rate; NR = not reported.

TABLE 5.5

Biomass Yield of Sulfate-Reducing Bacteria in Different Electron Donors and Acceptors[46]

Electron Acceptor	Electron Donor	Yield (g cells/mol Electron Donor)
SO_4^{2-}	H_2	1.9
$S_2O_3^{2-}$	H_2	4.5
SO_3^{2-}	H_2	4.0
SO_4^{2-}	Acetate	4.8
$S_2O_3^{2-}$	Acetate	10.4
SO_3^{2-}	Acetate	11.2
SO_4^{2-}	Propionate	0.086

TABLE 5.6
Summary of Sulfidogenic Reaction with Various Electron Donors

Electron donor	Reaction	Researchers
Hydrogen	$4H_2 + SO_4^{2-} + H^+ = HS^- + 4H_2O$ $8H_2 + 2SO_4^{2-} = H_2S + HS^- + 5H_2O + 3OH^-$	Annachhatre and Liamleam[47]
Formate	$SO_4^{2-} + 4HCOO^- + H^+ = HS^- + 4HCO_3^-$	
Methanol	$4CH_3OH + 3SO_4^{2-} = 4HCO_3^- + 3HS^- + 4H_2O + H^+$	
Ethanol	$C_2H_5OH + 0.5SO_4^{2-} = CH_3COO^- + 0.5HS^- + 0.5H^+ + H_2O$ $CH_3COO^- + SO_4^{2-} = 2HCO_3^- + HS^-$	
Molasses	$C_{12}H_{22}O_{11} + H_2O = 4CH_3CHOHCOOH$	
Lactate	$CH_3CHOHCOOH + 0.5H_2SO_4^{2-} = CH_3COOH + CO_2 + 0.5H_2S + H_2O$	
Acetate	$CH_3COO^- + SO_4^{2-} = HS^- + 2HCO_3^-$	
Propionate	$\text{Propionate} + SO_4^{2-} + H_2 = HS^- + HCO_3^- + \text{Acetate} + H_2O$ $\text{Propionate} + 2SO_4^{2-} + H_2 = 2HS^- + 3HCO_3^- + H_2O$	
Butyrate	$\text{Butyrate} + 3SO_4^{2-} + 2H_2 = 3HS^- + 4HCO_3^- + 5H_2O$ $\text{Butyrate} + SO_4^{2-} + 2H_2 + 6H_2O = HS^- + 2\ \text{Acetate}$	
Sugar	$\text{Glucose} + SO_4^{2-} = CH_3COO^- + 0.5HS\text{-} + 0.5H^+ + H_2O$ $\text{Glucose} + 3SO_4^{2-} = 3HS^- + HCO_3^- + 3H^+$	Thauer et al.[48]
Hydrocarbons	$CH_4 + SO_4^{2-} = HCO_3^- + HS^- + H_2O$ $CH_3COOH + SO_4^{2-} = 2CO_2 + S^{2-} + 2H_2O$ $C_6H_5COO^- + 0.75SO_4^{2-} + 4H_2O = 3CH_3COO^- + 0.75HS^- + HCO_3^- + 2.25H^+$ $C_6H_5COO^- + 3.75SO_4^{2-} + 4H_2O = 7HCO_3^- + 3.75HS^- + 2.25H^+$	Annachhatre and Liamleam[47]

Most reported studies on sulfate reduction have been limited to laboratory researches, which are essentially based on small and simplified reactor devices. Larger-scale operations with treatment capacities of above 15 tons of sulfur per day have also been reported in the literatures. The biological reduction process for both sulfate and heavy metals—eventually leading to metal precipitation—can be achieved in either one or few stages. In other words, the sulfate reduction and metal precipitation may occur simultaneously (in one unit or a single stage), or in a few units (separated operation mode). The single-stage system is less costly and may be efficient; however, it may be more prone to inhibition by the toxic substances in the wastewater. As such, the carefully designed balancing of the wastewater loading and the composition/control become crucial in order to maintain desirable operation efficiencies.

The SRBs are normally categorized into two groups: heterotrophic SRB and autotrophic SRB. The heterotrophic SRB uses organic compounds as substrates. On the other hand, the autotrophic SRB uses carbon dioxide as the carbon source and obtain electrons from oxidation of hydrogen gas.

Reduction of sulfate (SO_4^{2-}) to sulfide (S^{2-}) involves eight electrons as shown below (also Table 5.1).

$$\frac{1}{8}SO_4^{2-}+\frac{10}{8}H^{+}+e=\frac{1}{8}H_2S+\frac{1}{2}H_2O \tag{5.19a}$$

$$\frac{1}{8}SO_4^{2-}+\frac{19}{16}H^{+}+e=\frac{1}{16}H_2S+\frac{1}{16}HS^{-}+\frac{1}{2}H_2O \tag{5.19b}$$

Electron donors are essential for the treatment of sulfate-containing wastewater by the SRB. Some of sulfate-rich wastewaters may also contain highly concentrated organic matter, which may be utilized as electron donors (from carbons of the organic matter).

A suitable ratio of COD (organic matter providing electrons) to sulfate (which accepts electrons) is important to reduce sulfate from the water. A ratio of 0.67:1 is reportedly required in order to achieve complete reduction.[47] As the sulfate- and sulfuric-acid-containing wastewater from such industries as metal mining is normally deficient in organic matter, the sulfate reduction would become incomplete. In the event that such wastewater is to be treated, electron donors and carbon sources must be added. Two groups of organic matter can be employed to provide electron donors and carbon sources: (1) organic wastewater and (2) synthetic organic compounds. The requirement for organic matters, the advantages and disadvantages, the cost, and efficiency of biological sulfate reduction are generally similar to those for the biological denitrification process. Table 5.6 gives a list of sulfidogenic reaction.

Such waste organic matters as sewage, sewage sludge, animal manure, and industrial wastewater (e.g., soft drink and food wastewater) can be used as electron donors and carbon sources. The low-molecular-weight synthetic organic compounds that are less costly and by-products from certain biological processes (e.g., anaerobic digestion) can be used; typically they are lactate, acetate, propionate, pyruvate, butyrate, ethanol, and other alcohols, which may be from fermentation of such complex organic matter as carbohydrates, proteins, waste-activated sludge, and other dead biomass.

A list of kinetics rates in the biological sulfate reduction process by different electron donors is given in Tables 5.3 and 5.4. It is clear that the selection of electron donor has an important impact on the sulfate reduction rate. Higher sulfate removal rates can be obtained when H_2/CO_2, acetate, and ethanol are used.[22,24,49]

Table 5.5 gives the list of bacterial biomass yields of SRB as well as the electron donors and acceptors.[46] Different electron donors would cause a difference in biomass yield. H_2 as electron donor yields less biomass than acetate. The biomass yield is only 0.086 g cells/mol of electron donor when propionate is used. The substrate consumption rate of the sulfate reducers is dependent on the concentrations of electron donor and acceptor. There is competition between SRB and methanogens for the electron donor.

5.3.2.1 Thermodynamics

The overall process for microbial sulfate reduction can be splitted into two subprocesses: (1) energy generation and (2) cell synthesis. Both are essentially redox reactions. The first process provides energy for the second, as shown in Figure 5.11.

Reaction for energy generation (R_e):

$$R_e = R_a - R_d \tag{5.20}$$

Reaction for cell synthesis (R_s):

$$R_s = R_c - R_d \tag{5.21}$$

where R_a, R_d, and R_c represent half-reactions of electron acceptor, electron donor, and cell formation. The important reactions for the R_a, R_d, and R_c are listed in Tables 5.7 to 5.9.

The fraction for the first reaction is f_e, while that for the second reaction is f_s. Both are dependent on the operational conditions such as solid retention time (STR). The sum of f_e and f_s is 1, namely,

$$f_e + f_s = 1 \tag{5.22}$$

The overall reaction (R) is

$$R = f_e(R_a - R_d) + f_s(R_c - R_d)$$

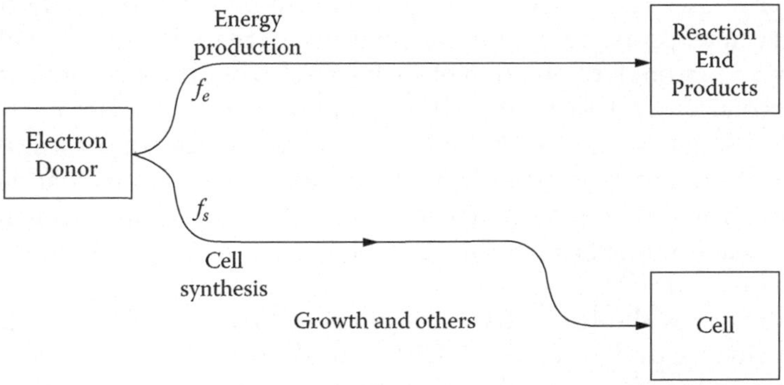

FIGURE 5.11
Illustration of biological reactions for the removal of sulfate from aqueous solution.

TABLE 5.7

Half-reactions and their Gibb's Standard Free Energy for Common Inorganic Electron Acceptors (R_a)[12]

No.	Reduced-oxidized Compounds	Half-reaction	$\Delta G^{o'}$ kJ/e eq
I-1	Ammonium-nitrate	$1/8\ NO_3^- + 5/4\ H^+ + e = 1/8\ NH_4^+ + 3/8\ H_2O$	−35.11
I-2	Ammonium-nitrite	$1/6\ NO_2^- + 4/3\ H^+ + e = 1/6\ NH_4^+ + 1/3\ H_2O$	−32.93
I-3	Ammonium-nitrogen	$1/6\ N_2 + 4/3\ H^+ + e = 1/3\ NH_4^+$	26.70
I-4	Ferrous-ferric	$Fe^{3+} + e = Fe^{2+}$	−74.27
I-5	Hydrogen-H^+	$H^+ + e = 1/2\ H_2$	39.87
I-6	Nitrite-nitrate	$1/2\ NO_3^- + H^+ + e = 1/2\ NO_2^- + 1/2\ H_2O$	−41.65
I-7	Nitrogen-nitrate	$1/5\ NO_3^- + 6/5\ H^+ + e = 1/10\ N_2 + 3/5\ H_2O$	−72.20
I-8	Nitrogen-nitrite	$1/3\ NO_2^- + 4/3\ H^+ + e = 1/6\ N_2 + 2/3\ H_2O$	−92.56
I-9	Sulfide-sulfate	$1/8\ SO_4^{2-} + 19/16\ H^+ + e = 1/16\ H_2S + 1/16\ HS^- + 1/2\ H_2O$	20.85
I-10	Sulfide-sulfite	$1/6\ SO_3^{2-} + 5/4\ H^+ + e = 1/12\ H_2S + 1/12\ HS^- + 1/2\ H_2O$	11.03
I-11	Sulfite-sulfate	$1/2\ SO_4^{2-} + H^+ + e = 1/2\ SO_3^{2-} + 1/2\ H_2O$	50.30
I-12	Sulfur-sulfate	$1/6\ SO_4^{2-} + 4/3\ H^+ + e = 1/6\ S + 2/3\ H_2O$	19.15
I-13	Thiosulfate-sulfate	$1/4\ SO_4^{2-} + 5/4\ H^+ + e = 1/8\ S_2O_3^{2-} + 5/8\ H_2O$	23.58
I-14	Water-oxygen	$1/4\ O_2 + H^+ + e = 1/2\ H_2O$	−78.72

Thus, we have

$$R = f_e R_a + f_s R_c - R_d \tag{5.23}$$

EXAMPLE

Develop an overall reaction for the biological reduction of sulfate. Assume that hydrogen gas is used as an electronic donor, CO_2 and NH_4^- are used as carbon and nitrogen sources, and f_e and f_s are 0.7 and 0.3.

SOLUTION

From Tables 5.7 to 5.9, we have:

$$R_d\text{: } H^+ + e = \frac{1}{2}H_2$$

$$R_a\text{: } \frac{1}{8}SO_4^{2-} + \frac{19}{16}H^+ + e = \frac{1}{16}H_2S + \frac{1}{16}HS^- + \frac{1}{2}H_2O$$

$$R_c\text{: } \frac{1}{5}CO_2 + \frac{1}{20}HCO_3^- + \frac{1}{20}NH_4^+ + H^+ + e = \frac{1}{20}C_5H_7O_2N + \frac{9}{20}H_2O$$

Overall reaction $= f_e R_a + f_s R_c - R_d$

TABLE 5.8

Half-reactions and their Gibb's Free Energy for Common Organic Electron Donors (R_d)[12]

No.	Compound	Half-reaction	$\Delta G^{o'}$ kJ/e eq
0-1	Acetate	$1/8\ CO_2 + 1/8\ HCO_3^- + H^+ + e = 1/8\ CH_3COO^- + 3/8\ H_2O$	−35.11
0-2	Alanine	$1/6\ CO_2 + 1/12\ HCO_3^- + 1/12\ NH_4^+ + 11/12H^+ + e = 1/12\ CH_3CHNH_2COO^- + 5/12\ H_2O$	−32.93
0-3	Benzoate	$1/5\ CO_2 + 1/30\ HCO_3^- + H^+ + e = 1/30\ C_6H_5COO^- + 13/30\ H_2O$	26.70
0-4	Citrate	$1/6\ CO_2 + 1/6\ HCO_3^- + H^+ + e = 1/18\ (COO^-)CH_2COH(COO^-)CH_2COO^- + 4/9\ H_2O$	−74.27
0-5	Ethanol	$1/6\ CO_2 + H^+ + e = 1/12\ CH_3CH_2OH + 1/4H_2O$	39.87
0-6	Formate	$1/2\ HCO_3^- + H^+ + e = 1/2\ HCOO^- + 1/2\ H_2O$	−41.65
0-7	Glucose	$1/4\ CO_2 + H^+ + e = 1/24\ C_6H_{12}O_6 + 1/4\ H_2O$	−72.20
0-8	Glutamate	$1/6\ CO_2 + 1/9\ HCO_3^- + 1/18\ NH_4^+ + H^+ + e = 1/18\ COOHCH_2CH_2CHNH_2COO^- + 4/9\ H_2O$	−92.56
0-9	Glycerol	$3/14\ CO_2 + H^+ + e = 1/14\ CH_2OHCHOHCH_2OH + 3/14H_2O$	20.85
0-10	Glycine	$1/6\ CO_2 + 1/6\ HCO_3^- + 1/6\ NH_4^+ + H^+ + e = 1/6\ CH_2NH_2COOH + 1/2\ H_2O$	11.03
0-11	Lactate	$1/6CO_2 + 1/12HCO_3^- + H^+ + e = 1/12CH_3CHOHCOO^- + 1/3H_2O$	32.29
0-12	Methane	$1/8CO_2 + H^+ + e = 1/8CH_4 + 1/4H_2O$	23.53
0-13	Methanol	$1/6CO_2 + H^+ + e = 1/6CH_3OH + 1/6H_2O$	36.84
0-14	Palmitate	$15/92CO_2 + 1/92HCO_3^- + H^+ + e = 1/92CH_3(CH_2)_{14}COO^- + 31/92H_2O$	27.26
0-15	Propionate	$1/7CO_2 + 1/14HCO_3^- + H^+ + e = 1/14CH_3CH_2COO^- + 5/14H_2O$	27.63
0-16	Pyruvate	$1/5CO_2 + 1/10HCO_3^- + H^+ + e = 1/10CH_3COCOO^- + 2/5H_2O$	35.09
0-17	Succinate	$1/7CO_2 + 1/7HCO_3^- + H^+ + e = 1/14(CH_2)_2(COO^-)_2 + 3/7H_2O$	29.09
0-18	Domestic Wastewater	$9/50CO_2 + 1/50NH_4^+ + 1/50HCO_3^- + H^+ + e = 1/50C_{10}H_{19}O_3N + 9/25H_2O$	*
0-19	General Organics	$(n-c)/dCO_2 + c/dNH_4^+ + c/dHCO_3^- + H^+ + e = 1/dC_nH_aO_bN_c + (2n-b+c)/dH_2O$ where $d = (4n+a-2b-3c)$	*
0-20	Cell Synthesis	$1/5CO_2 + 1/20NH_4^+ + 1/20HCO_3^- + H^+ + e = 1/20C_5H_7O_2N + 9/20H_2O$	*

Asterisks (*) indicate that $\Delta G^{o'}$ value is not available because the reduced species is not defined.

TABLE 5.9

Half-reactions for Cell Formation (R_c) and Common Electron Acceptor (R_a)

	Cell Synthesis Equations (R_c)
Ammonium as nitrogen source	$1/5CO_2 + 1/20HCO_3^- + 1/20NH_4^+ + H^+ + e = 1/20C_5H_7O_2N + 9/20H_2O$
Nitrate as nitrogen source	$1/28NO_3^- + 5/28CO_2 + 29/28H^+ + e = 1/28\ C_5H_7O_2N + 11/28H_2O$
Nitrite as nitrogen source	$5/26CO_2 + 1/26NO_2^- + 27/26H^+ + e = 1/26C_5H_7O_2N + 10/26H_2O$
Nitrogen as nitrogen source	$5/23\ CO_2 + 1/46N_2 + H^+ + e = 1/23C_5H_7O_2N + 8/23\ H_2O$
	Common Electron-Acceptor Equations (R_e)
Oxygen	$1/4O_2 + H^+ + e = 1/2\ H_2O$
Nitrate	$1/5NO_3^- + 6/5\ H^+ + e = 1/10\ N_2 + 3/5\ H_2O$
Sulfate	$1/8SO_4^{2-} + 19/16H^+ + e = 1/16H_2S + 1/16HS^- + 1/2\ H_2O$
CO_2	$1/8CO_2 + H^+ + e = 1/8CH_4 + 1/4\ H_2O$
Iron(III)	$Fe^{3+} + e = Fe^{2+}$

$$\frac{1}{2}H_2 = H^+ + e + \left(\frac{1}{8}SO_4^{2-} + \frac{19}{16}H^+ + e = \frac{1}{16}H_2S + \frac{1}{16}HS^- + \frac{1}{2}H_2O\right) \times 0.7$$

$$+\left(\frac{1}{5}CO_2 + \frac{1}{20}HCO_3^- + \frac{1}{20}NH_4^+ + H^+ + e = \frac{1}{20}C_5H_7O_2N + \frac{9}{20}H_2O\right) \times 0.3$$

Thus, we have:

$$0.0875\,SO_4^{2-} + 0.13125\,H^+ + 0.06\,CO_2 + 0.015\,HCO_3^- + 0.015\,NH_4^+ + 0.5\,H_2$$

$$= 0.04375\,H_2S + 0.04375\,HS^- + 0.485\,H_2O + 0.015\,C_5H_7O_2N$$

5.3.2.2 Type of Electron Donors

The most expensive operational cost in the biological reduction of sulfate is the electron donor. Soluble organic substrates as well as hydrogen are commonly used. However, organic wastes such as fermentation wastewater have attracted greater attention because of the low cost and wide availability.

5.3.2.2.1 Hydrogen

Sulfate reduction by hydrogen gas is thermodynamically more favorable than methanogenesis. Hence, hydrogen gas is considered to be one of the attractive electron donors for sulfate removal. However, there are competitive reactions among hydrogenotrophic sulfate reducers, homoacetogenesis microbes, and hydrogenotrophic methanogenesis microbes.

Hydrogenotrophic sulfate reducers (hydrogenotrophic sulfate reduction):

$$4H_2 + SO_4^{2-} + H^+ = HS^- + 4H_2O \tag{5.24}$$

Homoacetogenesis microbes:

$$4H_2 + 2HCO_3^- + H^+ = CH_3COO^- + 4H_2O \tag{5.25}$$

Hydrogenotrophic methanogenesis microbes:

$$4H_2 + HCO_3^- + H^+ = CH_4 + 3H_2O \tag{5.26}$$

Similar to other anaerobic reactions, the temperature of operation is important. Under thermophilic conditions (55°C), sulfate conversion rates of up to 7.5 g SO_4^{2-}/L d have been achieved using hydrogen. This is considerably less compared with mesophilic conditions.[50]

Hydrogen can be directly supplied to the reactors. It can also be generated on-site from other electron donors such as propionate, methanol, and glucose.

Consumption of hydrogen may lead to generation of hydroxide (due to consumption of hydrogen). However, this effect has been proved to be less obvious. The cells are often pH sensitive and may not function and reproduce well at low pH. pH control may be necessary. CO_2 may act to supply carbon themselves for the microbes. The presence of CO_2 may cause a decrease in the pH of the solution as the dissolution would lead to the formation of carbonic acid. It becomes more important during the start-up period. When acetate was used as a carbon source, cell production was one time higher than when carbon dioxide was used.[51]

5.3.2.2.2 *Acetate*

Acetate may act as both electron donor and carbon source in the sulfate reduction process. It may be generated in anaerobic digestion of organic waste (e.g. waste sludge) as an intermediate compound. As such, the cost of the compound may be low. However, not all microbes can use it as a substrate. For example, the species of the genus *Desulfotomaculum* can consume acetate; however, *Desulfotovibrio* cannot. The methanogens outcompete sulfate reducers because of their faster cell growth rate, even though sulfate reduction is thermodynamically favorable. In addition, sulfate removal is normally incomplete, and acetate is not fully utilized. The residual BOD and COD in the treated effluent may be high, which requires subsequent treatment.

5.3.2.2.3 *Ethanol*

Generally, a better sulfate removal efficiency up to 80% may be achievable when ethanol is used as an electron donor. Greater removal of ethanol was also reported. However, the cell growth rate is rather low when ethanol is used, which may limit its application. In addition, the effluent may have high BOD and COD as ethanol may not fully be oxidized.

5.3.2.2.4 Methanol

Methanol can directly participate in the microbial removal of sulfate. Its anaerobic reactions may produce H_2 (together with CO_2) and formate as the intermediate products, which are all efficient at providing electrons for the reduction of sulfate. As such, methanol has been considered to be one of the cost-effective organic substances for the biological removal of sulfate.

The process is temperature and pH-sensitive. Under thermophilic condition (>65°C), the sulfate-reducing bacteria outcompete methanogens. In addition, the operation under thermophilic conditions is more effective than under mesophilic conditions. In addition to temperature, pH plays a key role in the process. Methanogenesis is more depressed at low pH. As such, sulfate reduction becomes favorable under slightly acidic conditions.

5.3.2.2.5 Organic Waste

Organic wastes as electron donors for sulfate reduction include primary sewage sludge, sewage sludge, animal manure, compost, wood chips, cheese whey, leaf mulch, wood chips, sawdust, cellulose, and biodiesel manufacturing waste.[25] Good sulfide removal occurs only when a significantly large amount of organic matter is provided. It was reported that organic waste was successfully used as an electron donor and carbon source in the treatment of acid mine drainage.[52]

It is noted that the cellulose and hemicelluloses of the plant cell wall cannot be directly used by SRB.[44] The anaerobic degradation of organic matters (e.g., plant material) may convert polymeric material (e.g., cellulose and lignin) into smaller molecules (e.g., sugars and fatty acids) that are further fermented to carboxylic acids, alcohols, and hydrogen by fermentative microorganisms. These smaller organic molecules serve as electron donors for the reduction of sulfate (e.g., in the process of bioremediation of acid mine drainage). A few successful cases have been reported in the literatures. For example, a low-cost substrate, *Phalaris arundinacea* (acid hydrolysis of Reed Canary Grass hydrolysate) as an electron donor caused a good sulfate reduction with a rate of 2.2–3.3 g L^{-1} d^{-1}.[44] The acidic influent was neutralized, and the highest metal precipitation rates were 0.84 g Fe L^{-1} d^{-1} and 15 mg Zn L^{-1} d^{-1} respectively. The volatile fatty acids and hydrogen from the fermentation essentially resulted in the removal of heavy metal ions and sulfate from the synthetic mine wastewater. Wakeman and co-workers reported that silage was degraded into suitable electron donors (VFAs) for the SRB by the metal-tolerant fermentative microorganisms from the peat bog around the mine site.[45] During the treatment, Mn and Mg ions remained in the solution while Fe ions partially precipitated. Up to 87 mg of sulfate was reduced per gram of silage.

5.3.2.2.6 Molasses

Molasses from the sugar production process are widely available as a low-cost electron donor and carbon source for sulfate reduction. In its application, it is first fermented by such microorganisms as Lactobacilli to simple organic compounds, which are then consumed.

At a COD:S ratio of less than 2, good performance in the removal of sulfate in an upflow anaerobic sludge blanket reactor was reported. However, there are some substances left because of the poor biodegradation, which may lead to high residual BOD and COD in the treated effluent. In addition, high volatile fatty acids (VFAs) (e.g., acetate, propionate, and butyrate) generated in the fermentation limit the biological growth of cells (methanogens and sulfate reducers). Good monitoring systems for the pH of solution and possible addition of NaOH or $NaHCO_3$ are important to prevent the acidification of solution, for the microbial growth to be healthy.

5.3.2.2.7 Others

Formate can be used as an electron for most sulfate reducers. Lactate can be used as an electron and carbon source for several species of sulfate-reducing bacteria. Propionate, which can directly and indirectly participate in the sulfate reduction, is similar to acetate. Sugar is an effective electron donor and carbon source. *Desulfotomaculum antarcticum* can use glucose, while *Desulfovibrio nigrificans* and *Desulfotomaculum nigrificans* are able to grow on fructose.[51,53]

Hydrocarbons are considered less chemically reactive under anoxic conditions. Occasionally, they may serve as electron donors and carbon sources for the reduction of sulfate. As the changes in free energy are generally low, the reductions of the sulfate process are less efficient when hydrocarbons are involved.

5.3.2.2.8 Mixed Organics

Mixed organic compounds in wastewater can serve as electron donors and carbon sources. An ethanol and lactate mixture was reportedly operated successfully in a fluidized-bed reactor at 35°C for a period of 540 days for the mine wastewater treatment.[19] A sulfate reduction rate of 62–100 mmol SO_4^{2-} L^{-1} per day, iron and zinc precipitation rates of 11 mmol Fe L^{-1} per day and 1 mmol Zn L^{-1} per day, and a H_2S production rate of 73.2 mmol H_2S L^{-1} per day were achieved. The initial concentrations of iron and zinc were 0.1–3.2 mM and 0.1 mM, respectively; both were removed by a percentage of 99%. It was found that the control step of the process was acetate oxidation.

Vanbroekhoven et al. studied a bioprecipitation technology to remediate the groundwater in the vicinity of three nonferrous industrial sites historically contaminated with heavy metals and sulfate.[54] The characteristics of one site are given in Table 5.10, showing serious pollutions of zinc, sulfate, and cadmium. Such pure organic compounds as lactate as well as organic waste such as molasses and glycerol were used in the sulfate reduction, which were injected into the soil as illustrated in Figure 5.12. It was found that only glycerol caused efficient metal and sulfate removal as shown in Figure 5.13. Note that the sulfate and zinc concentrations were reduced to 200 mg/L and 0, respectively, after about 100 days.

Wine wastes containing ethanol as electron donor and carbon source were used for the bioremediation of acid mine drainage (AMD).[26] Anaerobic

TABLE 5.10

Water Characteristics at Different Points

Ground Water	pH	EC [mS/cm]	DO [mg/L]	DOC [mg/L]	SO_4^{2-} [mg/L]	NO_3^- [mg/L]	PO_4^{3-} [mg/L]	Zn [mg/L]	Cd [mg/L]
30 m-bg	4.18	17.0	0.6	<20	1280	2	25	139	5.2
65 m-bg	4.51	17.2	0.4	<20	1880	2	2	101	0.97
Aquifer (Metals in mg/kg dm)	**dm [%]**	**pH**	**TOC [mg C/kg dm]**	**Total S [mg/kg dm]**	**Mn**	**Fe**	**Cd**	**Zn**	**As**
30–32 m	85.22	4.18	0.022	219	16	6500	0.6	41	75
60–63 m	82.19	4.51	0.087	1540	160	120000	<0.1	48	50

Note: bg = belowground; dm = dry matter.[54]

FIGURE 5.12
Application of bioprecipitation technology to remediate the groundwater.[54]

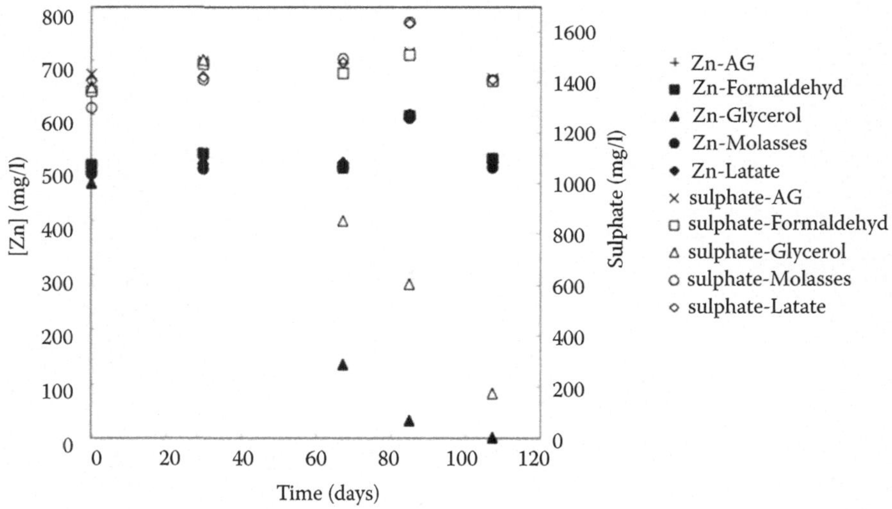

FIGURE 5.13
Effect of organic compound on removal of sulfate and zinc in groundwater.[54]

downflow packed bed reactors operated in a semicontinuous mode were used. Bioreactors I and II used ethanol and wine waste, respectively as electron donors. The overall performance of bioreactors is illustrated in Table 5.11. Efficient neutralization of water and sulfate removal (>90%) were achieved as shown in Figures 5.14 and 5.15; the treated effluent with

TABLE 5.11

Half Redox Reactions of Cr (VI) and Their Standard Redox Potential

	Bioreactor I		Bioreactor II	
Item	**Feed (AMD)**	**Treated Effluent**	**Feed (AMD)**	**Treated Effluent**
pH	2.4	6.5–7.5	2.4	5.0–6.7
sulfate	2300 mg/L	30 mg/L	2300 mg/L	200–905 mg/L
Fe	530 mg/L	9 mg/L	500 mg/L	35–290 mg/L
Cu	71 mg/L	3 mg/L	70 mg/L	2 mg/L
Zn	154 mg/L	4 mg/L	151 mg/L	5 mg/L

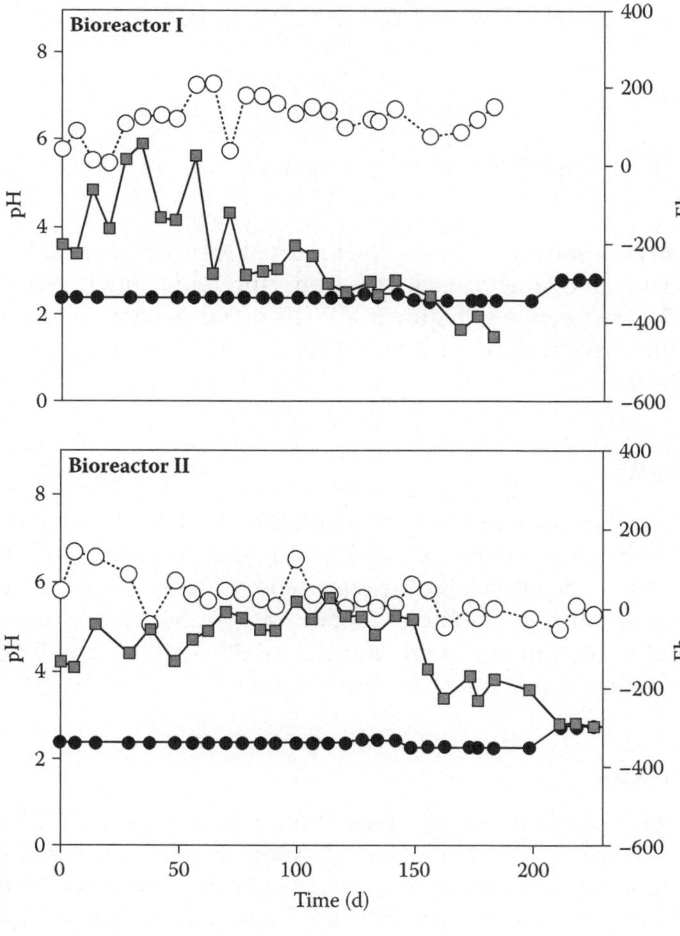

FIGURE 5.14
Change in pH and Eh in Bioreactors I and II: pH in AMD (●), pH in effluent from bioreactor (○) and Eh in effluent from bioreactor (■).

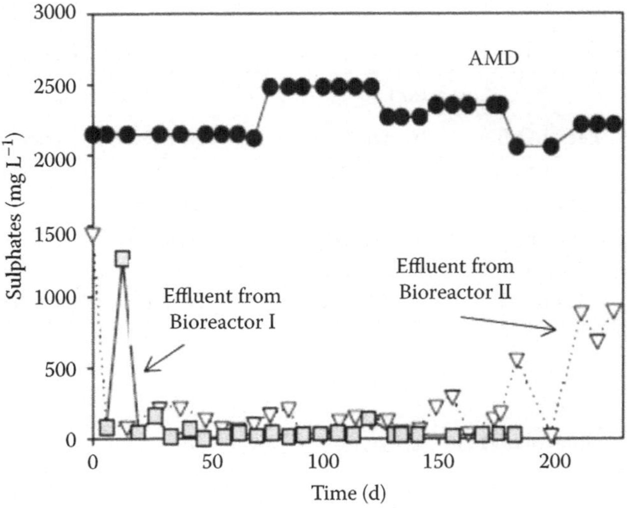

FIGURE 5.15
Performance of sulfate removal during operation of biological sulfate reduction from AMD.[26]

contaminant concentrations below the required legislation could be used for irrigation. The removal efficiencies for iron, zinc, and copper were 61%–91%, 97%, and 97%, respectively (Figure 5.16). The metal precipitates have significant amounts of sulfur, iron, silicon, calcium, and heavy metals, as shown in Figure 5.17.

5.3.2.3 Kinetics

A series of studies have been reported on the kinetics of biological sulfate removal. As shown in Tables 5.2 to 5.3,[44] the sulfate reduction rate ranges from 0.1 to 24 g L^{-1} d^{-1}. Typically, the rate is about 1 to 3 g L^{-1} d^{-1}. Temperature plays an important role. Sulfate reduction rates of 34, 22, and 6 mg/L/day were reportedly obtained at temperatures of 30°C, 20°C, and 9°C, respectively, when silage was used as an electron donor.[45]

5.3.3 Bioreactors

Batch reactor, fixed-bed reactor, filter, fluidized-bed reactor (FBR), upflow anaerobic sludge blanket reactor (UASB), gas-lift bioreactor (GLB), and moving sand bed reactor can be used in the biological removal of metal sulfate from wastewater (Tables 5.3 and 5.4). Selection of a reactor depends on cost, ease of operation, and nature of organic substances that are used as electron donor and carbon source. These reactors are generally suitable for soluble organic substrates. However, the solid organic

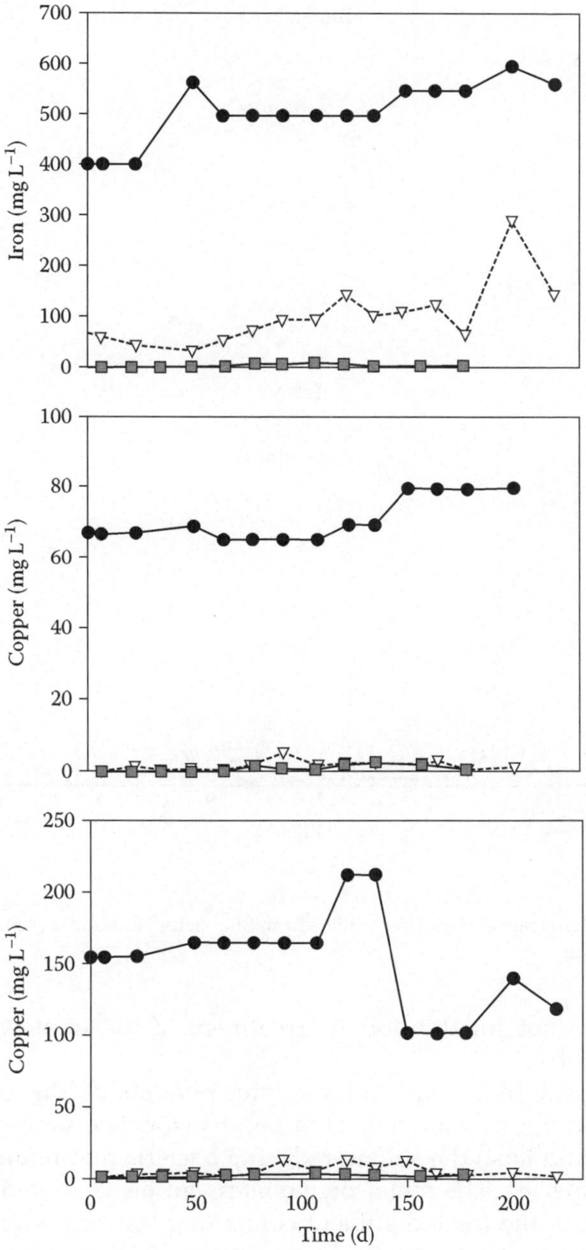

FIGURE 5.16
Removal of metals in the bioremediation of AMD: metal concentrations from AMD (●), from Reactor 1 (■) and from Reactor 2 (▽).

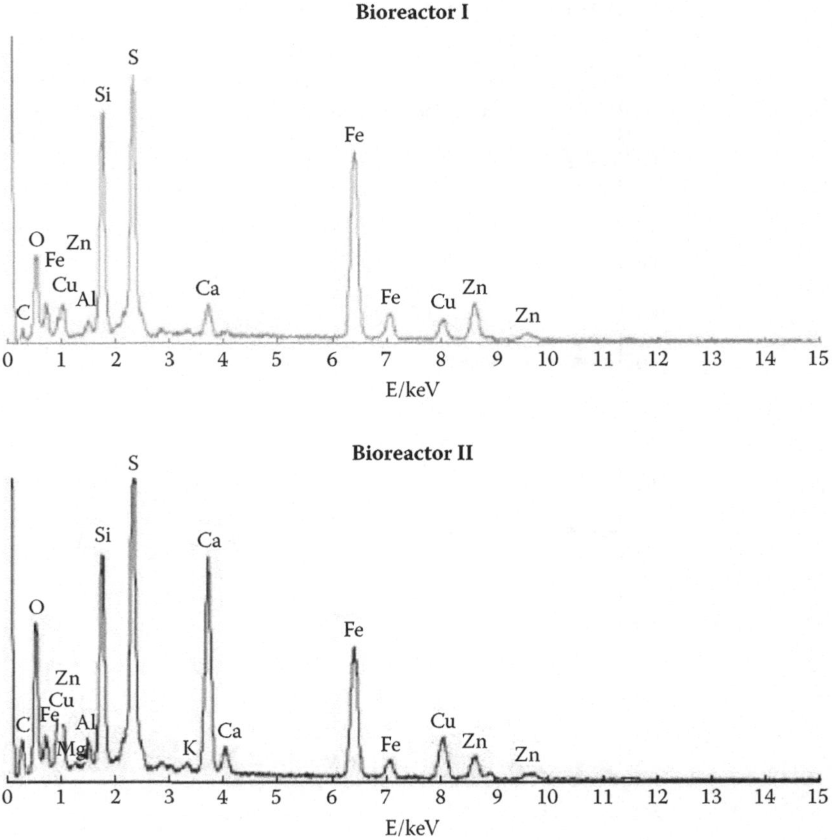

FIGURE 5.17
EDS and SEM micrographs of the precipitates from Bioreactor I (ethanol used) and Bioreactor II (wine waste used).

substrates are not suitable for the treatment of the wastewater by FBR, UASB, and GLB.

The biological reduction and the precipitation of the metal species can be carried out in a moving sand bed reactor. The sand particles may immobilize (and host) the sulfate-reducing bacteria and retain the precipitated metal species. The metal precipitates can be separated by the sand filter, leading to the treated effluent containing less or no metal particles. The precipitates can eventually be separated by back-washing, and the metal can be recovered by dissolving the precipitates in acidic solution and then chemical reduction. The moving sand bed reactor was used by Buisman et al.[55] Selenium and uranium were precipitated without sulfur components, whereas antimony, cadmium, copper, and zinc were precipitated as insoluble metal sulfides.

5.4 Reduction of Hexavalent Chromium

Chromium is a transition metal that is extensively used in various industries.[56–58] Chromium can exist in different oxidation states, from 2^- to 6^+. Cr(III) and Cr(VI) are the two thermodynamically stable oxidation states. Trivalent chromium is a naturally occurring element and is essential for adults when it is at a very low level. On the other hand, almost all hexavalent chromium (Cr(VI)) contaminant is produced because of human activities. Cr(VI) is an acute carcinogen and more mobile and toxic than Cr(III).

There are several methods employed in the conventional treatment of metal containing wastewater; these include chemical precipitation, coagulation and flocculation, oxidation, ultrafiltration, electrodialysis, reverse osmosis, and adsorption using activated carbon. The main strategies considered for Cr(VI) removal from wastewater are chemical reduction, chemical precipitation, ion exchange, adsorption, and membrane filtration.[59] A generally speaking they are costly and less efficient.

Typically, the treatment of hexavalent chromium involves first the reduction of Cr(VI) to the trivalent form by a reducing reagent with a high dosage. This is followed by the precipitation of Cr(III) by a strong base such as sodium hydroxide as a metal hydroxide at high pH, which leads to a large volume of metal precipitates. The watery precipitate sludge finally settles and is dewatered and disposed as discussed in Chapter 3.

It has been reported that at very low concentrations of Cr(VI), the redox reaction becomes kinetically slow and the residual level of Cr(VI) is still higher than the discharge limit. The high cost of the reduction chemicals, generation of toxic sludge, and incomplete reduction of Cr(VI) are the main driving factors for searching for more efficient new technologies.

5.4.1 Solution Chemistry of Chromium

5.4.1.1 Hexavalent Chromium

Hexavalent chromium is often associated with oxygen as chromate ($Cr_2O_4^{2-}$) or dichromate ($Cr_2O_7^{2-}$). The speciation of Cr(VI) in solution dependents on pH and the total concentration.

CrO_4^{2-} is the only species that can exist in solution above pH 8. The shift in pH results in a shift in the equilibrium; at pH values of 2–6, $HCrO_4^-$ is in equilibrium. The solution chemistry of hexavalent chromium is predicted using a chemical speciation prediction program, MINEQL and is shown in Figure 5.18.

It is also important to note that Cr(VI) is a strong oxidizing agent and its oxidation power is affected by the structure of reagent, as well as the nature and pH of the reaction medium.

The reduction of Cr(VI) to Cr(III) involving a three-electron change has a low probability of occurring via a single-step reaction. The formation of an

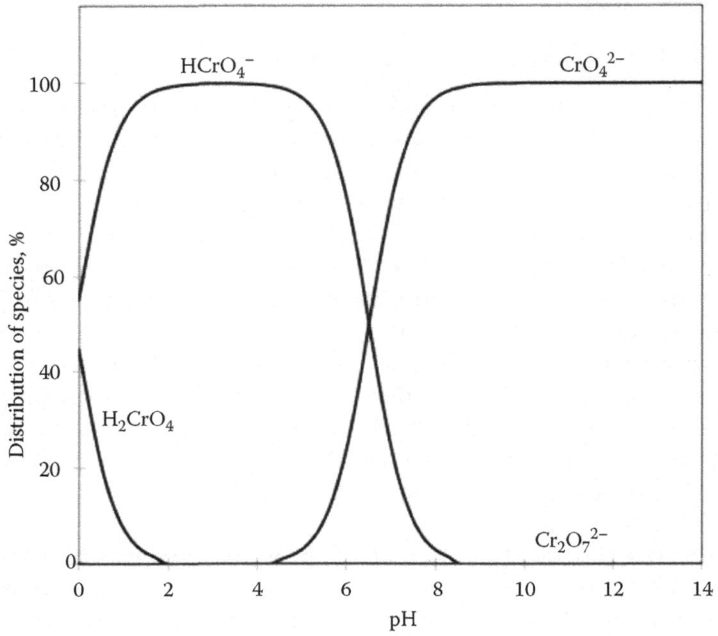

FIGURE 5.18
Hexavalent chromium speciation as a function of pH.

intermediate chromium oxidation state (i.e., Cr(V) and Cr(IV)) is possible for every redox reaction.

5.4.1.2 Trivalent Chromium

Among all the oxidation states, the most stable form is trivalent chromium. Free trivalent chromium essentially does not exist in aqueous solution, especially under natural conditions, because of its tendency to form coordination compounds, complexes, and chelates. At $10 > pH > 7.5$, the chromium is present as precipitates ($Cr(OH)_3$) indicating better removal for the chromium. Figure 5.19 shows the speciation of trivalent chromium in water solution.

5.4.2 Activated Sludge Process

Bioremediation can be conducted using strains of bacteria that are able to enzymatically reduce hexavalent chromium. Chromium-tolerant bacteria isolated from industrial effluents can detoxify hexavalent chromium. These microbial species normally function well in the laboratory under defined culture conditions. However, the bactericidal toxicants existing at many waste sites and wastewater would limit their growth and actual effectiveness.

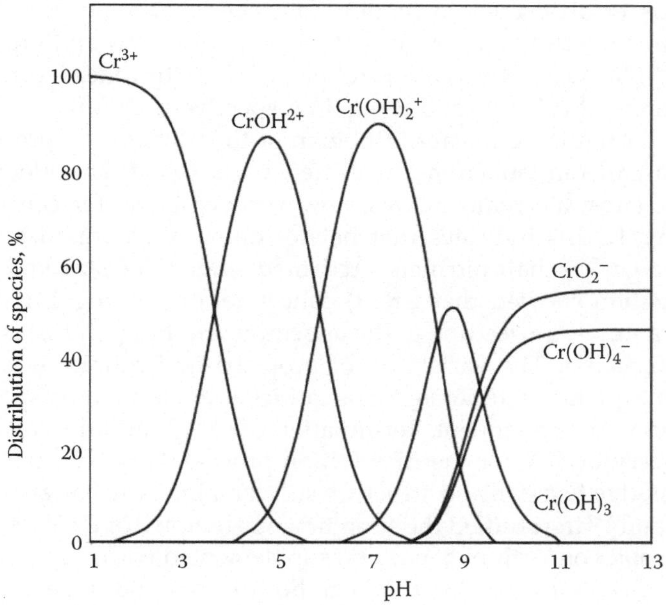

FIGURE 5.19
Trivalent chromium speciation as a function of pH.

The conventional activated sludge process (CASP) is widely used for treatment of domestic and industrial wastewater. Several studies were conducted on the effect of chromium on the performance of conventional activated sludge processes. Barth et al. reported that the aerobic biological treatment process could tolerate hexavalent chromium with concentrations up to 10 mg/L without significant loss in treatment efficiency.[60] Gokcay and Yetis demonstrated that an acclimatized activated sludge culture was fully capable of coping with the Cr(VI)-containing wastewater, and the culture was stimulated in terms of COD and BOD_5 removal by up to 1–25 mg Cr(VI)/L.[61] Dilek et al. claimed that Cr(VI) slightly affected the activated sludge performance in terms of COD removal efficiency, even up to 1–50 mg Cr(VI)/L.[62] Chen and Hao reported that Cr(III) was often accumulated in the cell membrane, while Cr(VI) was accumulated in the cells.[63] Cr(VI) ions could be reduced to Cr(III) and further adsorbed onto intracellular materials.

Caravelli et al. studied the reduction of Cr(VI) by *Sphaerotilus natans,* a filamentous bacterium usually found in activated sludge systems.[64] It was demonstrated that *S. natans* in aerobic conditions could efficiently reduce Cr(VI) to Cr(III) from dichromate solutions with 4.5–80 mg Cr(VI)/L in the presence of a carbonaceous source. The chromium(VI) reduction rate was dependent upon both Cr(VI) concentration and active biomass concentration. The study suggested that *S. natans* could be a potential microorganism to decontaminate Cr(VI) polluted wastewaters.

Stasinakis et al. used two parallel lab-scale continuous-flow activated sludge plants to study the removal efficiencies of Cr(VI) and organic compounds.[65,66] One was used as a control plant, while the other received Cr(VI) concentrations of 0.5, 1, 3, and 5 mg/L, respectively. Cr(VI) concentrations of 0.5 mg/L caused significant inhibition of the nitrification process; up to 74% decrease in ammonia removal efficiency was found. The effect of Cr(VI) on organic substrate removal was, however, negligible for concentrations up to 5 mg/L; this indicates that heterotrophic microorganisms are less sensitive to Cr(VI) than nitrifiers. Activated sludge floc size and structure characterization showed that Cr(VI) concentrations >1 mg/L reduced the filaments' abundance, leading to the appearance of pin-point flocs and free-dispersed bacteria. The variability of protozoa and rotifers was reduced. The effluent quality deteriorated as significant amounts of suspended solids escaped with the effluent. Termination of Cr(VI) addition resulted in a good recovery of 57% for the nitrification process. However, the damaged activated sludge floc size and structure still remained unchanged.

Chirwa and Wang studied the simultaneous removal of Cr(VI) and phenol in an anaerobic consortium of bacteria supplemented with a Cr(VI)-reducing organism, *Escherichia coli* ATCC 33456. Both *E. coli* and anaerobic phenol-degrading consortium culture were not acclimated to Cr(VI) prior to the experiments.[67] Optimum Cr(VI) reduction occurred at a phenol concentration of 200 mg/L at an initial Cr(VI) concentration of 2.0 mg/L. A higher phenol-degrader/*E. coli* ratio resulted in a higher rate of Cr(VI) reduction. A positive correlation between phenol degradation and Cr(VI) reduction indicated that *E. coli* utilized metabolites formed from phenol degradation as electron donors for Cr(VI) reduction.

5.4.3 Membrane Bioreactor

There are several unavoidable disadvantages in activated sludge processes, including production of large amounts of excessive sludge, sludge bulking, and requirement for large construction space. As such, numerous modifications have been conducted. In particular, the cost-effective ultra-/micro-filtration (UF/MF) combined with activated sludge processes prompted the emergence of the membrane bioreactor (MBR).

MBR has attracted increasing interest due to its advantages over activated sludge processes. Suspended solids are separated from the treated wastewater by a membrane filtration unit. MBR is able to achieve complete removal of solids, great nutrient and organic removals, a high loading rate, extremely low sludge production, and small land requirement.

One of the operational problems is the presence and toxicity of heavy metals due to industrial wastewater discharge. The metals originate from various industries such as semiconductor manufacturing, metal finishing, and battery manufacturing. Copper, cadmium, nickel, lead, chromium, and zinc with considerable concentrations are commonly found in influent of municipal sewage

TABLE 5.12

Summary of Experimental Conditions in the Four Operations of MBR

Code	Cr (VI) in Influent	Reasons for Selection
MBR-0	Cr-free	A metal-free synthetic wastewater was used as the feed for an MBR operation that lasted for 5 months.
MBR-0.4	0.4 mg/L	The yearly average Cr concentration in raw sewage in Singapore is 0.4 mg/L Cr(VI). Thus, it was selected to represent a typical value of Cr concentration in the domestic wastewater. An operation with an influent of 0.4 mg/L Cr(VI) was performed for 1 month.
MBR-10	10–11 mg/L	It has been reported that the addition of chromium above 11 mg/L will decrease the maximum growth rate (μ_{max}) and increase the Monod constant (K_s) in the CASP (Mazierski, 1995). 10–11 mg/L of chromium feed was used to test the toxic effect caused by the presence of Cr(VI) on the MBR. An operation with an influent of 10–11 mg/L was conducted for 1 month.
MBR-50	46–50 mg/L	A review of the literature indicated that the highest concentration of chromium used in the CASP was 50 mg/L. One-month continuous operation was conducted with 46–50 mg/L Cr(VI) in the influent.

treatment plants. Chromium, commonly found in the electroplating waste streams, has highly negative effects on wastewater treatment.

A study reported by Al-Malack on the toxicity of chromium on MBR performance is perhaps one of few available in the literatures.[68] A shock loading of chromium was applied to an MBR with an initial concentration of 20 mg/L on the second day followed by 40 and 50 mg/L on the third and fourth days. The COD removal initially dropped but a quick recovery was demonstrated.

A four-batch experimental study under the experimental conditions described in Table 5.12 was conducted by Zhao et al.[69] The major findings are given as follows, which can illustrate the performance of MBR on the simultaneous treatment of hexavalent chromium ion and organic wastewater. Figure 5.20 provides the process diagram of the MBR.

5.4.3.1 Effect of Metal on Membrane Flux

The flux rate of MBR is an important parameter in the operation of the MBR. It typically ranges from 0.12 to 7.2 $m^3/(m^2 \cdot day)$.[70] The flux depends on a number of complex interrelated parameters, including the characteristics of the wastewater, the TMP, the membrane pore size, and the biomass characteristics.

In the four-batch experimental study, a TMP of 720–740 mmHg was used, and the membrane module worked only at the first 15–20 min in every hour.[69] The membrane was in operation for approximately 8 months (5 month Cr-free, 1 month with 0.4 Cr mg/L, 1 month with 10–11 mg/L Cr, and 1 month with 46–50 mg/L Cr). As shown in Figure 5.21, a rapid flux decline in the initial

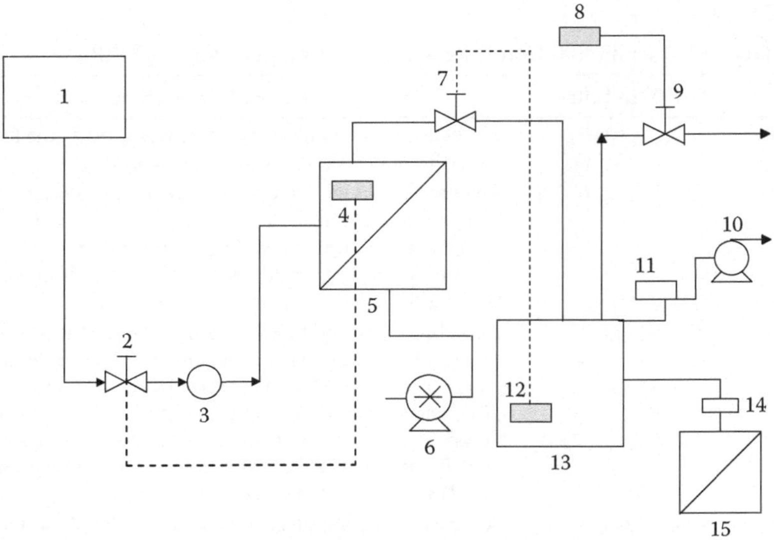

1. Influent Tank; 2. Solenoid Valve; 3. Influent Pump; 4. Level Controller;
5. Membrane Bioreactor; 6. Aeration Pump; 7. Solenoid Valve; 8. Time Controller;
9. Solenoid Valve; 10. Vacuum Pump; 11. Pressure Meter; 12. Level Controller;
13. Vacuum Chamber; 14. Flow Meter; 15. Auto-sampler.

FIGURE 5.20
Schematic diagram of MBR.

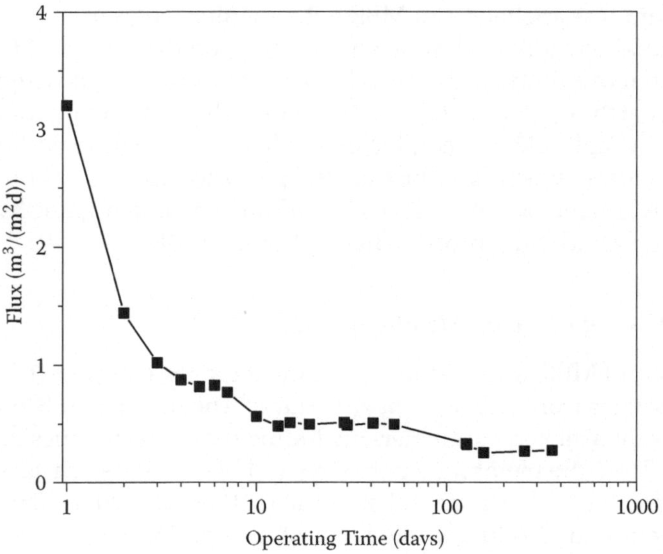

FIGURE 5.21
Flux decrease during MBR runs.

operation occurs. The flux drops from an initial value of 3.2 to 0.5 $m^3/(m^2 \cdot day)$ within 10 days after the start-up of the experiment. However, the flux is well maintained over the whole experimental period of around 8 months at a stable value of 0.25 $m^3/(m^2 \cdot day)$ without membrane cleaning.

The flux of membrane prior to usage is 3.2 $m^3/(m^2 \cdot day)$. The membrane after operation for approximately 8 months still has a flux of 0.7 $m^3/(m^2 \cdot day)$ compared with a flux of 2.8 $m^3/(m^2 \cdot day)$ of tap water. Comparison of these three values indicates that the fouling is not extremely severe. The fouled membrane may be easily recovered by simple physical cleaning.

5.4.3.2 Effect of Metal on Sludge Production

As hexavalent chromium is highly toxic, it may greatly affect biomass growth and sludge production. In addition, it is known that the solid retention time (SRT) can affect the microbial population and bacterial activities in MBR significantly. Canales et al. pointed out that biomass viability and sludge production in the absence of hexavalent chromium decreased when the SRT was increased.[71] It has been reported in the literature that sludge production of MBR is less than 0.3–0.5 kg-MLSS/kg-COD, much lower than conventional activated sludge processes.

Figure 5.22 shows the change in biomass concentrations versus operational time in the MBR that treated the organic wastewater with different hexavalent chromium concentrations.[69] The values of sludge production for MBR-0, MBR-0.4, MBR-10, and MBR-50 are listed in Table 5.13, together with the typical values reported for CASP and MBR. The sludge production in the MBR is approximately 10% of CASP. The sludge production of MBR-0 is higher than those of MBR-0.4 and MBR-10, while both MBR-0.4 and MBR-10 are similar to each other. The presence of Cr(VI) over a concentration of 0–10 mg/L does not adversely affect biomass growth. However, when the Cr(VI) concentration is increased to 50 mg/L, the biomass concentration decreases, as shown in Figure 5.22d, which indicates decomposition of biomass. In other words, the toxicity from Cr(VI) becomes significant when its concentration reaches 50 mg/L. The serious decomposition of cells is associated with higher nitrogen and phosphate concentrations in the effluent shown in Figures 5.24 and 5.26, which will be discussed.

5.4.3.3 Effect of Metal on Carbonaceous Pollutant Removal

The presence of chromium may affect the removal of carbonaceous contaminant (COD and BOD).[69] As demonstrated in Figure 5.23, great COD efficiency is achieved in the MBR when the chromium concentrations values are 0, 0.4, and 10 mg/L (MBR-0, MBR-0.4, and MBR-10). The effluent COD values are 1–20 mg/L (10 mg/L on average), 3–35 mg/L (17 mg/L on average), and 10–20 mg/L (14 mg/L on average) for MBR-0, MBR-0.4, and MBR-10, respectively. A COD removal efficiency of above 97% is obtained in MBR-0, MBR-0.4, and MBR-10.

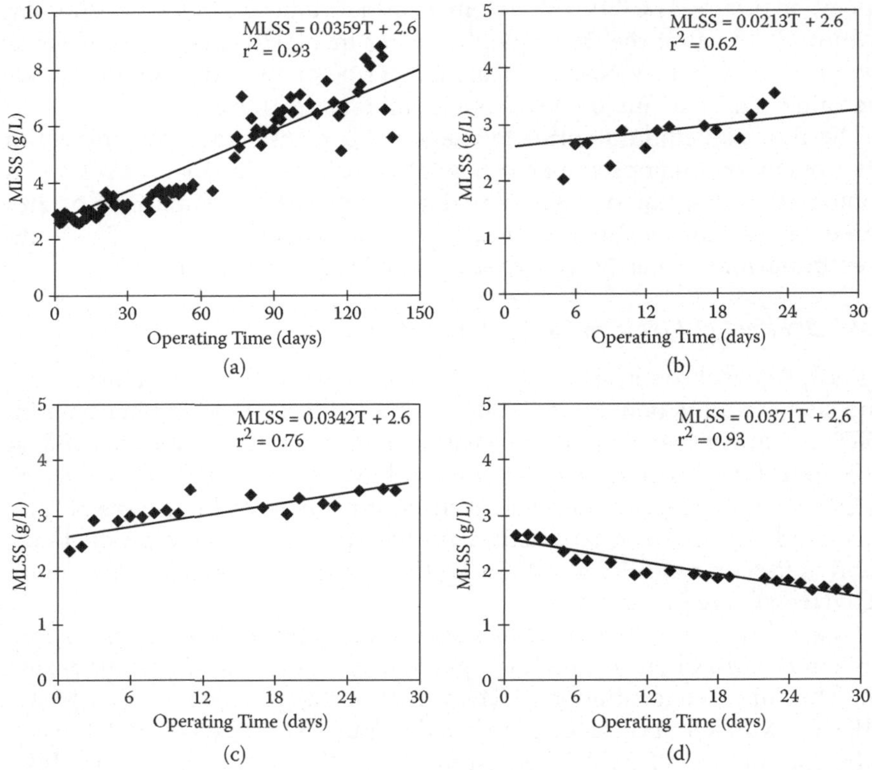

FIGURE 5.22
MLSS in MBR: (a) MBR-0, (b) MBR-0.4, (c) MBR-10, (d) MBR-50.

TABLE 5.13
Sludge Production in Different Runs

Process	Cr(VI) in Influent (mg/L)	Sludge Production kg-MLSS/kg-$COD_{utilized}$
CASP	—	0.3–0.5 (typical value)
MBR reported in literatures	—	0.03–0.2
MBR-0	0	0.06
MBR-0.4	0.4	0.03
MBR-10	10	0.05
MBR-50	50	−0.06

When the influent Cr(VI) concentration is 50 mg/L (MBR-50), the average COD concentration in the effluent increases to a level of 20–100 mg/L (52 mg/L on average). The removal efficiency for COD is 92%. The COD removal in MBR-50 is lower than MBR-0, MBR-0.4, MBR-10, and those reported in the literature.[72]

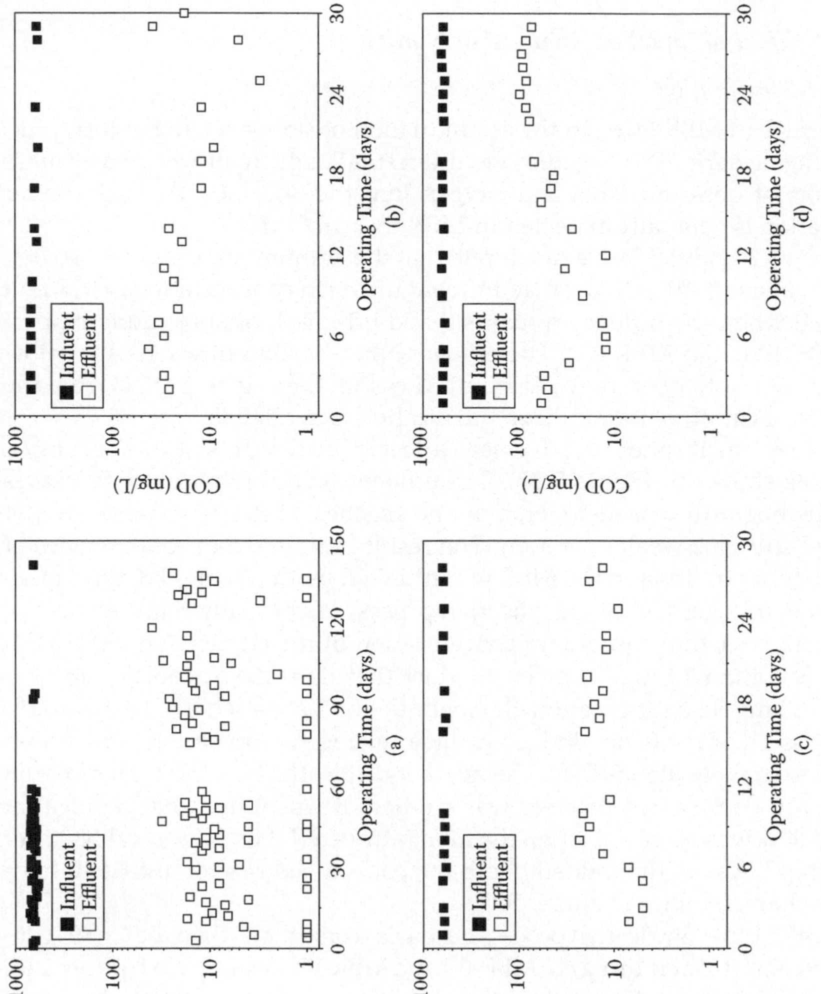

FIGURE 5.23
COD removal in MBR: (a) MBR-0, (b) MBR-0.4, (c) MBR-10, (d) MBR-50.

Generally speaking, the sludge culture in the MBR has a high resistance to Cr(VI) from the standpoint of carbonaceous removal. One reason may be the extremely long SRT (due to no wastage of sludge) and HRT (24 h) of the MBR, causing a relatively thorough carbonaceous oxidation. The carbonaceous uptake becomes slightly lower in MBR-50. It implies that the toxicity of metal would become much more severe when its concentration is 50 mg/L and higher.

5.4.3.4 Effect of Metal on Removal of Nutrient

5.4.3.4.1 Nitrification

Longer SRT in MBR leads to the accumulation of slow-growing autotrophic, nitrifying bacteria. The floccules are often small, which allows greater mass transport of both nutrients and oxygen into the floccules. As such, better nitrification is generally expected in MBR than in CASP.

Figures 5.24a and 5.24b demonstrate that the ammonium removal percentages are above 99% with average influent nitrogen concentrations of 27 and 28 mg/L when the influent contains 0 and 0.4 mg/L of chromium, respectively (MBR-0 and MBR-0.4). The effluent NH_4^+-N concentrations are below 0.3 mg/L for both operations. This indicates that the negative effect from the presence of Cr(VI) of 0.4 mg/L on nitrification is negligible.

However, nitrification is adversely affected in the presence of 10 mg/L Cr(VI), as shown in Figure 5.24c. The effluent NH_4^+-N gradually increases from the beginning of the operation and reaches a relatively stable level of 10 mg/L after a 2-week operation. The result indicates that some species of the nitrifying bacteria in the MBR are inhibited by the 10 mg/L Cr(VI) present in the influent, and the nitrification process is gradually retarded.

Nevertheless, the ammonium concentration in the effluent from the MBR is still less than 10 mg/L, given the fact that its influent concentration is around 20 mg/L; a nitrification efficiency above 50% is achieved. This could be because there was no sludge wastage during the operation, which created an extremely high SRT for the nitrifiers. The effect of Cr(VI) on nitrogen removal in packed bed reactors was studied. It was found that removal of around 80% was achieved when the total nitrogen (NH_4^+, NO_2^-, and NO_3^-) of 20–25 mg/L was in the influent. Both the packed bed reactor and MBR have similar characteristics of longer SRT.[69]

In MBR-50, the nitrification process becomes worse than when the Cr(VI) concentrations are 0.4 and 10 mg/L (MBR-0.4 and MBR-10), as shown in Figure 5.24d. The NH_4^+-N concentration in the effluent increases substantially after the start-up of the reactor. After 2 weeks, the NH_4^+-N concentration in the effluent almost reaches the same level as that in the influent. At the same time, the carbonaceous utilization efficiency also decreases as demonstrated in Figure 5.23. After 20 days from start-up, the NH_4^+-N and total nitrogen concentrations in effluent are the same or even higher than those in the influent. The decomposition of activated sludge in the Cr(VI) concentration at 50 mg/L may cause the higher total nitrogen concentrations in the effluent (Figure 5.25d).

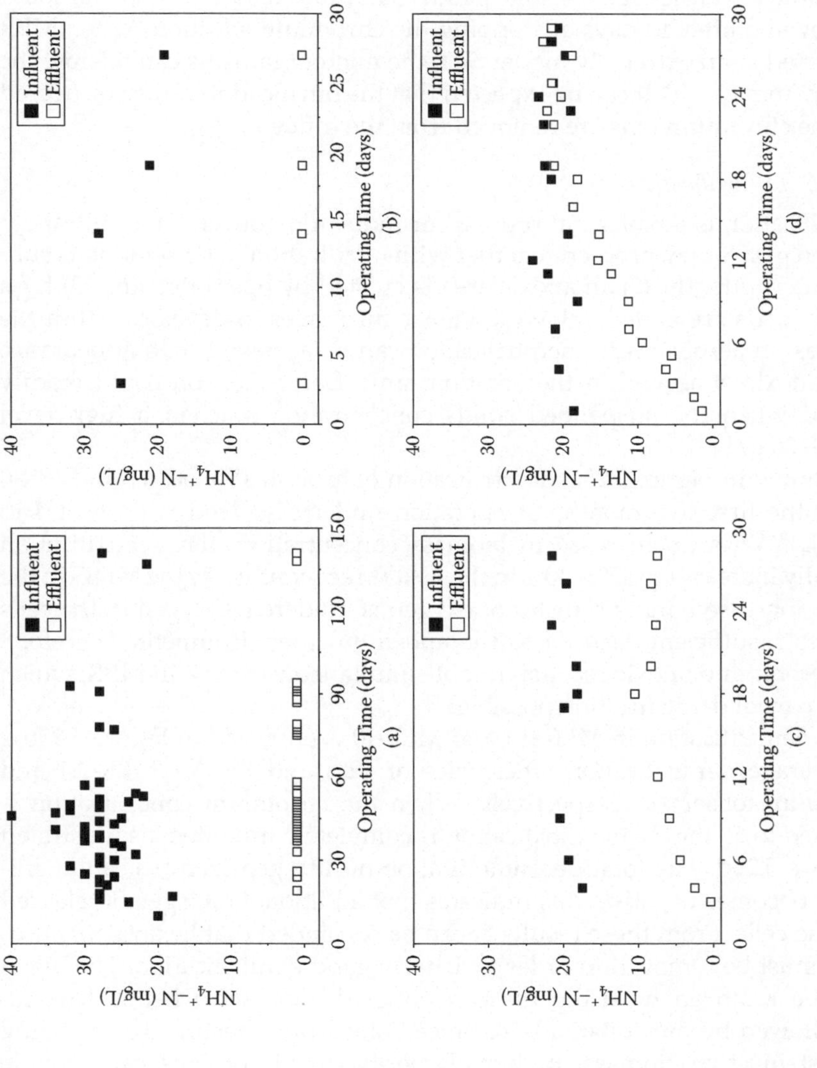

FIGURE 5.24
NH_4^+-N removal in MBR: (a) MBR-0, (b) MBR-0.4, (c) MBR-10, (d) MBR-50.

The results from the foregoing studies show that the nitrification bacteria are more sensitive than heterotrophs in the presence of Cr(VI); a Cr(VI) concentration up to 10 mg/L is able to inhibit the nitrification process partially, while a Cr(VI) concentration of 50 mg/L or higher completely inhibits the entire nitrification.

Stasinakis et al. studied the effect of hexavalent chromium on the removal of ammonia in an activated sludge plant.[65] They found that the removal could be recovered after 12 days of stopping the chromium addition. When MBR is operated at an extremely higher SRT, the microorganisms can tolerate the toxicity from Cr(VI). It can be expected that the nitrification can be recovered once the chromium ions are removed from the influent.

5.4.3.4.2 *Denitrification*

Denitrification is a biological process for reduction/conversion of nitrate to nitrogen gas. It may proceed together with nitrification if: (1) aeration is supplied intermittently, (2) an anoxic area is created by operation, and (3) high organic loads are added, allowing anoxic microsites to develop within the floccules. In aerobic MBR, denitrification can be achieved with appearance of an anoxic zone within the aeration tank. Denitrification could readily proceed when the suspended solids concentration reached a high level (e.g., 20–30 g/L).

As shown in Figure 5.25a, denitrification of around 15% occurs in MBR-0 during the first two months of operation under dissolved oxygen of 4 to 5 mg/L.[69] With an increase in biomass concentration, the denitrification gradually increases to 35%–40% in the next three months. As the MLSS in the MBR is increased, mass transfer of oxygen is hindered. Oxygen in the cells becomes insufficient, leading to the anoxic microenvironment. Therefore, both aerobic and anoxic reaction might simultaneously exist in MBR, which makes partial denitrification possible.

Poor denitrification in MBR-0.4 and MBR-10 is observed in Figure 5.25b,c. The average denitrification efficiencies of 12% and 5% for MBR-0.4 and MBR-10 are observed, respectively. When the chromium concentration is increased to 50 mg/L, the nitrification is completely inhibited, as illustrated in Figure 5.25d. The total decontamination of nitrogen (denitrification efficiency) becomes negative; this indicates that additional nitrogen is released from the cells. From these results, it can be concluded that hexavalent chromium must be removed in order to achieve good denitrification. The metal could be removed by using the conventional approach (chemical oxidation followed by precipitation). Advanced biosorption technology by using low-cost inactive biomass- or biopolymer-based biosorbents can provide another alternative for solving the problem.[73,74]

5.4.3.4.3 *Phosphorous Removal*

Phosphorous may be removed through biomass accumulation and sludge wasting in the activated sludge process. As shown in Figure 5.26a–c, the average

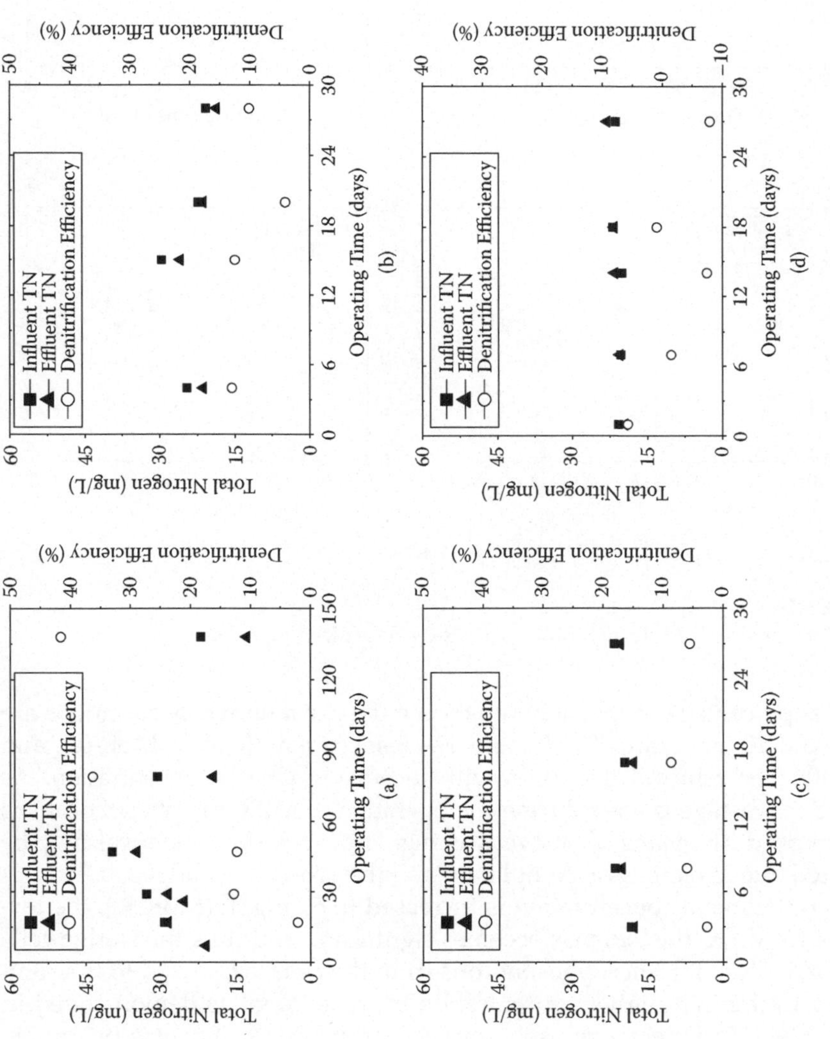

FIGURE 5.25
Denitrification in MBR: (a) MBR-0, (b) MBR-0.4, (c) MBR-10, (d) MBR-50.

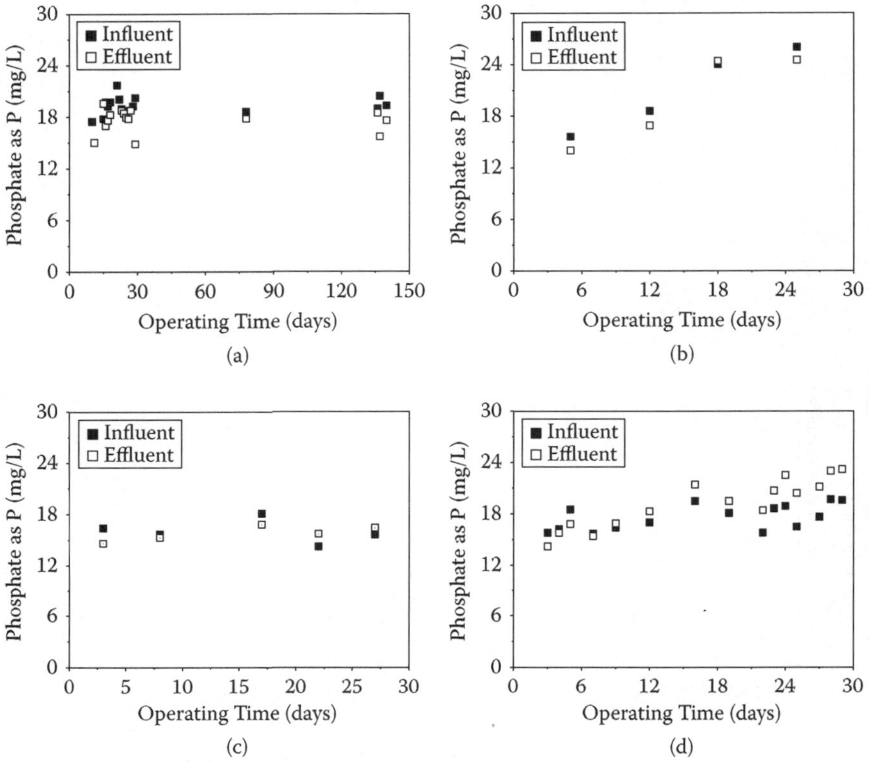

FIGURE 5.26
Phosphate removal in MBR: (a) MBR-0, (b) MBR-0.4, (c) MBR-10, (d) MBR-50.

removal percentages of phosphorous when the chromium concentrations are 0, 0.4, and 10 mg/L are 9%, 6%, and 1%, respectively (MBR-0, MBR-0.4, and MBR-10). The removal decreases with increasing Cr(VI) concentration. As no sludge wastage is used during the operation of MBR, the phosphorous is not removed by means of sludge wasting. However, some removal may be expected due to biomass accumulation and the growth of sludge.

When the metal concentration is increased to 50 mg/L (MBR-50), the toxicity of Cr(VI) to the biomass becomes significant, and thus the sludge cells begin to be lysed. Hence, phosphorous from the cells is released to the solution, resulting in a higher concentration in the effluent as demonstrated in Figure 5.26d. The average phosphorous concentration in effluent is 19.2 mg/L, while that in the influent is 17.6 mg/L. This observation is consistent with the decrease in biomass content (MLSS) in MBR-50 run illustrated in Figure 5.22d.

5.4.3.4.4 Change in Solution pH

pH plays an important role in biological wastewater treatment. From the viewpoint of biokinetics, the optimum pH for biomass growth ranges from

6.5 to 7.5. Outside of this range, biomass growth may be inhibited, causing a decrease in growth of biomass. At the same time, pH greatly affects both nitrification and denitrification. The nitrification process would cause consumption of alkalinity. On the other hand, nitrification may be inhibited if there's not enough alkalinity. When denitrification occurs, alkalinity is released to the water solution.

The change in pH in MBR is demonstrated in the study by Zhao et al.[69] In the operations, the influent pH was around 8.4. The effluent pHs for MBR-0, MBR-0.4, MBR-10, and MBR-50 are 5.8–7.3, 6.5–6.9, 7–7.3, and 7–7.3, respectively as shown in Figure 5.27. MBR-0 and MBR-0.4 show good ammonium removal (nitrification) (Figure 5.24a,b). The complete nitrification process causes a substantial pH decrease in the bioreactor due to the formation of NO_3^-. On the other hand, most nitrifying bacteria in MBR-10 and MBR-50 are inhibited as the influent contains higher Cr(VI) concentration (Figure 5.23c,d). As a result, the pH in effluent is higher than that of MBR-0 and MBR-0.4.

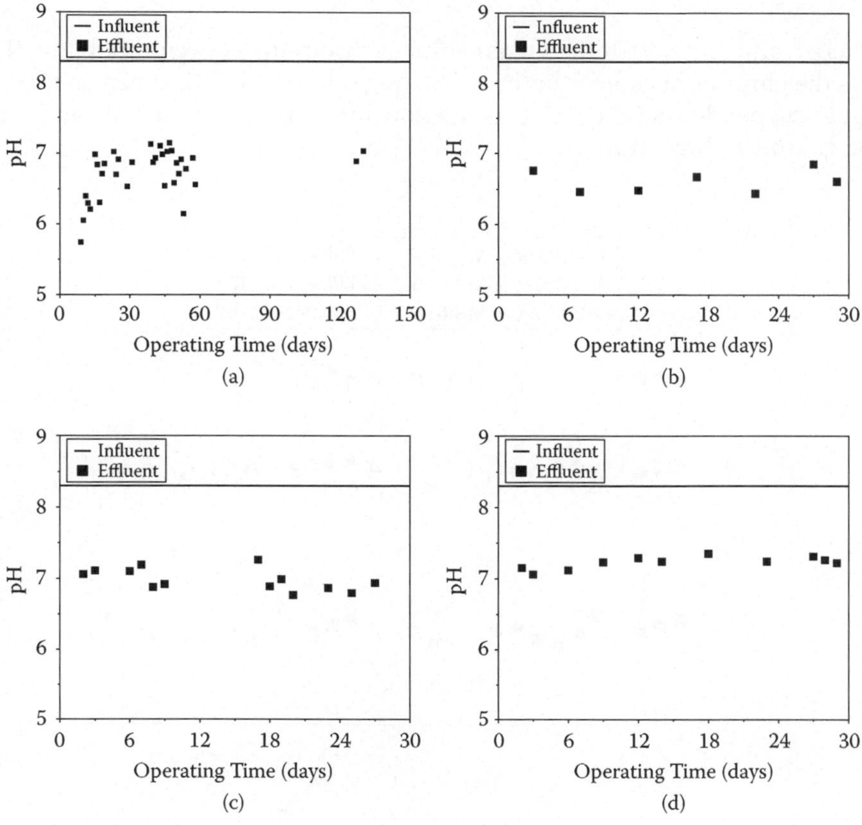

FIGURE 5.27
pH in MBR: (a) MBR-0, (b) MBR-0.4, (c) MBR-10, (d) MBR-50.[69]

5.4.3.4.5 Chromium Uptake

Hexavalent chromium in activated sludge culture may experience reduction and sorption. Thus, it can be anticipated that it can remain in the biomass, which was demonstrated by Zhao et al. and illustrated as follows.[69]

The changes in chromium concentration in the MBR are shown in Figure 5.28. In MBR-0.4, the chromium concentration in the effluent increases from the beginning and reaches a stable value of 0.1 mg/L on the fourth day. Since the metal concentration in the influent is very low, the saturation in uptake capacity is yet to be reached, and the metal concentration in effluent is still far below that in influent. In MBR-10 and MBR-50, the metal concentration increases and achieves a constant on the fifth and second day, respectively.

A mass balance can be established for the MBR with an influent with the chromium. The equation is as follows:

$$Q \times (c_i - c_e) \times dt = \overline{MLSS} \times V \times dc_s \tag{5.27}$$

where c_i and c_e are the influent and effluent chromium concentration (mg/L), c_s is the chromium content in the MLSS (mg/g), $\overline{MLSS}$ is the average mixed liquid suspended solid (g/L), Q is the flow rate (L), t is the time (d), and V is the volume of MBR (L).

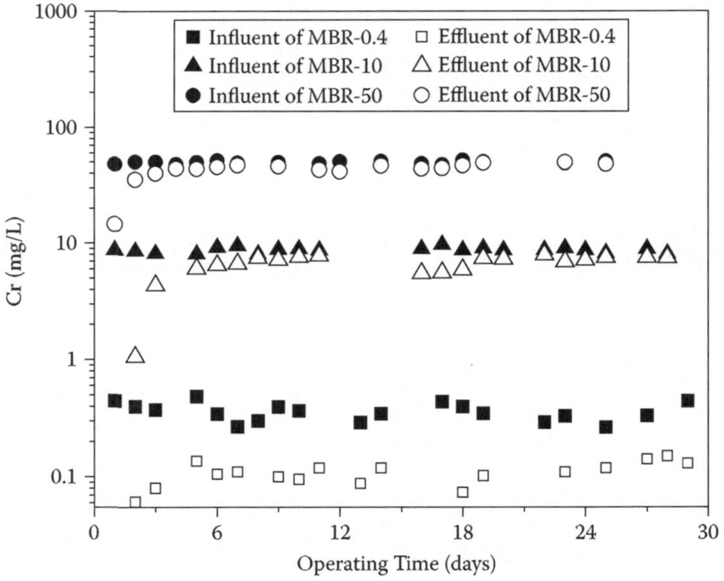

FIGURE 5.28
Changes in chromium concentrations of MBRs.[69]

Equation 5.27 can be integrated as follows:

$$\int_0^t Q\times(c_i - c_e)\times dt = \int_0^{c_c} \overline{MLSS}\times V\times dc_s \tag{5.28}$$

As Q, $\overline{MLSS}$, and V are constant, we can have

$$Q\times\int_0^t (c_i - c_e)\times dt = \overline{MLSS}\times V\times c_s \tag{5.29}$$

$$HRT = \frac{V}{Q} \tag{5.30}$$

Finally, we have

$$c_s = \frac{Q\times\int_0^t (c_i - c_e)\times dt}{\overline{MLSS}\times V} = \frac{\int_0^t (c_i - c_e)\times dt}{\overline{MLSS}\times HRT} \tag{5.31}$$

With the MLSS given in Figure 5.22, the HRT, and the metal concentration in the effluent shown in Figure 5.28, the chromium content can be estimated. The values are approximately 3, 7, and 10 mg-Cr/g-MLSS in MBR-0.4, MBR-10, and MBR-50, respectively. The chromium content obtained from the sludge digestion study shown in Figure 5.29 confirms the estimation for the case of MBR-50. It was reported that the hexavalent chromium biosorption capacity onto activated sludge was 10 mg-Cr(VI)/g-activated sludge at pH 6.[69]

Cr(VI) biosorption decreases as the solution pH is increased.[73,74] In the operation of MBR, the chromium concentration was as high as 50 mg/L. In such a toxic environment, the activated sludge still can remove organic carbon with great efficiency. This may be due to the low Cr(VI) biosorption in the operations. As shown in Figure 5.27, the solution pH ranges from 6.5 to 7.5. Under such a pH, less hexavalent chromium can be attached onto the cells. The bioaccumulation of the metal is not significant enough to cause malfunctions of heterotrophs. It can be concluded that great COD removal can still be achieved at the chromium content up to 10 mg-Cr/g-MLSS and pH of 7 to 8. Nitrogen and phosphorous removal efficiencies are adversely affected when chromium concentration is above 10 mg/L (Figures 5.25 and 5.26).

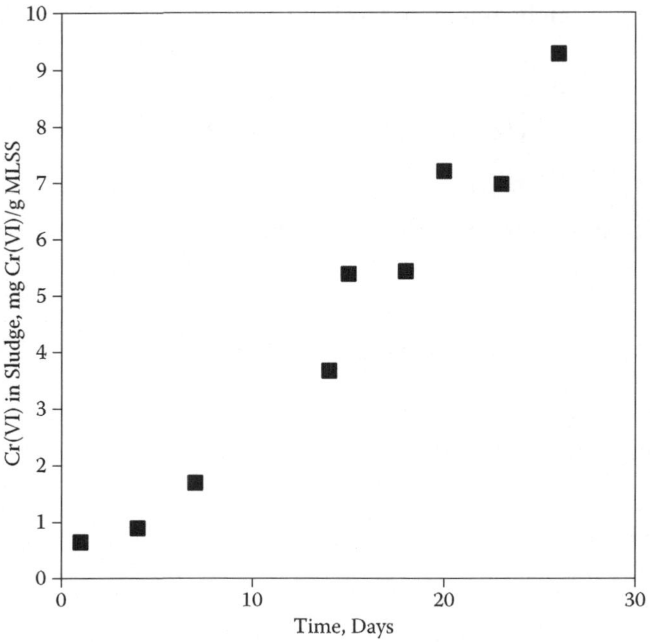

FIGURE 5.29
Chromium contents in activated sludge in MBR-10.

5.4.4 Inactive Biomass

Many dried/dead/inactive biosorbents, biomaterials, and biopolymers reportedly have the ability to cause chemical reduction of metals with higher valencies to lower valencies. Chromate can be reduced to trivalent chromium by several environmentally friendly seaweeds.[73,74] The findings from these researches are summarized as follows.[73]

Figure 5.30 shows that chromate in the solution is reduced as soon as two seaweeds (*Padina* and *Sargassum*) are in contact with the solution.[73] It is demonstrated that the instantaneous reduction is faster for *Padina* than for *Sargassum*. At 15 min, the ratio of Cr(VI):Cr(III) is 9.2:1 and 16.8:1, respectively. To reduce the ratio of Cr(VI):Cr(III) to 1.5, *Padina* takes 180 min, whereas *Sargassum* takes about 420 min. Both reduction processes occur until all the chromate is reduced to the trivalent form.

As untreated seaweed has high organic leaching during biosorption, surface modification/treatment for the seaweed becomes necessary. Yang and Chen reported that 0.2% formaldehyde could effectively reduce the organic leaching from the raw seaweed (RSW) *Sargassum sp.*[74] The resulted treated seaweed is termed *modified seaweed* (MSW).

Figure 5.31 shows the kinetics of Cr(VI) biosorption by the MSW and the RSW. At the MSW dosage of 3 g/L, the equilibrium time is 20 h when

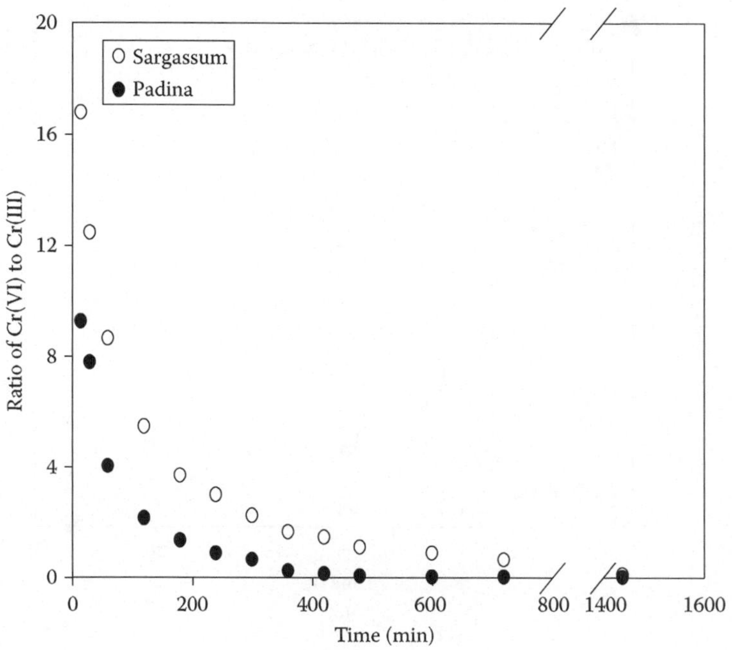

FIGURE 5.30
Ratio of Cr(VI) to Cr(III) in the supernatant during the progress of Cr(VI) biosorption (m = 1.0 g/L, C_0 = 1.0 mmol/L, pH = 2.0) solid points: *Padina;* hollow points: *Sargassum*.[73]

the initial concentration is 0.99 mM. Similarly, the biosorption equilibrium is achieved at around 20 h when the RSW is used. The chemical modification seems not to have an impact on the equilibrium time.

The concentrations of Cr(VI) and Cr(III) species in the solution during the biosorption are illustrated in Figure 5.31. The Cr(VI) concentration decreases, while the Cr(III) concentration increases during the operation. At the end of sorption, the concentration of Cr(VI) becomes undetectable for both MSW and RSW. From this experimental observation, one can see that biosorption is followed by two processes. The first process is chemical reduction that converts Cr(VI) to Cr(III), while the second process is metal coordination and ion exchange that lead to the accumulation of Cr(III) onto the solid. The sorption of cations onto the seaweed-based biosorbents is normally completed within 1 to 3 h as shown in Chapter 7 of this book. Thus, one can conclude that the reduction of Cr(VI) by biosorbents essentially controls the whole biosorption process.

Figure 5.32 shows the equilibrium concentrations of Cr(VI) and Cr(III) in the solution phase when the RSW and the MSW were used. The initial concentrations were set at around 1 and 2 mM, respectively, and the pH was varied. As shown the optimal biosorption pH at 2.0 is concentration

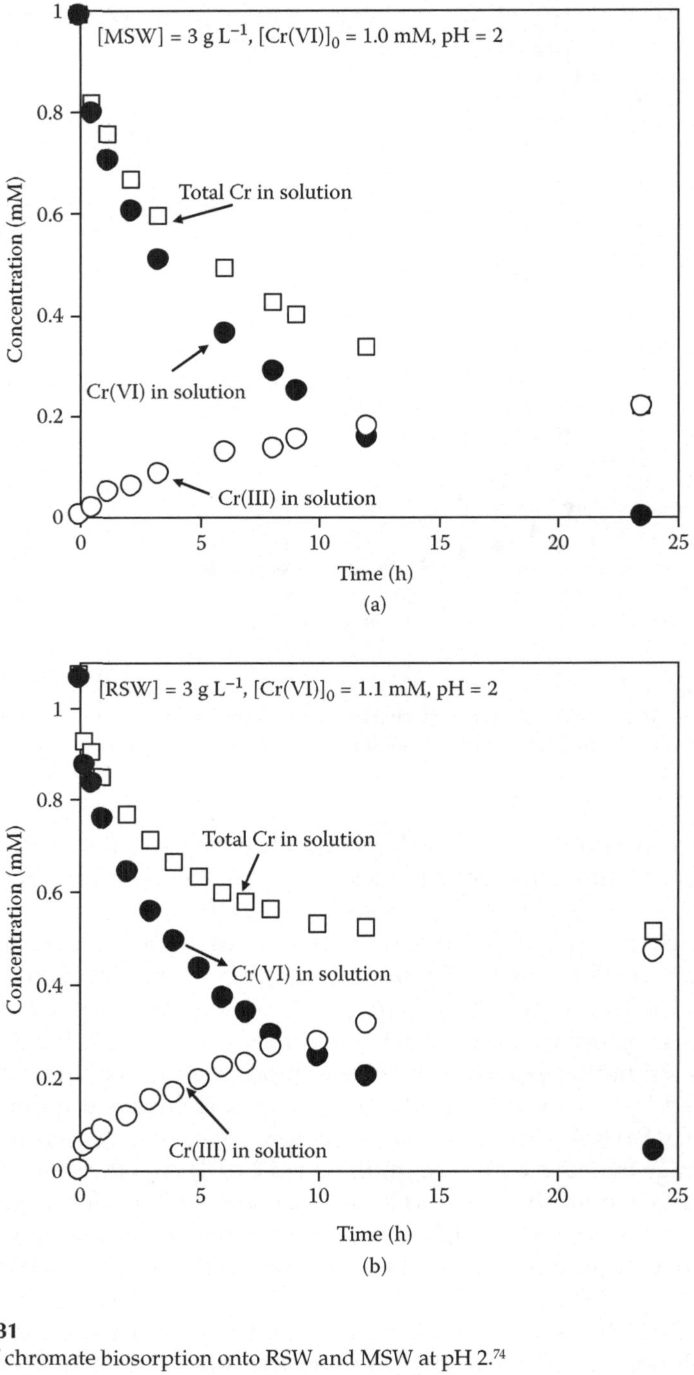

FIGURE 5.31
Kinetics of chromate biosorption onto RSW and MSW at pH 2.[74]

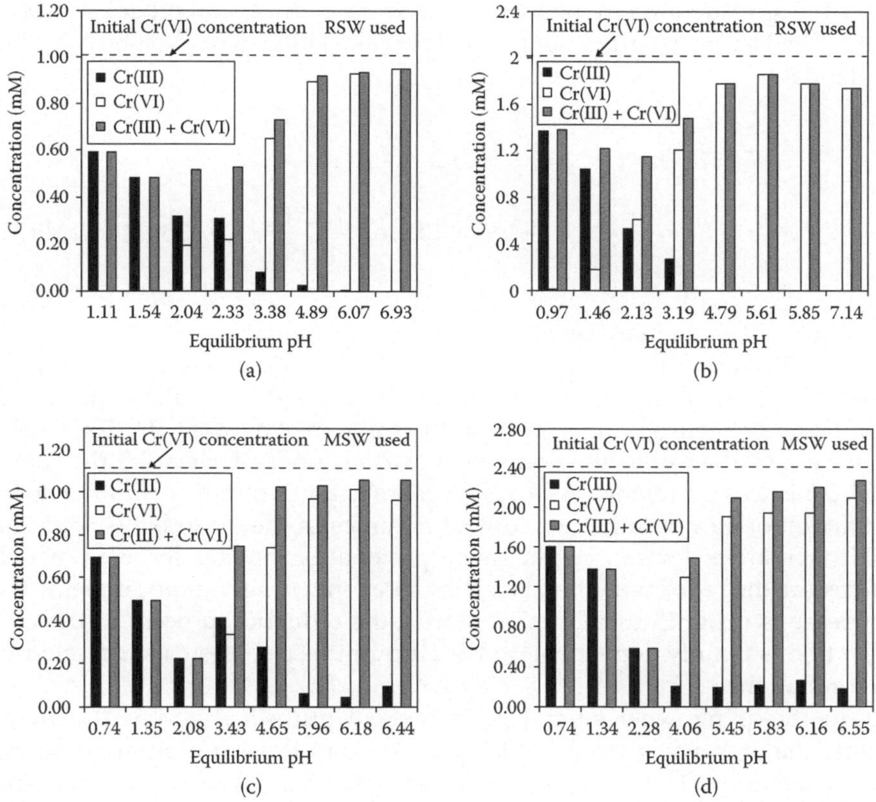

FIGURE 5.32
Distribution of chromium species in solution after biosorption under different pH. [RSW] = [MSW] = 3 g L^{-1}, contact time = 24 h. (a), (b): RSW; (c), (d): MSW.[74]

independent. The distribution of Cr(VI) and Cr(III) is strongly dependent on the solution pH; the initial metal concentration and the dosage of biosorbents also play some roles.

At pH < 2.0, all chromium species in the solution are in the form of trivalent chromium when the RSW was used, and the initial concentration of hexavalent chromium was 1 or 2 mM as illustrated in Figure 5.32a,b. When the pH was above 2.0, the concentration of Cr(VI) started to increase. A similar observation is seen in Figure 5.32c,d when the MSW was used and the initial concentrations of hexavalent chromium were 1.1 and 2.4 mM. A comparison of the foregoing four figures shows that the reduction strength increases when the formaldehyde is applied for the modification of seaweed. For example, at the initial concentration of 1.1 mM, Cr(III) can be observed at pHs ranging from 2 to 6 when the MSW is used as shown in Figure 5.32c. However, under similar conditions, the same species are not found when the RSW is used as demonstrated in Figure 5.32a. Reduction of Cr(VI) to

Cr(III) was also found in the biosorption process by several other biomaterials.[67] The Nernst equation can be used for calculation of oxidation-reduction potential:

$$E(HCrO_4^- / Cr^{3+}) = E_p(HCrO_4^- / Cr^{3+}) + 1.97 \times 10^{-2} \log \frac{[HCrO_4^-]}{[Cr^{3+}]} - 0.138\text{pH} \qquad (5.32)$$

where E_p is the standard oxidation-reduction potential.

As indicated in the foregoing equation, the redox potential increases as the pH is decreased. Figure 5.32 clearly indicates that trivalent chromium ion is the only chromium species in the solution at a lower pH. The Cr(III) can be adsorbed onto the sorbents; however, under lower pH, the metal uptake is less preferable. As a result, there is an optimum pH in the accumulation of chromium. The redox of organics in the biosorbents could be pH dependent; lower pH can cause protonation on the sorbents, which increases the redox potential. From the foregoing observations, it seems the increase is insignificant. In other words, the reduction process in the biosorption is mainly dependent on the chromium speciation, as indicated in the equation.

The foregoing experimental studies (e.g., Figures 5.31 and 5.32) demonstrate the presence of both Cr(III) and Cr(VI) in the solution phase due to the reduction effect from seaweed. An XPS analysis was used by Yang and Chen to further study the change in chemical structure in the solid phase.[74]

Figure 5.33 shows the XPS spectra of RSW and MSW exposed in Cr(VI) solution at pH 1 and 2 as well as in Cr(III) solution at pH 4. The binding energy values of chromium in Cr, Cr_2O_3, and $K_2Cr_2O_7$ are 574.4, 576.8~577.1, and 579.1–579.9 eV, representing Cr^0, Cr^{3+}, and Cr^{6+}, respectively.[75] Before Cr(VI) biosorption, there is no spectrum peak in the range from 570 to 590 eV, indicating that both RSW and MSW are free of chromium (data not shown).

Both Cr(VI) and Cr(III) peaks' areas on the XPS spectra are summarized and shown in Table 5.14. The result demonstrates that for the Cr(III) biosorption (also shown in parts a3 and b3 of Figure 5.33), there is no change in redox status on both RSW and MSW. For the Cr(VI) biosorption, the MSW reduces all Cr(VI) to Cr(III) on seaweed surface at both pH 1 and 2, whereas the RSW only reduces 48% Cr(VI) at pH 1 and 77% of Cr(VI) at pH 2. It is obvious that the MSW has greater reduction ability than the RSW.

Sorption of cation is enhanced at higher pH, while that of anion is improved at lower pH. Because of the presence of reducing functional groups, Cr(VI) is partially reduced, as demonstrated in Table 5.3 and Figure 5.9. At pH 1,

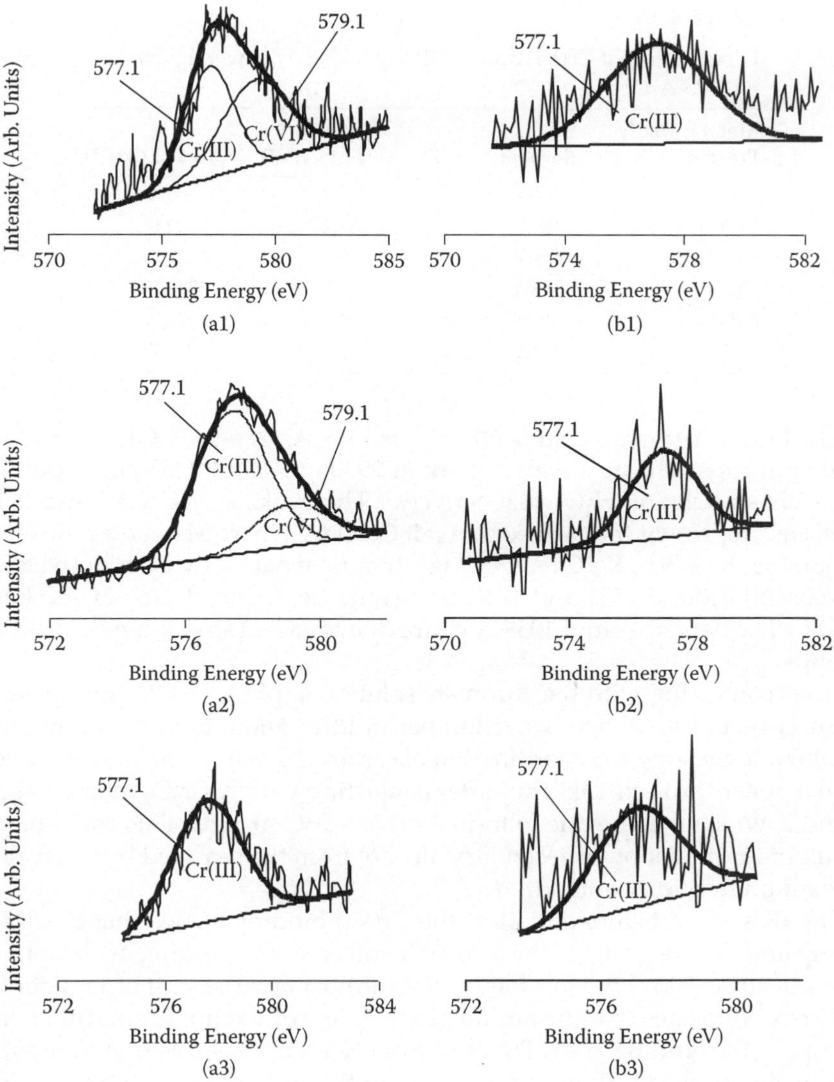

FIGURE 5.33
XPS spectra of RSW (a) and MSW (b). (a1, b1) contacted with Cr(VI) at pH 1; (a2, b2) contacted with Cr(VI) at pH 2; (a3, b3) contacted with Cr(III) at pH 4. Initial Cr concentration = 1 mM, biomass dosage = 3 g L^{-1}, contact time = 24 h.

nearly equal amounts of Cr(VI) and Cr(III) species were accumulated by the RSW. When the pH is increased to 2, more Cr(III) species are in the RSW.

The changes in surface functional groups in the biosorbents during the biosorption process are studied by FTIR technology. Figure 5.34 shows the infrared spectra of the biomass before and after the biosorption of $K_2Cr_2O_7$ or $Cr(NO_3)_3$.

TABLE 5.14
Distribution of Cr(VI) and Cr(III) Species on Seaweed Surfaces by XPS Analysis

Metal Initially Present	Sorbent	pH	Area of Cr(VI)	Area of Cr(III)
Cr(VI)	RSW	1	935.4	869.2
Cr(VI)	RSW	2	486.4	1622.0
Cr(III)	RSW	4	0	1532.2
Cr(VI)	MSW	1	0	1318.5
Cr(VI)	MSW	2	0	528.3
Cr(III)	MSW	4	0	713.7

The broad region around 3400 cm^{-1} can be assigned to –OH stretch and amine groups. The absorbance peak at 2930 cm^{-1} and 2363 cm^{-1} could be the –CH stretch and H_3O^+, respectively.[76] The peaks at 1645 cm^{-1} and 1427–1466 cm^{-1} represent the carboxylate salt COO-M, where M denotes the metal cations such as Na^+, K^+, Ca^{2+}, and Mg^{2+} that may naturally exist in the RSW. The brand around 1551 and 1246 cm^{-1} could be due to stretch of –NH and $-SO_3$.[77] The bands around 1038 cm^{-1} are due to –C–O stretching of alcoholic groups.

After contacting with the chromate solution at pH 1, $-COO^-$ group shifts from 1418 cm^{-1} to a lower wavenumber in RSW and a higher wavenumber in MSW, suggesting the coordination of carboxyl group to metal ions. A new peak appears around 1745 cm^{-1} after contacting with $K_2Cr_2O_7$ solution at pH 1 and 2, which may be due to formation of –C=O groups. This could be the result of the oxidation of secondary alcohol from the seaweed by hexavalent chromium in acidic solution.

The IR spectra before and after the Cr(VI) binding shows that the $-NH_2$ group and sulfate groups may also be involved in the binding of chromium. The shift from 3439 cm^{-1} to a lower wavenumber and the shift of 1551 cm^{-1} to 1549 cm^{-1} indicates that the amino group is one of the important functional groups in the binding of Cr. The shift from 1251 cm^{-1} to a lower wavenumber can be due to the binding of metal ions to the sulfate group, which mainly presents in sulfonic acids of polysaccharides, such as fucoidan.

A better understanding of the Cr(VI) biosorption mechanisms can help enhance the process performance. From the foregoing experimental results and the instrumental analysis, biosorption mechanisms are proposed (shown in Figure 5.35). The removal of anionic hexavalent chromium is a complicated process, in which reduction, surface complex formation, and ion exchange are involved.

Hexavalent chromium can form surface complexes with the protonated functional groups on the biosorbents, such as –COOH, $-NH_2$, and $-SO_3H$. The Cr(VI) anions can also oxidize the secondary alcohol groups on the biosorbents, while being reduced to Cr(III) cations.

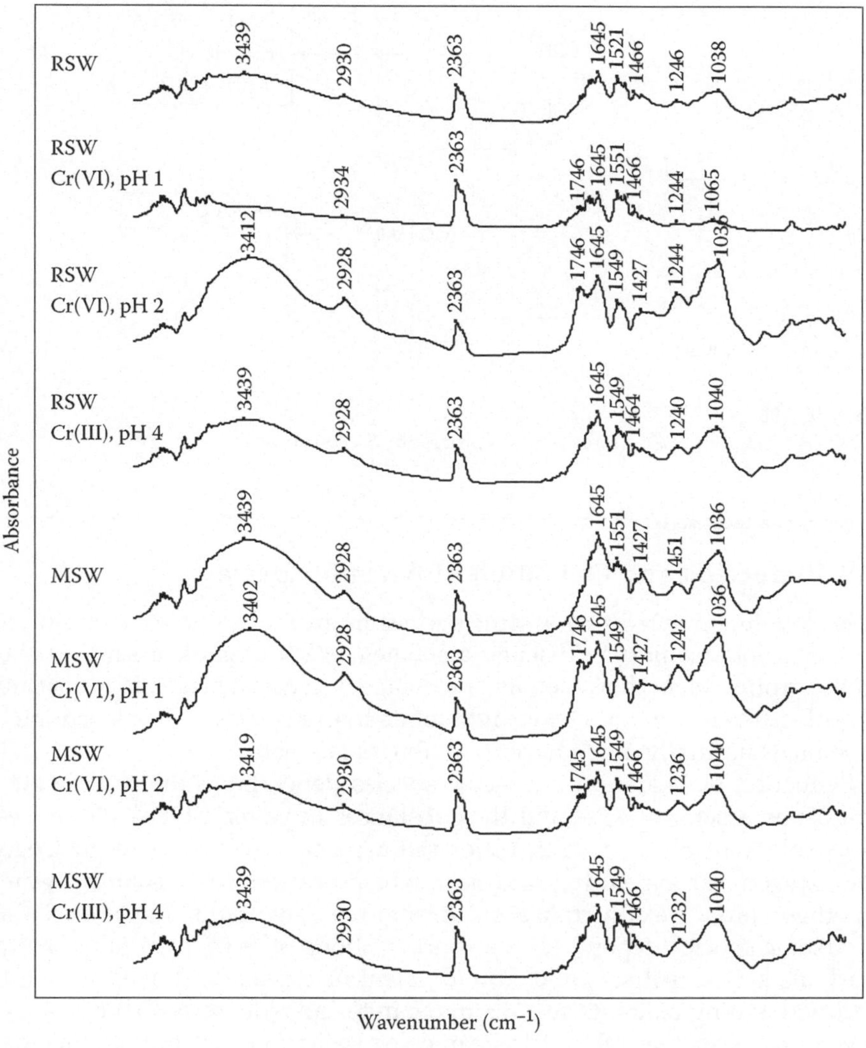

FIGURE 5.34
Fourier transform infrared spectra of RSW and MSW at the following conditions: before adsorption; after contacted with $K_2Cr_2O_7$ at pH 1.0; after contacted with $K_2Cr_2O_7$ at pH 2.0; after contacted with $Cr(NO_3)_3$ at pH 4.0.

The Cr(III) cations that formed from the redox reactions are then able to undergo ion exchange reactions with cationic alkaline metal ions (Alk in Figure 5.35) such as Ca^{2+}, which are initially bound onto the biosorbents. The experimental evidence reported in the literature has demonstrated that Cr(III) ions are accumulated onto seaweed biosorbents due to ion exchange. In addition, the carboxyl group can form metal complex due to coordination reactions.[73]

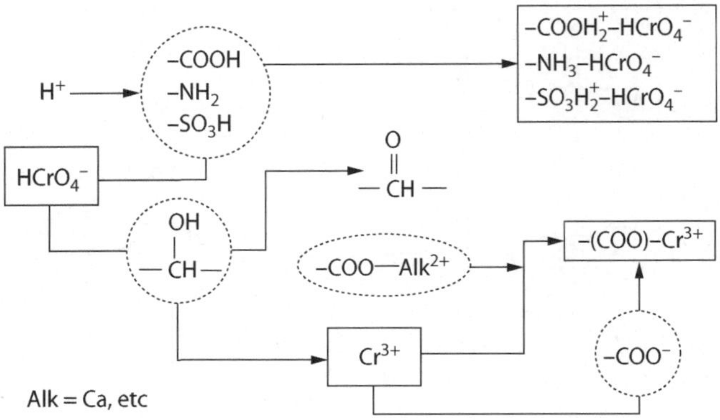

FIGURE 5.35
Schematic diagram of Cr(VI) biosorption process by *Sargassum sp.*

5.5 Reduction and Oxidation of Arsenic Species

It is well documented that arsenic species in groundwater occur mainly in the forms of arsenite (As(III)) and arsenate (As(V)). Organic arsenic may be in the groundwater; however, its percentage is generally less than inorganic arsenic. Arsenite is more toxic and mobile than arsenate. The adsorption of arsenite is normally much lower than that of arsenate.

Reduction or oxidation of arsenic species depends on the redox potential of the arsenic species and the substances in water. The whole process is greatly controlled by such major redox-sensitive elements as hydrogen ion, oxygen, carbon, sulfur, and iron, which are well distributed in water. In other words, the occurrence and strength of either reduction or oxidation of arsenic species depend on the concentrations of oxygen, organic matter (carbon), sulfur (sulfate), and iron (divalent or trivalent), as well as pH. In addition, strong oxidants such as manganite can oxidize As(III) to As(V).[78] Free radicals such as OH· radical can cause oxidation. Natural organic matter such as humic acid, on the other hand, can cause the reduction of As(V) to As(III). Figure 5.36 shows the Eh-pH stability diagram for the As-O-S system.[79]

5.5.1 Oxidation

As shown in Table 5.1, the standard potential for $H_3AsO_4/HAsO_2$ is 0.56 V. Thus As(III) may be oxidated to As(V). The oxidation rate of As(III) to As(V) can be increased by ozone, chlorine, hypochlorite, chlorine dioxide, or H_2O_2 as oxidants. The presence of manganese oxide can be reportedly used to oxidize As(III) to As(V). An advanced oxidation process such as TiO_2 may be able to cause oxidation. Some microbes may also cause the oxidation.

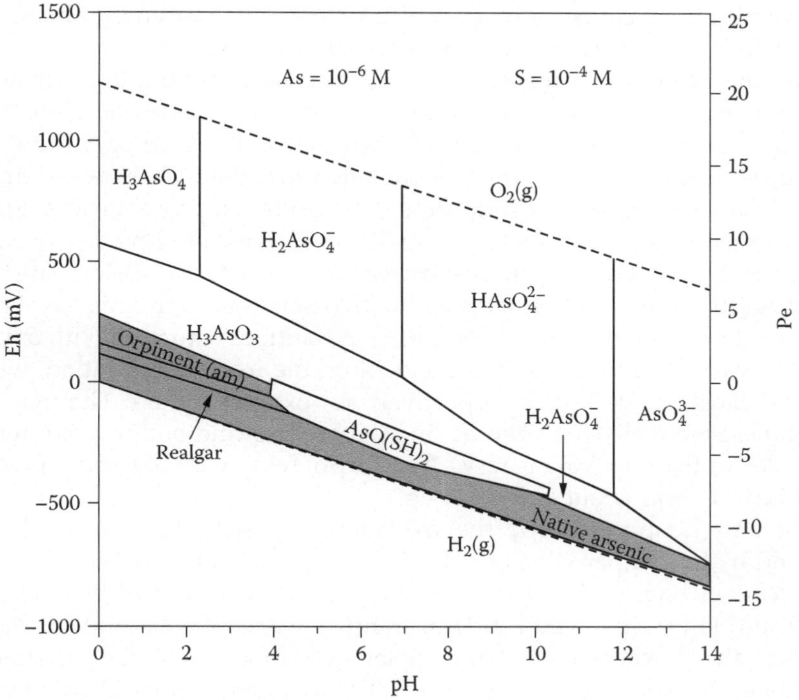

FIGURE 5.36
Eh-pH stability diagram for As-O-S system. The dashed lines indicate stability field for water. The gray area indicates a solid phase.[79]

5.5.1.1 Chemical Oxidation

Several oxidants can oxidate arsenite to arsenate, including air, ozone, chlorine, iron, and manganese compounds, H_2O_2, and Fenton's reagent. Such oxidation reactions may occur in both natural systems and treatment systems.

The oxidation of arsenite with manganese oxides in water treatment was investigated by Driehaus et al.[80] It was demonstrated that arsenite was easily oxidized by δ-modification of manganese dioxide. The kinetics study showed that the oxidation rate of As(III) into As(V) followed a second-order reaction kinetics with respect to As(III). The oxidation rate highly depends on the initial molar ratio of MnO_2 to As(III). Calcium had only a minor influence on the oxidation; the solution pH had no effect at pHs ranging from 5 to 10 with an initial molar ratio of MnO_2/As(III) of 14. It was reported that arsenite might be directly oxidized to arsenate by bacteria or react with biologically precipitated manganese oxides.

Chiu et al. reported that As(III) could be oxidized by manganite on the time scale of hours.[78] The overall conversion of As(III) to As(V) was slower at pH 6.3 than at pH 4. The presence of 200 μM phosphate (at pH 4) decreased the overall rate of conversion of As(III) to As(V). The presence of boric acid did not influence the conversion rate of As(III) to As(V).

Kim and Nriagu investigated the oxidation of arsenite into arsenate by using oxygen or ozone in groundwater samples, which contained 46–62 μg/L total dissolved arsenic (more than 70% as As(III)), 100–1130 μg/L iron, and 9–16 μg/L manganese.[81] It was demonstrated that the conversion of As(III) to As(V) became faster by using ozone. The oxidation process was sluggish with oxygen or air. The kinetics of As(III) oxidation could be described by using the modified pseudo-first-order reaction kinetics expression. The half-life of As(III) in the solution saturated with ozone was approximately 4 min. On the other hand, the half-life of As(III) in solutions saturated with oxygen and air was much longer and dependent on the iron concentration, which was 2–5 days and 4–9 days, respectively, for oxygen and air. The iron and manganese were also oxidized during the process, and both played important roles in the removal of As(V). The sorption capacity of freshly precipitated $Fe(OH)_3$ was about 15.3 mg As/g.

Hug and Leupin studied the oxidation of As(III) by dissolved oxygen and hydrogen peroxide at pH 3.5–7.5 in the presence of Fe(II, III) on a time scale of hours.[82] No oxidation of arsenite was observed by using O_2, 20–100 μM H_2O_2, dissolved Fe(III), or iron(III) hydroxides as single oxidants, respectively. However, partial or complete oxidation of arsenite occurred in parallel to the oxidation of 20–90 μM Fe(II) by oxygen and by 20 μM H_2O_2 in aerated solutions. At low pH, the addition of ·OH radical scavenger (2-propanol) quenched the As(III) oxidation. At neural pH, the addition of 2-propanol had a little influence on the oxidation of arsenite. High concentration of bicarbonate resulted in increased oxidation of arsenite. It was speculated that H_2O_2 and Fe(II) formed ·OH radicals at low pH, and an Fe(IV) species were formed at a higher pH. In addition, carbonate radicals might also be produced in the presence of bicarbonate. Jang and Dempsey reported that a significant oxidation was observed in the presence of As(V) and hydrous ferric oxide (HFO).[83]

Lee et al. reported that arsenite could be oxidized to arsenate by Fe(VI) with a stoichiometry of 3:2 [As(III):Fe(VI)].[84] The study showed that the reaction of As(III) with Fe(VI) fellowed a first-order reaction kinetics with respect to both reactants.

A manganese-loaded polystyrene matrix (R-MnO_2) was developed for the oxidation and removal of As(III).[85] The R-MnO_2 allowed the complete oxidation of As(III) in the solution even at high concentration. The oxidation and adsorption of arsenite onto the MnO_2 were involved during the removal process. The study showed that the oxidation of H_3AsO_3 by MnO_2 resulted in the formation of both $HAsO_4^{2-}$ and Mn^{2+}. A novel Fe-Mn binary oxide adsorbent was developed for effective As(III) removal by Zhang et al.[86] The results showed that As(III) were oxidized and adsorbed onto the binary oxide. The oxidation ability of the manganese oxide played an import role during the process.

The oxidation of arsenite by potassium permanganate ($KMnO_4$) was investigated under such conditions as pH, initial As(III) concentration, and dosage of Mn(VII) by Li et al.[87] The results demonstrated that potassium permanganate was an effective oxidant for the oxidation of As(III) into As(V) over a wide pH range. The performance of the oxidation of As(III) was not significantly influenced by the solution pH. The main reduction products of Mn(VII) were Mn(II) and $Mn(OH)_2$ under acidic and basic conditions, respectively. The ratio of Mn(VII) to As(III) was about 2:5 for the oxidation of As(III) to As(V). A pilot study on the enhancement in the As(III) removal was carried out with a process combined with oxidation, precipitation, and direct sand filtration as pretreatment before the ultrafiltration (UF).[88] The result showed that the pretreatment effectively facilitated As(III) removal, and residual arsenic concentration was below 10 μg/L in the effluent from the UF. Compared to chlorine, besides oxidation ability, the permanganate had positive seeding effects of in situ formed hydrous MnO_2 and the formation of larger floc. Thus, permanganate was more promising than chlorine in the treatment.

5.5.1.2 Catalytic Oxidation

As(III) can be oxidized to As(V) via catalytic oxidation. UV light is used in the processes. Photochemical oxidation of As(III) was investigated by Hug et al.[89] The water used contained 500 μg/L As(III), 0.06–5 mg/L Fe(II,III), and 4–6 mM bicarbonate at pH 6.5–8.0. It was found that dissolved oxygen and hydrogen peroxide did not oxidize As(III). In the dark, As(III) was partly oxidized after the addition of Fe(II) to the aerated water, likely due to the formation of reactive intermediates in the reduction of oxygen by Fe(II). It was observed that, under illumination with 90 W/m^2 UV-A light, over 90% of As(III) in solutions containing 0.06–5 mg/L Fe(II and III) could be oxidized photochemically within 2–3 h. The oxidation of As(III) could be strongly accelerated in the presence of citrate by forming Fe(III) citrate complexes.

The oxidation of As(III) by oxygen in the absence and in the presence of dissolved Fe(III) and illumination with near-ultraviolet light was study by Emett and Khoe.[90] It was demonstrated that the oxidation rate of As(III) to As(V) by oxygen was increased by several orders of magnitude in the presence of dissolved Fe(III) and irradiation with near-ultraviolet light. The free radicals played important roles: the rate of the initiation reaction was controlled by the rate of *photon* absorption by dissolved Fe(III)-hydroxo and Fe(III)-chloro species. The addition of arsenate or sulfate led to lower quantum efficiencies for the As(III) photooxidation process. In the absence of dissolved oxygen, two moles of Fe(III) could oxidize one mole of As(III), and the dissolved Fe(II) significantly hindered the oxidation of As(III). However, under oxic conditions, both Fe(II) and As(III) could be oxidized simultaneously, and the presence of Fe(II) and reducing solution pH increased the photon efficiency. The results demonstrated that iron compounds were a good photooxidant as ferric hydroxide was an excellent adsorbent for the resultant arsenate. The addition of ferrous salt in the presence of sunlight could be a practical method for the oxidation of As(III) in contaminated waters.

The photocatalytic oxidation of monomethylarsonic acid (MMA) and dimethylarsnic acid (DMA) using TiO_2 was studied by Xu et al.[91] The study demonstrated that MMA and DMA were readily degraded upon TiO_2 photocatalysis. DMA was oxidized to MMA as the primary oxidation product followed by oxidization to As(V). The pH of the solution affected the adsorption and photocatalytic degradation processes as the speciation of the arsenic substrates and surface charge of TiO_2 were pH dependent. The kinetics study indicated the removal of MMA and DMA by TiO_2 photocatalysis followed the Langmuir–Hinshelwood kinetics model. During the phtocatalysis, an addition of a hydroxyl radical scavenger, tert-butyl alcohol, greatly reduced the rate of degradation, which indicated that ·OH radical was the primary oxidant.

An advance oxidation process by ultraviolet light and a photo absorber (iron salts or sulfite) was studied by Zaw and Emett.[92] It was demonstrated that the iron-based photooxidation process oxidized and removed the arsenic in the mine water draining from a hard rock gold, silver, and lead mine in Montana. The results showed that the water treatment residues with and without cement were stable under 3-month testing.

The application of a sunlight-assisted process to oxidize and remove arsenic from the tube well water was studied in a village in Bangladesh. The process was simple to use for villagers in rural areas without electricity. The UV/sulfite process was preferred for use with UV as no solids were generated that might lead to the fouling of the lamps.

The presence of hydrogen peroxide could facilitate arsenite adsorption onto the ferrihydrite.[93] An attenuated total reflection-Fourier transform infrared spectroscopy study showed no oxidation of As(III) within minutes to hours in the absence of H_2O_2. In the presence of H_2O_2, the oxidation rate for adsorbed As(III) increased. The solution pH had an insignificant effect on As(III) oxidation. Iron was necessary to induce As(III) oxidation by catalytic H_2O_2 decomposition.

The TiO_2-photocatalyzed oxidation of As(III) to As(V) at neutral pH and over a range of As(III) concentrations was studied by Ferguson et al.[94] The complete oxidation of As(III) to As(V) was observed within 10 to 60 min of irradiation at 365 nm. The influence of phosphate at 0.5–10 μM on the photooxidation rate was negligible. The superoxide O_2· played an important role during the photooxidation.

Dutta et al. reported that the photocatalytic oxidation of As(III) to As(V) occurred in minutes, which followed zero-order kinetics.[95] The ·OH free radicals played an important role in the oxidation process.

The reactive intermediates produced by the corrosion of zerovalent iron (ZVI) in oxygen-containing water could be used to oxidize the arsenite to arsenate.[96] At pH 3–9, the half-life for the oxidation of 500 μg/L As(III) in the presence of 150 mg/L ZVI was 26–80 min. However, at pH 11, no oxidation of As(III) was observed in the first two hours. At pH values of 3, 5, and 7, the dissolved Fe(II) was determined as 325, 140, and 6 μM, and the peak concentration of H_2O_2 within 10 min was 1.2, 0.4, and less than 0.1 μM, respectively. The study suggested that oxidation of As(III) mainly occurred in solution by Fenton reaction, and was subsequently removed by sorption onto the freshly formed hydrous ferric oxides. The OH· radials were identified as the main oxidant at low pH.

The photocatalytic oxidation of arsenite and simultaneous removal of the resultant arsenate from aqueous solution using a municipal solid waste melted slag containing iron oxide and TiO_2 in the presence of UV light were investigated by Zhang and Itoh.[86] The oxidation of arsenite was rapid (3 h); the adsorption of the arsenate from the oxidation was slow (within 10 h). The arsenite could be oxidized to arsenate only by UV light at a slower rate (approximately 1/3 of that of the photocatalyzed reaction). Both alkaline and acidic conditions facilitated the oxidation reaction, and the optimum pH for the oxidation and adsorption was around 3.

The photocatalyzed oxidation in an upflow-through fixed-bed reactor with TiO_2-coated glass beads as packing materials was examined by Ferguson and Hering.[94] The beads could be repeatedly used for the As(III) oxidation. The competitively adsorbing anions, $NaNO_3$, did not significantly affect the catalyst activity.

Peroxydisulfate ions ($S_2O_8^{2-}$, KPS) were used as an oxidizing agent for the photochemical oxidation of As(III) to As(V) under UV light irradiation.[97] The rate of photochemical oxidation for As(III) by $S_2O_8^{2-}$ was exceptionally high. The UV light intensity was of primary importance for the dissociation of $S_2O_8^{2-}$ in producing sulfate anion radicals, which favored a higher reaction rate. The variation of pH from 3 to 9 had no influence on the reaction. However, the reaction rate was reduced by 20% by continuous purging of nitrogen, which indicated the key role of dissolved oxygen in the reaction. The presence of humic acid (up to 20 ppm) had no detrimental effect on the oxidation reaction.

Yoon et al. studied the usage of vacuum-UV (VUV) lamp (185 and 254 nm lights) for the oxidation of As(III).[98] The VUV lamp showed a higher performance for As(III) oxidation than other photochemical oxidation methods (e.g., UV-C/H_2O_2, UV-A/Fe(III)/H_2O_2, and UV-A/TiO_2). The presence of Fe(III) and H_2O_2 increased the As(III) oxidation efficiency; humic acid, however, did not have a significant effect on the reaction.

The photooxidation of As(III) under 254 nm irradiation in the presence of potassium iodide (100 μM), was studied.[99] The presence of iodide dramatically enhanced the oxidation rate, and the quantum yields of As(III) photooxidation ranged from 0.08 to 0.6, which depended on the concentrations of As(III) and iodide. The air- or N_2O-satuated solution enhanced the photooxidation of As(III). On the other hand, the photooxidation rate was markedly reduced in the N_2-saturated solution. The study suggested that the excitation of iodides under 254 nm irradiation resulted in the generation of iodine atoms and triiodides, which were involved in the oxidation process of As(III).

5.5.1.3 Biological Oxidation

Biological oxidation of As(III) to As(V) may occur in natural/contaminated subsurfaces systems. The microbes may directly and indirectly cause the oxidation. A new heterotrophic bacterial strain, ULPAs1, was isolated from arsenic-contaminated water.[100] The ULPAs1 showed a rapid and extensive oxidation of As(III) into As(V). The growth of the arsenite-oxidizing bacterium (ULPAs1) was independent of the presence of As(III) (1.33 mM) in the environment containing lactate as the sole organic carbon source. However, no growth took place in the absence of organic carbon source or in a rich medium (i.e., Luria–Bertani). The doubling time of the ULPAs1 was 1.5 h. The minimum inhibitory concentration of arsenic for the strain was 6.65 mM. The strain was demonstrated to be very effective for oxidation of arsenic. 16SrDNA sequence analysis showed that the strain belonged to β-Proteobacteria. This strain may be a good candidate for arsenic remediation in heavily polluted water.

The cultivation and application of arsenic-oxidizing bacteria (ULPAs1) for arsenic oxidation were further studied by Lievremont et al.[101] The strain was cultivated in batch reactors in the presence of chabazite and kutnahorite, which were used as microorganism-immobilizing materials. The results showed that the arsenite oxidative properties of ULPAs1 were maintained when cultivated in the presence of quartz or chabazite. It was found that the induced ULPAs1 oxidized As(III) in 2 days in the presence of chabazite, and As(V) was observed in the aqueous phase after 4 h. However, the oxidation rate of arsenite with the noninduced ULPAs1 was slower.

Two bacteria strains were isolated from acid waters originating from Carnoules mine tailings, and identified as *Thomonas* sp. by Lenoble et al.[102] The acid water contained high dissolved concentrations of arsenic and iron. It was found that the arsenic was rapidly precipitated with Fe(III). The precipitation rate was related to the oxidation of iron and enhanced by iron-oxidizing bacteria. Rapid arsenic oxidation was observed in the acid water because of the activity of arsenic-oxidizing bacteria.

Katsoyiannis et al. reported that indigenous iron- and manganese-oxidizing bacteria in groundwater could catalyze the oxidation of dissolved manganese (Mn(II)) to insoluble hydrous manganese oxide, subsequently removed by the filtration.[103] The process led to the formation of a natural coating on the surface of the filter medium. If arsenic was simultaneously present in the groundwater, it could subsequently be removed by the sorption onto manganese oxide. Rapid oxidation of As(III) to As(V) was observed prior to removal by sorption onto biogenic manganese oxide. The rate of As(III) oxidation was found to be significantly higher than the rates reported for abiotic As(III) oxidation by manganese oxides, indicating an important role for bacteria in both the oxidation of As(III) and the generation of reactive manganese oxide surfaces for the removal of As(III) and As(V) from the solution. The presence of phosphates at concentrations of around 600 μg/L did not affect the oxidation of As(III). It had an adverse influence on As(III) removal, which decreased the overall removal efficiency by 50%, though it did not affect the oxidation of As(III).

A bacterial strain, B2, was isolated from the biofilm growing in a biological groundwater treatment system used for Fe removal by Casiot et al.[104] It was demonstrated that bacteria strain could oxidize arsenite into arsenate. This strain was different from the genus *Leptothrix* commonly encountered in the biological iron oxidation processes. The study revealed that this isolated strain, B2, was the major constituent of the bacterial community in the biofilm. It was hypothesized that its presence contributed to arsenic oxidation.

The mineral-leaching archaebacterium *Sulfolobus acidocaldarius* strain BC has the ability to oxidize arsenite to arsenate.[105] The arsenite oxidation activity was eightfold higher for cells grown in the presence of arsenite when compared with cells grown without arsenite. The oxidation was sensitive to proteinase K and reached the highest activity level under acidic conditions.

5.5.2 Reduction

Reduction reactions occur when lakes, sediments, and aquifers become anaerobic. In the subsurface system, the decomposition of natural organic matter through a series of microbial activities would cause reduction and consumption of dissolved oxygen gas and production of carbon dioxide. Similar to the denitrification in well-known biological wastewater treatment process, the nitrate in these waters would be reduced to nitrite, N_2O, and N_2, leading to a decrease in the concentration of nitrate.

Insoluble manganese oxides and hydrous ferric oxides may be dissolved through reduction reactions to form Mn^{2+} and Fe^{2+}, respectively. These processes are followed by the reduction of SO_4^{2-} to S^{2-}; CH_4 can be produced through fermentation and methanogenesis; reduction of N_2 to NH_4^+ may occur if there are sufficient electrons available. The S^{2-} due to SO_4^{2-} reduction would react with the well-available Fe^{2+} to form FeS and ultimately FeS_2 (pyrite). It was reported that arsenic(V) reduction would occur after Fe(III) reduction but before SO_4^{2-} reduction.[79]

Three probable mechanisms were proposed to understand arsenic mobility in aquifers.[107] The release and mobilization of arsenic may be due to the oxidation of arsenic-bearing pyrite minerals. The dissolution of arsenic-contaminated iron oxy-hydroxides (FeOOH) due to the onset of reducing conditions in the subsurface may cause the mobilization of arsenic. Lastly, the phosphate ions that migrate into aquifers due to applications of fertilizers on soil could cause the release of arsenic that is originally adsorbed onto the soil because of the higher affinity of phosphate for the soil.

It was pointed out that the third mechanism could be the control step for the mobilization of arsenic in the aquifers.

The reduction of iron oxy-hydroxide to iron (II) may cause the reduction of As(V) to As(III), which results in the mobilization of arsenic according to the reactions as follows.[107] More than 50% of reduction in As(V) concentration occurred within 10 days in the presence of Fe(II); the concentration of arsenite increased. The redox reactions of iron and arsenic could occur in both biotic and abiotic pathways.

As natural organic matter is a prevalent constituent of natural waters, it is composed of various elements. The most important element is carbon, which exists in various valencies. The NOM is reactive toward both metals (e.g., arsenic and iron) and surfaces of metal oxides (e.g., iron oxides) and would greatly influence arsenic mobility.[107,108]

Tongesayi and Smart studied the reduction of As(V) with Suwannee River fulvic acid (FA).[109] It was demonstrated that the FA significantly reduced As(V) to As(III), which depended on solution pH. The reduction of As(V) to As(III) occurred under dark or light conditions. The presence of Fe(III) increased the reduction. The extent in the oxidation of As(III) to As(V) was greater at pH 2 than at pH 6. The reduction of As(V) by FA was hypothesized to occur via complexation, which was greatly enhanced by the presence of iron through intermetallic bridging. The reaction kinetics order could be two or higher.

Palmer et al. found that inorganic arsenates were reduced to arsenite by homogeneous aqueous solutions of humic and fulvic acids.[110] The fraction of arsenate initially reduced increased with an increase in humic concentration. The reoxidation of As(III) in the humic solutions order could be achieved by extended bubbling with air.

Some microbic species are reportely able to reduce As(V) to As(III). Several batch and column tests showed that As(V) was reduced to As(III) by Shewanella sp., with the reduced arsenic species subsequently removed by precipitation.[106] The precipitation reaction could be a major sink during microbially mediated arsenic reactive transport. The reduction of As(V) to As(III) may be due to the following reaction, where the hydrogen gas is produced by the microorganism:

$$H_2AsO_4^- + H^+ + H_2 = H_3AsO_3 + H_2O \qquad (5.33)$$

References

1. Gomezlahoz, C., Garciaherruzo, F., Rodriguezmaroto, J. M., and Rodriguez, J. J. Heavy-metal removal by chemical-reduction with sodium-borohydride—a pilot-plant study. *Separ. Sci. Technol.*, 27(12), 1569–1582, 1992.

2. Fanning, J. C., Brooks, B. C., Hoeglund, A. B., Pelletier, D. A., and Wadford, J. A. The reduction of nitrate and nitrite ions in basic solution with sodium borohydride in the presence of copper(II) ions. *Inorg. Chim. Acta,* 310(1), 115–119, 2000.
3. Simpson, D. K. Hydrazine—a powerful metal reductant. *Metal Finish.,* 83(4), 57–60, 1985.
4. Degen, A., and Macek, J. Preparation of submicrometer nickel powders by the reduction from nonaqueous media. *Nanostruct. Mater.,* 12(1–4), 225–228, 1999.
5. Ducampsanguesa, C., Herreraurbina, R., and Figlarz, M. Fine palladium powders of uniform particle-size and shape produced in ethylene-glycol. *Solid State Ionics,* 63–65, 25–30, 1993.
6. Nickel, U., Castell, A. Z., Poppl, K., and Schneider, S. A silver colloid produced by reduction with hydrazine as support for highly sensitive surface-enhanced Raman spectroscopy. *Langmuir,* 16(23), 9087–9091, 2000.
7. Tobe, M. L., and Burgess, J. *Inorganic Reaction Mechanisms.* Prentice Hall, NJ, 1999.
8. Audrieth, I. F., and Ogg, B. A. *The Chemistry of Hydrazine.* John Wiley & Sons, New York, 1951.
9. Chen, J. P., and Lim, L. L. Key factors in chemical reduction by hydrazine for recovery of precious metals. *Chemosphere,* 49(4), 363–370, 2002.
10. Schecher, W. D., and McAvoy, D. C. MINEQL+ chemical equilibrium modeling system, Version 4.5 for Windows. Environmental Research Software. Hallowell, ME, 2001.
11. Lim, V. W. L., Kang, E. T., and Neoh, K. G. Electroless plating of palladium and copper on polypyrrole films. *Synthetic Met.,* 123(1), 107–115, 2001.
12. Rittmann, B. E. and McCarty, P. L. *Environmental Biotechnology Principles and Application,* McGraw Hill, New York, 2001.
13. Haynes, W.M.M., CRC Handbook of Chemistry and Physics, 92nd Edition, Boca Raton, FL, Taylor, and Francis, 2012.
14. Lloys, T. *Formaldehyde on Trial.* James Lorimer & Company, Toronto, 1983.
15. El-Shazly, A. H., Mubarak, A. A., and Konsowa, A. H. Hexavalent chromium reduction using a fixed bed of scrap bearing iron spheres. *Desalination,* 185(1–3), 307–316, 2005.
16. Yurkow, E. J., Hong, J., Min, S., and Wang, S. Photochemical reduction of hexavalent chromium in glycerol-containing solutions. *Environ. Pollut.,* 117(1), 1–3, 2002.
17. Xu, X. R., Li, H. B., Li, X. Y., and Gu, J. D. Reduction of hexavalent chromium by ascorbic acid in aqueous solutions. *Chemosphere,* 57(7), 609–613, 2004.
18. Kaksonen, A. H., Franzmann, P. D., and Puhakka, J. A. Effects of hydraulic retention time and sulfide toxicity on ethanol and acetate oxidation in sulfate-reducing metal-precipitating fluidized-bed reactor. *Biotechnol. Bioeng.,* 86(3), 332–343, 2004.
19. Nevatalo, L. M., Makinen, A. E., Kaksonen, A. H., and Puhakka, J. A. Biological hydrogen sulfide production in an ethanol-lactate fed fluidized-bed bioreactor. *Bioresource Technol.,* 101(1), 276–284, 2010.
20. Bijmans, M. F. M., Peeters, T. W. T., Lens, P. N. L., and Buisman, C. J. N. High rate sulfate reduction at pH 6 in a pH-auxostat submerged membrane bioreactor fed with formate. *Water Res.,* 42(10–11), 2439–2448, 2008.
21. Vallero, M. V. G., Lettinga, G., and Lens, P. N. L. High rate sulfate reduction in a submerged anaerobic membrane bioreactor(SAMBaR) at high salinity. *J. Mem. Sci.,* 253(1–2), 217–232, 2005.

22. Vanhouten, R. T., Pol, L. W. H., and Lettinga, G. Biological sulfate reduction using gas-lift reactors fed with hydrogen and carbon-dioxide as energy and carbon source. *Biotechnol. Bioeng.*, 44(5), 586–594, 1994.
23. vanHouten, R. T., vanAelst, A. C., and Lettinga, G. Aggregation of sulphate-reducing bacteria and homo-acetogenic bacteria in a lab-scale gas-lift reactor. *Water Sci. Technol.*, 32(8), 85–90, 1995.
24. de Smul, A., Goethals, L., and Verstraete, W. Effect of COD to sulphate ratio and temperature in expanded-granular-sludge-blanket reactors for sulphate reduction. *Process Biochem.*, 34(4), 407–416, 1999.
25. Zamzow, K., Tsukamoto, T., and Miller, G. Waste from biodiesel manufacturing as an inexpensive carbon source for bioreactors treating acid mine drainage. *Mine Water Environ.*, 25, 163–170, 2006.
26. Costa, M. C., Santos, E. S., Barros, R. J., Pires, C., and Martins, M. Wine wastes as carbon source for biological treatment of acid mine drainage. *Chemosphere*, 75(6), 831–836, 2009.
27. Annachhatre, A. P., and Suktrakoolvait, S. Biological sulfate reduction using molasses as a carbon source. *Water Environ. Res.*, 73(1), 118–126, 2001.
28. Annachhatre, A. P., and Suktrakoolvait, S. Biological sulfide oxidation in a fluidized bed reactor. *Environ. Technol.*, 22(6), 661–672, 2001.
29. Maree, J. P., Gerber, A., and Hill, E. An integrated process for biological treatment of sulfate-containing industrial effluents. *J. Water Pollut. Control Fed.*, 59(12), 1069–1074, 1987.
30. Lo, K. V., Chen, A., and Liao, P. H. Anaerobic treatment of Baker's yeast wastewater. II. Sulfate removal. *Biomass*, 23, 25–37, 1990.
31. Maree, J. P., and Strydom, W. F. Biological sulfate removal in an upflow packed-bed reactor. *Water Res.*, 19(9), 1101–1106, 1985.
32. Maree, J. P., Hulse, G., Dods, D., and Schutte, C. E. Pilot-plant studies on biological sulfate removal from industrial effluent. *Water Sci. Technol.*, 23(7–9), 1293–1300, 1991.
33. du Preez, L. A., and Maree, J. P. Pilot-scale biological sulphate and nitrate removal utilizing producer gas as energy source. *Water Sci. Technol.*, 30, 275–285, 1995.
34. Stucki, G., Hanselmann, K. W., and Hurzeler, R. A. Biological sulfuric-acid transformation-reactor design and process optimization. *Biotechnol. Bioeng.*, 41(3), 303–315, 1993.
35. Vallero, M. V. G., Pol, L. W. H., Lettinga, G., and Lens, P. N. L. Effect of NaCl on thermophilic(55 degrees C) methanol degradation in sulfate reducing granular sludge reactors. *Water Res.*, 37(10), 2269–2280, 2003.
36. Hammack, R. W., Edenborn, H. M., and Dvorak, D. H. Treatment of water from an open-pit copper mine using biogenic sulfide and limestone—a feasibility study. *Water Res.*, 28(11), 2321–2329, 1994.
37. Polprasert, C., and Haas, C. N. Effect of sulfate on anaerobic processes fed with dual substrates. *Water Sci. Technol.*, 31(9), 101–107, 1995.
38. Henry, J. G., and Prasad, D. Anaerobic treatment of landfill leachate by sulfate reduction. *Water Sci. Technol.*, 41(3), 239–246, 2000.
39. Maree, J. P., and Strydom, W. F. Biological sulfate removal from industrial effluent in an upflow packed-bed reactor. *Water Res.*, 21(2), 141–146, 1987.
40. Whiteley, C. G., Enongene, G., Pletschke, B. I., Rose, P., and Whittington-Jones, K. Co-digestion of primary sewage sludge and industrial wastewater under

anaerobic sulphate reducing conditions: Enzymatic profiles in a recycling sludge bed reactor. *Water Sci. Technol.*, 48(4), 129–138, 2003.
41. Tsukamoto, T. K., Killion, H. A., and Miller, G. C. Column experiments for microbiological treatment of acid mine drainage: Low-temperature, low-pH and matrix investigations. *Water Res.*, 38(6), 1405–1418, 2004.
42. Zagury, G. J., Kulnieks, V. I., and Neculita, C. M. Characterization and reactivity assessment of organic substrates for sulphate-reducing bacteria in acid mine drainage treatment. *Chemosphere*, 64(6), 944–954, 2006.
43. Drury, W. J. Treatment of acid mine drainage with anaerobic solid-substrate reactors. *Water Environ. Res.*, 71(6), 1244–1250, 1999.
44. Lakaniemi, A. M., Nevatalo, L. M., Kaksonen, A. H., and Puhakka, J. A. Mine wastewater treatment using *Phalaris arundinacea* plant material hydrolysate as substrate for sulfate-reducing bioreactor. *Bioresource Technol.*, 101(11), 3931–3939, 2010.
45. Wakeman, K. D., Erving, L., Riekkola-Vanhanen, M. L., and Puhakka, J. A. Silage supports sulfate reduction in the treatment of metals- and sulfate-containing waste waters. *Water Res.*, 44(17), 4932–4939, 2010.
46. Speece, R. E. *Anaerobic Biotechnology for Industrial Wastewaters*. Arachae Press, TN, 1996.
47. Annachhatre, A. P., and Liamleam, W. Electron donors for biological sulfate reduction. *Biotechnol. Adv.*, 25(5), 452–463, 2007.
48. Thauer, R. K., Jungermann, K., and Decker, K. Energy-conservation in chemotropic anaerobic bacteria. *Bacteriol. Rev.*, 41(1), 100–180, 1977.
49. DeSmul, A., Dries, J., Goethals, L., Grootaerd, H., and Verstraete, W. High rates of microbial sulphate reduction in a mesophilic ethanol-fed expanded-granular-sludge-blanket reactor. *Appl. Microbiol. Biot.*, 48(3), 297–303, 1997.
50. vanHouten, R. T., Yun, S. Y., and Lettinga, G. Thermophilic sulphate and sulphite reduction in lab-scale gas-lift reactors using H-2 and CO_2 as energy and carbon source. *Biotechnol. Bioeng.*, 55(5), 807–814, 1997.
51. Widdel, F. Microbiology and ecology of sulfate-and sulfur-reducing bacteria. In *Biology of Anaerobic Micororganisms*. John Wiley & Sons, New York, 1988.
52. Gibert, O., de Pable, J., Cortina, J. L., and Ayora, C. Treatment of acid mine drainage by sulphate-reducing bacteria using permeable reactive barriers: A review from laboratory to full-scale experiments. *Rev. Environ. Sci. Biotechnol.*, 1, 327–333, 2003.
53. Klemps, R., Cypionka, H., Widdel, F., and Pfennig, N. Growth with hydrogen, and further physiological-characteristics of *Desulfotomaculum* species. *Arch. Microbiol.*, 143(2), 203–208, 1985.
54. Vanbroekhoven, K., Van Roy, S., Diels, L., Gemoets, J., Verkaeren, P., Zeuwts, L., Feyaerts, K., and van den Broeck, F. Sustainable approach for the immobilization of metals in the saturated zone: In situ bioprecipitation. *Hydrometallurgy*, 94(1–4), 110–115, 2008.
55. Buisman, C. J. N., and Jarich, N. Process for the Treatment of Waste Water Containing Heavy Metals. US Patent: US 6630071 B1, 2003.
56. Joel, B. Occurrences, uses, and properties of chromium. *Regul. Toxicol. Pharm.*, 26, 3–7, 1997.
57. Susan, P. F., and Dourson, M. L. Hexavalent chromium-contaminated soils: Options for risk assessment and risk management. *Regul. Toxicol. Pharm*, 25, 43–59, 1997.

58. Gode, F., and Pehlivan, E. A comparative study of two chelating ion-exchange resins for the removal of chromium(III) from aqueous solution. *J. Hazard. Mater.,* 100(1–3), 231–243, 2003.
59. Lazaridis, N. K., and Asouhidou, D. D. Kinetics of sorptive removal of chromium(VI) from aqueous solutions by calcined Mg-Al-CO_3 hydrotalcite. *Water Res.,* 37(12), 2875–2882, 2003.
60. Barth, E. F., Ettinger, M. G., and McDermott, G. N. Summary report on the effects of heavy metals on the biological treatment processes. *J. Water Pollut. Control Fed.,* 37, 86–96, 1965.
61. Gokcay, C. F., and Yetis, U. Effect of chromium(Vi) on activated-sludge. *Water Res.,* 25(1), 65–73, 1991.
62. Dilek, F. B., Gokcay, C. F., and Yetis, U. Combined effects of Ni(II) and Cr(VI) on activated sludge. *Water Res.,* 32(2), 303–312, 1998.
63. Chen, J. M., and Hao, O. J. Microbial chromium(VI) reduction. *Crit. Rev. Env. Sci. Tec.,* 28(3), 219–251, 1998.
64. Caravelli, A. H., Giannuzzi, L., and Zaritzky, N. E. Reduction of hexavalent chromium by *Sphaerotilus natans* a filamentous micro-organism present in activated sludges. *J. Hazard. Mater.,* 156(1–3), 214–222, 2008.
65. Stasinakis, A. S., Thomaidis, N. S., Mamais, D., Papanikolaou, E. C., Tsakon, A., and Lekkas, T. D. Effects of chromium(VI) addition on the activated sludge process. *Water Res.,* 37(9), 2140–2148, 2003.
66. Stasinakis, A. S., Thomaidis, N. S., Mamais, D., and Lekkas, T. D. Investigation of Cr(VI) reduction in continuous-flow activated sludge systems. *Chemosphere,* 57(9), 1069–1077, 2004.
67. Chirwa, E. N., and Wang, Y. T. Simultaneous chromium(VI) reduction and phenol degradation in an anaerobic consortium of bacteria. *Water Res.,* 34(8), 2376–2384, 2000.
68. Al-Malack, M. H. Performance of an immersed membrane bioreactor (IMBR). *Desalination,* 214(1–3), 112–127, 2007.
69. Zhao, W., Zheng, Y. M., Zou, S. W., Ting, Y. P., and Chen, J. P. Effect of hexavalent chromium on performance of membrane bioreactor in wastewater treatment. *J. Environ. Eng-Asce,* 135(9), 796–805, 2009.
70. Stephenson, T., Judd, S., Jefeerson, B., and Brindle, K. *Membrane bioreactors for wastewater treatment*. IWA, U.K., 2000.
71. Canales, A., Pareilleux, A., Rols, J. L., Goma, G., and Huyard, A. Decreased sludge production strategy for domestic waste-water treatment. *Water Sci. Technol.,* 30(8), 97–106, 1994.
72. Muller, E. B., Stouthamer, A. H., Vanverseveld, H. W., and Eikelboom, D. H. Aerobic domestic waste-water treatment in a pilot-plant with complete sludge retention by cross-flow filtration. *Water Res.,* 29(4), 1179–1189, 1995.
73. Sheng, P. X., Tan, L. H., Chen, J. P., and Ting, Y. P. Biosorption performance of two brown marine algae for removal of chromium and cadmium. *J. Dispers. Sci. Tech.,* 25(5), 679–686, 2005.
74. Yang, L., and Chen, J. P. Biosorption of hexavalent chromium onto raw and chemically modified Sargassum sp. *Bioresource Technol.,* 99(2), 297–307, 2008.
75. Moulder, J. F., Stickle, W. F., Sobol, P. E., and Bomben, K. D. *Handbook of X-ray Photoelectron Spectroscopy*. Perkin-Elmer Corp., MN, 1992.
76. Clothup, N. B., Daly, L. H., and Wiberley, S. E. *Introduction to Infrared and Raman Spectroscopy,* 3rd ed. Academic Press, London, 1990.

77. Sheng, P. X., Ting, Y. P., Chen, J. P., and Hong, L. Sorption of lead, copper, cadmium, zinc, and nickel by marine algal biomass: Characterization of biosorptive capacity and investigation of mechanisms. *J. Colloid Interface Sci.*, 275(1), 131–141, 2004.
78. Chiu, V. Q., and Hering, J. G. Arsenic adsorption and oxidation at manganite surfaces. 1. Method for simultaneous determination of adsorbed and dissolved arsenic species. *Environ. Sci. Technol.*, 34(10), 2029–2034, 2000.
79. Wang, Y. X., and Deng, Y. M. Environmental geochemistry of high-arsenic aquifer systems Chapter 5. In *Heavy Metals in the Environment*. Wang, L. K., Chen, J. P., Hung, Y. T., and Shammas, N. K. eds. CRC Press, Boca Raton, FL, 2009.
80. Driehaus, W., Seith, R., and Jekel, M. Oxidation of arsenate(Iii) with manganese oxides in water-treatment. *Water Research*, 29(1), 297–305, 1995.
81. Kim, M. J., and Nriagu, J. Oxidation of arsenite in groundwater using ozone and oxygen. *Sci. Total Environ.*, 247(1), 71–79, 2000.
82. Hug, S. J., and Leupin, O. Iron-catalyzed oxidation of arsenic(III) by oxygen and by hydrogen peroxide: pH-dependent formation of oxidants in the Fenton reaction. *Environ. Sci. Technol.*, 37(12), 2734–2742, 2003.
83. Jang, J. H., and Dempsey, B. A. Coadsorption of arsenic(III) and arsenic(V) onto hydrous ferric oxide: Effects on abiotic oxidation of arsenic(III), extraction efficiency, and model accuracy. *Environ. Sci. Technol.*, 42(8), 2893–2898, 2008.
84. Lee, Y., Um, I. H., and Yoon, J. Arsenic(III) oxidation by iron(VI)(ferrate) and subsequent removal of arsenic(V) by iron(III) coagulation. *Environ. Sci. Technol.*, 37(24), 5750–5756, 2003.
85. Lenoble, W., Laclautre, C., Serpaud, B., Deluchat, V., and Bollinger, J. C. As(V) retention and As(III) simultaneous oxidation and removal on a MnO_2-loaded polystyrene resin. *Sci. Total Environ.*, 326(1–3), 197–207, 2004.
86. Zhang, F. S., and Itoh, H. Photocatalytic oxidation and removal of arsenite from water using slag-iron oxide-TiO_2 adsorbent. *Chemosphere*, 65(1), 125–131, 2006.
87. Li, N., Fan, M. H., van Leeuwen, J., Saha, B., Yang, H. Q., and Huang, C. P. Oxidation of As(III) by potassium permanganate. *J. Environ. Sci-China*, 19(7), 783–786, 2007.
88. Li, G. B., Sun, L. H., Liu, R. P., Xia, S. J., and Yang, Y. L. Enhanced As(III) removal with permanganate oxidation, ferric chloride precipitation and sand filtration as pretreatment of ultrafiltration. *Desalination*, 243(1–3), 122–131, 2009.
89. Hug, S. J., Canonica, L., Wegelin, M., Gechter, D., and Von Gunten, U. Solar oxidation and removal of arsenic at circumneutral pH in iron containing waters. *Environ. Sci. Technol.*, 35(10), 2114–2121, 2001.
90. Emett, M. T., and Khoe, G. H. Photochemical oxidation of arsenic by oxygen and iron in acidic solutions. *Water Res.*, 35(3), 649–656, 2001.
91. O'Shea, K. E., Xu, T. L., and Cai, Y. Adsorption and photocatalyzed oxidation of methylated arsenic species in TiO_2 suspensions. *Environ. Sci. Technol.*, 41(15), 5471–5477, 2007.
92. Zaw, M., and Emett, M. T. Arsenic removal from water using advanced oxidation processes. *Toxicol. Lett.*, 133(1), 113–118, 2002.
93. Voegelin, A., and Hug, S. J. Catalyzed oxidation of arsenic(III) by hydrogen peroxide on the surface of ferrihydrite: An in situ ATR-FTIR study. *Environ. Sci. Technol.*, 37(5), 972–978, 2003.

94. Ferguson, M. A., Hoffmann, M. R., and Hering, J. G., TiO_2-photocatalyzed As(III) oxidation in aqueous suspensions: Reaction kinetics and effects of adsorption. *Environ. Sci. Technol.*, 39(6), 1880–1886, 2005.
95. Dutta, P. K., Pehkonen, S. O., Sharma, V. K., and Ray, A. K. Photocatalytic oxidation of arsenic(III): Evidence of hydroxyl radicals. *Environ. Sci. Technol.*, 39(6), 1827–1834, 2005.
96. Katsoyiannis, I. A., Ruettimann, T., and Hug, S. J. pH dependence of Fenton reagent generation and As(III) oxidation and removal by corrosion of zero valent iron in aerated water. *Environ. Sci. Technol.*, 42(19), 7424–7430, 2008.
97. Choi, H., Neppolian, B., and Celik, E. Photochemical oxidation of arsenic(III) to arsenic(V) using peroxydisulfate ions as an oxidizing agent. *Environ. Sci. Technol.*, 42(16), 6179–6184, 2008.
98. Yoon, S. H., Lee, J. H., Oh, S., and Yang, J. E. Photochemical oxidation of As(III) by vacuum-UV lamp irradiation. *Water Res.*, 42(13), 3455–3463, 2008.
99. Choi, W., and Yeo, J. Iodide-mediated photooxidation of arsenite under 254 nm irradiation. *Environ. Sci. Technol.*, 43(10), 3784–3788, 2009.
100. Weeger, W., Lievremont, D., Perret, M., Lagarde, F., Hubert, J. C., Leroy, M., and Lett, M. C. Oxidation of arsenite to arsenate by a bacterium isolated from an aquatic environment. *Biometals,* 12(2), 141–149, 1999.
101. Lievremont, D., N'negue, M. A., Behra, P., and Lett, M. C. Biological oxidation of arsenite: Batch reactor experiments in presence of kutnahorite and chabazite. *Chemosphere,* 51(5), 419–428, 2003.
102. Casiot, C., Morin, G., Juillot, F., Bruneel, O., Personne, J. C., Leblanc, M., Duquesne, K., Bonnefoy, V., and Elbaz-Poulichet, F. Bacterial immobilization and oxidation of arsenic in acid mine drainage (Carnoules Creek, France). *Water Res.*, 37(12), 2929–2936, 2003.
103. Katsoyiannis, I. A., Zouboulis, A. I., and Jekel, M. Kinetics of bacterial As(III) oxidation and subsequent As(V) removal by sorption onto biogenic manganese oxides during groundwater treatment. *Ind. Eng. Chem. Res.*, 43(2), 486–493, 2004.
104. Casiot, C., Pedron, V., Bruneel, O., Duran, R., Personne, J. C., Grapin, G., Drakides, C., and Elbaz-Poulichet, F. A new bacterial strain mediating as oxidation in the Fe-rich biofilm naturally growing in a groundwater Fe treatment pilot unit. *Chemosphere,* 64(3), 492–496, 2006.
105. Sehlin, H. M., and Lindstrom, E. B. Oxidation and reduction of arsenic by *Sulfolobus acidocaldarius* strain Bc. *Fems Microbiol. Lett.*, 93(1), 87–92, 1992.
106. Lim, M. S., Yeo, I. W., Prabhakar Clement, T., Roh, Y., and Lee, K. K. Mathematical model for predicting microbial reduction and transport of arsenic in groundwater systems. *Water Res.*, 41(10), 2079–2088, 2007.
107. Bose, P., and Sharma, A. Role of iron in controlling speciation and mobilization of arsenic in subsurface environment. *Water Res.*, 36(19), 4916–4926, 2002.
108. Redman, A. D., Macalady, D. L., and Ahmann, D. Natural organic matter affects arsenic speciation and sorption onto hematite. *Environ. Sci. Technol.*, 36(13), 2889–2896, 2002.
109. Tongesayi, T., and Smart, R. B. Arsenic speciation: Reduction of arsenic(v) to arsenic(III) by fulvic acid. *Environ. Chem.*, 3(2), 137–141, 2006.
110. von Wandruszka, R., Palmer, N. E., and Freudenthal, J. H. Reduction of arsenates by humic materials. *Environ. Chem.*, 3(2), 131–136, 2006.

6

Electrochemical Technologies for Heavy Metal Decontamination

6.1 Introduction

Electrochemical treatment techniques are becoming alternative wastewater treatment methods because such toxic wastewaters, as heavy-metal-containing wastewater, which are not easily biodegradable and require costly physical or physicochemical pretreatments. It has been demonstrated that electrochemical treatment technologies can be effectively used for heavy metal removal.[1–5] They can work directly or indirectly to achieve the objective of decontamination. There are at least two electrodes involved in the electrochemical treatment, namely, the anode and cathode. When a direct current is applied to the electrodes, oxidation and reduction reactions occur at the anode and cathode, respectively.

When a solution contains heavy metal ions, it may be reduced to its elemental form on the surface of cathode because of chemical reduction. The reduced elemental form of metal can deposit on the surface of the cathode. This is a direct approach to heavy decontamination, which is called electrodeposition.

On the other hand, when a sacrificial metal plate (normally iron or aluminum) is used at the anode, it can be oxidized and forms metal ions (Fe^{2+}, Fe^{3+}, or Al^{3+}), which subsequently form a series of metal hydroxides. The production of free metal ions and metal hydroxides can cause an effect such as chemical coagulation, which can remove heavy metal ions from water solution. This is an indirect approach (called electrocoagulation, EC) to heavy metal decontamination.

In this chapter, electrochemical technologies will be presented according to the nature of the operations. Reduction and oxidation reactions resulting from applied electricity are discussed in detail. A series of illustrative examples are provided throughout the chapter to ensure that readers have a better understanding of the technologies.

6.2 Electrodeposition Technology

Electrodeposition is an electrochemical technology in which metal ions are deposited on the surface of the electrode (cathode). An electroplating bath, an insoluble anode, and a suitable cathode are needed, which is much simpler than many other wastewater treatment technologies.

Cathodic removal of heavy metals by electrodeposition has several benefits in terms of costs, sludge produced, safety, recovery of pure metal, and versatility. Upon the treatment, elemental metal can be directly produced so as to obtain high-purity metal for industrial uses.

Other than recovering metals in their metallic form, the electrochemical treatment of metal ion pollutants has several advantages: (1) no extra chemical reagents required; (2) no sludge production; (3) high selectivity; (4) low operating cost; and (5) disinfection of the wastewater.[6,7] However, there are disadvantages: The deposition rate and the composition of the solution in some cases can cause the production of dendrites and loose or spongy deposits. The interference from the hydrogen evolution reaction or from oxygen gas production has to be minimized to ensure safety in operations.

6.2.1 Typical Reaction at Electrodes

The electrodeposition of metal from aqueous solution is based on the cathodic reduction of metal ions. Reduction reactions and oxidation reactions are involved in the cathode and anode, respectively.

6.2.1.1 Reduction Reactions at Cathode

At the cathode, the heavy metal ions may be reduced as shown:

$$M^{n+}\text{ (aq)} + ne^- = M\text{ (s)} \tag{6.1}$$

where M = metal; n = valence of the metal.

A few competing reactions may occur at the cathode. Among them, the most common reaction is the reduction of H^+ ions:

$$H^+\text{ (aq)} + e^- = \tfrac{1}{2}H_2\text{ (g)} \tag{6.2}$$

Nitrate can be reduced at the cathode, which results from a few reactions as follows:

$$NO_3^- + e^- = NO_2^- \tag{6.3a}$$

$$2NO_2^- + 8H^+ + 6e^- = N_2 + 4H_2O \tag{6.3b}$$

$$N_2 + 8H^+ + 6e^- = 2NH_4^+ \tag{6.3c}$$

6.2.1.2 Oxidation Reactions at Anode

A common reaction at the anode is as follows when chloride ions are absent:

$$4OH^- (aq) = O_2 (g) + 2H_2O + 4e^- \tag{6.4a}$$

In the presence of chloride ions, we have the following reaction:

$$2Cl^- (aq) = Cl_2 (g) + 2e^- \tag{6.4b}$$

When the anode is not inert, we have the oxidation reaction:

$$M_a^{m+} = M_a + me^- \tag{6.4c}$$

If the anode (M_a) dissolves into the solution during the electrochemical deposition process, any metal recovery efforts will then be futile. Careful selection of the anode is needed to ensure that it will be chemically stable in the electrolyte.

6.2.2 Factors Affecting Electrodeposition

Electrochemical technologies for metal removal and recovery have been widely studied. Kusakabe et al. (1986) studied the simultaneous electrochemical removal of copper and organic wastes by using a packed-bed electrode cell.[8] The final concentration of copper in the effluent was as good as 3 mg L^{-1}, and the removal of COD ranged from 25% to 41%. A study by Hwang et al. (1987) showed that in strong alkaline conditions (pH ≥ 12), cuprocyanide ions could directly be oxidized and copper oxide precipitates on the anode.[9] Kongsricharoern and Polprasert (1996) used an electrochemical precipitation process to treat an electroplating wastewater containing Cr concentrations of 570–2100 mg L^{-1}.[10] It was found that the Cr removal efficiencies were higher than 99%, and the Cr concentrations in the treated effluent were less than 0.5 mg L^{-1}.

In order to design an electrochemical deposition process for precious metal recovery, it is desirable to have a model to describe the metal recovery (or metal residual concentration in the bulk solution) as a function of operational parameters. The typical operation parameters in the studies are types of metals, the presence of other substances such as organic compounds (COD), pH, and initial metal concentrations. In addition, the effects of competitive factors such as presence of ions and organic compounds must be well understood as they are important in the design and operation.

Copper is the most important metal in the metal recovery in such industries as semiconductor, metal coating, and metal wiring.[11] Hence, the recovery of copper by electrodeposition reported by Chen and Lim[7] is selected and discussed in detail to illustrate the electrode position performance as a function of the key parameters.[7]

Copper generally forms compounds in two oxidation states, +1 and +2. The +3 oxidation state is also known, although trivalent copper tends to be unstable in the aqueous media. In the absence of complexing ligands, such as Cl^- ion, Cu^+ is unstable.[11] It had been generally accepted that electro-deposition of Cu^{2+} ions takes place through two steps. The first step is slow, while the second is rapid. The first step is controlled by the equilibrium between Cu^{2+} and Cu^+ at the electrode surface.

6.2.2.1 Effect of Initial Concentration

Initial metal concentration is important in the electrochemical deposition operation, which significantly affects the performance of the process. As shown in Figure 6.1, the residual copper concentration decreases dramatically in the first 6 h (e.g., recovery percentage of 57% for the initial concentration of 978 mg L^{-1}). The final copper concentrations of around 10 mg L^{-1} are observed after 24 h. This leads to a high percentage recovery of metal from 93% to 98%

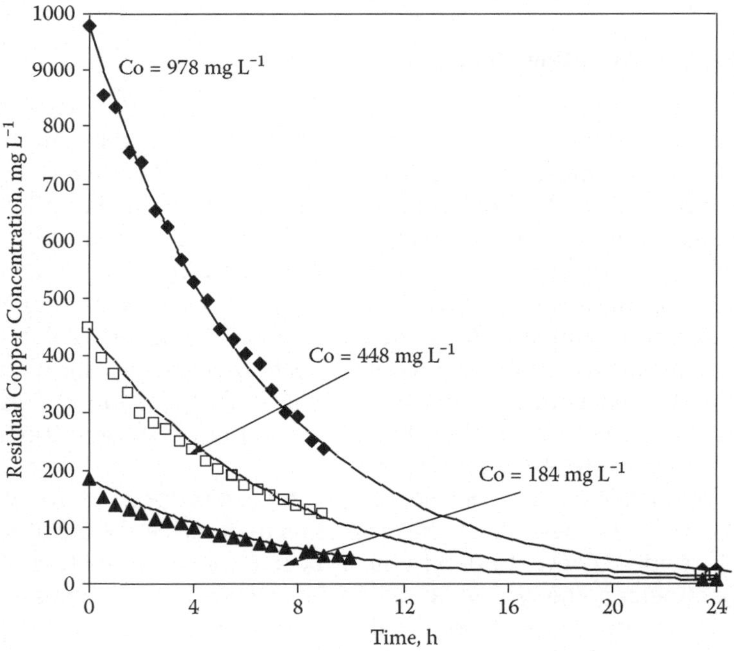

FIGURE 6.1
Effect of initial concentration on electrochemical deposition of copper, $C = C_0e^{-0.146t}$.

for the initial metal concentrations studied. It is also observed that the electroplating kinetics is faster when the initial copper concentration is higher. During the experiments, the solution pH decreased to 1.0 due to Equation 6.2. The temperature slightly increased to 23.5°C. Temperature would not change if the total dissolved solid (TDS) is higher. On the other hand, the temperature may be high when the TDS is low.

From the figure, it is noted that the decrease in concentration of copper for all initial concentrations of copper investigated exhibits an exponential trend. The following first-order chemical reaction kinetics shown may be used to describe the kinetics:

$$C = C_o e^{-kt} \tag{6.5}$$

where C is the concentration at time t, C_o is the initial concentration, and k is a constant associated with the operational condition.

As shown, the model fits the experimental data very well with a higher regression constant (r^2) of 0.99. As the pH is quite low, the metal is present as Cu^{2+} or $Cu(H_2O)_n^{+2}$, and no metal hydroxide species such as $CuOH^+$ are formed. The redox can then be considered as a sole reaction. Thus, the first-order reaction can describe the experimental observation well.

Numerous crystals with regular spherical and square shapes are observed in the SEM diagram given in Figure 6.2a. The values of binding energy (953.8 and 933.7 eV) in the XPS study shown in Figure 6.2b indicate that these crystals are Cu and CuO.[12] This is expected due to the oxidation of the elemental copper by the oxygen in the aqueous solution. The oxidation can be avoided if the copper plates are treated by antioxidation regents.

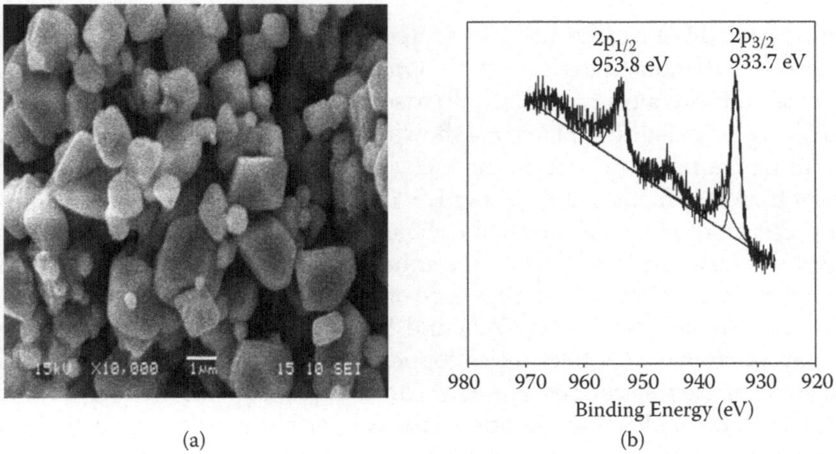

FIGURE 6.2
Microscopic study of electrochemical deposition of copper (initial copper concentration of 500 mg L^{-1}): (a) SEM micrograph; (b) XPS Cu spectrum.

6.2.2.2 Effect of Distance between Electrodes

The distance between the electrodes may be important in the operation. A narrower distance can reduce the metal transfer resistance. In order to achieve a narrower distance, more electrodes are needed, which may add to the cost of the operation. On the other hand, a wider distance may cause a reverse effect, which may be worsened when the TDS is at a lower level.

As shown in Figure 6.3a, the effect of the distance between electrodes can be neglected. The k values are independent of the distance between the electrodes when it is less than 22 cm.

6.2.2.3 Effect of Mixing

Mixing may be important in the operation as high mixing strength would facilitate the movement of metal ions, enabling speedy reduction. Figure 6.3b clearly shows that the recovery rate is higher when the bath is stirred at 800 rpm. On the contrary, the copper recovery, when the bath is stirred at 400 rpm, does not differ much from that in a stagnant bath. When the bath was stirred at 800 rpm, the final copper concentration was 2.78 mg L^{-1} with a percentage recovery of 99.4%. This is a tremendous improvement from the final copper concentration at 13.2 mg L^{-1} and percentage recovery of 97.2% when a stagnant bath was used. The significant enhancement is due to the increase in the mass transfer when the stirring is applied. There exists a threshold in the stirring speed in the metal recovery. Thus, it is recommended that the technique be used at a stirring speed above 800 rpm.

6.2.2.4 Effect of HA

Humic acid (HA) can be used to represent organic matters in wastewater. Figure 6.4a demonstrates that the copper recovery rate increases with an increase in HA concentration. Comparison of the k values at HA of 0, 10, 20, and 50 mg L^{-1} in Figures 6.1 and 6.4 shows that the presence of HA (< 20 mg L^{-1}) has an insignificant impact on the recovery kinetics; however, when the HA concentration is increased to 50 mg L^{-1}, the kinetics increases. The k value in the presence of HA concentration of 50 mg L^{-1} is 15% larger than that in the absence of HA. The HA effect on the enhancement of the copper recovery can be explained as follows. In the absence of HA, the copper ions are complexed with water molecules ($Cu(H_2O)_n^{+2}$) and hydroxides ($CuOH^+$, etc). Because of the introduction of HA, the species copper complexes are partially dissociated and some of the free copper ions (Cu^{2+}) are complexed with the HA.[13] Because the HA can be oxidized by the anode, the copper-HA species are then dissociated, which leads to a certain amount of free copper ions being available for the chemical reduction in the cathode.

In addition, the HA presence had an insignificant effect on the final recovery. As shown in Figure 6.4a, the percentage of copper recovery after

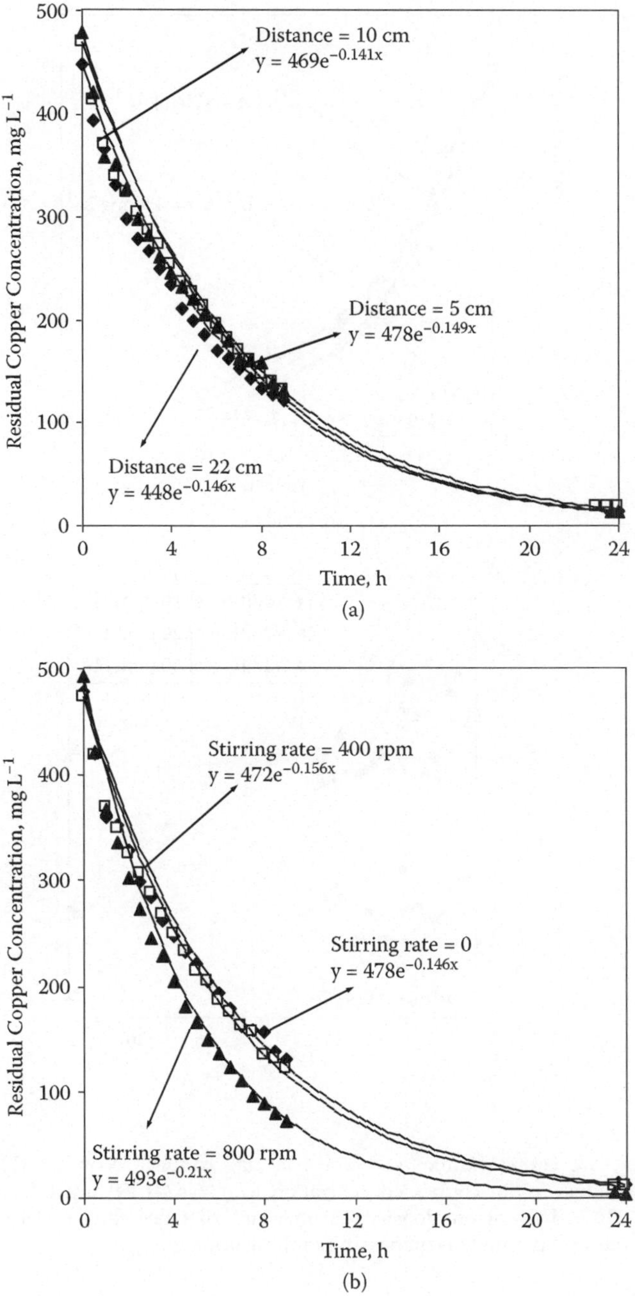

FIGURE 6.3
Physical operation parameters on copper recovery: (a) distance between electrodes; (b) stirring rate.

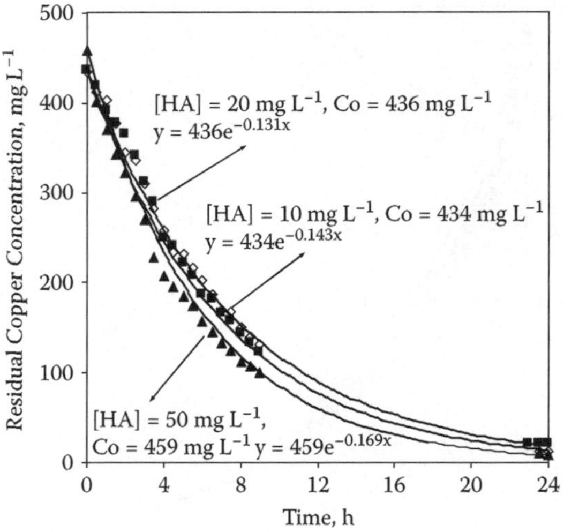

(a)

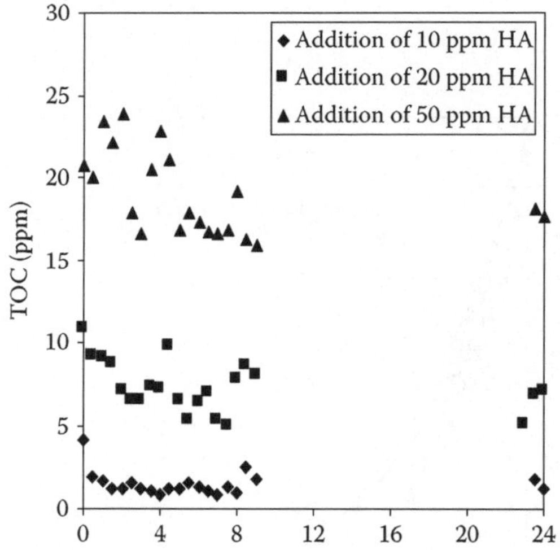

(b)

FIGURE 6.4
Metal recovery in copper-humic acid solution: (a) copper recovery; (b) humic acid removal. Conditions: initial copper concentration (Co) = 434~459 mg L^{-1}, initial pH: 1.4, temperature ~23°C–25°C, current density: 7.41 mA/cm^2, distance between electrodes: 22 cm, copper as cathode and platinum coated with titanium as anode.

(a) (b)

FIGURE 6.5
SEM micrographs of solids obtained for baths containing a fixed concentration of Cu^{2+} at ~450 mg/L with various concentrations of HA at (a) 20 ppm and (b) 50 ppm.

24 h for the bath containing HA reaches a value >95%. The residual copper (e.g., final copper concentration of 8.61 mg L^{-1} for HA concentration of 50 mg L^{-1}) can be easily treated by adsorption or biosorption after the pH of the residual solution is adjusted to 6–7.[14,15]

This electrochemical process also demonstrates some removal for the HA. The removal percentage was 50%, 50%, and 33% for initial HA concentrations of 10, 20, and 50 mg L^{-1}, respectively, as shown in Figure 6.4b. This is due to the anodic oxidation of the HA to simple compounds, such as H_2O and CO_2.

The SEM micrographs for the solids due to the reduction of copper in Cu-humic acid solution are shown in Figure 6.5. There is not a great difference in the morphology of the solids obtained when the solution has different concentrations of humic acid. Comparison of Figure 6.2a with Figure 6.5 shows that the particles formed in the absence of HA are larger and the edges are better defined than those formed in the presence of HA.

The XPS spectra in Figure 6.6 demonstrate that copper oxide is formed in the presence of humic acid. This finding is in a good agreement with that in the absence of humic acid illustrated in Figure 6.2b.

6.2.2.5 Effect of EDTA

EDTA has been widely used in the semiconductor industry. Its presence may affect metal recovery.[7] Figure 6.7a shows that the recovery of copper slightly decreases as the EDTA:copper ratio increases. The *k* value decreases by 16% when EDTA is present. EDTA inhibits copper recovery efficiency as

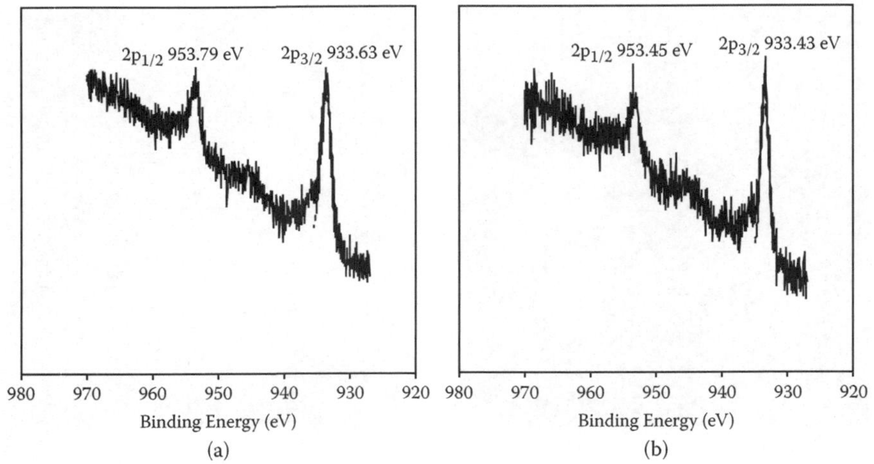

FIGURE 6.6
XPS Cu spectra on the solid obtained for baths containing a fixed concentration of Cu^{2+} and various concentrations of HA at (a) 20 ppm of HA and (b) 50 ppm of HA.

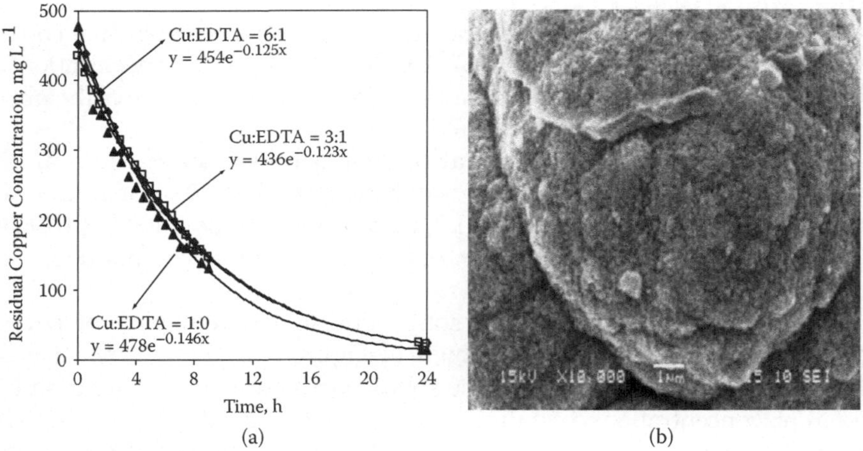

FIGURE 6.7
Effect of EDTA on copper recovery of Cu: (a) recovery percentage; (b) SEM micrograph, Cu:ETDA = 6:1.

it is a strong chelating ligand and can form stronger complexes with metal ions in solution.

There is no obvious degradation of EDTA for the first 4 h (figure not shown). The degradation becomes more rapid at a Cu:EDTA ratio of 3:1 than at the ratio of 6:1. The final removal efficiencies of EDTA are 40% and 55% at ratios of 6:1 and 3:1, respectively.

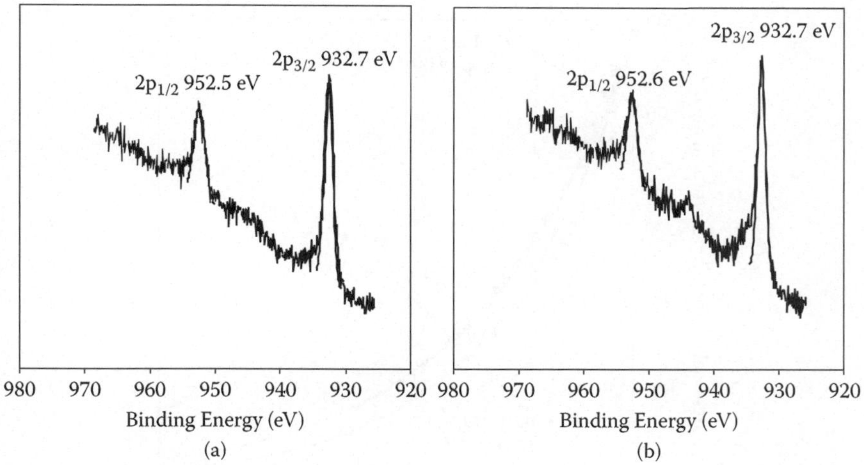

FIGURE 6.8
XPS Cu spectra of solids obtained from baths containing various molar ratios of Cu to EDTA: (a) 6:1 and (b) 3:1.

At a Cu:EDTA molar ratio of 3:1, there is more EDTA than at 6:1. As time passes, the copper concentration decreases and hence more EDTA becomes available, which may be subjected to anodic oxidation. This explains the finding of more rapid degradation of EDTA in the solution with a Cu:EDTA ratio of 3:1. For the solution with less EDTA, at the end of the operation time, the limited amount of EDTA would still be bound to the copper ions and cannot undergo anodic oxidation. Therefore, a higher removal percentage of EDTA occurs at a lower Cu:EDTA ratio.

The morphology in Figure 6.7b shows that the solid particles are of a much larger size than the solids obtained in an additive-free solution (Figure 6.2a). The change is due to the slow electroplating kinetics in the presence of EDTA.

Furthermore, the binding energies of 952.5 – 952.6 eV and 932.7 eV in Figure 6.8 clearly indicate the presence of copper oxides (CuO) and Cu^0.[12,16,17] This finding is the same as what we just observe in Figure 6.2.

6.2.2.6 Effect of Ionic Strength

In a metal solution, some inert ions such as sodium and chloride always exist. They can be considered as ionic strength (I). Figure 6.9a shows that its presence in the solution slightly inhibits the rate of copper recovery.[7] The decrease in copper recovery is likely due to electronic repulsion among the ions. As a result, the mass transfer of copper ions to the cathode was hindered. The presence of ionic strength seems to have less impact on the final concentration. As shown, the removal after 24 h operation in all the cases is as high as 95%.

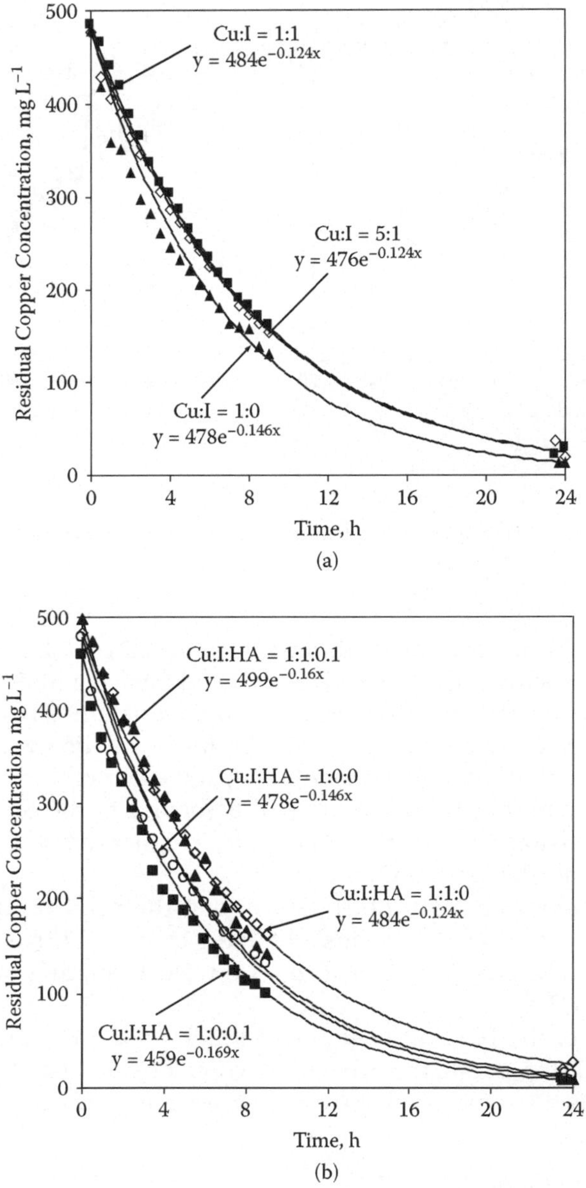

FIGURE 6.9
Effect of ionic strength and HA on copper recovery of Cu: (a) ionic strength; (b) ionic strength and HA.

The coeffect of ionic strength and HA on the copper recovery was also investigated. As shown in Figure 6.9b, addition of HA into the copper solution with ionic strength enhanced the recovery (see Cu:I:HA = 1:1:0.1 versus Cu:I:HA = 1:1:0 in Figure 6.9b). The k value increases by 28% (from 0.124 to 0.16 h^{-1}). The HA seems to overcome the encumbrance because of the presence of ionic strength.

In the case of TOC removal for the bath containing copper, ionic strength and HA reached 50%, compared to 33% in the bath containing only copper and HA. It has been speculated that fewer copper ions are adsorbed onto the HA because of the presence of ionic strength, which leads to more HA for its oxidation. As a result, more HA removal is observed. It was reported that an increase in ionic strength would result in a decrease in the copper adsorption onto the HA.[13]

The morphology of the solids obtained demonstrates that the surfaces are quite similar to each other in the presence and absence of ionic strength (Figure 6.10). An increase in ionic strength seems to result in an increase in the number of smaller crystals on the surfaces.

6.2.3 Recovery of Multicomponent Metal Ions

Industrial wastewater commonly contains various kinds of heavy metal. Hence, it is important to evaluate the feasibility of application of electrodeposition for multicomponent metal recovery. A case study on the recovery of copper and silver was conducted. It was demonstrated that the electrochemical approach is able to recover multicomponent metal ions from the solution. From Figure 6.11a, the rates of recovery for both metal ions are slower under competing conditions than when in a single component metal ion solution. The k values show that the rate of recovery for silver is faster than for copper. This is due to the competition between the two different metal ions for the same limited cathode surface area under constant current density. More important, this is because silver has a more favorable standard electrode potential than copper, as shown.

$$Ag^+ + e^- \Leftrightarrow Ag^0 \qquad E^o = +0.80\ V \tag{6.6a}$$

$$Cu^{2+} + 2e^- \Leftrightarrow Cu^0 \qquad E^o = +0.34\ V \tag{6.6b}$$

The morphology of the solids obtained in an Ag-containing solution and a Cu-Ag solution is shown in Figure 6.11b,c. The morphology of copper solid exhibits a more angular structure shown previously (e.g. Figure 6.2a), while silver solids have a branch-like structure (Figure 6.11b). A mixture of copper and silver crystals results in the morphology shown in Figure 6.11c.

(a)

(b)

(c)

FIGURE 6.10
SEM micrographs of solids obtained from the solutions: (a) mass ratio of Cu:Na = 1:1; (b) mass ratio of Cu:Na = 5:1; (c) mass ratio of Cu:Na:HA = 1:1:0.1.

FIGURE 6.11
Effect of competition on bimetal recovery: (a) silver and copper residual concentration as a function of time; (b) SEM micrographs, silver only; (c) SEM micrographs, Ag:Cu = 1:1; (d) XPS Ag spectrum, Ag:Cu = 1:1; (e) XPS Cu spectrum, Ag:Cu = 1:1.

The valence states of the solids obtained were ascertained using the XPS spectra as shown in Figure 6.11d,e. The binding energies of 954.4 and 933.2 eV correspond to CuO and Cu.[12,18] The binding energy of 367.7 eV indicates the presence of Ag_2O.[19] The results clearly demonstrate that oxidation overnight lead to the formation of these metal oxides.

The competitiveness of metal ions (silver, copper, and lead ions) for electrochemical recovery is shown in Figure 6.12a. It can be observed that silver showed the highest rate of recovery followed by lead, with copper having the slowest rate. The result is rather surprising since the reduction potential of copper is positive, while that of lead is negative.

Figure 6.12b show the morphology of the solids obtained at the cathode, in which lead solids exhibit a stick-like structure (lead only). Figure 6.12c demonstrates that the silver and lead were the predominant species on the surfaces. The XPS spectrum analysis shows that lead and silver exist in the forms of PbO and Ag/Ag_2O, respectively (Figure 6.12d,e).[18,20]

6.2.4 Industrial Application

The recovery of metals by electrodeposition can be used in various industries, such as surface finishing industry and printed circuit board manufacturing industry. Application of metal recovery should be highly useful in metal mining industry, especially in the production of precious metals such as gold.[14] The recovery of metals by electrodeposition here involves two steps: deposition of heavy metals and stripping of the collected metals. The collection step involves electrically reduction, and the stripping can be accomplished chemically or electrochemically. For dilute effluent, an ion-exchange unit can be used to concentrate the metal concentration. High-concentration streams can be treated directly using a recovery system as in the metal surface finishing industry. Metal powders can be formed on the surface of carbon cathodes. Therefore, physical separation is sufficient. The metals recovered can be of quite high purity. An industrial application of electrodeposition for metal recovery is now described as follows.

An electrochemical device shown in Figure 6.13 was installed in a metal finishing company in Baltimore, to remove cadmium.[14] A carbon fiber cathode was used. It had a larger surface-area-to-volume ratio, approximately 1,000 times larger than other types of cathodes. The rinse water, consisting of cadmium, is passed through the cathode, where the metal (i.e. the metallic contaminant, cadmium) is coated/plated on its surface. The coated metal is removed from the cathode electrode by reversing the current or rinsing it with a stripping solution. The "regenerated" electrode is then returned to the plating batch for the next operation's cycle.

As shown in Figure 6.13, the wastewater from the production process is pumped through a filter for the removal of particulate matter and then to the electrochemical reactor. The treated water is returned to the rinse tank. Note that the power supply for the reactor is from a rectifier that converts AC to DC.

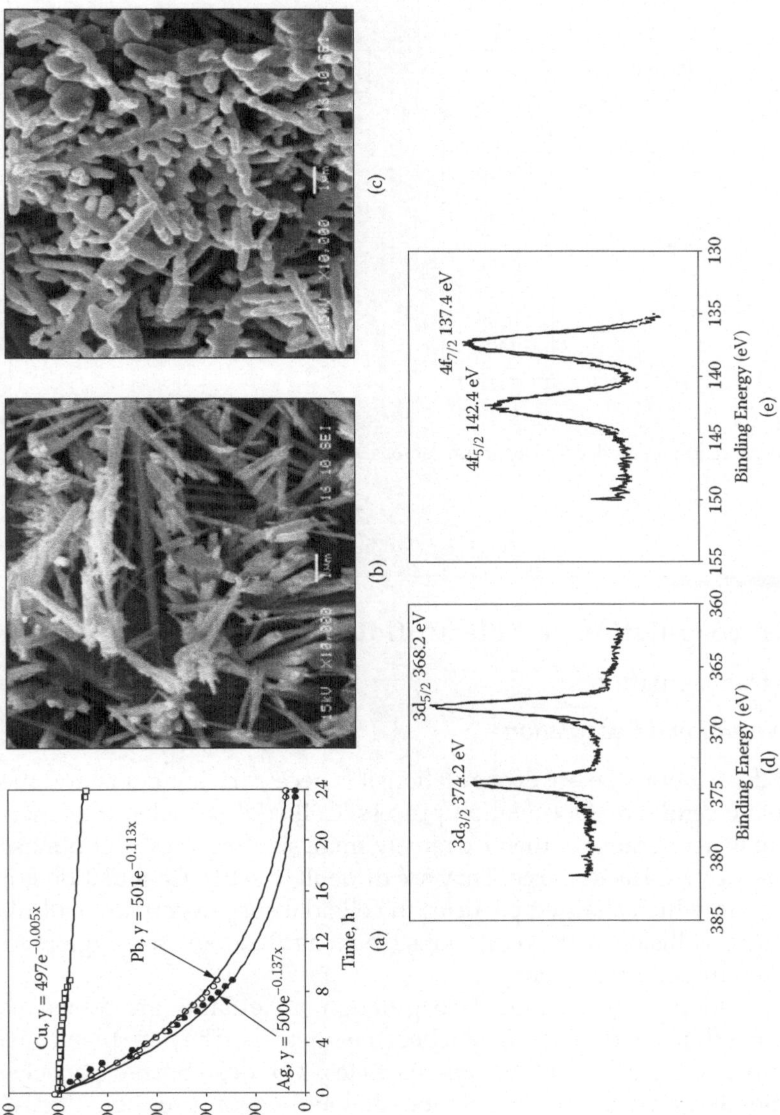

FIGURE 6.12
Effect of competition on trimetal recovery: (a) recovery of respective metal species; (b) SEM micrographs, $[Pb]_0 = 500$ mg L^{-1}; (c) SEM micrographs, Cu]:Ag:Pb = 1:1:1; (d) XPS Ag spectrum, Cu:Ag:Pb = 1:1:1; (e) XPS Pb Spectrum, Cu:Ag:Pb = 1:1:1.

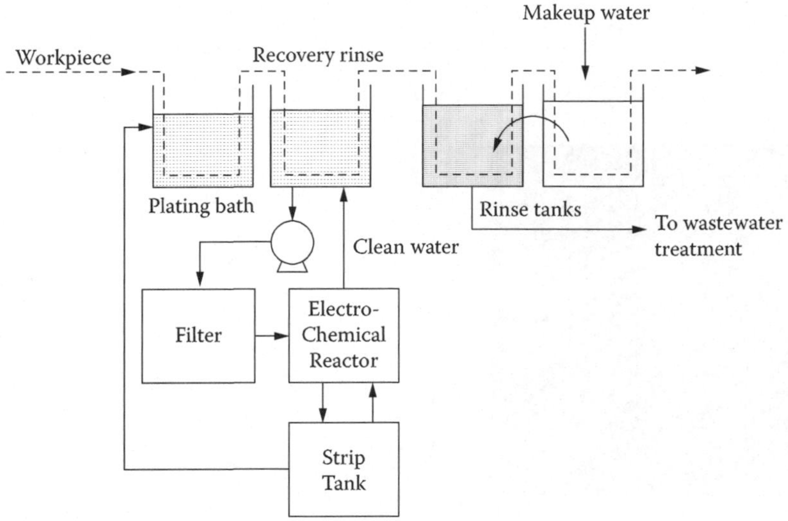

FIGURE 6.13
Electrochemical treatment device for cadmium removal.[14]

6.3 Electrocoagulation and Electroflotation

6.3.1 Electrocoagulation

6.3.1.1 Conventional Coagulation

Before our description of electrocoagulation, it is necessary to understand the conventional coagulation/flocculation process. Colloidal particles are generally stable in water because of their extremely small size, high surface-to-mass ratio, and negative surface charge. They are difficult to settle. Coagulation is a phenomenon in which charged particles in colloidal suspension are neutralized by mutual collision with counterions (i.e., coagulant) and are agglomerated, followed by sedimentation.[21]

In the operation of conventional coagulation, coagulants are added in order to destabilize the colloids, which become unstable. They are then combined together to form larger/unstable particles. The destabilized particles are agglomerated during the flocculation. Agglomeration produces "flocs" that are settleable. Flocculation is carried out under low mixing. Typically, coagulation occurs in a reactor, where rapid mixing is applied; flocculation occurs in another reactor, where slow mixing/stirring is applied. The reaction time for the coagulation is only about 1–5 min, while the flocculation takes about 30 min. In most of the discussion in the literature, the term *coagulation* is often used to describe both coagulation and flocculation.

6.3.1.2 Definition of EC

EC is the process of removal/treatment of suspended, emulsified, or dissolved contaminants (e.g., oil, dye, heavy metals) in an aqueous solution by introducing an electric current into the medium.[21,22] In the EC process, the coagulant (such as Al, Fe ions hydroxides) is electrochemically generated in situ by electrolytic oxidation of an appropriate anode material, which brings about neutralization of electric charges and formation of metal hydroxides for removing pollutants. For example, aluminum anodes are used to produce aluminum cations that have the same effect as the addition of Al-based coagulants in conventional treatment systems. In the process, formation of metal hydroxides and coagulation occur simultaneously. This process has been demonstrated to be very effective in removing contaminants from water without using chemicals and is also easy to operate.

It is generally accepted that the EC process involves three successive stages: (a) formation of coagulants by electrolytic oxidation of the anodes; (b) destabilization of the contaminants, particulate suspension, and breaking of emulsions; (c) aggregation of the destabilized suspended solids to form flocs. A simplest form of EC reactor may be made up of an electrolytic cell with one anode and one cathode, on which oxidation and reduction reactions occur respectively. The conductive anodes are commonly known as "sacrificial electrodes" and may commonly be made of iron or aluminum.

6.3.1.3 Typical Electrode Connection

In an EC reactor, the electrode can be connected in monopolar and bipolar forms. A simple arrangement of the electrode connections is shown in Figure 6.14, where the electrodes are arranged by monopolar or bipolar connections in the EC reactor. With a monopolar connection, the same electric potential is connected between *n* pairs of anodes and cathodes.

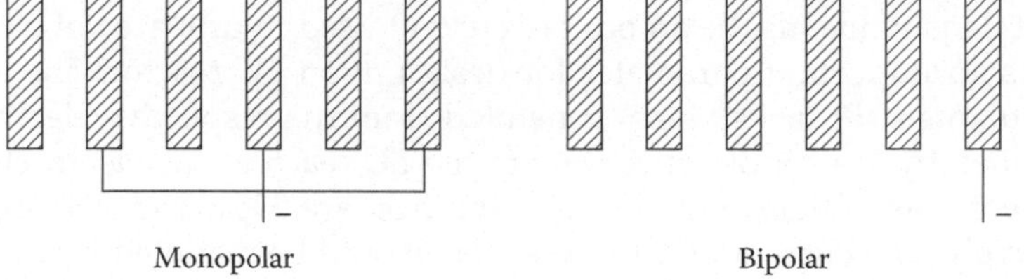

FIGURE 6.14
Monopolar and bipolar electrode connections in the EC reactor.

However, the same current passes through *n* pairs of electrodes in bipolar connection mode.

6.3.1.4 Electrode Reactions

In the anode, the following oxidation reactions occur:

$$Me_1 \text{ (insoluble1)} = Me_1{}^{m+} \text{ (soluble)} + me^- \quad (6.7a)$$

$$4OH^- = 2H_2O + O_2 + 4e^- \quad (6.7b)$$

$$2Cl^- = Cl_2 + 2e^- \quad (6.7c)$$

where Me_1 is the metal in the anode and can be sacrificial metal such as iron or aluminum.

In the cathode, the following reduction reactions occur:

$$2H^+ + 2e^- = H_2 \text{ (g)} \quad (6.8a)$$

$$Me_2^{m+} \text{ (soluble)} + me^- = Me_2 \text{ (insoluble)} \quad (6.8b)$$

where Me_2^{m+} is the soluble metal ion in the solution.

The EC offers the anodic oxidation and in situ generation of metal hydroxide and hydroxides of aluminum. For instance, in the process of removal of As(III) and As(V) by EC, the As(III) can be oxidized to As(V), and then both species are removed via adsorption and coagulation by the metal hydroxides produced. Noted that element that has higher valency may be reduced by cathode. For example, nitrate NO_3^- may be reduced to nitrogen gas (N_2) by cathode.

6.3.1.5 Factors Influencing EC

The most important factors affecting EC performance include current density, solution pH, electrolyte concentration, and temperature. A detailed discussion of the influence of the important factors on the EC process is now presented.

6.3.1.5.1 *Current Density*

Current density means the amount of current used in unit area of electrode, which is an important parameter for design of an EC reactor. The amount of coagulants (Al^{3+} or Fe^{2+} ions) generated from the respective electrodes is determined by the supply of current to the EC reactor. The electrochemical equivalent mass values of aluminum and iron are 335.6 and 1041 mg/(Ah), respectively. The current density selection should be made with other operating parameters such as pH, temperature, as well as flow rate to ensure high current efficiency. A large current density means a small EC reactor. However, very high current density would lead to a significant decrease in current efficiency and waste electrical energy in heating up the water. A current density of 20–25 A/m^2 is recommended for an EC reactor to operate for a long period of time without maintenance.

6.3.1.5.2 *pH Effect*

The effects of solution pH on EC are reflected by the current efficiency as well as the solubility of metal hydroxides. It is generally found that the aluminum current efficiencies are higher at either acidic or alkaline condition than at neutral.[22] The treatment performance depends on the nature of the pollutants, with the best pollutant removal found near pH 7. The power consumption is, however, higher at neutral pH due to the variation of conductivity. When conductivity is high, the pH effect is not significant. The effluent pH after EC treatment would increase for acidic influent but decrease for alkaline influent, which is one of the advantages of the EC process.

6.3.1.5.3 *Presence of Sodium Chloride*

Sodium chloride is usually used to increase the conductivity of the wastewater to be treated. The addition of sodium chloride would lead to a decrease in power consumption because of the increase in conductivity. Moreover, it has been found that chloride ions could significantly reduce the adverse effect of other anions such as HCO_3^- and SO_4^{2-}. The existence of the carbonate or sulfate ions would lead to the precipitation of Ca^{2+} or Mg^{2+} ions that forms an insulating layer on the surface of the electrodes, which would lead to a significant increase in power consumption and a significant decrease in the current efficiency. Moreover, the electrochemically generated chlorine from chloride was found to be effective in water disinfections.

6.3.1.5.4 *Temperature*

Although EC has been widely used for decades, the information on the effect of temperature is quite limited. In general, the current efficiency of aluminum increases with increasing solution temperature. The increase of current efficiency with temperature is attributed to the increased activity of destruction of the aluminum oxide film on the electrode surface. However, as the temperature is too high, there is a shrinkage of the large pores of the $Al(OH)_3$ gel, resulting in more compact flocs that are more likely to deposit on the surface of the electrode. Hence, there is an optimal temperature for the EC process, which should be experimentally determined. Moreover, the water conductivity is affected by temperature. It is generally found that higher temperature gives a higher conductivity, and hence results in a lower energy consumption.

6.3.2 Electroflotation

Electroflotation (EF) is a simple unit process that floats pollutants to the surface of a water solution by tiny bubbles of either hydrogen or oxygen gases produced via water electrolysis. Therefore, hydrogen evolution and oxygen evolution electrochemical reactions occur at the cathode and anode, respectively. EF was first proposed by Elmore in 1904 for flotation of valuable minerals from ores.

6.3.2.1 Selection of Electrodes

The selection of anode and cathode materials is based on their stability, effectiveness, and chemical properties over the gas evolution reactions. According to those reactions, the cathode produces hydrogen while the anode produces oxygen.

In order to select the cathode, the selection can be based on the hydrogen overpotential of the material. Electrode materials such as nanocrystalline Ti–Ru–Fe–O coated on Ti or Fe[23] and NiW and TiO_2 composite layers on carbon fibers[24] can be used effectively in a hydrogen evolution reaction with lower overpotential. However, it has been reported that the materials with lower overpotential produce larger hydrogen bubbles than materials with higher overpotential.[25] Therefore, selection of material for fabrication of electrode should be done based on an optimization process. Generally speaking, cathodes made of iron, aluminum, or titanium can provide hydrogen bubbles suitable for EF.

Oxygen evolution anodes produce oxygen gas bubbles to enhance the flotation of sludge, suspended solids, and oils and grease from the water. Lower oxygen evolution overpotential and stability in the oxygen evolution reaction are the main criteria in the selection of suitable electrodes. Although it is possible to use conventional anodes such as stainless steel, graphite, and iron, they are either soluble or show high overpotential to oxygen evolution reaction. Lead and cobalt oxides can be coated on to Ti or Ni substrates to develop stable oxygen evolution anodes. However, the release of lead and cobalt ions to the solution may be possible if their oxides are not firmly attached onto the substrates, which may cause contamination of the heavy metals in the water.

A wide variety of DSA-typed anodes are stable at the oxygen evolution reaction. Most of them consist of IrO_2 as one of the coating materials; the most common substrate is Ti. Ghany and co-workers used coating oxides of Mn, Mo, and Fe onto the Ti substrate with an intermediate IrO_2 layer.[26] This IrO_2 layer prevents the formation of titanium oxide (insulating layer) between the coating and Ti substrate. Addition of iron enhances electrode stability at elevated temperatures. A combination of Ta_2O_5 and IrO_2 as coating layer on the Ti substrate was used successfully in the oxygen evolution.[27,28]

6.3.2.2 Typical EF Cell

The electrochemical cell arrangement for EF is generally with two horizontal electrodes: one anode and one cathode. These two inert electrodes are arranged near the bottom of the reactor and preferably are of mesh

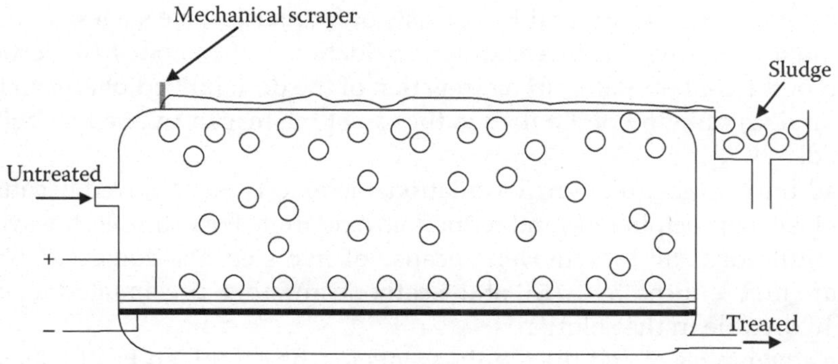

FIGURE 6.15
Electrochemical reactor cell for electroflotation.

configuration. The flow of wastewater can be sent through a perforated feeder, which is placed above the electrodes. Then, the floated waste containing the contaminants from the influent and bubbles can be mechanically removed into the sludge collector. The cell arrangement is illustrated in Figure 6.15.

6.3.2.3 Factors Affecting EF

The performance of an EF system is reflected by the power consumption and the pollutant removal efficiency. The pollutant removal efficiency mainly depends on the size of the bubbles formed.

Power consumption is related to the cell design, electrode materials, as well as operating conditions such as solution pH, current density, and so forth.

The size variation of the bubbles depends on the water pH and current density, as well as the electrode. It has been found that the hydrogen bubbles are smallest at neutral pH. For oxygen bubbles, their sizes increase with pH. Some researchers have demonstrated that the size of the gas bubbles depends on the current density. A decrease of gas bubble size was found with increase of current density at the low end of current density. However, when the current density is too high, no clear trend would be observed. It should be noted, however, that the cathode materials affect the size of the hydrogen bubbles, and so do the anode materials.

6.3.3 Combination of EC and EF

6.3.3.1 Introduction

EC combined with the EF process is an electrochemical technology that can be used for heavy metal removal. It can also be used for the removal of such contaminants as suspended solids (SS), color, oil and grease (O & G) and organic compounds.

The combination of EC and EF consists of four successive stages: (a) in situ generation of coagulants by electrolytic oxidation of the anodes; (b) destabilization of the contaminants; (c) aggregation of the destabilized phases to form flocs; and (d) carrying of the flocs to the top of the reactor by the fine bubbles formed.

In addition to coagulation/flocculation, metal ions are removed because of hydrolysis, coprecipitation, and reduction/oxidation. For example, hexavalent chromium ions can be removed because of two step reactions. The Cr(VI) ions are first reduced to Cr(III) at the cathode and then precipitated together with hydroxide in the solution.

The advantages of and the requirements for the combined EC/EF technology can be summarized as follows:

a. Higher efficiencies for treatment of organics, solids, colors, and heavy metal ions
b. No chemical addition needed
c. Lower amount of sludge production compared to conventional coagulation/flocculation
d. Disinfection of microorganisms
e. Easy to use for treatment of low-volume wastewater
f. No pH adjustment requirement
g. No additional requirement for mechanical agitation
h. Shorter contact time and smaller footprint of system
i. Higher current efficiency
j. The electrodes need to be regularly replaced because the "sacrificial electrodes" are dissolved in the wastewater streams.
k. Trace amounts of aluminum and iron content may exist in the effluent. The effluent may have yellow color when iron is used.
l. Passivation of anodes may occur because of the formation of oxide film, which leads to an increase in power consumption.

6.3.3.2 Electrodes

In the process of EC/EF, the most important factor is the release of metal ions from the anode and subsequent metal hydroxide formation. The electrode that releases metal ions is termed as the *sacrificial electrode*. They are iron (Fe) and aluminum (Al). The reactions of anode and cathode are given in Equations 6.7 and 6.8. The typical reactions are shown in Table 6.1.

Hydrogen gas produced at the cathode creates turbulence in the system, leading to enhancement of the mixing. The gas also serves to transport the insoluble coagulated particles to the surface of the solution. Thus, a floating

TABLE 6.1
List of Typical Chemical Reactions in EC/EF

Location	Reaction
Anode	$4OH^- = 2H_2O + O_2 + 4e^-$
	$2Cl^- = Cl_2 + 2e^-$
	$Al = Al^{3+} + 3e^-$ (when aluminum is applied)
	$Fe = Fe^{2+} + 2e^-$ (when iron is applied)
	$Fe^{2+} = Fe^{3+} + e^-$ (when iron is applied)
Cathode	$2H^+ + 2e^- = H_2\,(g)$
	$O_2 + 2H_2O + 4e^- = 4OH^-$
Solution phase	$nAl^{3+} + mOH^- = Al_n(OH)_m^{3n-m}$ (when Al is applied)
	$iFe^{2+} + jOH^- + kH_2O = Fe_i(H_2O)_k(OH)_j^{2i-j}$ (when Fe is applied)
	$\alpha Fe^{3+} + \beta OH^- + \gamma H_2O = Fe_{i\alpha}(H_2O)_\gamma(OH)_\beta^{3\alpha-\beta}$ (when Fe is applied)
	(i, j, k, m, n, α, β, and $\gamma = 0, 1, 2, \ldots$)

layer is formed at the liquid surface consisting of both hydrogen bubbles and entrapped suspended matter.

For an individual electrode, the net current flowing through the wire to the electrode is the difference between the anodic and cathodic currents developed by the electrode reactions at the electrode interface. The utilization of electrolytic precipitation for isolating a number of elements from aqueous solution has been practiced by analytical chemists since the 19th century.

The principles can be grouped into two basic categories: (a) electrolysis at constant applied cell potential for the deposition of certain elements, and (b) electrolysis at variable potential to maintain convenient current through the cell for the deposition of a large group of elements. It is obvious that such techniques can be applied to the area of wastewater treatment for the removal of undesirable elements.

In addition to the dissolution of anodic metals or generation of oxygen, other strong oxidizing agents, for example, chlorine, can be generated in the anode if the appropriate type of electrolytes are present in solution. For the generation of chlorine, a high level of chloride is needed for the process to be economically feasible.

6.3.3.3 Cell Arrangements

Electrochemical cells for EC/EF can be arranged as shown in Figure 6.16. The simplest form of this type of arrangement consists of two electrodes, one anode and one cathode. The electrical connection can be bipolar or monopolar; however, a bipolar connection is preferred. Other than that, the flow of wastewater can be manipulated into a single or multiple channels. If enough H_2 is produced at the cathode, EF may take place.

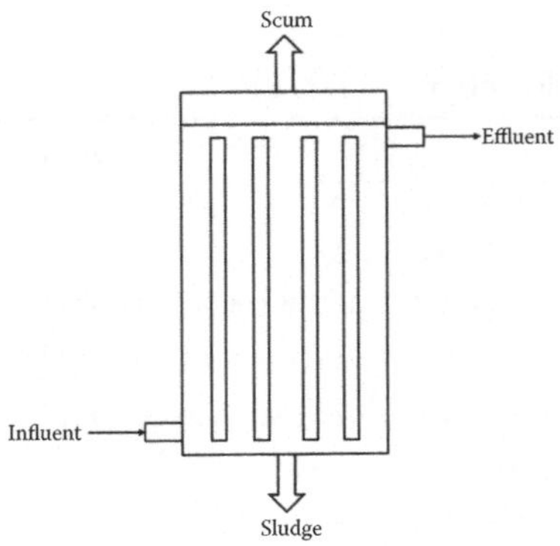

FIGURE 6.16
Electrochemical reactor for EC/EF.

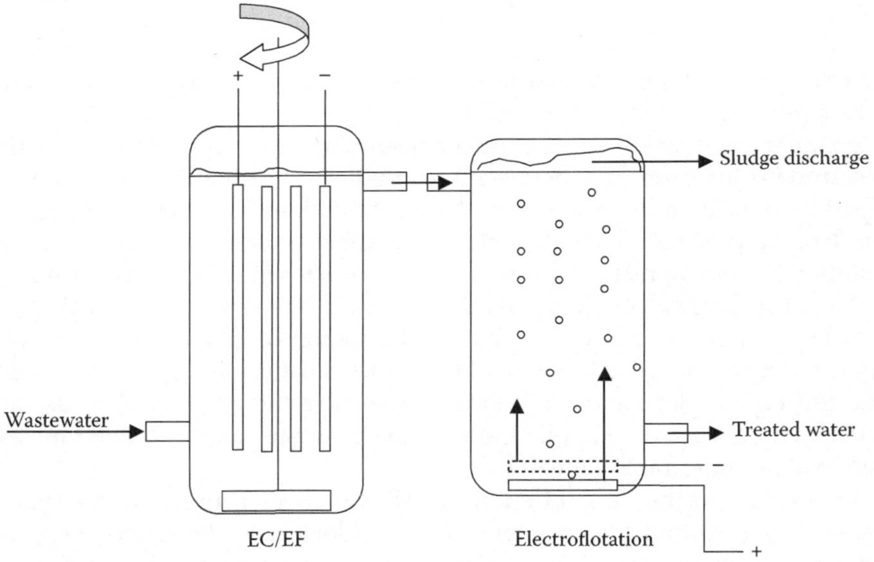

FIGURE 6.17
Combined EC/EF and electroflotation technologies.

If the hydrogen gas bubbles in the EC/EF given in Figure 6.16 are insufficient, one more EF would be added after the EC/EF, and the whole system would then become a combined EC/EF–electroflotation process. The combined system illustrated in Figure 6.17 would improve the efficiency of heavy metal removal. The EC/EF is placed before the EF. In the EC/EF process, heavy metals will be coagulated/precipitated, and some of the flocs are able to float due to the production of gas bubbles. Slight agitation is needed in order to minimize the settling of sludge. The treated water can then be sent to and through the EF. The separation of sludge from the water can be enhanced by selecting good oxygen evolution electrodes for use in the EF. Note that the electrodes used in the electroflotation must be inert in order to avoid oxidation at the anode, because the oxygen that may be produced through the oxidation of hydroxide ions.

6.3.4 Case Studies

To better understand the application of EC and EF for the removal of heavy metals (copper, zinc, chromium, cadmium, nickel, etc.) from aqueous solution, some case studies are now discussed as follows.[29–42] The most recently development of EC and EC/EF technologies is the so-called hybrid treatment systems.[42,43] They can remove various pollutants such as dyes, metals and virus.

6.3.4.1 Copper Removal

Escobar et al. performed a study on EC for the removal of copper, lead, and cadmium in natural waters and simulated wastewater.[29] In the study, commercial laminate steel was used as an electrode, and the simulated wastewater contained 12 mg/L of copper, 4 mg/L of lead, and 4 mg/L of cadmium. It was found that copper removal was achieved primarily by adsorption on to the coagulated product. The optimum conditions for the process were identified as pH 7, flow rate 6.3 cm^3/min, and a current density between 31 and 54 A/m^2. At a critical electrode geometric area of 35 cm^2 and electrolysis time of 6 min, a maximum copper removal rate of 80% was observed. A linear relationship was identified between the current density and the mass of generated sludge. In addition, specific energy consumption and the current density were linearly related.

Continuously operated EC tests were performed to explore their treatment efficiencies and to identify the optimum operating conditions for treatment of copper chemical mechanical polishing (Cu-CMP) wastewaters.[30] In the testing, the EF cell was a 1-L Pyrex glass reactor, which was equipped with a temperature control. Cast aluminum (Al) and iron (Fe) plates (20 cm L × 3 cm W) were used as the anode/cathode pair, and the electrodes were situated approximately 1 cm apart. Three pairs of electrodes were adopted in the experimental tests, as shown in Figure 6.18. The electrode pairs were dipped in the CMP wastewater with an effective electrode surface area of 72 cm^2. A magnetic stirrer was employed in the reactor to maintain good mixing of the CMP wastewater during the EC process. It was found that EC of the CMP wastewater was able to realize a 96% reduction efficiency of turbidity (NTU) and more than 96%–99% copper ions removal efficiency with an applied potential of 30 V in 30 min (Figure 6.19). The Copper removal and time are linearly related. The effluent wastewater was clear and can be considered for recycling in the Cu-CMP processes.

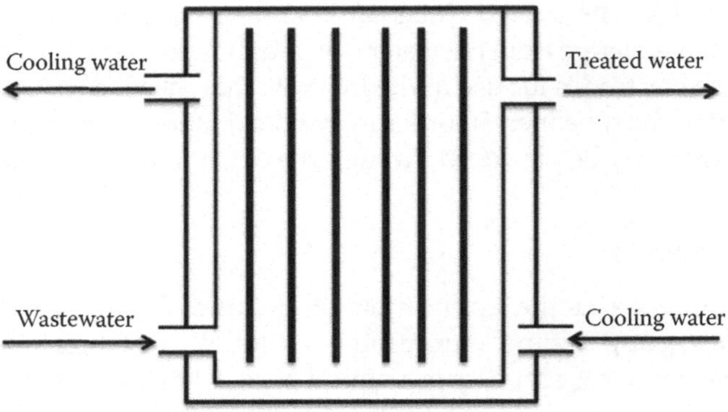

FIGURE 6.18
Schematic diagram of electrocoagulation system for treatment of Cu-CMP wastewater.[30]

6.3.4.2 Zinc Removal

Application of EC for treatment of zinc-containing rinse wastewater from an automotive assembly plant was investigated.[31] The EC experiments were carried out in batch and continuous modes with a single-compartment electrochemical reactor. Aluminum (99.53% purity) and iron (99.50% purity) as sacrificial electrodes were used as both anode and cathode, respectively. The total effective electrode area for batch and continuous modes was 146.3 and 660 cm^2 and an electrode gap of 11 and 20 mm, respectively. In the study, effects of the process variables such as pH, current density, electrode material, and operating time were studied with respect to zinc removal efficiencies, electrical energy, and electrode consumptions.

As illustrated in Figure 6.20, it was found that the optimized pH values were 3–5 and 2.5–7 for the removal of zinc with iron and aluminum electrodes, respectively. The results showed that zinc removal was more efficiently achieved by Al electrode than by Fe electrode at the optimal pH. The zinc removal was significantly affected by current density as shown in Figure 6.21. The zinc removal efficiency was sharply increased from 28% to 99% when the current density was increased from 10.0 to 60.0 A/m^2 with aluminum electrodes, and the further increase of current density did not enhance the zinc removal. When iron was used as electrodes, the zinc removal efficiency was obviously increased from 5% to 99% when current density was increased from 10.0 to 60.0 A/m^2, and the further increase of current density did not enhance the zinc removal. The energy consumptions were 0.18–11.29 kWh/m^3 for Al electrode and 0.24–8.47 kWh/m^3 for Fe electrode in the same current density range. The electrode consumptions increased from 0.01 to 0.35 kg electrode/m^3 for Al electrode and from 0.20 to 0.62 kg electrode/m^3 for Fe electrode with increasing current density from 10.0 to 100.0 A/m^2.

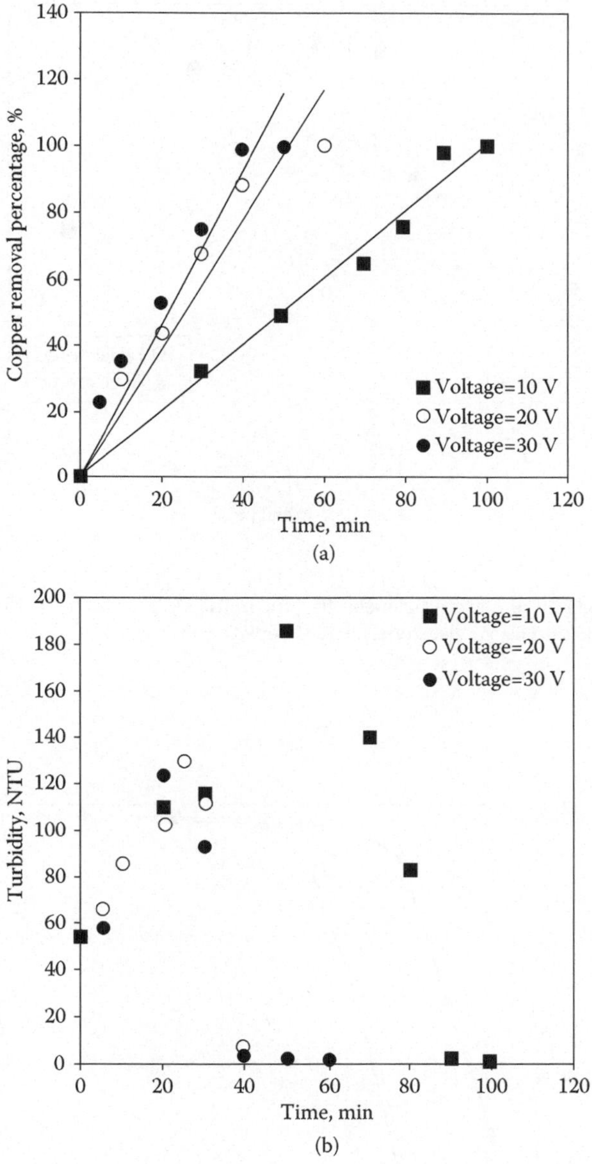

FIGURE 6.19
Effect of applied potential on the removal of copper ions and final turbidity of Cu-CMP wastewaters: (a) removal efficiency of copper ions and (b) turbidity of treated water. Characteristics of wastewater: turbidity = 180 NTU, initial copper concentration = 83 mg/L and COD = 277 mg/L.[30]

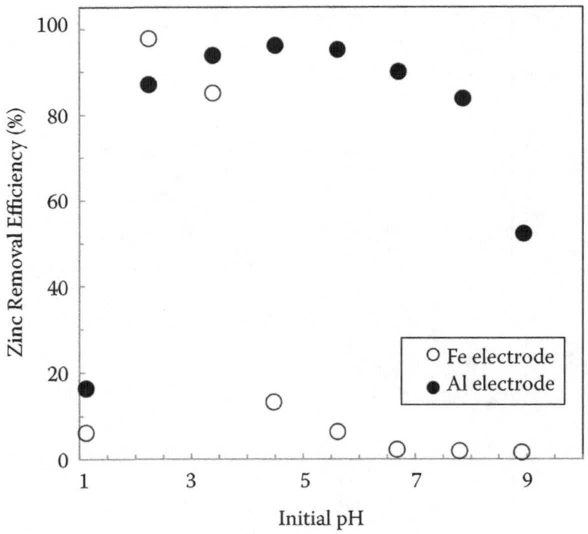

FIGURE 6.20
Effect of initial pH on removal efficiencies for zinc in the EC process (t_{EC} = 15 min for Fe electrode and t_{EC} = 25 min for Al electrode, current density = 60 A/m^2).[31]

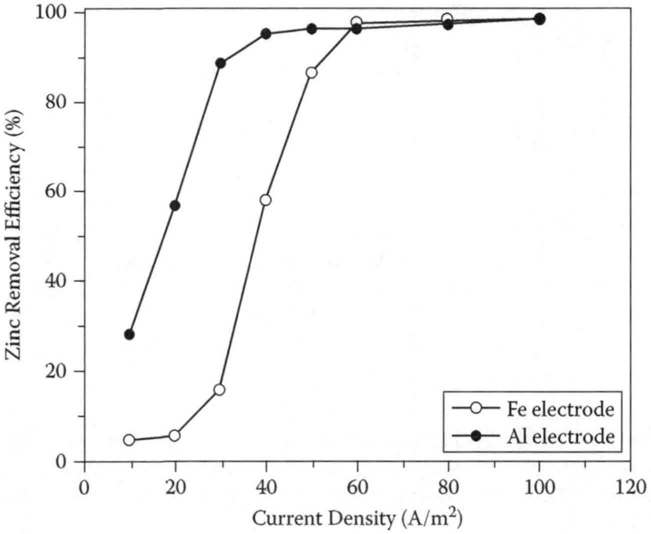

FIGURE 6.21
Effect of current density on zinc removal efficiencies in the EC process (t_{EC} = 15 min and pH 3 for Fe electrode, t_{EC} = 25 min and pH 5 for Al electrode).

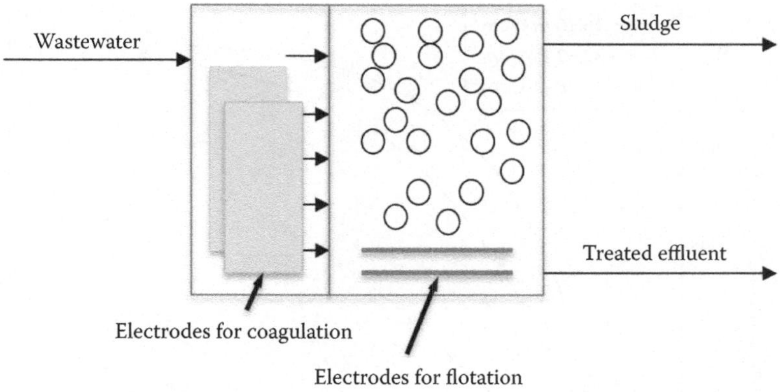

FIGURE 6.22
Schematic diagram of the EC/EF system for treatment of zinc wastewater.

A combination EC and EF cell was developed and investigated for removing zinc from synthetic solution by Casqueira and co-workers.[32] In the study, the influence of some relevant parameters/variables, such as collector concentration, tension, and current density variation on the removal efficiency, was studied in detail. The schematic diagram of the EC/EF apparatus was illustrated in Figure 6.22. The obtained results demonstrated that the combined EC and EF process up to 96% of the zinc from the wastewater. The predominant zinc ionic species, Zn^{2+}, in the aqueous solution interacted with the sodium dodecyl sulfate (SDS) collector by electrostatic interaction. The optimum conditions for zinc removal were as follows: voltage of 15 V current of 400 mA, 150 mg/L Na_2SO_4 as an electrolytic support, inlet pH around 7.0, influent concentration of zinc 20 mg/L, 0.1% ethanol (as a frother), 8 mA/cm^2 current density, and SDS:zinc molar ratio of 3:1.

6.3.4.3 Chromium Removal

Godler and co-workers reported a study on trivalent chromium (Cr^{3+}) removal by EC/EF technology.[33] The influences of important operation parameters such as cell current density, pH, and contact time on the removal efficiency were investigated. In the study, a 1.0-L Perspex electrolytic reactor (102 × 102 × 124 mm) equipped with a flat plate Al electrode was used for the batch EC experiments, and NaCl was added to provide enough conductivity. The results showed that a higher stirring speed enhanced Cr^{3+} removal due to the enhanced contact with Al floc. It was observed that the applied current density significantly affected the removal efficiency. The effects of contact time and current density on the removal efficiency are illustrated in Figure 6.23. It was found that almost complete removal of Cr^{3+} was observed at current density and contact time of 48.78 mA/cm^2 and 60 min, respectively. For the pH effect study, the experiments were carried out at pH 3.4, 4.2 and 5.0 under identical

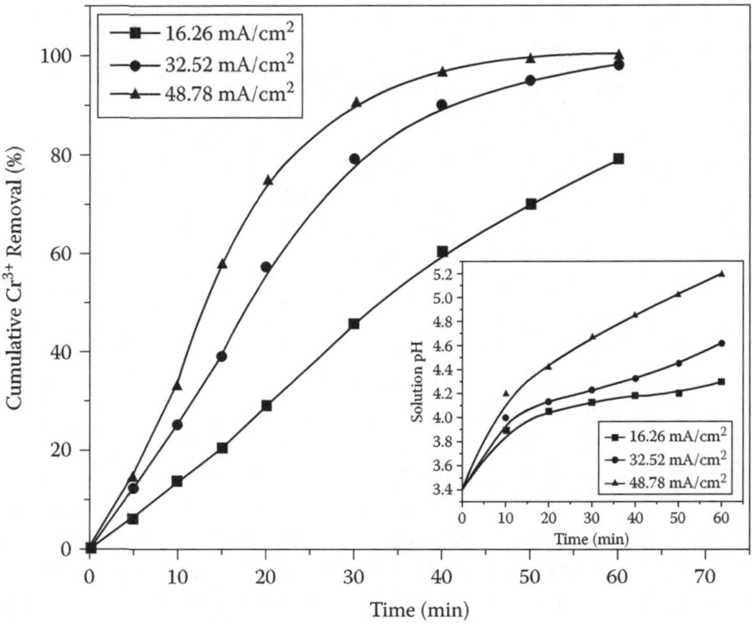

FIGURE 6.23
Progress of Cr(III) removal for different current densities. Experimental conditions: initial Cr^{3+} concentration = 1000 mg/L, initial pH = 3.4, and supporting electrolyte NaCl 1000 mg/L.

conditions, and the results are shown in Figure 6.24. It can be seen that the time required for the same level of Cr^{3+} removal is less for a higher initial pH.

Gao and co-workers designed a combined EC and EF process to reduce CrO_4^{2-} to Cr^{3+} and then to remove the total Cr from wastewater to below 0.5 mg/L.[34] The designed experimental facility for the Cr(VI) removal process is schematically shown in Figure 6.25. The apparatus consists of three major parts: production of Fe^{2+} for reduction of Cr^{6+} to Cr^{3+}, coagulation of the newly formed precipitates with pH adjustment, and separation of the sludge via flotation. It was found that Cr(VI) was successfully reduced to Cr(III) by the in situ formation of Fe^{2+} electrochemically. In the EC cell, the resulting Fe^{3+} ions were found to be good coagulants together with Al^{3+}, which was produced electrochemically from the hybrid Al–Fe electrodes. The gas bubbles generated from water electrolysis–EF can effectively float up the sludge formed. The flotation efficiency could be improved by addition of a small amount of surfactant. After treatment, the suspended solids was found to be below 3 mg/L and the total Cr content less than 0.5 mg/L at the optimal conditions. The power consumption was less than 1 kWh/m³ water at the conductivity of 1.5 mS/cm.

Arroyo and co-workers carried out a study on the effect of pH and chloride concentration on the removal of hexavalent chromium in a batch

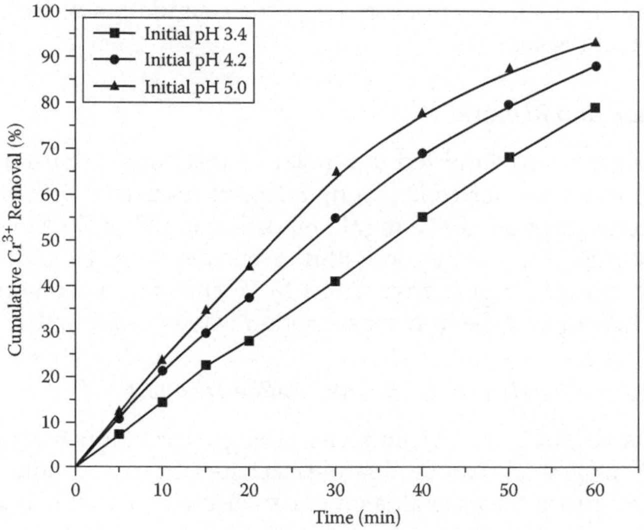

FIGURE 6.24
Effect of initial pH on cumulative removal of Cr(III).

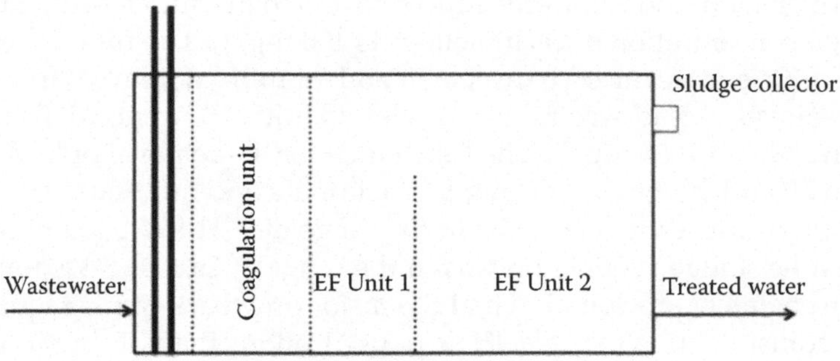

FIGURE 6.25
Schematic diagram of combined EC and EF system for treatment of hexavalent chromium wastewater.

EC reactor.[35] In the work, the effects of pH and chloride ion concentration on the removal of Cr(VI) from wastewater by batch EC using iron plate electrodes have been investigated. It was observed that the presence of chloride ions enhanced anode dissolution due to pitting corrosion. Fe^{2+} ions generated during anode dissolution led to the reduction of Cr(VI) to Cr(III), which were subsequently coprecipitated with Fe^{3+} ions at relatively low pH. The reduction of Cr(VI) to Cr(III) and the solubility of both chromic and iron hydroxides were highly pH dependent. At lower pH, the reduction of Cr(VI) to Cr(III) by Fe^{2+} ions was preferred, but the coagulation of Fe^{3+} and Cr(III) was favored at higher pH.

6.3.4.4 Cadmium Removal

Four iron electrodes connected in bipolar connection were used in a 1.56 L EC/EF cell to remove cadmium ions from an industrial effluent that had an initial concentration of 500 mg/L and solution pH of 7.[36] More than 99% removal for the Cd was achieved within a reaction time of 20 min and under an applied voltage ranging from 20 to 40 V. The energy consumption was about 0.36 kW/kg and the iron consumption was 2.61–8.45 g/L.

6.3.4.5 Removal of Heavy Metals from Saline Leachate

A comparative study of EC and chemical precipitation technologies for removing heavy metals from soil saline leachate (SSL) was conducted.[37] In the study, the EC process was evaluated via an electrolytic cell using mild steel as electrodes, whereas chemical precipitation was evaluated using either calcium hydroxide or sodium hydroxide as precipitant. It was observed that, at a pH varying between 7 and 8, EC was more effective than chemical precipitation in removing metals from SSL having 124 mg Pb/L and 38 mg Zn/L. Except cadmium, the concentrations of heavy metals in the mixture, including Cr, Cu, Ni, Pb, and Zn, were reduced to less than 4 mg/L using EC while the initial concentration of each metal was 100 mg/L. The time to reach the removal efficiencies varied between 30 and 90 min. Moreover, at least 94% of the removal of lead was achieved with its initial concentrations ranging between 250 and 2000 mg/L. The time taken for the removal of lead ranged between 20 and 90 min, according to the initial concentrations. The energy consumption was estimated to be in the range of 9–11 kWh/ton of dry soil. The metallic sludge production was in the range of 29.8–41.3 kg/ton of dry soil. The chemical precipitation had the major drawback of the requirement for pH adjustment. When NaOH was used, more than 15 mg/L of each element was found at the end of the treatment for the case of heavy metal mixture. Even if $Ca(OH)_2$ was used, cadmium and lead were not removed effectively. It can, therefore, be concluded that EC/EF overperforms chemical precipitation.

6.3.4.6 Nickel and Zinc Removal

Kabdasli and co-workers[38] conducted a study on the removal of Ni and Zn by EC/EF technology. Stainless steel electrodes were used as both anodes and cathodes, which were connected in the monopolar mode. The cell volume was 1800 mL. The metal finishing wastewater with a chloride content of approximately 1500 mg/L and pH in the range of 6.5–7.5 was treated. The optimized removal percentages for Ni and Zn occurred at contact times of 150 and 60 min, respectively. The current density to achieve these results was 9 mA/cm^2.

6.3.4.7 Arsenic Removal

Kumar and co-workers conducted a laboratory-scale study to evaluate EC as a treatment technology for arsenite and arsenate removal from water.[39] In the study, three electrode materials, namely, iron, aluminum, and titanium, were tested to assess their efficiencies for arsenic removal. It was found that arsenic removal efficiency with different electrode materials followed the sequence: iron > titanium > aluminum. The EC was able to bring down the arsenic concentration to less than 10 μg/L with iron as electrodes. In the direct current density range of 0.65 to 1.53 mA/cm^2, higher current density resulted in faster arsenic removal. In the pH range from 6 to 8, the effect of pH on arsenic removal was not significant. It should be noted that EC was found to be more efficient for As(III) removal than chemical coagulation (CC), whereas the As(V) removal performance of both EC and CC was nearly same. In the EC process, the As(III) removal mechanism seemed to be oxidation of As(III) to As(V) and then surface complexation with iron hydroxides. A similar phenomenon was observed by other researchers. Gregor reported that the As(III) removal by EC was much better than by chemical coagulation, while As(V) treatment by both approaches showed virtually the same results.[40] The improvement by EC was probably due to the oxidation of As(III) to As(V).

Parga and co-workers conducted a study on removal of As(III) and As(V) from the contaminated groundwater by using EC/EF technology.[41] In the study, the electrocoagulation system was modified. Before entering the EC reactor, the arsenic-containing water was passed through a porous tube medium where air was injected. Carbon steel electrodes were used as anodes and cathodes in the EC reactor. The electrodes were placed vertically; the interelectrode gap was 6 mm. A bench-scale study was carried out by using a reactor with a volume of 1.2 L. The arsenic-contaminated groundwater from La Comarca Lagunera, Mexico, was used after adding NaCl (1 g/L). It was found that more than 99% of arsenic was removed within 92 s. The study was further extended to a pilot-scale study with a capacity of 30 L/min. The well water with conductivity of 600–4000 μS/cm and arsenic concentration of 0.025–0.05 ppm was used for the study. The pH of the well water was 5.5–7.1. Within 1 min of HRT, the arsenic concentration dropped to 0.002 ppm (2 ppb). It was estimated that the power requirement for a well producing 2500 m^3/day is 150 W/min.

6.3.4.8 A Hybrid EC/EF–Membrane Process

A hybrid EC/EF–membrane process was studied for treatment of industrial Cu production wastewater containing high concentrations of heavy metal cations such as Cd, Cu, Zn, Pb, and Fe as well as heavy metal anions such as As and Se.[42] The schematic diagram for the experimental setup is shown in Figure 6.26, which mainly consists of primary lime neutralization

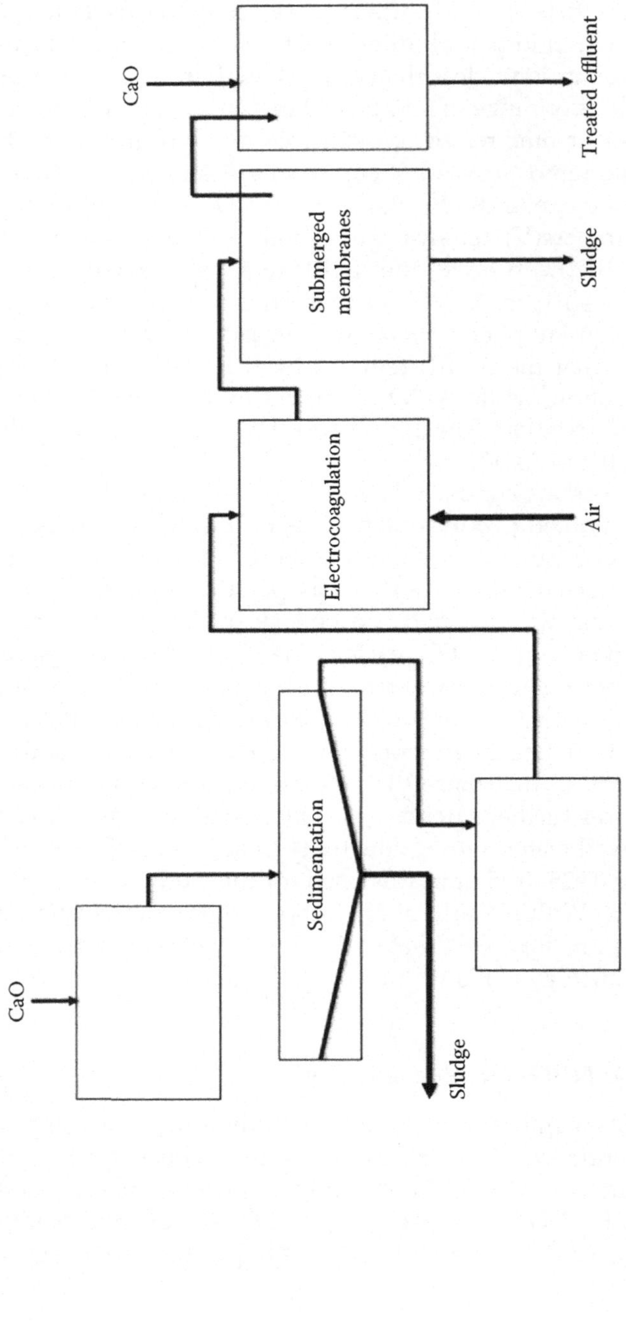

FIGURE 6.26
Schematic diagram of hybrid EC/EF and membrane process for treatment of concentrated metal wastewater.[42]

with sedimentation, EC, microfiltration, and final lime neutralization with sedimentation. In the EC/EF reactor, iron plate and stainless steel mesh were used as anodes and cathodes, respectively. A microfiltration membrane with 0.3-μm pores was employed. The size of reactor volume was 1 L, and the flow rate was in the range of 3–10 L/h. The hybrid process of EC/membrane filtration demonstrated very promising results for removal of Se as well as other metals (especially As, Cu, and Pb) in the treatment of industrial wastewater. It was found that, at a contact time of 20 min and under a current density of 4.8 mA/cm^2, the concentration of Se was reduced by 98.7%. For similar treatment conditions, the removal efficiencies of As, Cu and Pb, and Cd and Zn were reported as 99.9%, 98%, and 99.9%, respectively. The sludge from the microfiltration could be used as an adsorbent.

References

1. Gutierrez, C., Hansen, H. K., Nunez, P., Jensen, P. E., and Ottosen, L. M. Electrochemical peroxidation as a tool to remove arsenic and copper from smelter wastewater. *J. Appl. Electrochem.*, 40 (6), 1031–1038, 2010.
2. Pociecha, M., and Lestan, D. Using electrocoagulation for metal and chelant separation from washing solution after EDTA leaching of Pb, Zn and Cd contaminated soil. *J. Hazardous Mater.*, 174(1–3), 670–678, 2010.
3. Camarillo, R., Llanos, J., Garcia-Fernandez, L., Perez, A., and Canizares, P. Treatment of copper (II)-loaded aqueous nitrate solutions by polymer enhanced ultrafiltration and electrodeposition. *Sep. Purif. Technol.*, 70(3), 320– 328, 2010.
4. Dermentzis, K. I., Davidis, A. E., Dermentzi, A. S., and Chatzichristou, D. An electrostatic shielding-based coupled electrodialysis/electrodeionization process for removal of cobalt ions from aqueous solutions. *Water Sci. Technol.*, 62(8), 1947–1953, 2010.
5. Britto-Costa, P. H., and Ruotolo, L. A. M. Electrochemical removal of copper ions from aqueous solutions using a modulated current method. *Sep. Sci. Technol.*, 46(7), 1205–1211, 2011.
6. Li, X. Y., Fan, X. J., Gu, J. D., Ding, F., and Tong, A. S. F. Electrochemical wastewater disinfection: Identification of its principal germicidal actions. *J. Environ. Eng., ASCE*, 130, 1217–1221, 2004.
7. Chen, J. P., and Lim., L. L. Recovery of precious metals by an electrochemical deposition method. *Chemosphere.*, 60, 1384–1392, 2005.
8. Kusakabe, K., Nishida, H., Morooka, S., and Kato Y. Simultaneous electrochemical removal of copper and chemical oxygen demand using a packed-bed electrode cell. *J. Appl. Electrochem.*, 16, 121–126, 1986.
9. Hwang J. Y., Wang Y. Y., and Wan C. C. Electrolytic oxidation of cuprocyanide electroplating waste waters under different pH conditions. *J. Appl. Electrochem.*, 17, 684–694, 1987.
10. Kongsricharoern, N., and Polprasert, C. Chromium removal by a bipolar electro-chemical precipitation process. *Wat. Sci. Technol.*, 34, 109–116, 1996.

11. Rajeshwar, K., and Ibanez, J. G. Environmental electrochemistry—fundamentals and applications in pollution abatement. Academic Press, CA, 1997, 361–497.
12. Maroie, S., Haemers, G., and Verbist, J. J. Surface oxidation of polycrystalline "alpha" (75% Cu et 25% Zn) and "bêta" (53% Cu et 47% Zn) brass as studied by XPS: Influence of oxygen pressure. *Applications Surface Sci.*, 17, 463–476, 1984.
13. Liu, A., and Gonzalez, R. D. Modeling adsorption of copper(II), cadmium(II) and lead(II) on purified humic acid. *Langmuir*, 16, 3902–3909, 2000.
14. U.S. Environmental Protection Agency, *The Electroplating Industry: Environmental Regulations and Technology*. EPA/625/10-85/001, Washington, DC, 1985.
15. Chen, J. P., and Lim, L. L. Key factors in chemical reduction by hydrazine for recovery of precious metals. *Chemosphere*, 49, 363–370, 2002.
16. Rochefort, A., Abon, M., Delichere, P., and Bertolini, J. C. Alloying effect on the adsorption properties of Pd50Cu50(111) single crystal surface. *Surface Sci.*, 294, 43–52, 1993.
17. Marcus, P., and Bussel, M. E. XPS Study of the passive films formed on nitrogen-implanted austenitic stainless steels. *Appl. Surf. Sci.*, 59, 7–21, 1992.
18. Wagner, C., Riggs, W., Davis, L., Moulder, J., and Muilenberg, G. E. *Handbook of X-ray Photoelectron Spectroscopy*. Physical Electronics Division. Perkin Elmer Corporation, Eden Prairie, MN, 1979.
19. Briggs, D., and Seah, M. P. *Practical Surface Analysis*, 2nd ed., Vol. 1. John Wiley & Sons, New York, 1993.
20. Kanai, H., Yoshiki, M., Hayashi, M., Kuwae, R., and Yamashita, Y. Grain-boundary-phase identification of a lead-based relaxor by X-ray photoelectron spectroscopy. *J. Am. Ceramic Soc.*, 77, 2229–2231, 1994.
21. Mollah, M. Y. A., Schennach, R., Parga, J. R., and Cocke, D. L. Electrocoagulation (EC)—science and applications. *J. Hazardous Mater.*, B84, 19–41, 2001.
22. Chen, G. Electrochemical technologies in wastewater treatment. *Sep. Purif. Technol.*, 38, 11–41, 2004.
23. Irissou, E., Blouin, M., Roue, L., Huot, J., Schulz, R., and Guay, D. Plasma-sprayed nanocrystalline Ti-Ru-Fe-O coatings for the electrocatalysis of hydrogen evolution reaction. *J. Alloys Compounds*, 345, 228–237, 2002.
24. Rashkov, R., Arnaudova, M., Avdeev, G., Zielonka, A., Jannakoudakis, P., Jannakoudakis, A., and Theodoridou, E. NiW/TiO_x composite layers as cathode material for hydrogen evolution reaction. *Int. J. Hydrogen Energy*, 34, 2095–2100, 2009.
25. Rajeshwar, K., and Ibanez, J. G. *Environmental Electrochemistry: Fundamentals and Applications in Pollution Abatement*. Academic Press, CA, 361–497, 1997.
26. Ghany, N. A. A., Kumagai, N., Meguro, S., Asami, K., and Hashimoto, K. Oxygen evolution anodes composed of anodically deposited Mn-Mo-Fe oxides for seawater electrolysis. *Electrochim. Acta*, 48, 21–28, 2002.
27. Xu, L., Xin, Y., and Wang, J. A comparative study on IrO_2-Ta_2O_5 coated titanium electrodes prepared with different methods. *Electrochim. Acta*, 54, 1820–1825, 2009.
28. Scialdone, O., Randazzo, S., Galia, A., and Filardo, G. Electrochemical oxidation of organics at metal oxide electrodes: The incineration of oxalic acid at IrO_2-Ta_2O_5 (DSA-O_2) anode. *Electrochim. Acta*, 54, 1210–1217, 2009.

29. Escobar, C., Soto-Salazar, C., and Toral, M. I. Optimization of the electrocoagulation process for the removal of copper, lead and cadmium in natural waters and simulated wastewater. *J. Environ. Manage.*, 81, 384–391, 2006.
30. Lai, C. L., and Lin, K. S. Sludge conditioning characteristics of copper chemical mechanical polishing wastewaters treated by electrocoagulation, *J. Hazardous Mater.*, B136, 183–187, 2006.
31. Kobya, M., Demirbas, E., Dedeli, A., and Sensoy, M. T. Treatment of rinse water from zinc phosphate coating by batch and continuous electrocoagulation process. *J. Hazardous Mater.*, 173, 326–334, 2010.
32. Casqueira, R. G., Torem, M. L., and Kohler, H. M. The removal of zinc from liquid streams by electroflotation. *Miner. Eng.*, 19, 1388–1392, 2006.
33. Golder, A. K., Samanta, A. N., and Ray, S. Trivalent chromium removal by electrocoagulation and characterization of the process sludge. *J. Chem. Technol. Biotechnol.*, 82, 496–503, 2007.
34. Gao, P., Chen, X. M., Shen, F., and Chen, G. H. Removal of chromium(VI) from wastewater by combined electrocoagulation-electroflotation without a filter. *Sep. Purif. Technol.*, 43, 117–123, 2005.
35. Arroyo, M. G., Perez-Herranz, V. Montanes, M. T., Garcia, J., and Guinon, J. L. effect of pH and chloride concentration on the removal of hexavalent chromium in a batch electrocoagulation reactor, *J. Hazardous Mater.*, 169, 1127–1133, 2009.
36. Bazrafshan, E., Mahvi, A. H., Nasseri, S., Mesdaghinia, A. R., Vaezi, F., and Nazmara, S. Removal of cadmium from industrial effluents by electrocoagulation process using iron electrodes. *Iran. J. Environ. Health Sci. Eng.*, 3(4), 261–266, 2006.
37. Meunier, N., Drogui, P., Montane, C., Hausler, R., Blais, J. F., and Mercier, G. Heavy metals removal from acidic and saline soil leachate using either electrochemical coagulation or chemical precipitation. *J. Environ. Eng.*, May 2006, 545–554, 2006.
38. Kabdasli, I., Arslan, T., Olmez-Hanci, T., Arslan-Alaton, I., and Tunay, O. Complexing agent and heavy metal removals from metal plating effluent by electrocoagulation with stainless steel electrodes. *J. Hazardous Mater.*, 165, 838–845, 2009.
39. Kumar, P. R., Chaudhari, S., Khilar, K. C., and Mahajan, S. P. Removal of arsenic from water by electrocoagulation. *Chemosphere,* 55(9), 1245–1252, 2004.
40. Gregor, J. Arsenic removal during conventional aluminium-based drinking-water treatment. *Water Res.*, 35(7), 1659–1664, 2001.
41. Parga, J. R., Cocke, D. L., Valenzuela, J. L., Gomes, J. A., Kesmez, M., Irwin, G., Moreno, H., and Weir, M. Arsenic removal via electrocoagulation from heavy metal contaminated groundwater in La Comarca Lagunera Mexico. *J. Hazardous Mater.*, B124, 247–254, 2005.
42. Mavrov, V., Stamenov, S., Todorova, E., Chmiel, H., and Erwe, T. New hybrid electrocoagulation membrane process for removing selenium from industrial wastewater. *Desalination,* 201, 290–296, 2006.
43. Tanneru, C. T., and Chellam, S., Mechanisms of virus control during iron electrocoagulation–Microfiltration of surface water. *Water Res.*, 46, 2111–2120, 2012.

7

Adsorption: Materials, Chemistry, and Applications

7.1 Introduction

Sorption and adsorption can be defined, respectively, as gas–solid and liquid–solid phenomena in which the particular components are accumulated or concentrated at the surface between the two phases. Unbalanced forces of attraction between the gas–solid or liquid–solid phases result in an increase of concentration of the particular components in the solid phase. In this chapter, discussion on sorption/adsorption focuses on the solid–liquid system.

Sorption can be categorized as either physical sorption or chemisorption, based on the strength of these forces. Physical sorption involves only relatively weak forces, while in chemisorption a chemical bond is formed between the sorbate component(s) and the sorbent components on the solid surface. Metal adsorption occurs according to chemisorption or chemical adsorption reactions.

In the past several decades, a dramatic increase in quantity of metal contaminants has posed many serious environmental problems. The most common treatment processes like precipitation and ion exchange are usually effective in reducing the extent of contamination but are not economical. The novel absorption by various sorbents, such as biosorbents and nano-sized sorbents, has emerged as one of the most effective technologies for removing organic and inorganic pollutants from water and wastewater. According to a literature survey, metals and adsorption are the most commonly studied subjects of research in the past 10 years.[1] Numerous researches have been carried out during this time on the subjects of metal adsorption, metal sorption, and metal biosorption. A large number of papers have been published that provide a good understanding of the process and various methods for preparation of effective sorbents.

There are several types of adsorbents that can be used for decontamination of heavy metals. The most important adsorbents are activated carbon, biosorbent, metal oxide, and synthesized sorbent. They can be from both natural and synthesized/man-made sources.

Activated carbon is undoubtedly the most popular and widely used adsorbent in wastewater treatment applicants because of its large surface area and its high degree of surface reactivity.[2] Activated carbons, in the form of powder-activated carbon (PAC), granular-activated carbon (GAC), activated carbon fibrous (ACF), and activated carbon cloth (ACC), have been utilized in water treatment. Although activated carbon is designed for the treatment of organic contaminants, it can still remove some heavy metals.

Biosorbent is an adsorbent that is based on active or inactive biomaterial or its derivative. Biosorption of heavy metal onto biosorbent has shown increasing importance in recent years.[3–4] Bacteria, fungi, yeast, and algae have been shown to remove heavy metals in reasonably good amounts from aqueous solutions.[5–10] Normally, pretreatment of raw biomaterial must be performed prior to the application because it can reduce leaching of unwanted substances such as organic compounds and light metal ions during the biosorption and improve performance.

Metal oxides can be from natural sources. Adsorption onto soil clays (metal oxides) constitutes one of the most important sinks for heavy metals in soil.[11–13] There are three basic species of clays: smectites (e.g., montmorillonite), kaolinite, and micas, out of which montmorillonite has the highest cation adsorption capacity.

Important natural oxides include iron oxide (goethite), aluminum oxide, and manganese oxide. Iron oxide is an abundant component in the earth, which has been studied extensively and is considered an important adsorbent for trace heavy metals. The alumino–silicate mineral zeolite, which consists of a wide variety of species such as clinoptilolite and chabazite, has the potential of providing precise and specific separation of heavy metal ions. The aluminum substitution of silicon provides a negative framework favorable for a high cation adsorption capacity.

In addition to the natural metal oxides, synthesized metal oxides have increasingly become important for metal removal from aqueous solutions. They can be prepared by engineered approaches and demonstrate very good adsorption capacities for metal decontamination. For example, Fe-Mn binary oxide developed by Zhang et al. has demonstrated extremely high arsenic uptake from groundwater.[14]

Synthesized organic sorbent has also attracted great attention as most commercially available sorbents such as GAC cannot effectively remove heavy metals. The sorbent is fabricated through complex chemical pathways. As such, this type of sorbent is often costly. However, the sorbent that is fabricated with a designed approach would have specific functional groups can effectively remove heavy metals.

In the past 10 years, great effort has been made in the development of cheaper and/or more effective adsorbents. The number of low-cost adsorbents increases steadily.[15–51] The adsorbents are either metal oxides or biosorbents.

The low-cost adsorbents include industrial wastes such as iron(III) hydroxide waste from fertilizer industries,[15] chabazite,[16] siderite,[17] peat,[18] sewage sludge,[19] apple residual, bagasse fly ash,[21] sugarcane bagasse,[22] grape bagasse,[23] and by-product wastes such as cactus pulp,[24] sunflower stalks,[25] groundnut husks,[26] crab shell,[27] sawdust,[28–29] rice husk,[30–32] bark,[33–35] plant leaves,[36–38] coconut shells,[39] textiles,[40] chitin,[41] chicken feathers,[42] fulvic acid,[43] wetland plants,[44] red mud,[45] vermicompost,[46] citrus reticulate,[47] and parthenium.[48] Various seaweeds and other raw biomaterials have also been studied and reported in the literature. These materials from industrial waste, bioproducts, or natural sources can serve as low-cost alternatives to expensive adsorbents for treatment of heavy metal waste streams.[49–51]

In this chapter, the aforementioned adsorbents are discussed for their adsorption chemistry, treatment capacities, and factors that affect the adsorption processes. A series of reactors that are used for the metal waste treatment is presented. Several case studies are included to illustrate the adsorption performance and mechanisms.

7.2 Activated Carbon

Depending on the operational conditions in the manufacture of carbons, the commercial available activated carbons can be divided into H- and L-types.[52–54] Those that are activated at high temperatures and adsorb H^+ are called H-type carbons. Examples of H-type activated carbons are coconut shell based and dust coal carbons such as Filtrasorb 400 and 200.[55,56] Those low-temperature oxidized carbons that can adsorb OH^- ions primarily are called L-type carbons, such as wood-based carbons.

Adsorption by activated carbon has been used for removing the toxic metal pollutants from aqueous solution. GAC is generally used in batch and fixed-bed columns. Occasionally, PAC is used, which is not recovered for regeneration; however, such an application is usually limited. Numerous studies have demonstrated that many factors, such as pH, ionic strength, initial metal ion concentration, competing metal ions, adsorbent dosage and type, background electrolyte, and complexing agent, influence metal ion adsorption.

7.2.1 Surface Properties

Important surface properties of activated carbons include surface charge, presence, and effect of functional groups, porosity, and size. The surface charge and functional groups are related to each other and determine metal adsorption performance of activated carbon as illustrated in Figure 7.1.

A titration study can be used to measurement the functionality of activated carbon. Typical titration results of activated carbon at different ionic strengths are shown in Figures 7.2 to 7.4. The pH under which the surface

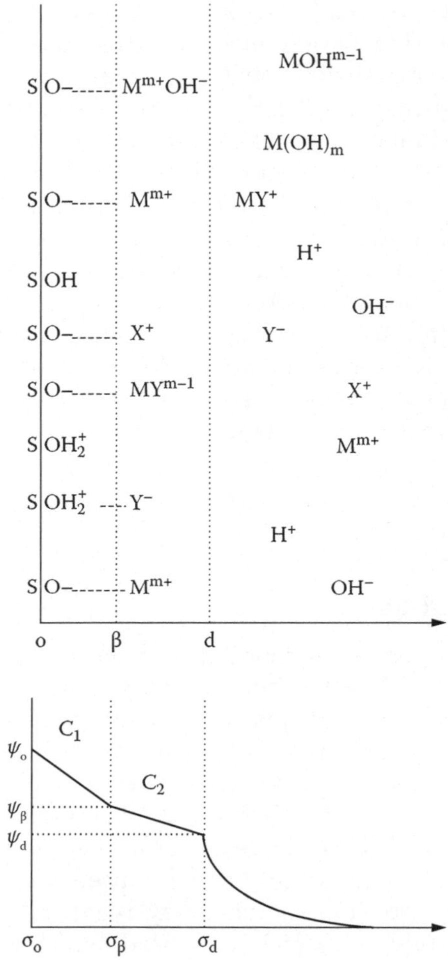

FIGURE 7.1
Schematic illustration of two-pK triple-layer surface complex formation model. m, M, Y, X defined in Table 7.1.

charge is zero is termed the pH of point of zero charge (pH_{PZC}). The values of pH_{pzc} range from 5 to 11 for activated carbons. The figures show the pH_{pzc} of 7.2 for the Filtrasorb 200 granular activated carbon.

Surface charge density of activated carbon (σ_0, C/cm^2) is defined by the uptake of protons by the surface and is determined by potentiometric titration as follows:

$$\sigma_o = \frac{(C_A - C_B + [OH^-] - [H^+])F}{Sa} \tag{7.1}$$

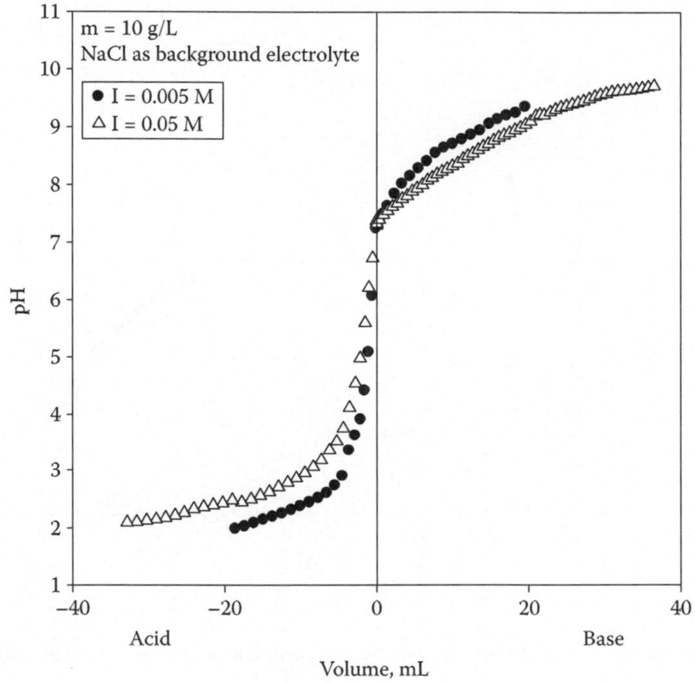

FIGURE 7.2
Potentiometric titration of Filtrasorb 200 granular-activated carbon at different ionic strength: pH curve.

where C_A and C_B are the concentrations of acid and base needed to reach a point on the titration curve, and mol/L, $[H^+]$ and $[OH^-]$ are the concentrations of H^+ and OH^-, in mol/L, F is the Faraday constant (96,485 C/mol), S is the specific surface area, in m^2/g, and a is the concentration of activated carbon in g/L. Note that this equation can also be used for other adsorbents such as metal oxides.

At a pH less than pH_{pzc}, a positively charged surface prevails. At a pH above pH_{pzc}, a negatively charged surface is formed. The surface charge as a function of solution pH for an activated carbon (Filtrasorb 200) is shown in Figure 7.2. The pH-dependent surface charge property implies that at a higher pH, adsorption of metal ions will be higher.

The surface charges result from the chemical reactions on the activated carbon surface, given in Equations (a) and (b) of Table 7.1, according to the two-pK triple-layer surface complex formation model (SCFM) illustrated by Figure 7.1. The surface contains an ionizable functional group, –SOH. The surface charges of activated carbon depend on the degree of the ionization and consequently on the solution pH according to the reactions.

Ion strength may affect the surface charge of adsorbent. Its effect, however, is less obvious than pH. As shown in Figure 7.3, the surface charge of GAC decreases with an increase in ionic strength when the pH value of the

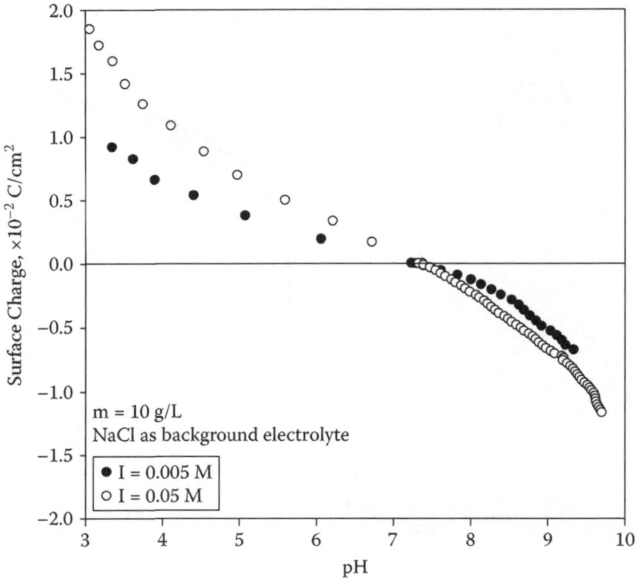

FIGURE 7.3
Surface charge of Filtrasorb 200 granular activated carbon as a function of pH at different ionic strengths.

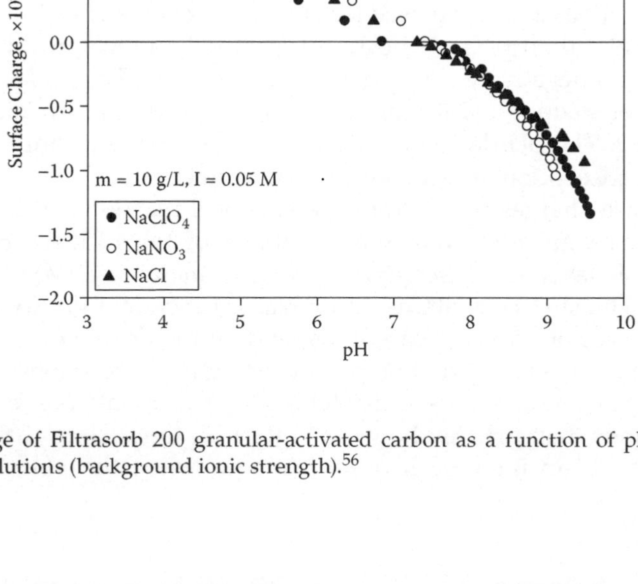

FIGURE 7.4
Surface charge of Filtrasorb 200 granular-activated carbon as a function of pH in various Electrolyte solutions (background ionic strength).[56]

TABLE 7.1

Two-pK Triple-Layer Complex Formation Model for Metal Adsorption

Reaction	Reaction Equations
Surface protolysis reaction	a. $SOH + H^+ + \exp(-y_o) \overset{K_{H1}}{\Leftrightarrow} SOH_2^+$
	b. $SOH - H^+ - \exp(-y_o) \overset{K_{H2}}{\Leftrightarrow} SO^-$
Electrolyte (XY) surface protolysis reaction	c. $SOH + Y^- + H^+ + \exp(-y_o) - \exp(-y_\beta) \overset{K_X}{\Leftrightarrow} SOH_2^+Y^-$
	d. $SOH + X^+ - H^+ - \exp(-y_o) + \exp(-y_\beta) \overset{K_{Na}}{\Leftrightarrow} SO^-X^+$
Metal adsorption reaction	e. $SOH + M^{2+} - H^+ - \exp(-y_o) + 2\exp(-y_\beta) \overset{K_{Cu}}{\Leftrightarrow} SO^-M^{2+}$
	f. $SOH + M^{2+} - 2H^+ - \exp(-y_o) + \exp(-y_\beta) \overset{K_{CuOH}}{\Leftrightarrow} SO^-MOH^+$
	g. $SOH + M^{2+} - H^+ + Cl^- - \exp(-y_o) + \exp(-y_\beta) \overset{K_{CuCl}}{\Leftrightarrow} SO^-MCl^+$

where $y_o = e\psi_o/kT$ referred to o-layer and $y_\beta = e\psi_\beta/kT$ to β-layer, which are illustrated in Figure 7.1; M = heavy metal ions such as Cu, Zn, or Co; X = less active cations such as sodium; Y = anions such as Cl, NO_3, or ClO_4.

solution is below pH_{PZC}. When the solution pH is above pH_{PZC}, however, ionic strength has little effect on surface charge. The surface charge is a higher at the higher ionic strength. The difference is not as large as that below pH_{PZC}.

Besides the effect of magnitude ionic strength, different types of electrolytes may affect the property of the surface charge on activated carbon. Figure 7.4 demonstrates the surface charge as a function of pH under three different electrolytes, NaCl, $NaNO_3$, and $NaClO_4$, which were obtained from the potentiometric titration studies of an activated carbon (Filtrasorb 200).

Comparison of the surface charge densities in Figure 7.4 indicates that the values in NaCl are slightly higher than those in $NaNO_3$ and $NaClO_4$ when the pH is lower than pH_{PZC}. A reverse trend, however, can be observed at $pH > pH_{PZC}$. The values for $NaNO_3$ and $NaClO_4$ are quite close to each other. The pH at point of zero charge is slightly different, which is in the order of $NaClO_4 < NaCl < NaNO_3$. This is possibly due to the similar surface complex formation reactions between the functional groups and nitrate and perchlorate, as both of them are inert anions in the solutions. At the pH above pH_{PZC}, the surface charge in $NaClO_4$ solution is almost the same as that in $NaNO_3$ solution.

7.2.2 Effect of pH

A series of studies has shown that adsorption of metal ion onto activated carbons from aqueous solution is strongly pH dependent. Typically, cationic metal adsorption increases from low pH (e.g., 1 to 2) to slightly higher pH (e.g., 5 to 7), after which the adsorption reaches its maximum demonstrated in Figure 7.5.[55,56]

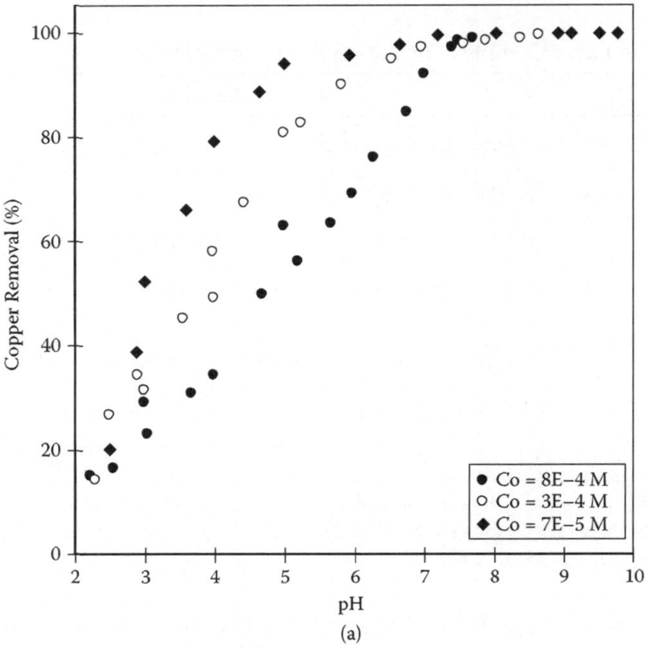

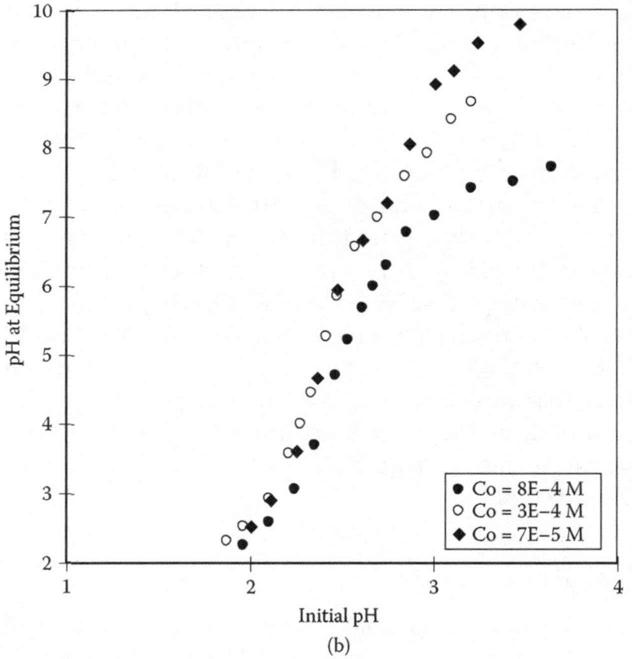

FIGURE 7.5
Effect of pH on adsorption of copper: (a) adsorption as a function of pH; (b) variation of pH in sorption.[60]

Sigworth and Smith were among the first to report that adsorption of an inorganic by activated carbon depended on solution pH.[57] The researchers showed that lead showed little adsorption at pH 2 but fairly good removal at pH 5. The onset of adsorption occurs before hydrolysis and precipitation of metals, and generally coincides with the loss of the outer hydration sheaths of the metal ion.

The adsorption density of Calgon Filtrasorb 400 increased with increasing pH to a maximum value and then declined rather rapidly with any further increase in pH for removal of chromium.[58] When the pH was greater than 10, no appreciable adsorption was observed. Nezter and Hughes found that adsorption of lead, cobalt, and copper was insignificant at pH < 2, but adsorption dramatically increased in the pH range from 4 to 10.[59] Copper adsorption equilibrium experiments with Filtrasorb 400 were carried out by Chen et al.[60] The removal efficiency increased from 10% to 95% in the equilibrium pH values ranging from 2.3 to 7 as shown in Figure 7.5a.

In addition, a dramatic increase in pH and emission of small gas bubbles were observed during the adsorption as shown in Figure 7.5b.[60] Corapcioglu and Huang reported that the adsorption of copper ion by activated carbon increases dramatically from pH 3 to 5, and that adsorption edge of lead is between pH 3 and 6.[53,61] The researchers suggested that pH was a dominant parameter because it affected the charge distribution of various species and the hydroxyl group distribution at the carbon surface. Reed and Matsumoto showed that the adsorption edge for cadmium ion using Darco HDB or Nuchar SN (PAC) ranges from pH 3 to 9.[62]

An increase of solution pH results in a decrease of surface charge, an increase of negatively charge sites, and eventually an increase of metal ion binding. More important, the adsorption reactions given in Table 7.1 play key roles in the metal uptake. As shown in Equations (e) to (g), higher pH (i.e., a lower concentration of hydrogen ions) can enhance the adsorption. The binding sites in activated carbon for metal removal are carboxylate, amine, phosphate, sulfate, hydroxyl, and other organic functional groups.[63]

The presence of organic matters (chelating agents) such as humic acid and ethylenediaminetetraacetic acid (EDTA) can greatly change heavy metal speciation. This may affect metal adsorption. On the other hand, organic matter may be adsorbed onto activated carbons, which could facilitate metal adsorption. The pH effect on heavy metal adsorption in the presence of organic matters may be different from that in the absence of organic matter.[64,65]

7.2.3 Types of Metal Ions

Different metal ions behave differently in the adsorption process. At the same pH, the adsorption of one metal may be different to that of another metal, even though both metals have the same valency and are under the same pH. As such, the hypothesis of adsorption due to abstraction that many researchers argue is fairly incorrect can be better illustrated by an experiment reported by Chen and Lin.[66]

During the experiment, only one kind of metal ion is added into the solution. It is shown in Figure 7.6 that the adsorption of copper, zinc, and cobalt ions on activated carbon strongly depends on the solution pH. Higher solution pH results in higher metal ion removal. The adsorption curves of zinc and cobalt ions are almost identical and significantly different from the copper adsorption curve. The adsorption of zinc and cobalt starts at pH 4, while adsorption of copper begins at pH 2. The amount of the metal ions absorbed by the activated carbon is in the following descending order: $Cu^{2+} > Zn^{2+} \approx Co^{2+}$. For example, at pH 5, the metal ion removal percentages for copper, zinc, and cobalt are 94%, 30%, and 28%, respectively.

At a fixed pH, the surface charge density is always independent of the types of metal ions in the solutions. The role of electrostatic repulsion or attraction is, therefore, the same for metal ions. The difference in adsorption capacities for copper, zinc, and cobalt must be caused by other mechanisms.

In the experiment, pH is in the range of 1.7 to 7.5. According to the calculation using MINEQL+, copper ions exist as Cu^{2+} and $CuCl^+$ at pH 1 to 5.8, while tenorite is formed after 5.8. At pH 1 to 7.5, zinc and cobalt ions are in the form of Zn^{2+} and $ZnCl^+$, and Co^{2+} and $CoCl^+$, respectively. It can be fairly

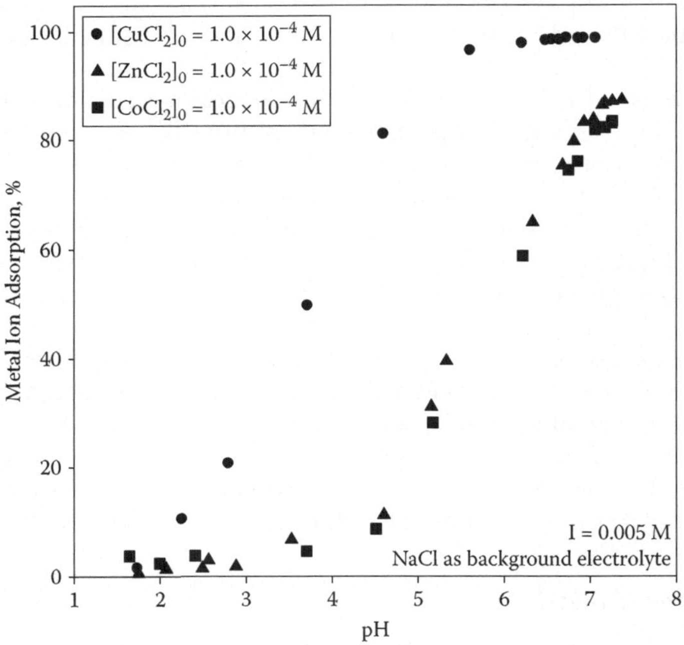

FIGURE 7.6
Comparison of copper, zinc, and cobalt ion adsorption equilibrium. Note that single-species metal is in the adsorption.[66]

reasonable to assume that each metal ion removal is due to the adsorption of the free metal ions and the metal chlorides.

The Pauline ionic radii of these three metal ions are as follows: 0.069 nm for Cu^{2+}, 0.074 nm for Zn^{2+}, 0.074 nm for Co^{2+}.[67] As copper ions have the smallest radius, they may easily penetrate into smaller pores in activated carbon and thus have greatest access to the interior carbon surfaces. In addition, the radius of hydrated ion (e.g., Zinc hydroxi, 0.44 nm) is much larger than that of the free ion (0.074 nm),[68] which may affect the adsorption.

Surface complex formation between the metal ions and functional groups can play an important role. On the basis of the observations in the figure, one may conclude that the functional groups have higher affinity for copper than zinc and cobalt.

The different metal adsorption onto the activated carbon may further be illustrated by comparing the affinities of these metals for organic compounds. The constants of reactions between organic compounds and metal ions are typically in the order of $Cu^{2+} > Zn^{2+} \approx Co^{2+}$. For example, constants of reaction between DL (phenylethylene)-dinitrilotetraacetic acid ($C_{16}H_{20}N_2O_8$) and copper, cobalt, and zinc are $10^{18.7}$, $10^{16.9}$, and $10^{17.05}$ M^{-1}, respectively.[69] l-2-aminopropanoic acid (alanine) can be complexed with these three metal ions with the reaction constants of $10^{8.11}$, $10^{4.31}$, and $10^{4.58}$ M^{-1}. The metal ion adsorption order observed in the above figure is in agreement with that of the reaction constants for metal–organics complexes reported in the literature. As the activated carbon and organics (such as alanine) have similar functional groups, such as carboxyle and hydroxyl groups,[70,71] the trends for the reaction constants for copper, zinc, and cobalt are similar.

7.2.4 Effect of Ionic Strength

Ionic strength may affect heavy metal adsorption onto activated carbons. This may be important as metal-bearing wastewater streams often contain high levels of total dissolved solid (TDS). In addition to heavy metal ions in the streams, inert or less active ions such as sodium, chloride, calcium, and carbonate are present, which contribute to the TDS and ionic strength.

Ionic strength is a gross measure of the amount of background electrolytes in aqueous solution. It plays a complicated role in metal ion adsorption from the aqueous solution. The ionic charges of the background electrolytes, as well as the specific species affect the chemistry of both solution phase and adsorbent surface. On the one hand, alteration of ionic strength changes the aqueous solubility of metal species and, slightly, the species distribution. On the other hand, the surface charges on the adsorbent can cause counter ions closely attached near the charged surface, forming an EDL on the adsorbent surface. These complicate the effect of ionic strength on metal ion adsorption.

A few studies demonstrated lower metal adsorption at higher ionic strength. Huang and Smith reported that cadmium removal by activated carbon decreases as ionic strength is increased from 0.01 to 0.1M.[72] Reed and Nonavinakere observed that adsorption of Ni^{2+} and Cd^{2+} by activated carbon decreased by 10% to 20% with an increase of ionic strength from 0.01 M to 0.1 M.[73] Pesavento et al. reported that copper ion adsorption by activated carbon is independent of ionic strength.[74] It was claimed that competition for adsorption sites in activated carbon between added ionic strength and copper ions was insignificant.

However, many studies show that the metal ion adsorption increases with an increase in ionic strength when activated carbon is used, which is similar to the adsorption by metal oxide. Figure 7.7 illustrates slightly better adsorption of heavy metal when the ionic strength is increased from 0.005 M to 0.05 M.

These observations suggest two mechanisms, that is, competition and EDL compression. Competition from increased background electrolytes depresses heavy metal ion adsorption, while enhanced ELD compression facilitates its adsorption. Therefore, the apparent ionic strength effect is due to the results from these two mechanisms. The enhancement in the adsorption results from the compression of EDL, while the competition for the functional groups between metal ions and other ions in solution is less important.

It should be noted that the effect of pH is much more important than ionic strength. In addition, ionic strength may greatly reduce the metal sorption onto biosorbents.

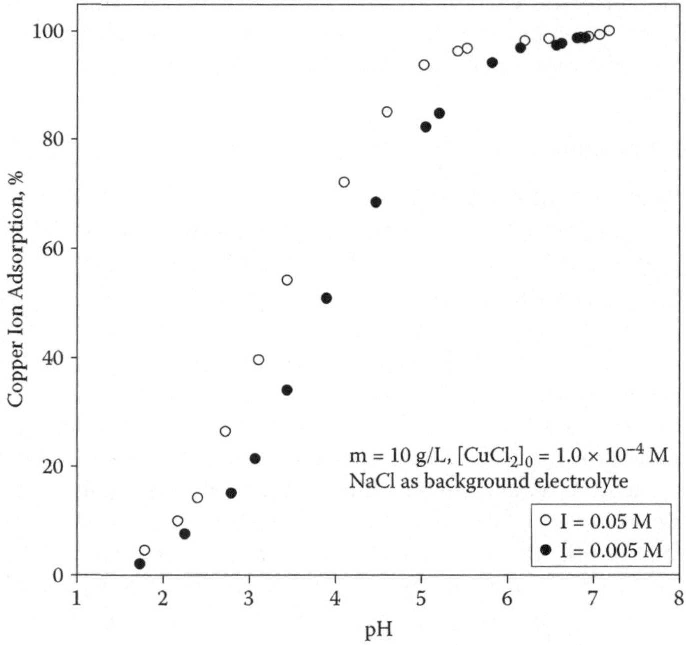

FIGURE 7.7
Copper ion adsorption equilibrium: effect of ionic strength.[66]

7.2.5 Effect of Background Electrolyte

In addition to ionic strength, background electrolyte composition may affect heavy metal adsorption onto activated carbons. Reed and Nonavinakere investigated the Cd^{2+} removal by Darco KB carbon in 0.01 M $NaNO_3$ and $NaClO_4$ solutions and concluded that the type of background electrolyte has no effect on metal removal.[74]

However, the adsorption behaviors of $CuCl_2$, $Cu(ClO_4)_2$, and $Cu(NO_3)_2$ are different. In the study by Chen and Lin, copper nitrate or perchlorate was used as the copper solution, while, correspondingly, sodium nitrate or perchlorate was added as background electrolyte.[66] Figure 7.8 shows that the results for copper nitrate and percholates are almost the same. The adsorption efficiencies of $Cu(ClO_4)_2$ and $Cu(NO_3)_2$ are higher than $CuCl_2$ (sodium chloride as background electrolyte).

Lower removal of copper (in $CuCl_2$) could result from the competition between the surface functional groups and chloride ions (Cl^-) for copper ions. The calculation using MINEQL+ shows that no complex species containing NO_3^- or ClO_4^- exists in the solution. Chloride is a complexing anion, and nitrate and perchlorate are noncomplexing anions. Regardless of the formation of the surface complex containing these complexing anions, the concentration of free metal ion is reduced because of the complexation of free metal ion with the complexing anion (Cl^-), which has a higher concentration

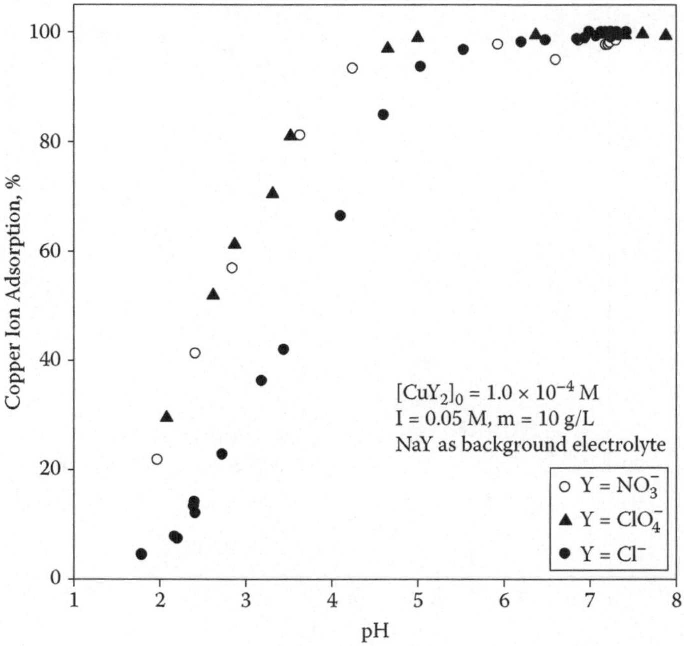

FIGURE 7.8
Effect of background electrolytes on copper adsorption (NaCl, $NaNO_3$, and $NaClO_4$).[66]

in the solution. Nevertheless, the formation of the surface complex of free copper ion with surface functional group results in removal of most copper ion. Thus, the copper ion removal is higher in the case of noncomplexing anion than complexing anion.

7.2.6 Effect of Initial Concentration and Dosage

Metal removal by activated carbon depends on the adsorption capacity under a given condition. The concentration of solution per gram of carbon (adsorbate/adsorbent ratio) plays an important role in the treatment efficiency.

Tan and Teo reported that the initial lead concentration greatly affected the lead removal by an activated carbon.[75] Namasivayam and Kadirvelu concluded that mercury loading per gram of coir pith carbon (mg metal/g carbon) decreased with a decrease in adsorbate concentration.[76]

It was reported that the adsorption edge of copper shifted to a higher pH when the initial metal ion concentration became higher.[60,66] Similar trends have been observed by Corapcioglu and Huang and Reed and Matsumoto.[53,61,62] Figure 7.9 illustrates the effect of initial metal concentration on the adsorption efficiency typically reported in the literature.[66] As shown, the metal removal decreases when the initial concentration is increased.

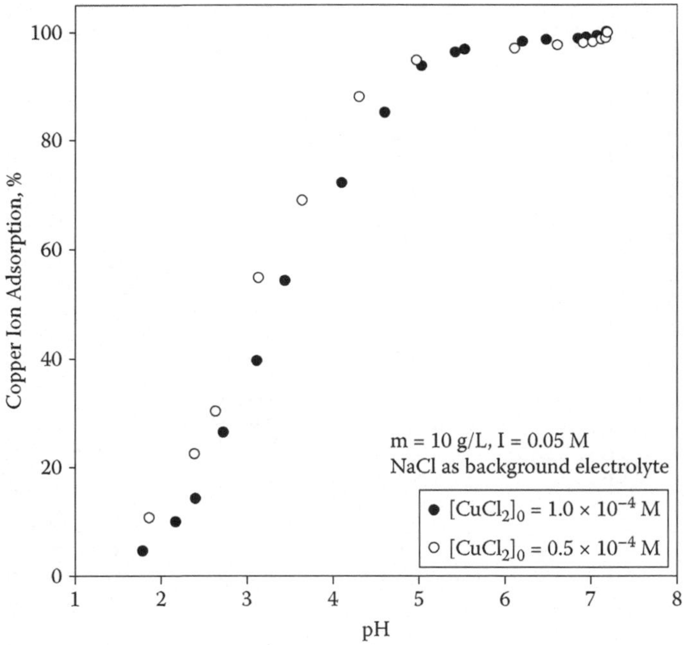

FIGURE 7.9
Copper ion adsorption equilibrium; effect of initial copper ion concentration.[66]

As adsorption capacity (or adsorption sites) of any adsorbent (e.g., activated carbon) is limited, the total amount of metal ions adsorbed is limited. Therefore, the relative adsorption amount, that is, the metal removal percentage (adsorption efficiency), is lower for the initial higher concentration. In addition, the higher surface coverage, resulting from the increase of metal ion concentration, raises the activation energy for reaction, thereby making it more difficult for the surface to bind metal ions.

7.2.7 Adsorption Isotherm

Heavy metal ion adsorption capacities of various activated carbons at a fixed pH and under a constant temperature can be described by traditional adsorption isotherms. The typical isotherms used are the Langmuir and Freundlich equations with a detailed description given in Chapter 8. The equation constants can be calculated by best fitting the experimental results with least-squares regression to the equation(s). Table 7.2 lists some adsorption isotherm constants for adsorption of metals by different carbons.

7.2.8 Presence of Industrial Organic Matters

Removal of specific contaminants is usually accomplished in the presence of a large amount of other pollutants. A series of studies have demonstrated that natural organic matter is the primary source of materials that hinder the effectiveness of activated carbon in removing metal contaminants. With the increasing application of complexing ligands in the industrial and agricultural fields, more complexing ligands are discharged into the environment. It is found that complexing ligands have an important effect on the metal ion distribution between solid phase and liquid phase because metal complexes have different adsorption characteristics compared with the uncomplexed metal ions.

EDTA is reported in various studies in the literature as an important industrial organic waste. EDTA has been widely applied in the industries as a strong complexing agent of most metals. As a result, the EDTA species complexed with metal ions could exist in many waste streams. It is also considered as a cocontaminant in some cases. Adsorption becomes one of the few treatment techniques for removal of both metal and EDTA.

Adsorption of EDTA (in the absence of metal ions) by an activated carbon at different solution pHs was investigated by Chen and Wu.[64] It is shown in Figure 7.10a that EDTA adsorption fluctuates between 55% and 95% in the pH range of 2.5 to 7.3. When solution pH is low, the EDTA adsorption is rather high, and reaches a maximum of 95% at pH 3.3. As solution pH is increased, EDTA adsorption decreased gradually.

Figure 7.10b shows that final solution pH increases sharply when the initial solution pH is low. When the initial solution pH is above 4, the final solution pH reaches the maximum value of pH 7.3. This indicates that

TABLE 7.2
List of Constants for Adsorption Isotherms

No.	Metal	Carbon	pH	Ionic Strength	Complexing Agent	q_{max}	b	K	1/n	Reference
1	Cu(II)	Filtrasorb 200	6	0.05 M	No	0.1107 mol-Cu/kg	130.28 m^3/mol			77
2	Cu(II)	Filtrasorb 200	6	0.005 M	No	0.0527 mol-Cu/kg	162.05 m^3/mol			
3	Cu(II)	Filtrasorb 200	6	0.005 M	EDTA=2×10^{-4} M	0.02 mol-Cu/kg	2.39×104 m^3/mol			
4	Cu(II)	Fitrasorb 200	4.9	~0.01 M	No	6.14 mg/g	0.18 L/mg			78
5	Cu(II)	CAC	4.9	~0.01 M	No	14.92 mg/g	0.08 L/mg			12
6	Cu(II)	SAC	4.9	~0.01 M	No	11.85 mg/g	0.2 L/mg			12
7	Cu(II)	Activated carbon from *Eucalyptus camaldulensis* Dehn bark	5			29.12 mg/g	0.096 L/mg			79
	Pb(II)		5			110.5 mg/g	0.0317 L/mg			
8	Mn(II)	Activated carbon granular from coconut shell	5.8			75.65 mg/g	0.010 L/g			80
	Fe(II)		5.8			81.89 mg/g	0.016 L/g			
	Ni(II)		5.8			83.16 mg/g	0.005 L/g			
	Cu(II)		5.8			86.01 mg/g	0.983 L/g			
9	Cu(II)	Activated carbon obtained by pyrolysis of potato peel	6			62.76 mg/g	0.087 L/mg			81

10	Pb(II)	Commercial-activated carbon	5.8					18.98	0.28	82
	Pb(II)	Cotton stalk–activated carbon	4.4					26.83	0.326	
11	Pb(II)	Carbon nanofibers grown on powdered activated carbon	5.5			100 mg/g	0.4 L/mg			83
12	Ni(II)	Activated carbon from *Hevea brasiliensis*	6			16.89 mg/g	0.3178 L/mg			84
	Zn(II)		6			22.42 mg/g	0.2131 L/mg			
13	Pb(II)	*M. baccifera* raw charcoal	5			53.76 mg/g	1.317 L/mg			85
	Pb(II)	*M. baccifera* activated charcoal	5			10.66 mg/g	0.231 L/mg			
14	As(III)	Supported nano zero-valent iron on activated carbon	6.5			18.19 mg/g	8.952 L/mg			86
	As(V)		6.5			12.02 mg/g	71.35 L/mg			
15	Cr(VI)	Acid modified waste activated carbons	3			10.93 mg/g	4.2087 L/mg			87
16	Cu(II)	Cattle-manure-compost based activated carbons	4.1			6.06 mg/g	0.189 L/mg			88
	Pb(II)		4.1			9.004 mg/g	0.594 L/mg			
17	Cu(II)	Chestnut shell activated carbon	5			98.04 mg/g	0.0037 L/mg			88
	Cu(II)	Grapeseed activated carbon	2					0.19	0.91	
18	Hg(II)	Carbonaceous sorbents derived from walnut shell	5			151.5 mg/g	0.009 L/mg			89

(*Continued*)

TABLE 7.2 (CONTINUED)

List of Constants for Adsorption Isotherms

No	Metal	Carbon	pH	Ionic Strength	Complexing Agent	q_{max}	b	K	1/n	Reference
19	Pb(II)	Activated carbon prepared from Phaseolus aureus hulls	6			21.8 mg/g	0.28 L/mg			90
	Zn(II)		7			21.2 mg/g	0.26 L/mg			
	Cu(II)		7			19.5 mg/g	0.27 L/mg			
	Cd(II)		8			15.7 mg/g	0.19 L/mg			
20	Pb(II)	Activated carbon prepared from *Spartina alterniflora*	5.6					31.57	0.27	91
21	Cu(II)	Poly(N,N-dimethylaminoethyl methacrylate) modification of activated carbon	5.5			31.46 mg/g	0.889 L/mg			92
22	Cr(VI)	Acid-treated *C. albidum* seed shells carbon	5			59.63 mg/g	1695 L/mg			93
23	Hg(II)	Activated carbon prepared from the hulls of *C. pentandra*	6					11.24	0.243	94
	Hg(II)	Activated carbon prepared from the hulls of *P. aureus*	7					9.51	0.268	
	Hg(II)	Activated carbon prepared from the hulls of *C. arietinum* waste	7					8.36	0.275	

24	Cu(II)	Activated carbon from *Eucalyptus camaldulensis* Dehn. bark	5		28.8 mg/g	0.19 L/mg			95
	Pb(II)		5		113.85 mg/g	0.23 L/mg			
25	Cu(II)	Activated carbon from a new precursor hazelnut husks	6.7				1.5517	0.3187	96
	Pb(II)		6.7		13.05 mg/g	0.255 L/mg			
26	Cd(II)	Activated carbon prepared from olive stone	7		1.851 mg/g	0.388 L/mg			97
27	As(V)	Granulated activated carbon	6.5	0.01 M NaCl			0.09	0.85	98
	Cd(II)		6.5	0.01 M NaCl			1.45	0.91	
	Cr(VI)		6.5	0.01 M NaCl			0.22	1.08	
	Cu(II)		6.5	0.01 M NaCl			0.37	1.25	
	Ni(II)		6.5	0.01 M NaCl			0.97	1.06	
	Zn(II)		6.5	0.01 M NaCl			0.91	1.01	

Note: CAC: Citric acid modified activated carbon Fitrasorb 200[78]; SAC: Sodium hydroxide-extracted CAC.[12]

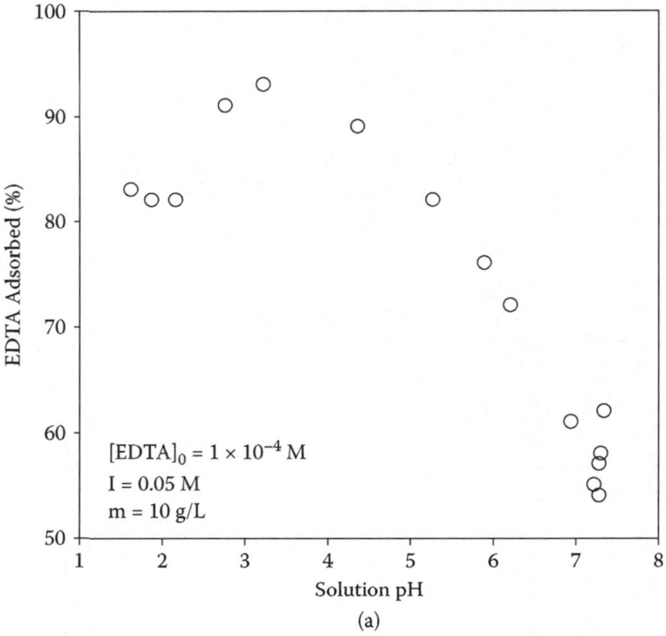

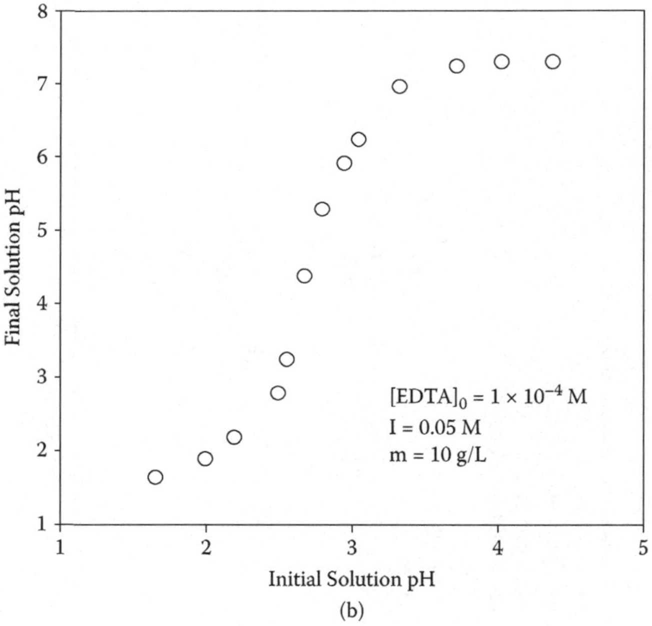

FIGURE 7.10
Effect of solution pH on EDTA adsorption: (a) variation of EDTA adsorbed, (b) variation of solution pH.[64]

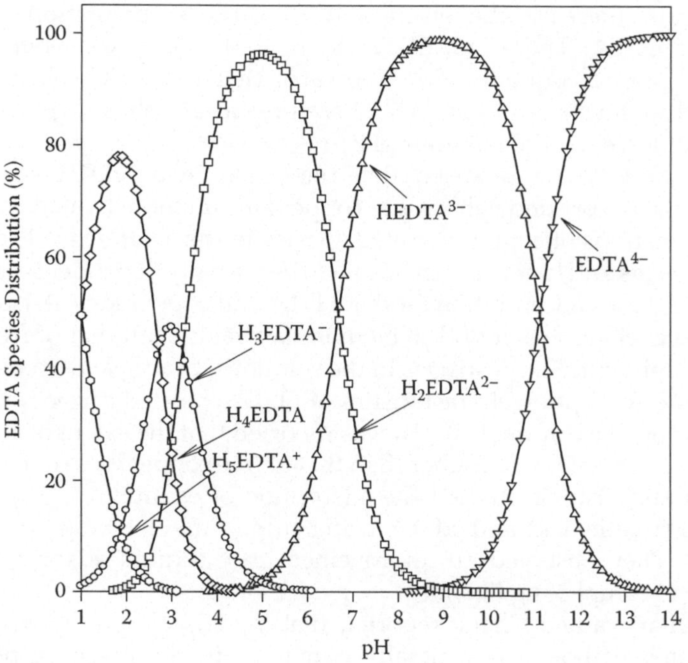

FIGURE 7.11
Variation of EDTA species distribution with pH.

there is an adsorption of hydrogen ions during EDTA adsorption by the carbon.

The EDTA species distribution calculated by MINEQL at different pHs is illustrated in Figure 7.11. It can be seen that the negatively charged H_2EDTA^{2-} and $HEDTA^{3-}$ are the dominant species in the pH range in Figure 7.9. Therefore, the EDTA uptake is due to adsorption of these two major species (i.e., H_2EDTA^{2-} and $HEDTA^{3-}$) on the carbon. Based on experimental observations, the EDTA adsorption is hypothesized to be due to the following reactions:

$$SOH + EDTA^{4-} + H^{+} \rightarrow SOH - HEDTA^{3-} \tag{7.2a}$$

$$SOH + EDTA^{4-} + 2H^{+} \rightarrow SOH - H_2EDTA^{2-} \tag{7.2b}$$

A decrease in solution pH can increase the EDTA adsorption; the solution pH increases after the EDTA adsorption. The surface positive charges of activated carbon decrease when the solution pH is increased. This decreases its electrostatic attractive interactions with EDTA. At a higher solution pH, the EDTA species contains higher negative charges.

This would increase the electrostatic repulsive interactions between adsorbed EDTA. Therefore, EDTA adsorption decreases when solution pH is increased. Equation 7.2b indicates that more hydrogen ions are adsorbed at lower solution pH. Therefore, final solution pH increases sharply at lower initial solution pH.

Chung and Young reported that the presence of EDTA significantly changed the adsorption behavior of copper ions on the activated carbon due to the formations of copper chelated species in the solutions.[99] The adsorption isotherms and kinetics were found to be strongly dependent on the solution pH and the concentration ratio of EDTA and copper ions. Adsorption of EDTA-Cu chelates was found to be more favorable than that of free copper ions and unbound EDTA species in the solutions. Chung and Young further studied the adsorption characteristics of EDTA-chelated copper ion by activated carbon in a column test.[100] It was reported that the extent of adsorption of EDTA-Cu chelates was higher than that of free copper ions in the solution.

Huang and Ostovic studied the adsorption of cadmium in the presence of nitrilotriacetic acid and EDTA using three different kinds of activated carbon.[101] They observed an improvement in cadmium adsorption in the presence of complexing ligands.

Bhattacharyya and Cheng reported that the adsorption of heavy metals by activated carbon was generally enhanced in the presence of organic ligands.[102] The adsorption of Cd-Ni-EDTA increased as pH was increased from 3 to 7 and then decreased as pH was further increased. The maximum adsorption capacity was found at pH 7 for Cd-Ni-EDTA and pH > 9 for the Cd-Ni-TRIEN system. They proposed that the adsorption capacity could be related to the electrostatic interactions between the charges of solutes and the surface of activated carbon.

The effects of pH and initial copper concentrations on the adsorption were studied by Chen and Wu.[64] The concentration ratio values of copper to EDTA varied from 1:2.5 to 0.5:1. As shown in Figure 7.12, the copper adsorption in the presence of EDTA increases with an increase in pH. The removal increases from 10% to 95% as solution pH is increased from 2 to 7.5. Compared to the single-species metal ion adsorption (e.g., Figure 7.2), the adsorption percentage of EDTA-chelated copper is slightly higher. As illustrated in Figure 7.12a,b, the variation in concentrations ratio values of copper ions to EDTA appears to have insignificant impact on copper ion adsorption.

In the presence of EDTA, copper precipitation is completely inhibited. When the concentration of EDTA is equal to that of copper ions, the calculation by the MINEQL shows that free copper ions (Cu^{2+}) are the dominant species at a pH less than 2, while the negatively charged $CuHEDTA^-$ and $CuEDTA^{2-}$ species are dominant for a pH greater than 2. The pH_{pzc} of the carbon is pH 7.2.[56] Hence, the activated carbon surface is positively charged when solution pH is less than 7.2. The negatively charged $CuEDTAH^-$ and $CuEDTA^{2-}$ chelates, therefore, is expected to exhibit much higher affinity for the positively charged activated carbon than the Cu^{2+} and $CuOH^+$ present in the single copper ion adsorption.

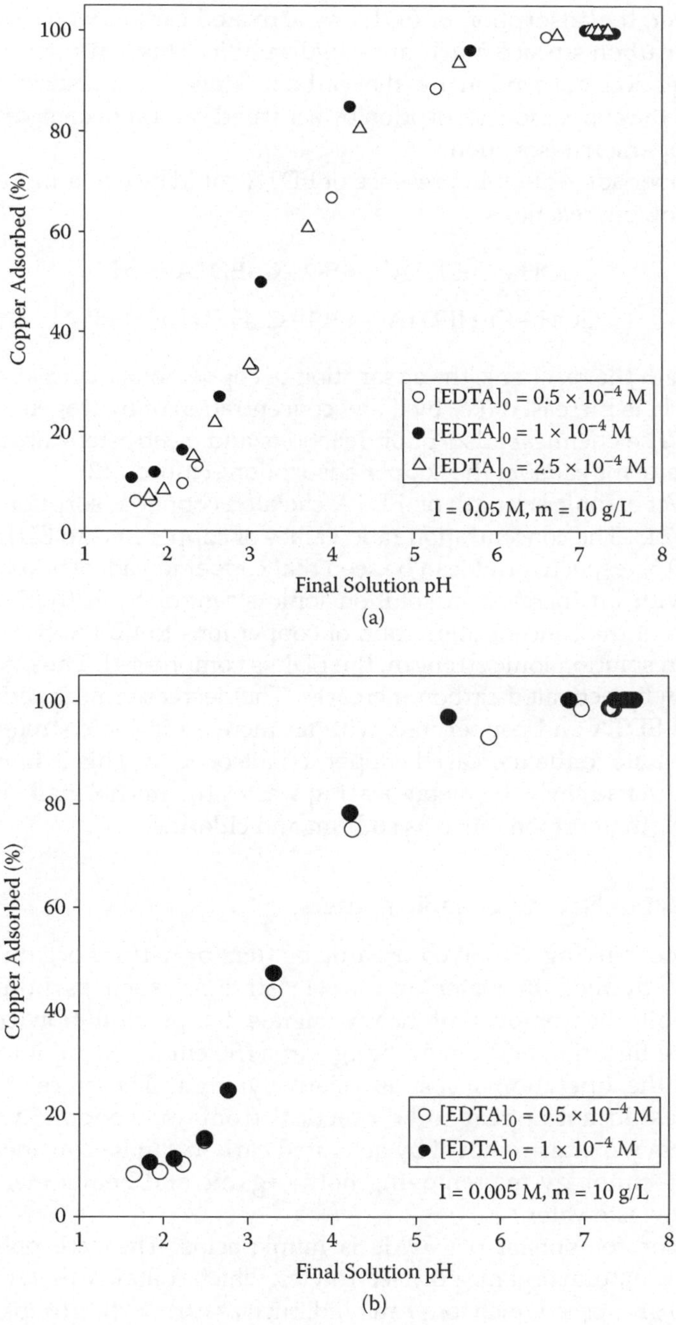

FIGURE 7.12
Effect of concentration ratio of copper ions to EDTA on EDTA-chelated copper adsorption: (a) $[Cu]_0 = 1 \times 10^{-4}$ M, (b) $[Cu]_0 = 0.5 \times 10^{-4}$ M.[65]

Moreover, the adsorption of EDTA by activated carbon would render the activated carbon surface much more hydrophilic. This can further increase the copper ion adsorption by the carbon. Thus, the presence of EDTA increased the copper ion adsorption by activated carbon in comparison with single copper ion adsorption.

The copper adsorption in presence of EDTA could be explained according to the following reactions:

$$SOH + CuEDTA^{2-} \rightarrow SO - CuEDTA^{3-} + H^{+} \quad (7.3a)$$

$$SOH + CuHEDTA^{-} \rightarrow SO - CuHEDTA^{2-} + H^{+} \quad (7.3b)$$

As shown in the reactions, the adsorption of copper ions increases when the solution pH is increased (i.e., by a low concentration of hydrogen ions in the solution). The chemical adsorption reactions and electrostatic attraction are the mechanisms causing the copper adsorption (Figure 7.12).

The effect of ionic strength on EDTA-chelated copper adsorption is shown in Figure 7.13. The concentration ratio values of copper ions to EDTA were at 1:1 and 1:0.5, respectively. It can be seen that copper ion adsorption increases slightly with an increase in solution ionic strength from 0.005 to 0.05M, regardless of the concentration ratio of copper ions to EDTA. Because of an increase in solution ionic strength, the EDL is compressed. Thus, copper ion adsorption by activated carbon increases. The decrease in the activity coefficients of EDTA and copper ions with an increase in ionic strength might also contribute to the increased copper ion adsorption. This is important in a practical sense since the metal-bearing waste streams normally have high ionic strength (inert ions such as sodium and chloride).

7.2.9 Effect of Natural Organic Matters

Naturally occurring dissolved organic matters or natural organic matters (NOMs) contained in water and waste streams, such as humic acids, may inhibit the removal of heavy metals by precipitation treatment. Membrane filtration and ion exchange are efficient for metal ion removal; however, the operational cost is relatively high. Therefore, these two approaches do not appear to be practical from an economic viewpoint. Hence, heavy metals removal by activated carbon can be considered as an effective technology for removing both organic and heavy metal pollutants from wastewater.

The major component of NOMs is humic acids. They are polydisperse mixtures of natural organic polyelectrolytes, which contain different types of functional groups to which ions can bind, such as carboxylic groups, phenolic groups, amino, sulfhydryl, and quinone groups. Because of these functional groups, the humic acids behave as heterogeneous ligands, which have strong binding properties and electrostatic interactions with metal ions. It is indicated that they can control metal ion concentrations in surface and groundwaters.

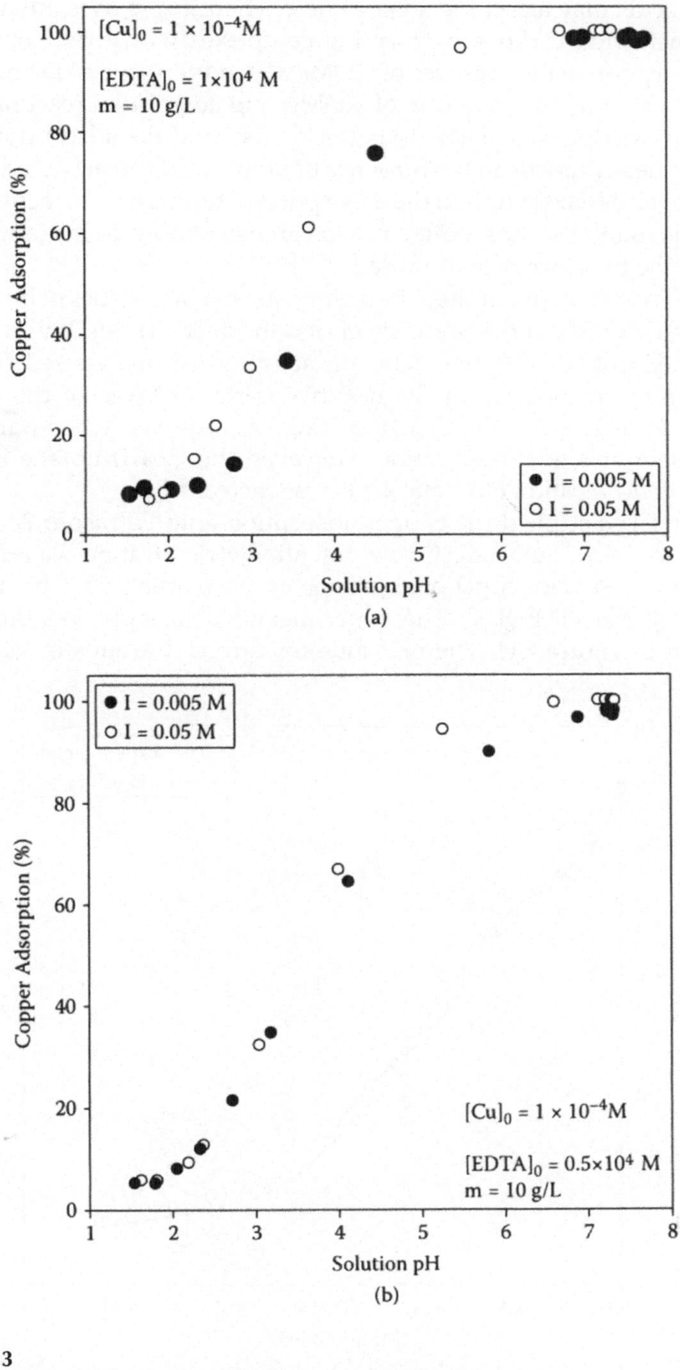

FIGURE 7.13
Effect of ionic strength on EDTA-chelated copper adsorption.

Humic acids may affect the adsorption of chemical species, mainly metal ions, on activated carbons by forming complexes with these ions. Heavy metal adsorption in the presence of NOMs depends on several factors such as the surface charge, functionality of NOMs, and adsorption reactions. There limited knowledge is available to better understand the adsorption characteristics of heavy metals in the presence of humic acids on activated carbons.

Lafrance and Mazet studied the adsorption of humic acid on activated carbon. It was found that adsorption caused an increase in the surface negative charge in the presence of sodium salts.[103]

Ferro-Garcia et al. researched the adsorption characteristics of humic acids on activated carbon in the presence of dissolved Cr(III) salt. It was reported that the adsorption of humic acids on activated carbon caused both pore blockage and an increase in the negative surface charge of the activated carbon.[104] A decrease in the Cr(III) uptake was observed when humic acids were present at low concentration. However, the Cr(III) uptake increased when the concentration of humic acids was increased.

Chen and Wu studied the copper adsorption onto Filtrasorb 200 GAC in the presence of a humic acid.[63] The potentiometric titration was conducted to identify the surface functional groups of the humic acid. The data were modeled by the FITEQL.[105] The experimental data and modeling results are shown in Figure 7.14. The pK value for carboxyl groups is 4.4, and the

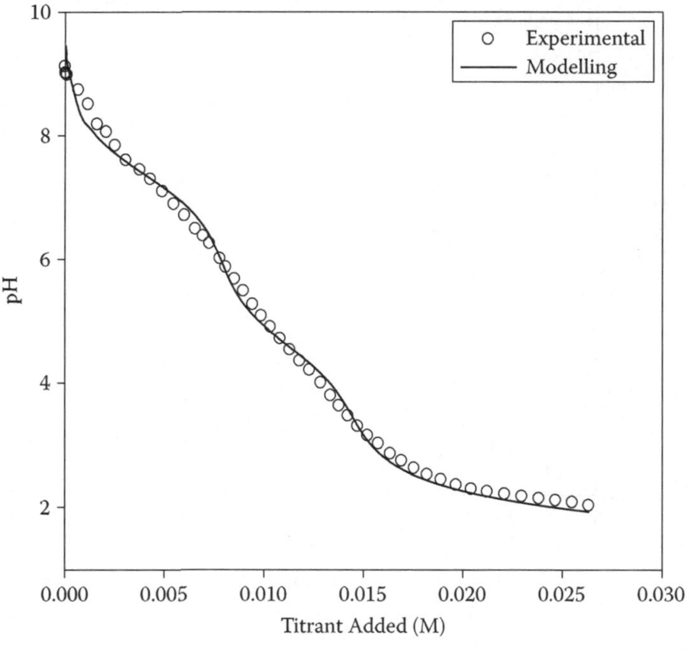

FIGURE 7.14
Potentiometric titration of humic acid in 0.1 M $NaNO_3$ solution.

concentration of carboxyl groups is 1.7 mmol per gram humic acid. The pK value for phenolic groups is 7.2, and the calculated concentration of phenolic groups is 1.2 mmol per gram humic acid.

Comparison of the pK values for the humic acid and the Filtrasorb 200 shows that adsorption of humic acid on activated carbon may reduce the positive charges of activated carbon surface in addition to increase the surface functional groups. This is expected to increase cationic heavy metal adsorption by activated carbons in presence of humic acid.

The adsorption of humic acid onto Filtrasorb 200 activated carbon as a function of solution pH is shown in Figure 7.15. It is seen that humic acid adsorption is strongly pH dependent. A decrease in solution pH from 6.7 to 2 substantially increases humic acid adsorption from 50% to 80%.

As pH plays an important function regarding NOM adsorption into carbon, the adsorption isotherm for NOM must be determined at a fixed pH value that can be controlled by a buffering solution such as phosphorous solution. Figure 7.16a shows the adsorption isotherm of humic acid by the Filtrasorb 200 activated carbon.[65] The adsorption is in good agreement with the Langmuir equation. Figure 7.16b demonstrates the change in the specific surface area of the carbons due to humic acid adsorption. The specific surface area of the carbon without humic acid adsorption is 934 m^2/g. The carbon surface area significantly decreases to about 600 m^2/g after 0.5 mg humic acid per gram of activated carbon is adsorbed, that is, a reduction by one-third due to the adsorption. The adsorbed humic acid causes the blockage of porous structures of the carbon, leading to the decrease in the specific surface area.

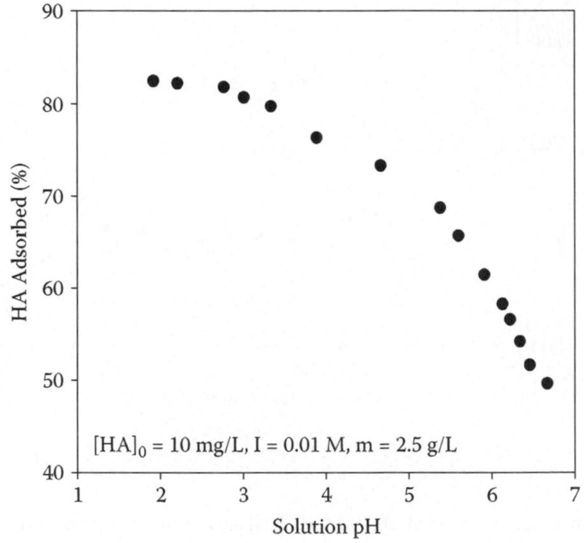

FIGURE 7.15
Effect of pH on humic acid adsorption onto Filtrasorb 200 activated carbon.[63]

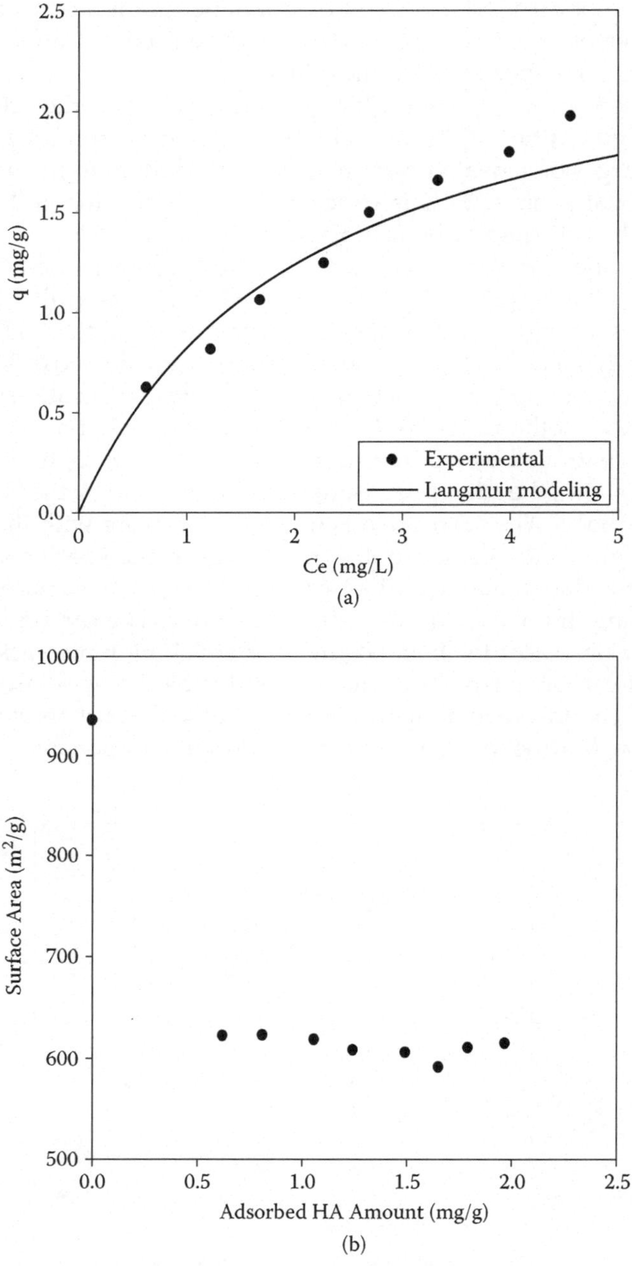

FIGURE 7.16
Adsorption of humic acid by Filtrasorb 200 activated carbon in phosphate buffering solution of pH 5 with I = 0.01 M: (a) humic acid adsorption isotherm, (b) variation of carbon surface area with adsorbed humic acid amount.[63]

The adsorption isotherm of copper ions onto the Filtrasorb 200 carbon under different solution conditions is shown in Figure 7.17.[65] The Langmuir equation well describes the isothermal adsorption at different pH conditions. The maximum adsorption capacities are listed in Table 7.3. The presence of humic acid enhances the copper adsorption. In addition, the highest adsorption occurs at pH 5.5 to 6.0 (not controlled by the phosphorous buffering solution).

The effect of humic acid concentration on copper ion adsorption is shown in Figure 7.18a. The humic acid significantly changes the behaviors of copper ion adsorption onto the carbon. Copper ion adsorption distinctly reduces in

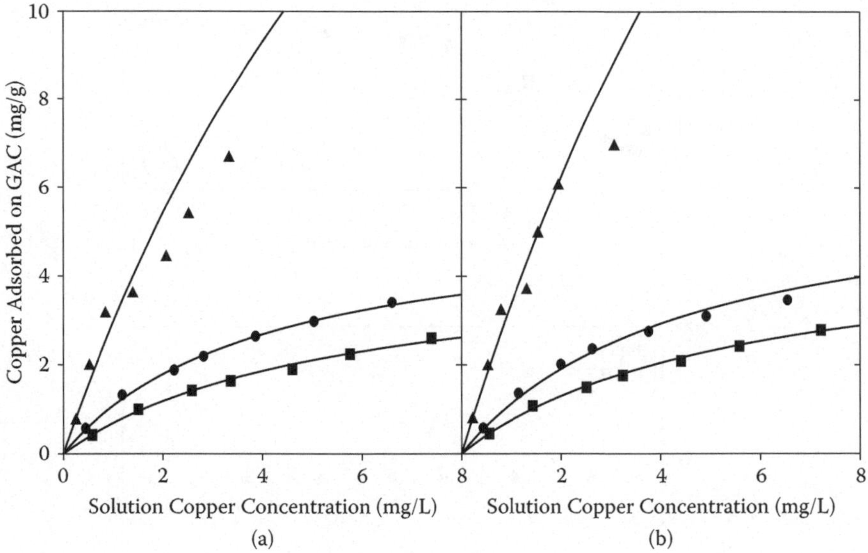

FIGURE 7.17
Copper adsorption isotherms of Filtrasorb 200, (a) no presence of HA and (b) 5 mg/L HA: (■) pH 4, (●) pH 5, (▲) pH not controlled. — fitting by Langmuir isotherm. pH 4 and pH 5 are controlled by phosphate-buffering solution.[63]

TABLE 7.3

Copper Adsorption Capacities of the Filtrasorb 200 Carbon[63]

Solution pH Condition	q_{max} without HA (mg/g)	q_{max} with HA (mg/g)
pH 5.5–6.0 (not controlled)	33.21	42.11
pH 5 (controlled by buffering)	5.37	6.28
pH 4 (controlled by buffering)	4.40	4.83

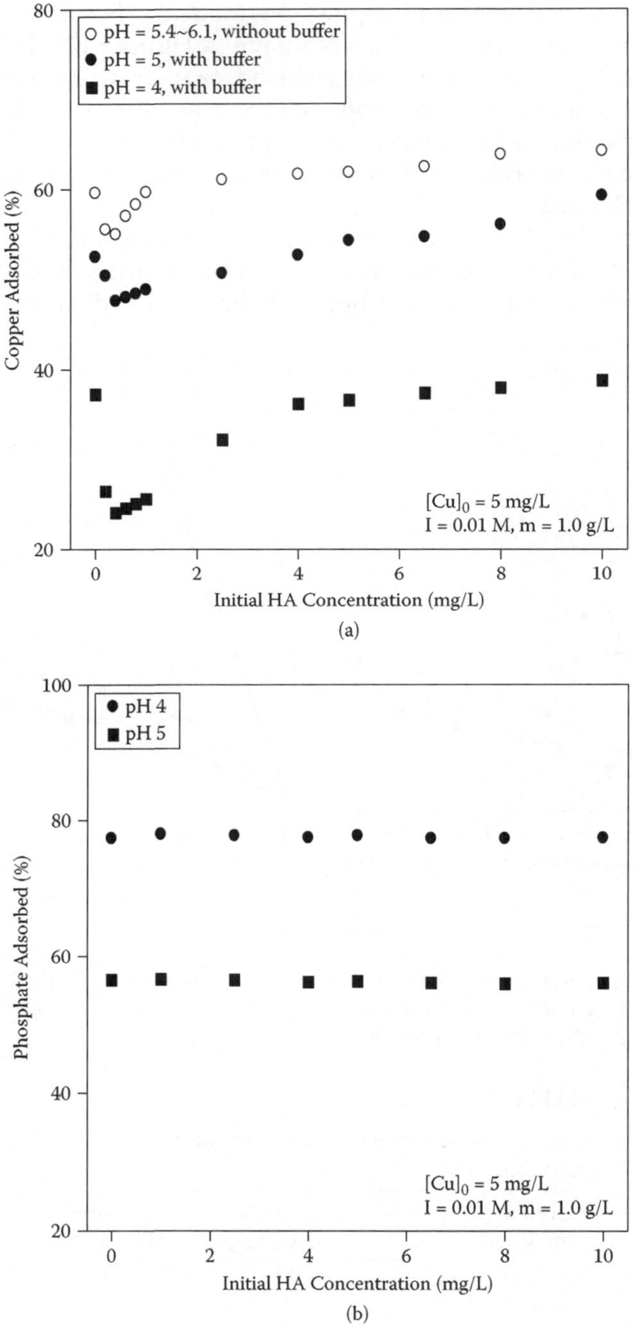

FIGURE 7.18
Effect of humic acid with different initial concentrations on copper ion adsorption (a) and phosphate adsorption (b).[63]

the presence of a low concentration of humic acid, but enhances in the presence of highly concentrated humic acid. There exists a critical concentration of humic acid (0.25 mg/L). Figure 7.18b shows that the absorption of phosphate is not affected by the humic acid.

The heavy metal adsorption can be affected by the presence of other heavy metal ions. The adsorption of copper ions with competitive zinc and cobalt ions in the presence of 5 mg/L humic acid is shown in Figure 7.19. It can be seen that addition of zinc ions and/or cobalt ions in the solution slightly diminished copper ion adsorption. This is attributed to the different affinities of these metal ions to the adsorption sites provided by the carbon and/or the absorbed humic acid.

It is common that heavy metal ions, complexing ligands, and NOM simultaneously exist in waste streams. Figure 7.20 shows the effect of citric acid, EDTA, and humic acid on copper uptake. Copper ion adsorption greatly increases in the presence of citric acid, whereas the addition of EDTA slightly decreases copper ion uptake.

Several mechanisms may be used to explain copper adsorption in the presence of the ligands. The content of adsorption sites of activated carbon may become increased due to the adsorption of a complexing ligand, which leads to more uptake of metal ions. On the other hand, as the complexing ligand may not be strongly adsorbed onto the carbon and remains in the aqueous solution, it may compete with the metal ion for the binding sites, which decreases the metal ion adsorption.

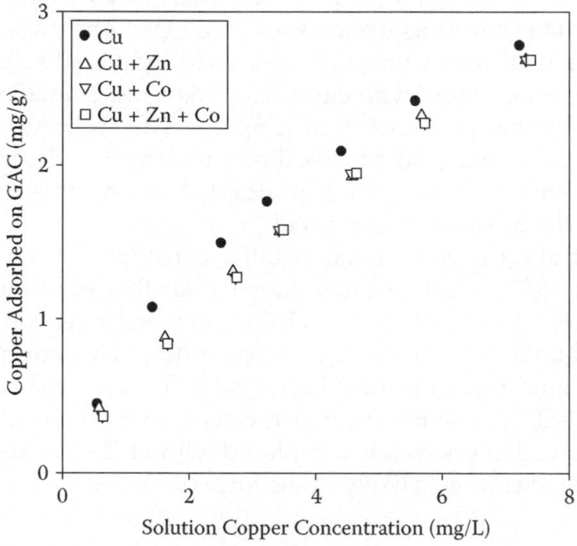

FIGURE 7.19
Effect of competitive metals on copper adsorption in phosphate buffering solution of pH 4 with 5 mg/L humic acid.[63]

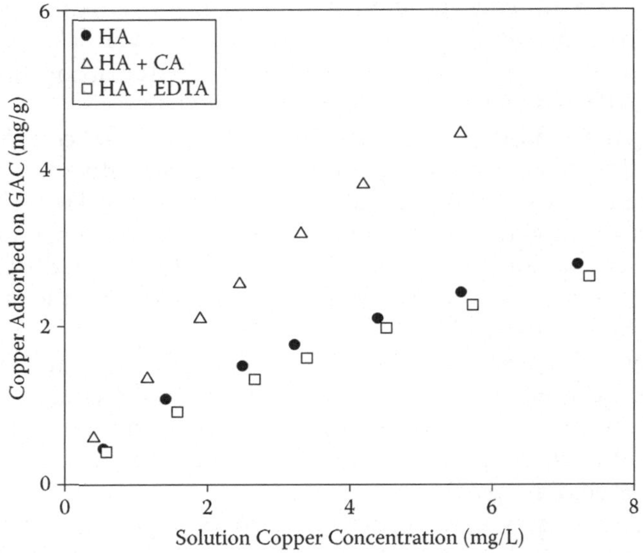

FIGURE 7.20
Effect of citric acid (CA) and EDTA on copper adsorption in the presence of 5 mg/L humic acid (HA). The pH is controlled at 4 by phosphate buffering solution.[63]

As the citric acid has a much smaller molecular size, it can be greatly adsorbed by the activated carbon. The adsorption of citric acid provides "newly enhanced adsorption sites" for the metal adsorption. The adsorbed citric acid provides more adsorption sites for copper ions since the citric acid molecule contains three carboxyl groups. The conditional stability constant of the copper complexation with citric acid is 4.67 much smaller than that of the humic acid (ranging from 4.89 to 10.5).[106–109] This indicates that competition between the humic acid adsorbed on the carbon and citric acid in the solution for the metal binding is insignificant. As such, more copper adsorption occurs in the presence of citric acid.

On the contrary, the conditional stability constant of Cu^{2+} complexation with EDTA is 18.72, which is much larger than that of humic acid.[106] The presence of EDTA in the solution can bring about stronger competition with the adsorbed humic acid for the copper ion binding. This inhibits copper ion adsorption by binding to adsorbed humic acid. The activated carbon can also adsorb the Cu-EDTA complexes, which causes some removal of the metal ions. As the metal ion is strongly complexed with EDTA, the metal adsorbed onto the activated carbon is likely in the form of Cu-EDTA.

7.2.10 Effect of Surfactant

Surfactants (surface active agents) are amphiphilic compounds that contain two regions of very different characteristics, one part being polar and hydrophilic and the other being nonpolar and hydrophobic, which are referred to as the head and tail groups, respectively. Since the two parts have very different properties, the molecules have limited solubility in any solvent and tend to accumulate at the interface between two phases (air–water, oil–water, and solid–liquid), where the polar part can immerse itself into the more polar phase, and the nonpolar part can do likewise.

Surfactants are classified according to the nature of the hydrophilic portion of the molecules. The head group may carry negative charge (anionic), positive charge (cationic), both negative and positive charges (zwitterionic), or no charge (nonionic). Differences in the chemistry of surfactants due to the nature of the hydrophobic tails (degree of branching, carbon number, and aromaticity) are usually less pronounced than those due to the hydrophilic head groups.

At sufficient concentration, surfactants in solution begin to aggregate into micelles containing some 50–100 individuals. Micelle formation occurs above a critical concentration of surfactant monomers referred to as the *critical micelle concentration* (cmc), typically ranging between 0.1 and 1 mM. In a micelle, the individual monomers are oriented with their hydrophilic moieties in contact with the aqueous phase, and their hydrophobic moieties are embedded in the interior of the aggregate.

Surfactants are widely used in many industrial and commercial products and processes throughout the world. They have been used industrially as adhesives, flocculating, wetting and foaming agents, de-emulsifiers, and penetrants. Petroleum users have traditionally been the major users. Surfactants increase the solubility of petroleum components or lower the interfacial tension to enhance mobility of the petroleum. They have been used for mineral flotation as well as pharmaceutical industrials. They also have received extensive application in environmental prevention and remediation owing to their advantageous properties, such as solubility enhancement, surface tension reduction, critical micelle concentrations, wetting ability, and foaming capacity.

The wide application of surfactants produces severe environmental pollution and creates a series of problems for wastewater treatment. The adsorption process has been one of the few treatment methods that can simultaneously remove heavy metals and surfactants from wastewater streams.

The presence of surfactants in aqueous solution can significantly change the surface property of the adsorbent and the metal adsorption behaviors. Lee and Koopal observed that the surface charge of rutile was altered due to the screening by adsorbed cationic and anionic surfactants.[110] Al-Asheh et al. recently reported that copper removal by an adsorbent increased when sodium dodecyl sulfate (SDS) concentration in the solution increased up to certain value.[42] Activated carbon allows for a practically complete elimination of surfactants from wastewater. Little is reported in the literature on metal ion removal by activated carbons in the presence of surfactants.

The effect of surfactants on metal adsorption can be illustrate by the following studies that used three surfactants as the adsorbates. SDS,

TABLE 7.4
The cmc Values of the Studied Surfactants in 0.01M NaCl Solution

Surfactant	Charge Property	cmc (g/L)
HDTMA	Cationic	0.18
Tergitol	Nonionic	0.11
SDS	Anionic	1.61

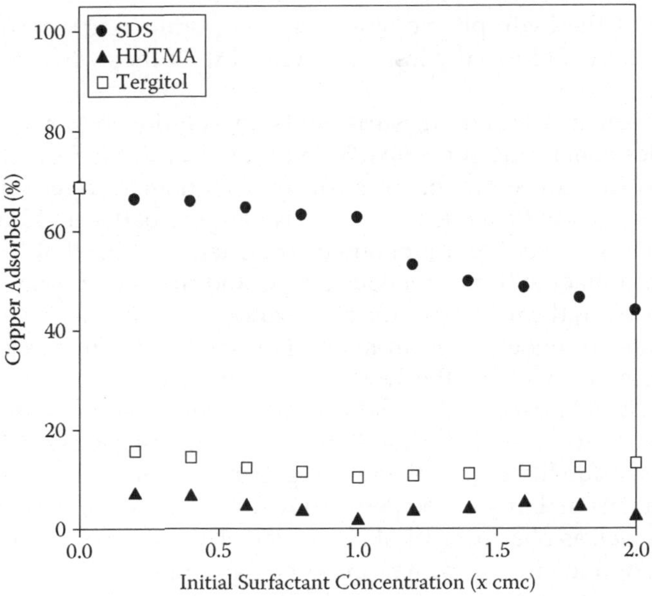

FIGURE 7.21
Effect of concentration of surfactant on copper ion adsorption by activated carbon.

hexadecyltrimethylammonium (HDTMA), and Tergitol (type 15-S-5) were selected as anionic, cationic, and nonionic surfactants. The cmc values of three surfactants are listed in Table 7.4.

The behaviors of copper ion adsorption by Filtrasorb 200 activated carbon in the presence of three surfactants are illustrated in Figure 7.21. The adsorption behavior strongly depends on the type of the surfactants.

Figure 7.21 shows that the presence of SDS with an initial concentration below 1 cmc slightly decreases copper ion adsorption. However, when the initial SDS concentration increased above 1 cmc, an apparent copper ion adsorption decrease occurred.

The effects of HDTMA and Tergitol on the copper ion adsorption are significantly different from the SDS. Both remarkably decrease the copper adsorption. In the absence of HDTMA or Tergitol, the adsorption is 80%.

However, in the presence of the surfactants, the adsorption sharply reduces to 15% and 7% for Tergitol and HDTMA, respectively.

Any change in the concentration of Tergitol and HDTMA does not largely affect the metal adsorption. When their initial concentrations are below 1 cmc, an increase of the concentration of HDTMA or Tergitol causes a decrease in the adsorption. The adsorption gradually increases with a further increase in the concentration above the cmc.

The pH_{pzc} of the activated carbon is 7.2.[56] At the studied solution pH of around 5.5, the carbon surface is positively charged. Adsorption of negatively charged SDS can cause neutralization of the surface positive charges on the carbon. This reduces the electrostatic repulsive interactions between copper ions and the carbon surface, resulting in a higher copper adsorption.

On the other hand, the SDS adsorption may occupy the adsorption sites of the activated carbon, leading to less copper adsorption. Owing to its large molecular size, the adsorbed SDS is expected to bring about considerable blockage of the porous structures of the carbon. Both competition and pore blockage may result in a decrease in the copper ion adsorption. The experimental results in Figure 7.21 indicate that the beneficial reduction of electrostatic repulsive force between the surface and copper ions is overwhelmed by the adverse competition and pore blockage from the SDS adsorption.

When the micelles are formed in the aqueous solution after the SDS concentration reaches above 1 cmc, the micelle microenvironment is expected to compete with the activated carbon for SDS molecules with strong hydrophobic attraction. This may reduce the SDS adsorption. However, since the SDS micelle microenvironment is highly negatively charged, it will compete for copper ions with activated carbon. Consequently, a distinct decrease of copper ion adsorption is observed after the initial concentration of SDS reaches above its cmc.

Owing to its intrinsic hydrophobicity, the activated carbon can adsorb the positively charged HDTMA. This enhances the surface positive charges of the carbon, which increases the electrostatic repulsive interaction between the carbon and copper ions. The HDTMA adsorption brings about the blockage of porous structures of the carbon, which reduces the available adsorption sites for the copper ions. These two adverse effects sharply decrease the copper ion adsorption.

When the initial HDTMA is above its cmc, micelles form in the aqueous solution. The micelle microenvironment competes with the activated carbon for the HDTMA and decreases its adsorption into the activated carbon. This reduces the electrostatic repulsive interactions between the carbon surface and copper ions, and decreases the pore blockage in activated carbon. Therefore, a gradually recovered copper ion adsorption was observed, as shown in Figure 7.21.

Unlike SDS and HDTMA, the nonionic Tergitol cannot impart apparent influence on surface charges of activated carbon. The reduced copper ion removal shown in Figure 7.21 may result from the reduction of adsorption sites and the blockage of porous structures in the carbon by the adsorbed Tergitol. When the initial Tergitol concentration is above its cmc, micelles compete with activated carbon for Tergitol molecules via hydrophobic attractions, which reduces the blockage of porous structures in activated carbon. Therefore, copper ion adsorption gradually increases as the Tergitol concentration increases above its cmc.

The change in the pH in the experiment of adsorption (data given in Figure 7.21) is demonstrated in Figure 7.22. It can be seen that the solution pH increases with an increase in the initial concentration of surfactant SDS, which indicates consumption of hydrogen ions. Conversely, the solution pH decreases with an increasing concentration of HDTMA and Tergitol. This suggests that hydrogen ions are released with the adsorption of HDTMA and Tergitol.

In summary, all the surfactants cause pore blockage in activated carbon to reduce metal adsorption. Negative-charged surfactant (e.g., SDS) neutralizes the surface positive charges of carbon while positive-charge surfactant (e.g., HDTMA) enhances surface positive charges. Nonionic surfactant (e.g., Tergitol) does not have a significant influence on the surface charge of activated carbon. Therefore, the effect of surfactants on metal adsorption mainly depends on the hydrophilic heads of the surfactants.

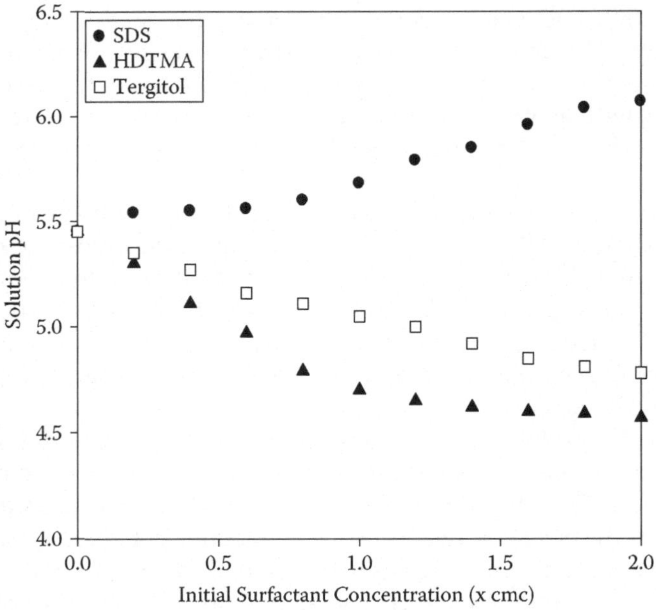

FIGURE 7.22
Change in pH during copper adsorption in the presence of surfactants.

7.2.11 Effect of Competing Ions

Multicomponent metal ion adsorption is normally involved in treatment systems, as well as in natural systems. Compared to the amount of literature concerning monometal adsorption, relatively few data are available on multicomponent metal ion adsorption.

Generally, the adsorption capacity for metals by activated carbon is slightly or greatly reduced when more than one metal is present. The adsorption of the metal that has a higher affinity for adsorption sites of activated carbon is less affected by other metal(s). On the other hand, the metal adsorption is greatly reduced if it has a lower affinity for adsorption sites.

Netzer and Hughes reported that copper greatly affects the adsorption of cobalt. The adsorption capacity for lead is twice as much as that of copper and 10 times more than that of cobalt.[59]

Budinova et al. analyzed the adsorption of the individual metal ions from an aqueous solution containing four ions together in equal concentration and found that the presence of other metal ions in the solution diminishes the adsorption of each of the ions. This effect is greatest for the lead ions and smallest for the copper ions.[111]

Simultaneous cadmium and zinc adsorption experiments were carried out by Gabaldon et al.[112] They concluded that removal percentages for both metals were lower in the competitive process, and the adsorption increased as the pH or molar metal/carbon ratio was increased, which was greater for zinc.

Reed and Nonavinakere reported that the presence of a second heavy metal does not affect removal of the primary metal ion.[74] If such a conclusion is reached, the decision to use activated carbon to treat a multiple metal waste stream could be based on single metal adsorption isotherms. However, it isn't common that two metals have the same affinity for adsorption sites.

Figures 7.23 to 7.25 show competitive metal adsorption together with single-species metal adsorption. The adsorption behavior for individual metal in the mixtures is similar to that for a single metal.

The adsorption removal of copper ions is slightly hindered by the presence of zinc and zinc/cobalt ions as illustrated in Figure 7.23.[66] As copper affinity for activated carbon is the highest among these three metal ions, the inhibitory effect of added cations (zinc and/or cobalt) is not significant, which was also observed by Netzer and Hughes.[59]

On the other hand, Figures 7.24 and 7.25 demonstrate that zinc and cobalt ion adsorption capacities are significantly reduced in the presence of copper ion because the affinities between the functional groups and these two metal ions are relatively weak. When the strongly binding metal (Cu) is added, the adsorption edge for weaker-binding metal (Zn or Co) shifts to higher pH. In addition, the removal for zinc and cobalt in the triple metal ion adsorption is almost the same, mainly due to the similarity in the affinity between the functional groups and these two metal ions. This order is

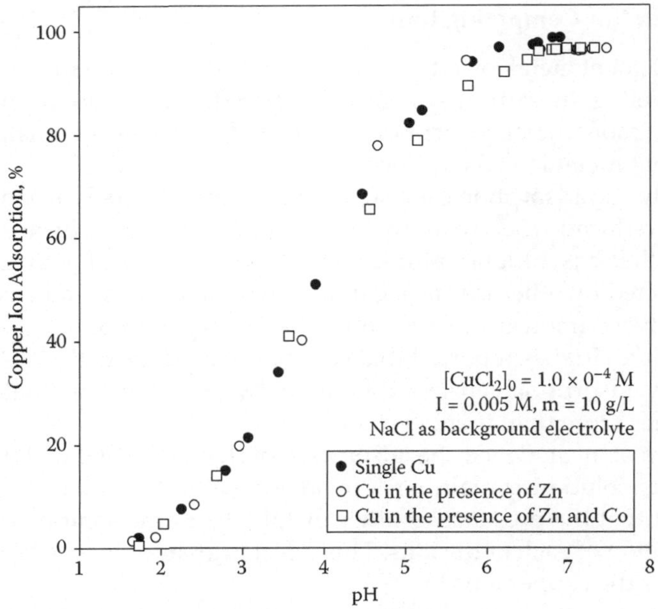

FIGURE 7.23
Copper ion adsorption equilibrium: effect of competitive metal ions.

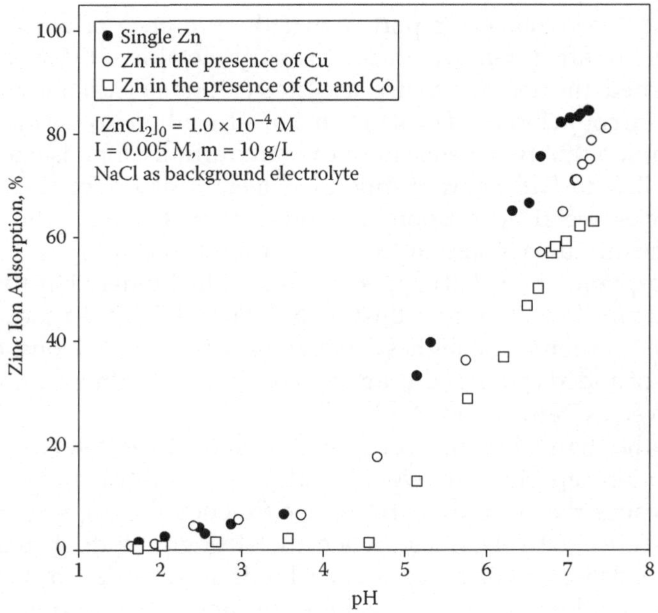

FIGURE 7.24
Zinc ion adsorption equilibrium: effect of competitive metal ions.

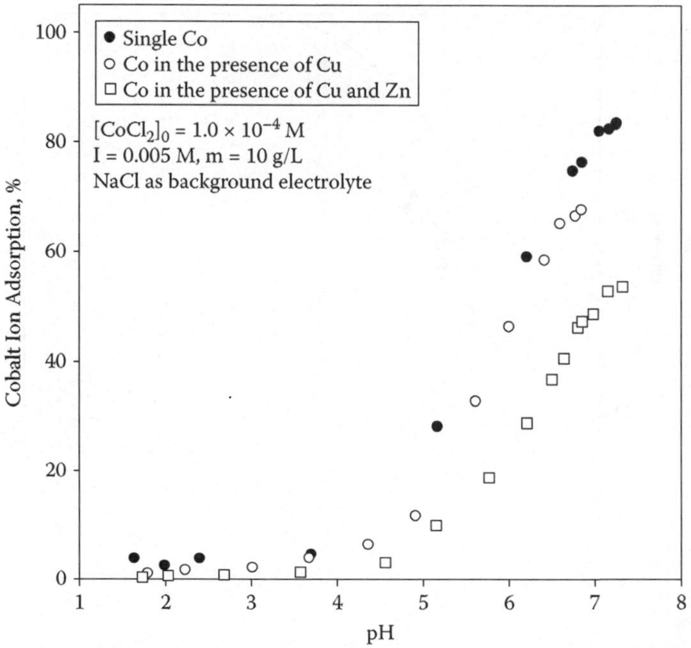

FIGURE 7.25
Cobalt ion adsorption equilibrium: effect of competitive metal ions.[66]

in agreement with the observation in the experiments of single metal ion adsorption (Cu > Zn ≈ Co).

7.2.12 Temperature Effect

The adsorption process may be exothermic or endothermic; however, no conclusive trend is established for metal ion adsorption by various adsorbents. Exothermic reaction is indicated by positive enthalpy change value ΔH_0; consequently, metal ion adsorption is favored at lower temperature. Endothermic reaction is indicated by the negative value of enthalpy change; therefore, metal ion adsorption is favored at a higher temperature. Generally speaking, heavy metal adsorption onto activated carbons is endothermic.

Leyva-Ramos et al. observed that the amount of Cd^{2+} adsorbed on GAC was reduced about three times by increasing the temperature from 10°C to 40°C.[113] Huang and Blankenship studied the effect of copper adsorption onto Filtrasorb 400.[114] As shown in Figure 7.26, the adsorption increases as the temperature is increased from 25°C to 102°C. Take the adsorption at pH 3, the removal increases from 15% to 60% when the temperature is increased from 25°C to 102°C.

Heavy metal removal may be due to metal precipitation and adsorption. Metal precipitation is an endothermic process. The temperature effect

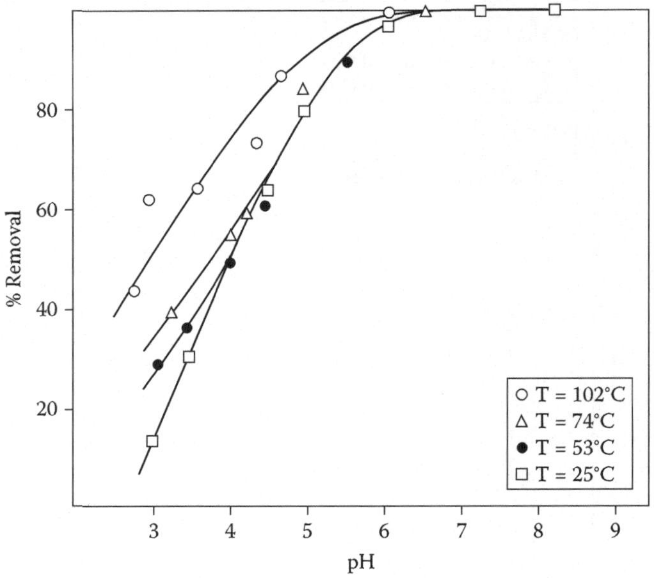

FIGURE 7.26
Effect of temperature on copper removal by Filtrasorb 400 activated carbon. Condition: 10 g/L GAC, ionic strength of 0.05 M, $[Cu(ClO_4)_2] = 1 \times 10^{-4}$ M.[114]

is more obvious at pH less than 4 as shown in Figure 7.26. Under such a solution condition, the metal precipitation (endothermic) does not occur, and the adsorption is the only mechanism for metal removal. The experimental observations in the figure further confirm that the adsorption is endothermic.

7.2.13 Effect of Carbon Type

The adsorption capacities of different carbons are different. Depending on the operational conditions during manufacture, commercially available activated carbon can be divided into H- and L-type.[52–54,56]

Huang and co-workers have done a series of researches on the measurement of the adsorption capacities of different commercial activated carbons. The surface properties of activated carbons from major carbon producers are given in Table 7.5. The specific surface areas of activated carbons range from 500 to 2000 m^2/g. Most of them have the specific surface areas above 1000 m^2/g.

Huang and Blankenship showed that the Hg^{2+} adsorption capacity of the Filtrasorb (H-carbon) is less than Nuchar SA (L-carbon).[114] Corapcioglu and Huang reported that Nuchar SA, Nuchar SN, and Darco KB were the most effective for copper, lead, nickel, and zinc removal.[53] The Darco carbons given in Table 7.5 had the least adsorption capacities for heavy metal. Darco

TABLE 7.5

Summary of Physical and Chemical Properties of Commercial Activated[53]

Type of Carbon	pK_{a1}^{int}	Surface Acidity[a] pK_{a2}^{int}	N_B	pH_{ZPC}	ZPC[b] Washed	IEP[b] Unwashed
Nuchar						
SA	2.3	5.7	2.04	4.0	~2.0	~2.0
SN	2.6	5.4	1.06	4.0	~1.8	~1.8
WVG	7.7	10.4	0.71	9.0	~1.8	~9.2
WVW	6.4	8.5	0.62	7.3	~2.2	~2.2
WVL	5.6	10.2	0.33	7.9	~2.0	~2.0
Filtrasorb						
100	7.0	11.0	0.77	9.0	~2.0	~2.0
200	6.8	9.6	1.53	8.2	~1.8	~1.8
300	7.6	12.1	1.12	9.8	~7.0	~1.8
400	8.6	12.2	0.88	10.4	~3.0	~2.0
Darco						
HDC	9.7	11.8	4.34	10.6	–8.5	~4.0
HD3000	6.1	11.0	3.33	8.5	~2.0	~2.0
S51	8.6	9.0	3.33	8.8	~8.8	~2.0
G60	4.4	7.2	2.40	6.2	~2.0	~2.0
KB	4.2	6.1	0.66	5.2	~2.0	~2.0

Note: N_B = Total number of surface acid sites, in units of $\mu C\ cm^{-2}$.
ZPC = Zero point of charge.
IEP = Isoelectric point.

[a] From alkalimetric titration.

[b] From electrophoretic mobility measurements.

HD3000 was the least effective for copper and lead removal at pH 2.5 to 5.5. Darco HDC displayed the lowest adsorption capacities for nickel and zinc. The adsorption capacities of the Filtrasorb activated carbons fell between Nuchar and Darco series.

To search for more effective carbon, many researchers developed new carbon using various raw materials, such as agricultural by-products, or prepared carbon by physical and chemical treatment. Budinova et al. studied the adsorption of lead, copper, zinc, and cadmium by the activated carbons obtained from apricot stones, coconut shells, and lignite coal by pyrolysis in the presence of water vapor.[111] They found there was a considerable difference in the adsorption capacities of the three activated carbons with lead and copper.

7.2.14 Modification of Activated Carbon

Activated carbon is generally produced for the removal of organic compounds in water and wastewater. As such, its capacity for heavy metal removal is limited. However, it can be further modified to develop desirable physicochemical properties by the adequate choice of activation procedures for heavy metal removal. It is even possible to prepare carbons with designated proportions of micro-, meso-, and macro-pores.

7.2.14.1 Chemical Approaches

Carbons can be treated by acids, bases, or oxidizing agents in order to produce favorable chemical and physical properties for different applications (e.g., separation and catalysis). Gas phase oxidation increases mainly the content of hydroxyl and carbonyl surface groups, while liquid phase oxidation especially enhances that of carboxylic groups.[115]

Surfaces properties of activated carbons can be modified by inorganic acids. The modification mainly changes the surface chemistry of the carbons and sometimes alters their specific surface area and porosity. For example, treatment of carbon by phosphoric acid can cause a high surface area and high degree of porosity.[116–118] Both strong and weak acidic oxygen functional groups can be introduced by acid modification.[119]

Oxidation of carbon by a H_2O_2 solution increases hydroxide groups in the carbon when the solution pH is not controlled. However, when the solution pH is fixed at 2.5 or 11.5, the content of hydroxide groups decreases while the contents of other oxygen groups (e.g., carbonyl and carboxyl) increase.[120]

HNO_3 treatment increases the contents of acidic surface functional groups.[121] Various surface oxygen groups and structures containing N-O bonds (nitro-groups and nitrate complexes) may be further developed.[122] The effect of nitrate acid treatment on the surface area of carbons is not conclusive. Mazet et al. showed a significant increase in the surface area,[123] while Gomez-Serrano et al. observed a slight variation.[122]

Polymers and chelating agents have been selected to modify activated carbons. The density of positive surface charges on the carbons can be increased after they are grafted by cationic polymers.[124,125] The quantity of surface functional groups is also changed.[126]

7.2.14.2 Physical Approaches

Carbons can be treated physically for improvement of their properties for different applications. It is reported that heat treatment in an inert atmosphere can selectively remove some of these functional groups. Surface carboxylic acid groups of carbons disappear after treatment in H_2 at 723 K.[119] De la Puente et al. observed the removal of carboxylic acid groups at a temperature of 400 to 623 K.[127] The elimination of stable ether groups was found at temperatures above 823 K. At higher temperatures, the concentrations of carboxylic acid groups decrease, and subsequently ketone and quinone groups disappear.[128] With a further increase in temperature, elimination of both ether groups and aliphatic structures occurs.[129]

Menendez et al. modified an activated carbon by a microwave device in a nitrogen flow. Most acidic oxygen groups are removed from the carbon surface, resulting in a significant increase in the carbon pH.[130]

7.2.14.3 Metal Performance of Modified Activated Carbons

Heavy metal adsorption onto activated carbons is due to a series of adsorption reactions between the metal ions and the organic functional groups in the activated carbons. The adsorptive behaviors may be significantly altered after the surfaces of carbons are modified.[131–133]

Ravindran et al. impregnated 8-hydroxyquinoline (oxine) on an activated carbon.[134] It was found that oxine-impregnated carbon had a much higher adsorption of cadmium, lead, and zinc. Lakov et al. modified activated carbons by 3-methyl-1-phenyl-pyrazolone-5 (HMPP).[135] The copper adsorption capacity was enhanced by 50%, even though the specific surface area of the carbon decreased by 80%. More recently, Monser and Adhoum reported that a substantial improvement in Cu, Zn, and Cr removal capacity was achieved when tetrabutyl ammonium iodide or sodium diethyl dithiocarbamate was applied onto an activated carbon.[136]

Heat treatment can improve metal adsorption capacity. Anodic oxidation of carbon results in enhancement of both uptake capacity and kinetics for Cr(VI).[131] Air oxidation can cause an improvement in copper adsorption.[132] In the study of copper adsorption onto a sulfuric acid modified carbon, it was observed that the concentration of acidic surface oxides on the carbon surfaces increased, and cation exchange reactions occurred.[133] The copper adsorption onto the carbon treated in an ammonia atmosphere was due to the formation of surface complexes with the nitrogen- and oxygen-containing functional groups.[137]

Case Study I: Modification of Activated Carbon by Acid and Base

Filtrasorb 200, a commercial GAC from Calgon Corporation (Pittsburgh, PA), was modified by several acid and base approaches given in Table 7.6 for the enhancement of heavy metal removal.[65] Table 7.7 gives the specific surface area, the carbon pH, and the total acidity capacity of the activated carbons. The variations among the specific surface areas of various activated carbons are within 2.5%, indicating no change in the surface area of carbons. Similar findings were reported by Wang and Lu.[138] These observations are different from those of activated carbons modified by organic compounds. For example, surface modification by citric acid reduces the surface area of activated carbon by 34%.[78] As simple inorganic acids (HCl and HNO_3) have small molecular sizes, they do not significantly change specific surface area of carbons.

Carbon pH and total acidity capacity (TAC) can be used as a more straightforward approach to compare chemical properties of carbons. Carbon pH can be treated as an approximate measure of the pH at the point

TABLE 7.6

List of Chemical Approaches for Modification of Activated Carbon[65]

Carbon	Brief Description of Modification
DI-AC	Filtrasorb 200 washed by DI water
HA37-AC	DI-AC treated by concentrated HCl and washed by DI water
NA10-AC	DI-AC treated by 10% HNO_3 at 0.1g/mL and washed by DI water
NA30-AC	DI-AC treated by 30% HNO_3 at 0.1g/mL and washed by DI water
NA50-AC	DI-AC treated by 50% HNO_3 at 0.1g/mL and washed by DI water
NA69-AC	DI-AC treated by concentrated HNO_3 at 0.1g/mL and washed by DI water
SH-AC	DI-AC treated by 1 M NaOH and washed by DI water
HA37-SH-AC	HA37-AC treated by 1 M NaOH and washed by DI water
NA69-SH-AC	NA69-AC treated by 1 M NaOH and washed by DI water

Note: AC: activated carbon; DI: deionized; HA: hydrochloric acid; NA: nitric acid; SH: sodium hydroxide. Value in symbol indicates volume percentage.

TABLE 7.7

Changes in Surface Properties of Activated Carbons After Modifications[65]

Carbon Type	Specific Surface Area	Carbon pH	Total Acidity Capacity (mmol/g)
DI-AC	648 m^2/g	6.88	0.15
SH-AC	664 m^2/g	10.71	0.16
HA37-AC	659 m^2/g	5.71	0.19
HA37-SH-AC	662 m^2/g	8.49	0.19
NA69-AC	636 m^2/g	3.74	0.51
NA69-SH-AC	647 m^2/g	7.49	0.37

of zero charge (pH_{PZC}). It is shown in Table 7.7 that the carbon pH values for the carbons are different, ranging from 3.74 of the NA69-AC to 10.71 of the SH-AC. The DI-AC has a carbon pH of 6.88, close to the determined value 7.2 for the same activated carbon by a potentiometric titration.[56]

The carbon pH indicates that the surface of the DI-AC is positively charged when the solution pH is below 6.88. It becomes negative at pH > 6.88.

It is known that a stronger acidic functional group is deprotonated at lower pH, while a weaker acidic functional group is deprotonated at higher pH. The treatment of DI-AC by both HNO_3 (NA69-AC) and HCl (HA37-AC) greatly reduces the carbon pH. The NA69-AC and the HA37-AC are negatively charged when solution pH is above 3.74 and 5.78, respectively. The NaOH treatment of the carbon (SH-AC) distinctly increases the carbon pH values; this can result in positive charges on the surfaces of the carbons at rather higher solution pH.

Sodium hydroxide can be used to target various oxygen functional groups, including carboxylic acid, lactone, and phenolic groups.

Therefore, the total acidity capacity determined by sodium hydroxide titration can quantify total oxygen functional groups that can effectively form complexes with metal ions.

As shown in Table 7.7, NA69-AC has a TAC of 0.51 mmol/g, 240% more than that of DI-AC. This indicates that HNO_3 treatment introduces a large number of oxygen-containing functional groups. As a result, a higher amount of sodium hydroxide is required to neutralize the functional groups in the activated carbon. The TAC of HA37-AC is slightly higher than that of DI-AC and much lower than that of NA69-AC. The HCl treatment does not significantly oxidize the functional groups in the DI-AC carbon.

NaOH treatment does not alter the TAC values of activated carbons (SH-AC versus DI-AC and HA37-SH-AC versus HA37-AC). The TAC of the NA69-SH-AC is 147% higher than that of DI-AC, but 27% lower than that of NA69-AC. This could be due to the elimination of some acidic functional groups by NaOH.

The XPS and FT-IR studies indicate that the HNO_3-modification generates a significantly large number of surface functional groups such as carbonyl, carboxyl, and nitrate groups. NaOH causes an increase in the content of hydroxyl groups. The HCl-treatment results in an increase in the amount of single-bonded oxygen functional groups such as phenols, ethers, and lactones.

The Langmuir constants (q_{max} and b) for different activated carbons are listed in Table 7.8. The maximum copper adsorption capacity (q_{max}) at a constant pH 4.9 follows a descending order of NA69-AC > NA69-SH-AC > NA50-AC > NA30-AC > NA10-AC > SH-AC > DI-AC > HA37-SH-AC > HA37-AC. The q_{max} for DI-AC is close to that for Filtrasorb 200. Both HNO_3– and NaOH-treated activated carbons have the best adsorption performance among all the activated carbons.

TABLE 7.8

List of Constants of Langmuir Equation for Copper Adsorption onto Activated Carbons[65]

Carbon type	q_{max} (mg/g)	b $(mg/L)^{-1}$
DI-AC	6.15	0.18
SH-AC	6.75	0.27
NA10-AC	6.93	0.91
NA30-AC	9.23	0.91
NA50-AC	9.91	0.74
NA69-AC	15.34	0.23
NA69-SH-AC	14.97	0.20
HA37-AC	5.08	0.20
HA37-SH-AC	5.38	0.38

Note: pH 4.9 was controlled by 0.07 M sodium acetate and 0.03 M acetic acid; sodium chloride of 0.01 M was added.

Case Study II: Modification of Activated Carbon by Citric Acid

Citric acid ($HOOC-CH_2-COH(COOH)-CH_2-COOH$) has a higher binding affinity to heavy metal. It is biodegradable and widely used in food and pharmaceutical industries. It is a small organic compound with a rich amount of carboxylic groups.[12]

Filtrasorb 200 activated carbon from Calgon was modified by citric acid for the enhancement of heavy metal removal.[12] It was washed with deionized water to remove fine powdered-activated carbon left after sieving. The carbon denoted by WAC reacted with a 1-M citric acid solution at a ratio of 4-gram carbon to 25-mL citric acid solution for 30 minutes at 50°C overnight. The citric-acid-modified carbon (CAC) was then washed with DI water until no turbidity could be observed when a 0.1-M lead nitrate solution was added to the washed liquid.

As shown in Table 7.9, the modification by citric acid significantly improves its copper ion adsorption capacity. The maximum adsorption capacity increased from 6.14-mg Cu/g-WAC to 14.92-mg Cu/g-CAC. After the modification, the activated carbon surface became more homogeneous. The specific surface area of carbon decreases, which results from the pore blockage by adsorbed citric acid molecules. However, the adsorption kinetics seems not to be affected. The adsorption equilibrium can be established within 400 min.

Case Study III: Modification of Activated Carbon by Polyaniline

Filtrasorb 200 from Calgon was ground and then washed by deionized water to remove the fine particles (referred to as RGAC).[20] Aniline was made up to 100-mL solution with ethanol. $FeCl_3{\cdot}6H_2O$ and aniline with a molar ratio of 3:1 was then added, and the aniline solution was stirred for 3 h. Subsequently, RGAC was added to this solution and stirred. The entire mixture was vacuum filtered and thoroughly washed with DI water to remove the excess reactants until the washed liquid became colorless. The carbon after the modification was referred to as MGAC.

TABLE 7.9

List of Properties of Activated Carbons[12]

Carbon Type	Langmuir Constants		Specific Surface Area, m^2/g
	q_{max}, mg/g	b, $(mg/L)^{-1}$	
WAC	6.14	0.18	648
CAC	14.92	0.08	431

Note: WAC: Activated Carbon after washing by water.
CAC: Citric Acid-modified Carbon.

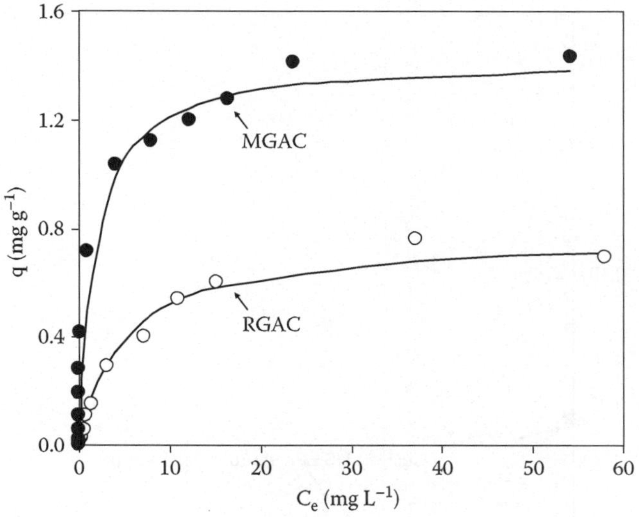

FIGURE 7.27
Adsorption isotherms of RGAC and MGAC at pH 5.5.[20]

The optimal pH range is 5.5 to 6.5 for the MGAC. The adsorption follows a typical Langmuir-type adsorption isotherm as shown in Figure 7.27. The best-fitted maximum adsorption capacities (q_{max}) of the RGAC and MGAC are 0.78 and 1.43 mg g^{-1}, respectively. The polyaniline modification of RGAC enhances the arsenate adsorption capacity by 84%. The corresponding affinity (b) values of the RGAC and the MGAC are 0.17 and 0.50 L mg^{-1}, respectively. The higher affinity of MGAC indicates that As(V) is more affinitive to the MGAC surface.

Figure 7.28 shows an arsenic uptake kinetics when an adsorbent dosage is 5 g L^{-1}, and initial As(V) concentration is 0.15 mg L^{-1} (150 ppb). MGAC can reduce the As(V) to 0.0013 mg L^{-1} (1.3 ppb) in 30 min, while RGAC only reaches an equilibrium As(V) concentration of 0.086 mg L^{-1} after 2 h. This clearly demonstrates that MGAC is able to treat trace-level arsenic in the contaminated water below the current US EPA standard. Better adsorption performance of modified activated carbon for possible arsenate wastewater treatment is further illustrated by a continuously operated fixed-bed reactor. As shown in Figure 7.29, the carbon is able to reduce the arsenate concentrations ranging from 0.12 to 1.91 mg/L to less than 10 ppb. Higher influent concentration leads to earlier breakthrough of the fixed-bed reactor. The curves in the figure are from the modeling simulation by a diffusion-control model. The best fitting between the experimental observations (points) and the modeling simulation (curves) indicates that the adsorption is controlled by mass transfer (diffusion).

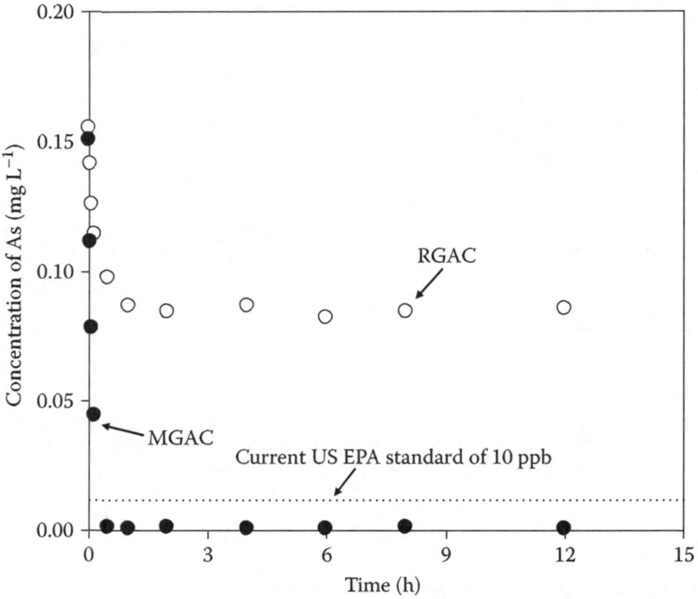

FIGURE 7.28
Compare of arsenate adsorption kinetics on RGAC and MGAC at initial arsenate concentration of 0.15 mg L^{-1} and adsorbent dosage of 5 g L^{-1}.[20]

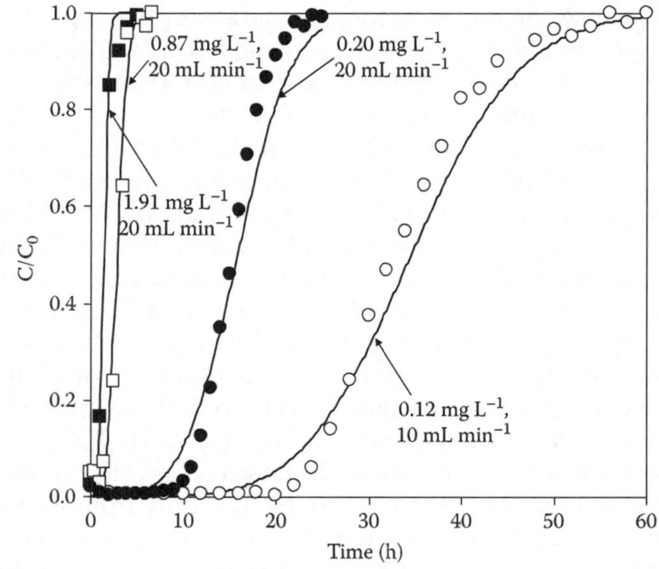

FIGURE 7.29
Performance of fixed-bed column for arsenate adsorption by a modified activated carbon.[20]

7.3 Biosorbents

One promising technology for removal of heavy metals is biosorption, where biopolymers and nonliving organisms are as biosorbents. Biopolymers are extracted from, and have common chemical properties with, nonliving organisms. The nonliving organisms include fungi, bacteria, and fermentation waste.

It has been documented that biosorbents have a high potential to sequester and accumulate rapidly heavy metal ions present in aqueous solutions. The cost of biosorbents is normally lower than activated carbons and other synthesized adsorbents as shown in Table 7.10. If one considers the cost per kg of metal removed, the advantage of using biosorbent becomes much more obvious. Furthermore, the used biosorbents, after coming into contact with a weakly acidic solution, can be reused, and their effectiveness for metal ion removal is comparable to that of fresh biosorbents.

Originated from living organisms living in sea and freshwaters, biosorbents are rich in carbon, hydrogen, oxygen, nitrogen, and phosphorus. The organic ligands or the functional groups in biosorbents play a dominant role in removal of various heavy metal contaminants. These representative functional groups are carboxyl, hydroxyl, sulfate, phosphate, and amine groups. Among them, the carboxyl group is the most important for metal binding. The biosorption mechanisms include complexation, coordination, surface precipitation, chelating, and ion exchange.

Among various biosorbents reported in the literature, marine algae are identified as good biosorbents because of their low cost, renewability, and high metal biosorption capacity. They can effectively remove heavy metal ions from concentrations ranging from few parts per million to several hundred parts per million. Maximum metal biosorption capacity can be as high as a few millimoles per gram biosorbent. The biosorption of cations is more effective than that of anions. The biosorption capacity is normally much higher than commercial adsorbents and ion exchange resins.

7.3.1 Preparation of Biosorbents

Several criteria must be considered in the selection of an appropriate biosorbent. First, it must be low cost in both acquisition (e.g., harvesting and

TABLE 7.10

Comparison of Price of Different Sorbents

Type of Sorbent	Price of Sorbent
Biosorbent	\$0.2 to \$ 5 per kg
Activated carbon	\$0.5 to \$7 per kg
Ion exchange resin	\$10 to \$ 40 per kg

manufacturing) and recycle/reuse. Second, the biosorbent must have a high sorption capacity, great selectivity, and fast sorption kinetics. These properties ensure the compact reactor system ease of operation and sustainability. Third, the biosorbent must have stable mechanical properties in order to be used in various reactors such as column operation. Any change in the size (shrinking or swelling) during the operation is highly undesirable and must be avoided.

Raw biomass can be collected from natural or industrial sources. Figure 7.30 shows two seaweeds, *Padina* sp. and *Sargassum* sp., which have been reported for heavy metal biosorption in the literature. Normally, raw biomass cannot be directly used as a biosorbent for metal removal. First, several impurities such as organic matters, light metal ions, and particular matters are released from the raw biomass when metal wastewater is treated. Additional cost is required for the removal of these substances, even though heavy metal ions are removed. Among the impurities, the organic matters leaching from the biosorbents are often considered as a secondary pollutant and retard the biosorption technology in water and wastewater treatment. Second, metal removal capacity may be enhanced if modification is conducted on raw biomass. The modified biomass may have longer recycles of application.

Take *Saragsium* sp. as an example. The total organic carbon level in the solution becomes quite high after the biosorption process is completed. Kratochvil and Volesky reported that total organic carbon of the effluent from a Sargassum-packed column was around 24 mg/L at the early phase of operation.[139] The TOC during the desorption can be as high as 55 mg/L. It is common to observe that the color of water after biosorption turns yellowish or green. Figure 7.31 shows the chemically modified *Sargarssium* sp., reported by Chen and Yang.[140] It is clearly demonstrated that the chemical modification greatly reduces the organic leaching.

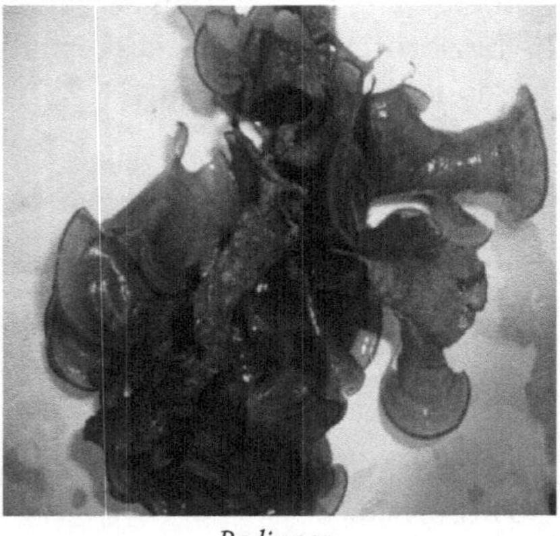

Padina sp.

Sargassum sp.

FIGURE 7.30
Illustration of typical seaweeds for heavy metal biosorption.

FIGURE 7.31
Effect of chemical modification for *Sargassum* sp.: raw biomass (right); modified biomass (right).

Depending on the requirement for quality of the treated water and the influent characteristics, different preparation approaches for biosorbents are available for selection. If treated water free of organic substances is required, chemical modification of raw biomass must be used, which can greatly minimize organic leaching during the operation. However, this will not have to be adopted if the treated streams are discharged to sewage work as the organic substances can subsequently be treated biologically. Figure 7.32 describes one approach for biosorbent preparation.

Case Study: Modification of Sargassum Sp.
Like many other marine algae, *Sargassum* sp. (a brown algae) has a rich content of polysaccharides in cell walls, which is mainly responsible for higher metal biosorption. Such important function groups as carboxyl, sulfate, and amino are identified in the biomass. It has a high amount of organic substances such as carbohydrates, protein, lips, and pigments. Some of them can inevitably become dissolved in the aqueous solutions during the biosorption operation.

A total of 13 chemical reagents were made for the modification of *Sargassum* sp.[140] 10 g/L raw seaweed (RSW), *Sargassum* sp., was reacted with the chemical solutions for 24 h. The resulted modified seaweeds (MSWs) were then filtered from the mixture, washed, and dried in an oven overnight at 60°C.

The weight loss percentage (W_{cm} %) due to the chemical modifications can be determined by:

$$W_{cm}\,\% = \frac{W_i - W_f}{W_i} \times 100\,(\%) \tag{7.5}$$

where W_i is the dry weight of RSW, and W_f is the dry weight of MSW.

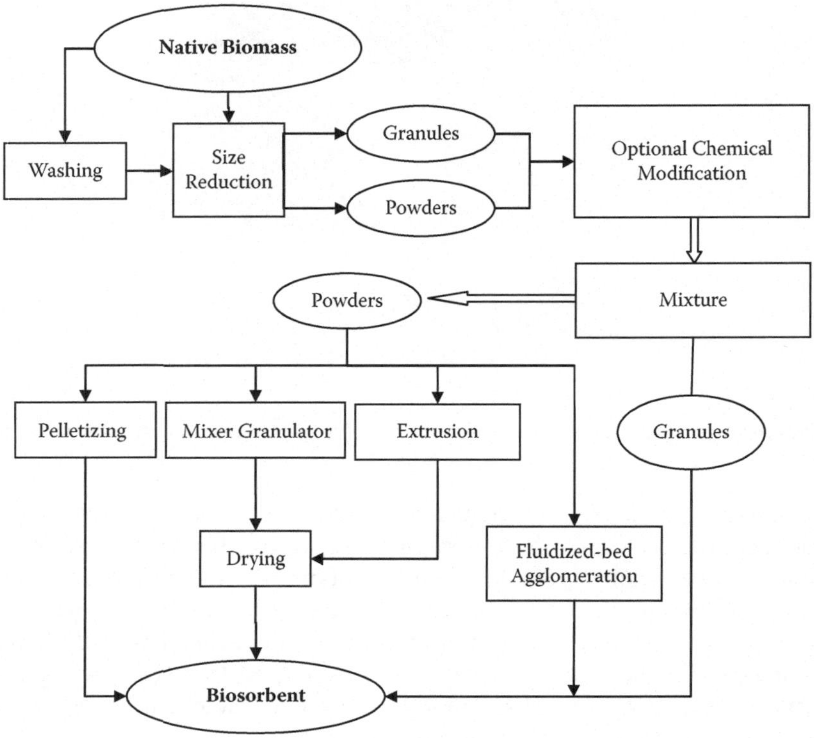

FIGURE 7.32
Schematic diagram of biosorbent preparation.

It is observed that the organic leaching from RSW can be decreased if it is prewashed by the DI water. In order to compare this approach with the chemical modification approaches, RSW with a weight of one gram was washed by DI water several times, filtrated, and dried. The weight loss was then determined.

The MSW with a dosage of 1 g/L was used to remove copper that had an initial concentration of 6 mM and pH 5.0. A higher initial concentration of copper ions was used in order that the MSW would become saturated with metal ions. The contact time was controlled at 24 h. Weight loss and organic leaching due to the biosorption, and metal removal efficiency of different MSWs were determined and used to screen chemical modification methods.

Considering a weight loss due to the modification, the biosorption capacities of RSW and MSWs should be compared based on the mass of seaweed before the modifications. Therefore, the metal biosorption capacity with this as a consideration can be expressed as

$$q_{e-OSW} = q_e \times (1 - W_{cm}\%) \tag{7.6}$$

where q_{e-OSW} is the biosorption capacity based on the weight of seaweed before modifications (mmole metal/g OSW) and q_e is the biosorption capacity based on the weight of MSWs or RSW (mmole metal/g MSW or RSW) and can be determined. W_{cm} % can be obtained by Equation 7.5.

Since the heavy metals are strongly adsorbed onto the RSW and the MSWs, the amount of heavy ions sorbed on the solids must be taken into account in the calculation of weight loss due to the biosorption. Some light metals (e.g., calcium ions) are released into the solution because of the ion exchange discussed later. The weight loss due to the biosorption ($W_{biosorption}$ %) with such a consideration can be determined as follows:

$$W_{biosorption}\ \% = \frac{(wt_i - wt_f) + wt_m - wt_{mo}}{wt_i} \times 100\ (\%) \tag{7.7}$$

where, wt_i is weight of biosorbents before biosorption, wt_f is weight of biosorbents after biosorption, wt_m is weight of heavy metal ions (e.g., Cu^{2+}) adsorbed onto the biosorbents, and wt_{mo} is weight of light metal ions (e.g., Ca^{2+}) released from the biosorbents.

It has to be recognized that there is a weight loss during the chemical modifications as stated in Equation 7.5. Thus, a weight loss based on the mass of the seaweeds before modification can be calculated:

$$W_{biosorption-OSW}\ \% = \frac{W_{biosorption}}{1 - W_{cm}} \times 100\ (\%) \tag{7.8}$$

The simplest approach to modify seaweeds is to wash them by using the DI water. It was found that, the weight loss of seaweed due to the washing was 29.5%. A total of 13 different chemical reagents by acid, base, aldehydes, and their combinations were used for modification of RSW in order to reduce the weight loss and enhance the biosorption capacity.

The weight loss percentage of approximately 24.5 % for $CaCl_2$, glutaraldehyde, and formaldehyde-modified *Sargassum* sp. is demonstrated in Figure 7.33. A more serious weight loss percentage can be observed when acid or base is used. Higher weight loss percentages of 33.8% and 43.3%, respectively, are caused by 0.1 M HCl and 0.1 M NaOH. The weight loss percentages of 30% and 28.1% are observed when 0.1 M HCl followed by $CaCl_2$ or 10% formaldehyde is used.

As shown in Figure 7.34, all of the MSWs show higher copper adsorption capacity (q_e) than the RSW. When the weight loss (W_{cm} %) is considered, the metal biosorption capacities (q_{e-OSW}) for MSWs and RSW are virtually similar to each other, except in Cases a-d when acid or base is used for the modification. This indicates that the portion of biomass removed during the chemical modifications is less essential for the heavy metal biosorption.

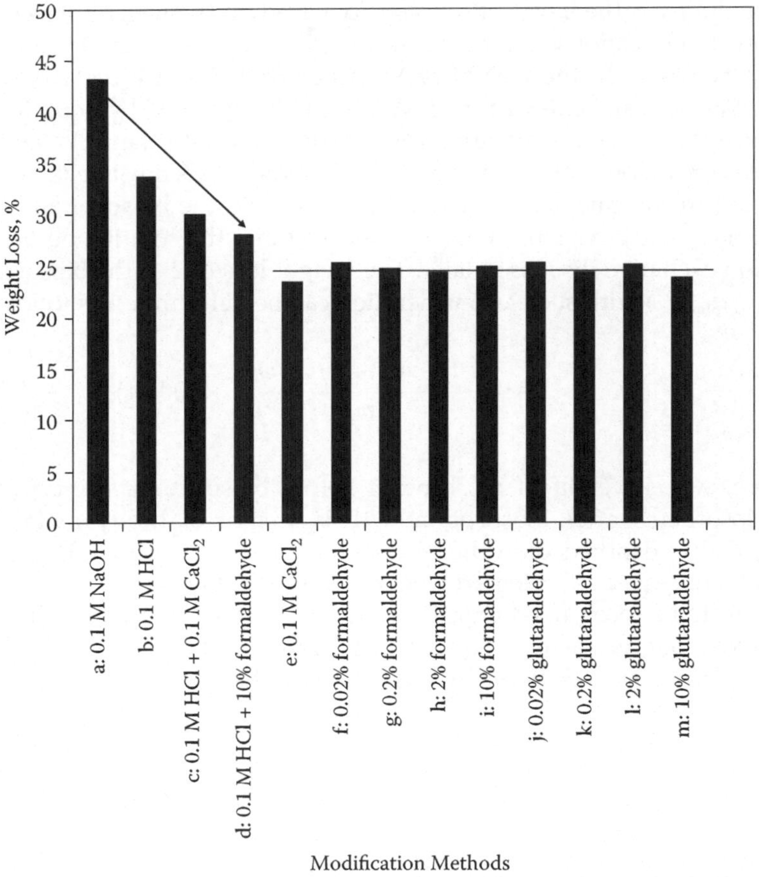

FIGURE 7.33
Comparison of weight loss percentages of different modification approaches. S/L = 10, contact time = 24 h.[140]

It is observed that the lowest biosorption capacity ($q_{e\text{-}OSW}$) occurs when 0.1 M hydrochloric acid is used. The biosorption capacity of the HCl MSW is 73% of the RSW, which is due to the strong hydrolysis reactions of organic substances from the seaweeds by the acid.

The weight loss in the MSWs and the RSW during the biosorption is illustrated in Figure 7.35. The weight loss of RSW ($W_{biosorption}$ %) during the copper biosorption is the highest (24.1%) among all the biosorbents. The second highest weight loss ($W_{biosorption}$ % of 19.1%) occurs when the 0.1 M HCl MSW is used. The averaged weight loss percentage of the aldehydes MSWs is 11.3%, far below the highest (RSW). When the weight loss during the chemical modifications is considered, most of the MSWs have lower weight loss ($W_{biosorption\text{-}OSW}$ %) than the RSW (24.1%). The weight

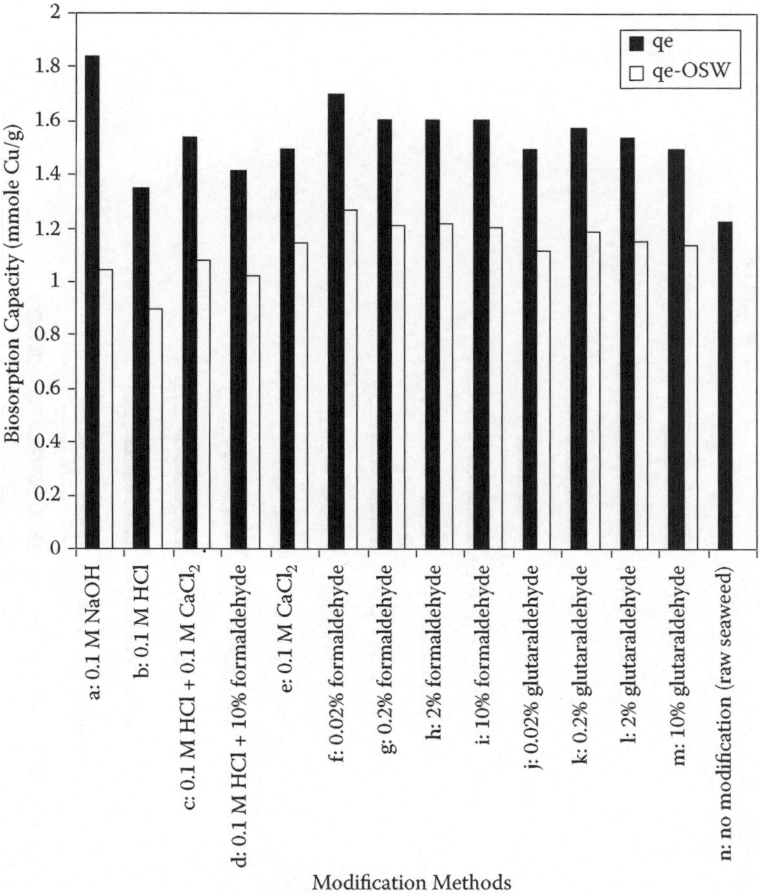

FIGURE 7.34
Comparison of copper biosorption capacities of raw and modified *Sargassum* sp.; biosorption: MSW or RSW = 1 g/L; initial pH = 5, $[Cu]_0$ = 6 mM, contact time = 24 h.[140]

loss of HCl MSW ($W_{biosorption\text{-}RSW}$ % of 28.8%) is higher than that of the aldehydes MSWs (15.1%) and the RSW.

It was observed during the experiment that, when the RSW was used, the solutions became yellow or brown, which resulted from the pigments leaching from the raw seaweeds. There are many types of pigments in brown seaweeds, including chlorophyll II and carotene. Most of these pigments are soluble in organic solvents. A solution with 1 g RSW/L being contacted with DI water for 24 h has a TOC of 110.9 and 186.3 mg/L at pH 5.0 and 2.0, respectively.

Figure 7.36 shows organic leaching during copper biosorption by both MSWs and RSW. As shown, the leaching from MSWs pretreated by aldehydes (0.02% to 2%) and $CaCl_2$ is much lower than that from RSW.

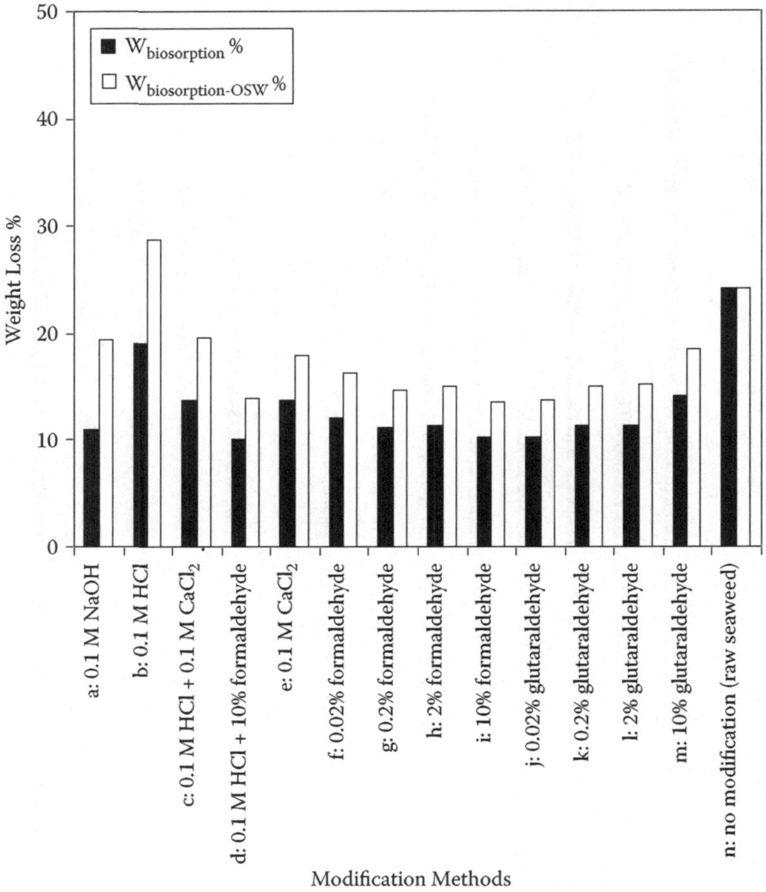

FIGURE 7.35
Comparison of weight loss of different modification approaches during copper biosorption: MSW or RSW = 1 g/L; initial pH = 5, $[Cu]_0$ = 6 mM, contact time = 24 h.[140]

The solutions have TOC less than 5 mg/L, much lower than the leaching from RSW (19 mg/L TOC). Higher organic leaching can be observed when acid or 10% aldehydes is used, which may be due to organic dissolution by acid and high dosages of aldehydes applied in modifications.

Based on weight loss during the modification and the metal biosorption, organic leaching and heavy metal removal capacity, the chemical modification by 0.2% formaldehyde is the best for copper biosorption.

Aldehydes such as formaldehyde used in this study are commonly used for preservation of plant and animal tissues. Formaldehyde can cause polymerization. Polymerized aldehydes can react with primary amine groups in seaweeds as illustrated:

$$R_1-CHO+R_2-NH_2 \rightarrow R_1-CH=N-R_2+H_2O \quad (7.9)$$

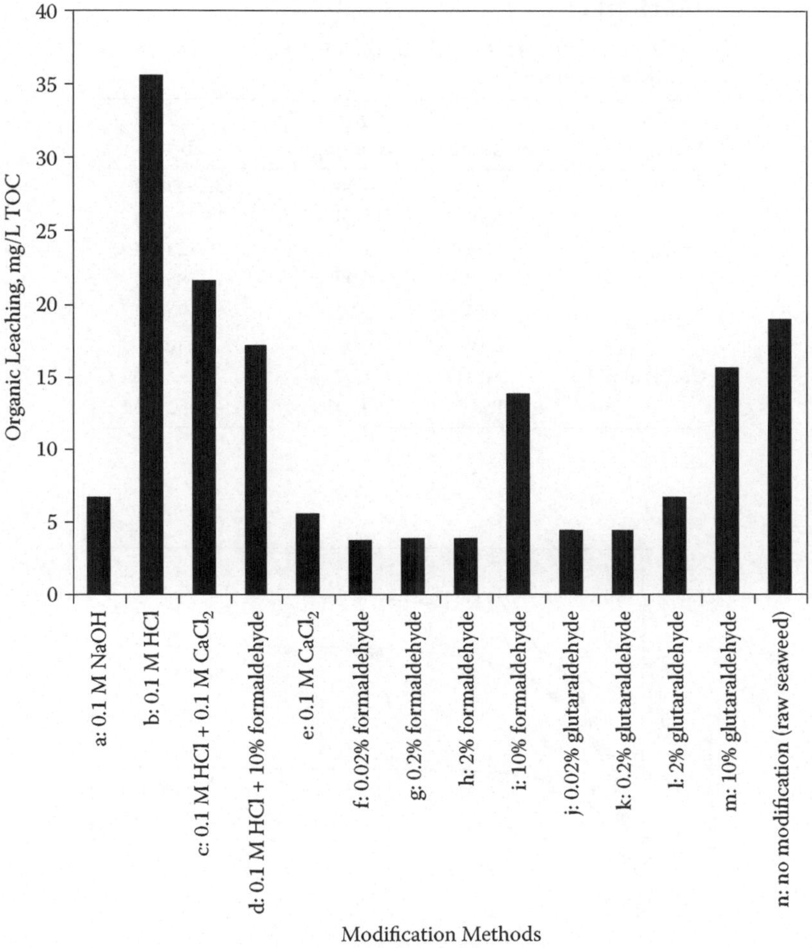

FIGURE 7.36
Organic leaching from raw and modified *Sargassum* sp. during copper biosorption. Biosorption: MSW or RSW = 1 g/L; initial pH = 5, $[Cu]_0$ = 6 mM, contact time = 24 h. Water samples were filtrated by using 0.45 μm membrane filters.[140]

The metal sorption isotherm experiments of RSW and MSW were conducted for cationic and anionic heavy metal ions.[140,141] Values of q_{max} and b for the metal biosorption onto RSW and 0.2% formaldehyde MSW are listed in Table 7.11. Comparison of the q_{max} values of MSW and RSW shows that the metal biosorption is enhanced by 26% to 200% after the chemical modification is applied. This clearly demonstrates that chemical modification significantly improves the biosorption capacity.

Figure 7.37 shows the kinetics of metal biosorption by the 0.2% formaldehyde MSW. More than 90% of the maximum metal uptake can be

TABLE 7.11

Langmuir Constants of Metal Biosorption by Raw and Modified *Sargassum* sp.

Metals	Biosorbent	q_{max} (mmol/g)	b
Pb(II)	MSW	1.46	132.84 mM^{-1}
	RSW	1.16	14.23 mM^{-1}
Cu(II)	MSW	1.37	10.42 mM^{-1}
	RSW	0.99	8.78 mM^{-1}
Ni(II)	MSW	1.22	5.69 mM^{-1}
	RSW	0.61	4.6 mM^{-1}
Cr(VI) at pH 2	MSW	1.123	2.162 mM^{-1}
	RSW	0.601	0.652 mM^{-1}

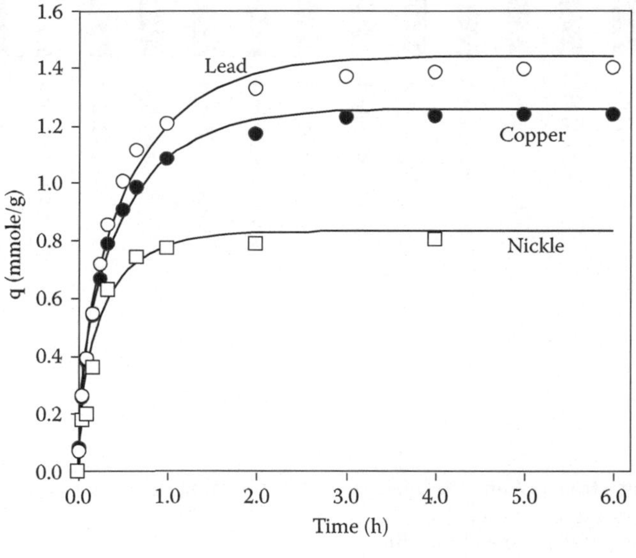

FIGURE 7.37
Kinetics of heavy metals biosorption by modified *Sargassum* sp. [MSW] = 1 g/L; $[Cu]_0$ = 1.0 mM, $[[Ni]_0$ = 1.20 mM, $[Pb]_0$ = 1.20 mM. Noted that the points represent experimental data while the lines represent results from modeling simulation by an intraparticle diffusion model.[140]

achieved within 40 min. No further heavy metal removal is observed after 3 h for all the cations. Comparison of this finding with those reported in the literature shows that the modification does not alter the biosorption kinetics.[9,139,140] The biosorption kinetics involves a rapid process followed by a slightly slow process.

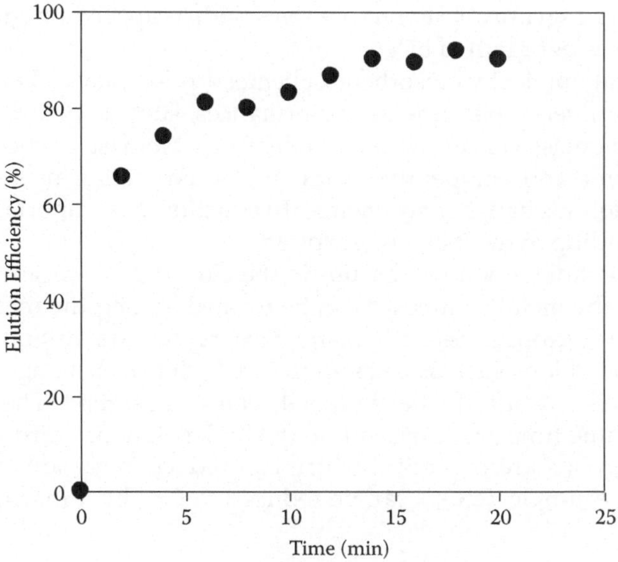

FIGURE 7.38
Elution efficiency of copper as a function of time. Copper ions were adsorbed onto the MSW with a biosorption capacity of 1.23 mmol/g; 0.1 M HCl was used for the metal elution.[140]

Copper desorption kinetics was studied with the date shown in Figure 7.38.[140] The desorption sharply increases during the first several minutes; nearly 70% of maximum desorption occurs in 2 min. It is completed within 20 min. The kinetics of desorption is much faster than that of biosorption (equilibrium time of 4 h) as shown in Figure 7.37.

It was found that a five-cycle operation of metal sorption and desorption confirms that the MSW was much better than the RSW. The concentration of eluted copper solution was above 700 mg/L, which could be further recovered by chemical reduction technologies.

7.3.2 Biosorption Chemistry

Generally, the performance of a biosorption system is determined by the characteristics of raw wastewater (e.g., pH, ionic strength, temperature, and metal species), type of biomass (e.g., capacity, kinetics, selectivity, immobilization, size, and structure), and operation factors of process (e.g., stirring speed).

The physisorption (i.e., physical interactions) such as electrostatic abstraction may cause some removal of heavy metals. It plays an important role in sorption of organic substances. It is reported that biosorbents can remove organics, such as dyes, phenols, and pesticides. For example, active sludge, bacteria, and fungi are able to remove dyes with a capacity ranging from 100 to 300 mg dyes per gram of biosorbents.

In the metal uptake by biosorbents, chemosorption plays a key role. Many factors can influence biosorbents performances, such as pH, pollutant concentration, biomass concentration, temperature, biomass particle size, mixing conditions, and competitive components. Formation of metal surface complexes, ion exchange, and chemical reduction are important chemical reactions, leading to the metal biosorption.

First, the organic functional groups in the biosorbents are similar to weak acids. Thus, the metal complexes can be formed among the metal ions and the functional groups. Second, many biosorbents are initially saturated with light metal ions such as calcium ions, which can exchange with heavy metal ions. As a result, the heavy metals can be adsorbed. The light metal ions at the same time are released into the bulk solution. Third, some of the functional groups are reductive, which can reduce metal ions with higher valency. For example, CrO_4^{2-} can be reduced to Cr^{3+} by a marine algal biomass (Chapter 5).

Case Study: Modeling of Metal Biosorption onto Calcium Alginate Biosorbent

A calcium alignate biosorbent developed by Chen and coworker has demonstrated excellent Biosorption capacities for different heavy metal ions.[142] It was found that the biosorbent contained such alkaline earth metal ions (light metals) as calcium and magnesium ions existed in the biosorbents. They could exchange with the heavy metal ions such as copper and lead ions, leading to the removal of the heavy metal ions. In addition, the formation of metal complexes might play an important role.[142,143]

There may be several organic functional groups for metal binding. It can be assumed that there is one generalized type of functional group (e.g., R^{2-}) in order to simplify the simulation of the process. The functional groups are complexed with alkaline earth metal ions such as calcium, leading to the formation of metal–organic complexes (e.g., CaR). Ion exchange is the main pathway through which heavy metal ions are stripped away.

When the biosorbent is immersed in an aqueous solution, alkaline earth metal ions (e.g., calcium) can be replaced by hydrogen ions according to the following reaction:

$$2H^+ + [Ca\mathit{R}] = Ca^{2+} + [H_2\mathit{R}] \tag{7.10a}$$

When a heavy metal ion is present in the solution, an exchange between metal ions and alkaline earth metal ions inside the biosorbent takes place:

$$M^{2+} + [Ca\mathit{R}] = Ca^{2+} + [M\mathit{R}] \tag{7.10b}$$

TABLE 7.12

List of Reactions for Modeling of Metal Uptake onto Calcium Alginate-Based Biosorbent

No.	Reaction	Equilibrium Constant
a	$2H^+ + R^{2-} = H_2R$	$10^{8.000}\ M^{-2}$
b	$Ca^{2+} + R^{2-} = CaR$	$10^{5.657}\ M^{-1}$
c	$Pb^{2+} + R^{2-} = PbR$	$10^{7.257}\ M^{-1}$
d	$Cu^{2+} + R^{2-} = CuR$	$10^{6.205}\ M^{-1}$

Content of functional group (R^{2-}) = 3.338×10^{-3} mol/g.

where M represents heavy metal ion to be removed, such as copper and lead.

The higher affinity of R^{2-} for M^{2+} is the driving force for pushing the equilibrium to the right.

In addition, some functional groups that do not participate in the binding of alkaline earth metal ions may react with the heavy metal ions to form an ion-pair ($M^{2+}R^{2-}$), leading to the metal uptake.

$$M^{2+} + R^{2-} = [MR] \tag{7.10c}$$

Table 7.12 is used for the modeling of the copper and lead uptake by a calcium alginate-based biosorbent.[142] The equations listed are different to Equations (7.10a–b). However, these equations can be easily obtained via a combination of the equations in Table 7.12. For example, Equation c minus Equation b can yield Equation 7.10b.

In modeling, the equilibrium constants of reactions in Equations a and b of Table 7.12 can be obtained based on the titration experimental data of the biosorbent, as shown in Figure 7.39. The equilibrium constants of reactions in Equations c and d can then be estimated by using the sorption isotherm data. Figure 7.40 shows the fitting for the lead biosorption. After all the equilibrium constants are obtained, modeling simulation can be conducted for the pH effect and competitive sorption experiments shown in Figures 7.41 and 7.42, respectively. It is demonstrated that the model can be successfully used for the simulation of metal biosorption.

7.3.3 Biosorption Performance

Many studies point out that pH is the most critical factor that can affect biosorbents' capacity.[144–190] The pH effect is often determined by metal species and different functional groups on biosorbents. The uptake of cationic heavy metal ions increases when the solution pH is increased

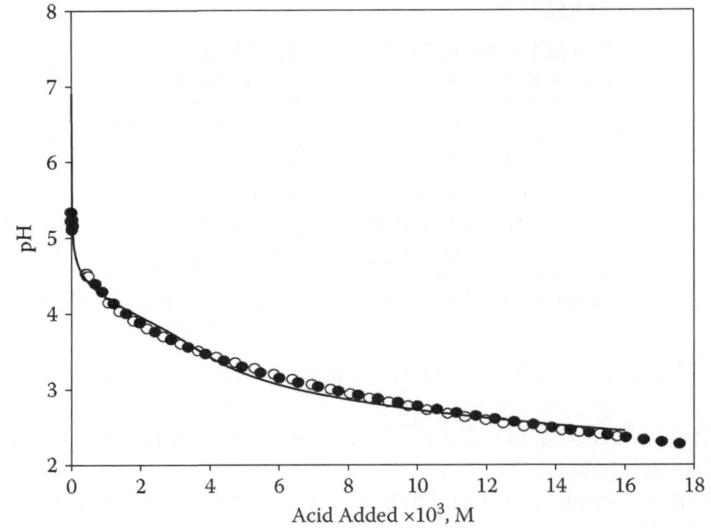

FIGURE 7.39
Modeling of titration experimental data of calcium alginate-based biosorbent. $NaNO_3$ added as ionic strength, m = 5 g/L, HNO_3 used as titration reagent, solid points (I = 0.05 M), hollow points (I = 0.1 M), curve represents modeling results.[142]

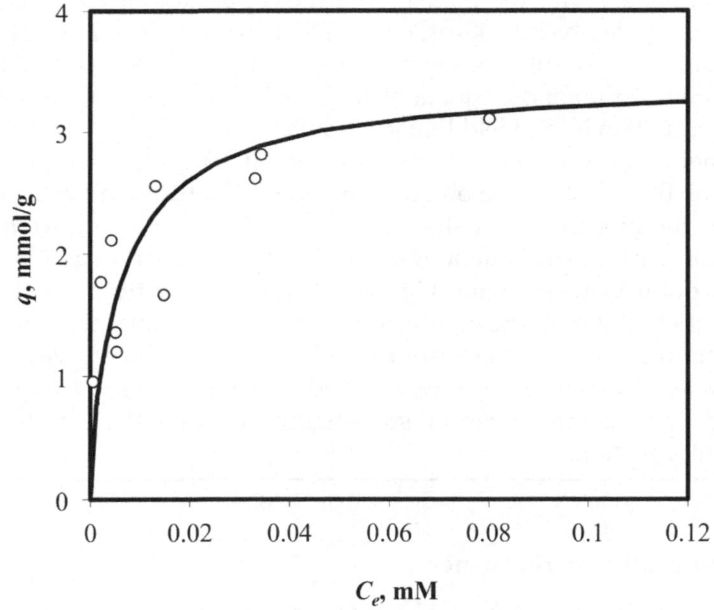

FIGURE 7.40
Lead sorption isotherm of calcium alginate-based biosorbent. Illustration: ○, experimental; —, modeling.[142]

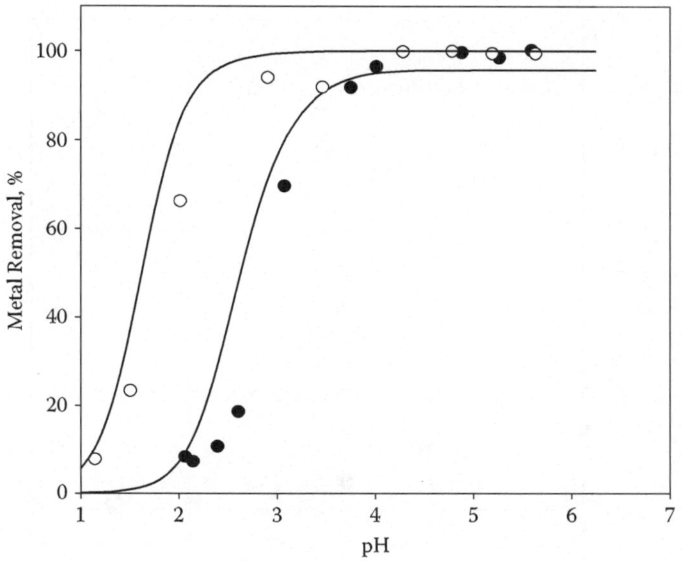

FIGURE 7.41
pH effect on metal removal by calcium alginate biosorbent. Condition: $[Pb]_o = 1.0 \times 10^{-4}$ M, m = 1 g/L, $[Cu]_o = 1.0 \times 10^{-4}$ M, [calcium alginate] = 0.15 g/L. Illustration: ● experimental (Cu), ○ experimental (Pb), — modeling.[142]

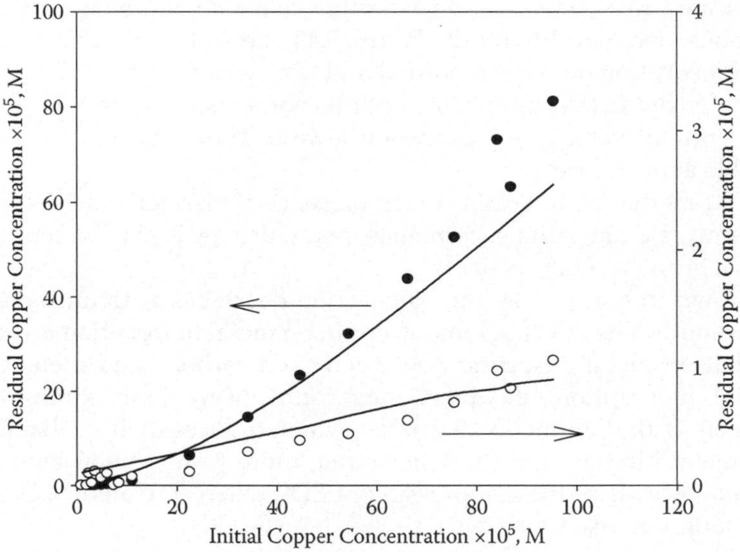

FIGURE 7.42
Competitive biosorption by calcium alginate biosorbent. pH 5.5–7.0, m = 0.167 g/L, $[Pb]_o$ = 1.0×10^{-4} M. Points and curves represent experimental and modeling results, respectively.[142]

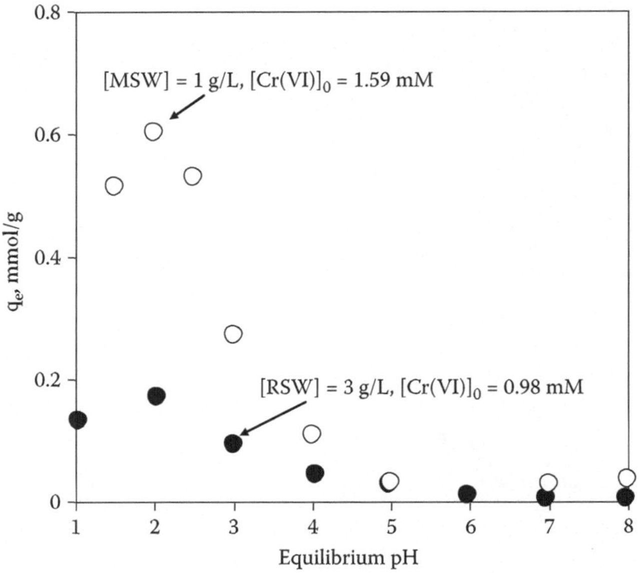

FIGURE 7.43
Effect of pH on biosorption onto RSW and MSW. pH was controlled and contact time was 24 h. RSW represents raw *Sargassum* sp. MSW represents the 0.2% formaldehyde-treated *Sargassum* sp.

as illustrated in Figure 7.41. On the other hand, the anionic heavy metal biosorption behaves differently. Figure 7.43 shows the pH dependence of Cr(VI) biosorption onto the modified and raw seaweeds. pH 2.0 is the optimal pH for the Cr(VI) uptake for both biosorbents.[141] When the pH is out of this optimal value, the biosorption lowers. The metal uptake becomes negligible at pH above 6.

Ionic strength is an indication of the presence of inert ions such as sodium, potassium, calcium, sulfate, carbonate, and chloride. It plays a certain role, though it is not as much as pH.

As shown in Figure 7.44, the biosorption decreases with an increase in ionic strength. The effect of ionic strength on metal biosorption is different from that on metal adsorption onto activated carbons and metal oxides. In metal biosorption, the inert ions contributing ionic strength may participate with the functional groups, which reduces their availability for heavy metal binding. On the other hand, more heavy metal ions can be adsorbed because of the compression of EDL when activated carbons and metal oxides are used.

Any biomass can adsorb heavy metal ions from aqueous solutions. For example, activated sludge has good metal-binding capacity. As shown in Figure 7.45, good metal sorption can be observed, which explains the reason for heavy metal removal in the activated sludge process.[144]

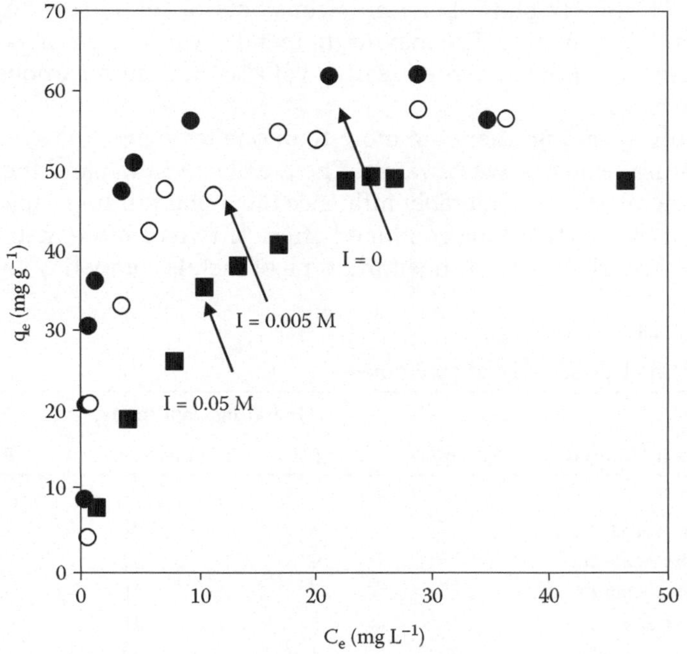

FIGURE 7.44
Effect of ionic strength on copper biosorption.

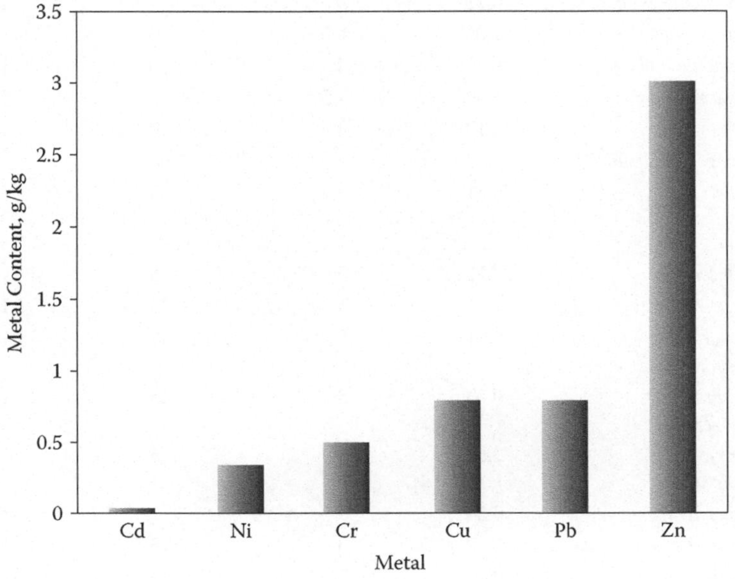

FIGURE 7.45
Heavy metal biosorption onto activated sludge.

Tables 7.13 to 7.17 show Langmuir constants of metal sorption of four important heavy metals. The maximum metal sorption capacities (q_{max}) are much larger than most commercially available activated carbons and ion exchange resins.

Generally speaking, there are more than one toxic metallic species existing in natural water or wastewaters. The metal speciation and the presence of other metals can considerably influence the metal-binding capacities and kinetics of biosorption. The combined effect of two or more metals on biosorption depends on the number and type of metals competing for binding

TABLE 7.13

Biosorption Uptake of Lead by Biomass

Biosorbent Material	q_{max} (mmol/g)	Operating Conditions pH	Temperature (°C)	Reference
Algae				
Ascophyllum nodosum	1.27	5	21	145
Cladophora glomerata	0.35	4.5	30	146
Durvillaea potatorum	1.55	5	21	145
Ecklonia radiata	1.26	5	21	145
Gracilaria canaliculata	0.20	4.5	30	146
Gracilaria corticata	0.26	4.5	30	146
Laminaria hyperbola	1.35	5	21	145
Lessonia flavicans	1.45	5	21	145
Padina gymnospora	0.31	3.5		147
Padina pavonia	1.05	4.5	30	146
Polysiphonia violacea	0.49	4.5	30	146
Sargassum hystrix	1.38	4.5	30	146
Sargassum natans	1.13	4.5	30	146
Schizomeris leibleinii	0.31	4.5	30	148
Ulva lactuca	0.61	4.5	30	146
Fungus				
Absidia coerula	0.71	5	25	149
Agaricus bisporus	1.05	5	25	149
Cladosporium cladosporioides	0.81	5	25	149
Penicillium chrysogenum	0.63	5	25	149
Phycomyces blakesleeanus	0.60	5	25	149
Rhizopus arrhizus	0.24	5		150
Rhizopus oryzae	1.07	5	25	149
Saccharomyces cerevisiae	0.49	5	25	149

TABLE 7.14
Biosorption Uptake of Copper by Biomass

Biosorbent Material	q_{max} (mmol/g)	Operating Conditions pH	Temperature (°C)	Reference
Algae				
Ascophyllum nodosum	1.19	5	21	145
Cyclotella cryptica	0.41	6		151
Durvillaea potatorum	1.31	5	21	145
Ecklonia maxima	1.22	5	21	145
Ecklonia radiata	1.11	5	21	145
Laminaria japonica	1.20	5	21	145
Lessonia flavicans	1.25	5	21	145
Lessonia nigresense	1.26	5	21	145
Porphyridium purpureum	0.27mg/g	6		151
Sargassum filipendula	0.89	4.5	22	152
Sargassum vulgare	0.93	4.5	22	152
Scenedesmus subspicatus	0.21	6		151
Bacterium				
Arthrobacter sp.	0.16	5		153
Arthrobacter sp.	0.73	5.4		153
Arthrobacter sp.	0.72	6		153
Sphaerotilus natans	0.91	6	25	154

sites, initial metal concentration, order of metal addition, and exposure time. The uptake of heavy metals will not be greatly affected if it has a higher affinity for biosorbent. However, less removal occurs for the metal that has a lower affinity for biosorbent.

The competitive biosorption of hexavalent chromium, lead, and copper ions on *Bacillus* sp. in a batch-stirred system was investigated by Nourbakhsh et al.[162] The results showed that lead biosorption increased widely, while chromium and copper biosorption decreased in comparison with the biosorption of only one kind of metal ion.

The effect of the presence of multiple metal ions on metal removal efficiency by two distinct (flocculent and nonflocculent) yeast wastes was studied by Marques et al.[163] It was observed that only copper was significantly affected by the presence of the other metals (lead and cadmium) in solution and only when the nonflocculent yeast biomass was used as biosorbents. Cadmium was only slightly affected by the presence of both copper and lead; the lead removal was not affected by the presence of any or both of the interfering metals for the two biosorbents used.

TABLE 7.15

Biosorption Uptake of Cadmium by Biomass

Biosorbent Material	q_{max} (mmol/g)	Operating Conditions		References
		pH	Temperature (°C)	
Algae				
Ascophyllum nodosum	1.03	5	21	145
Durvillaea potatorum	1.12	5.4	21	155
Ecklonia maxima	1.15	5	21	145
Ecklonia radiata	1.04	5	21	145
Laminaria hyperbola	0.82	5	21	145
Laminaria japonica	1.11	5	21	145
Sargassum filipendula	0.66	4.5	22	152
Sargassum fluitans	0.71	4.5	22	152
Sargassum vulgare	0.79	4.5	22	152
Scenedesmus subspicatus	0.06	6		151
Schizomeris leibleinii	0.44	5	30	148
Spirulina platensis	0.87	7	26	156
Fungus				
Rhizopus arrhizus	0.56		23	157
Rhizopus arrhizus	0.13	5		150
Rhizopus arrhizus	0.26	6		150
Rhizopus arrhizus	0.39	7		150
Rhizopus oligosporus	0.37	*NA	23	157
Rhizopus oryzae	0.31	*NA	23	157
Trametes versicolor	0.165	5.5	25	158
Trametes versicolor	0.141	5.5	25	158
Bacterium				
Arthrobacter sp.	0.119	6		153
Sphaerotilus natans	0.225	6	25	154

* NA: not available.

El-Naas and co-workers (2007) used *Chlorella vulgaris* for the biosorption of heavy metals.[164] As shown in Figure 7.46, the uptake of lead is adversely affected by the presence of copper ions, while zinc ions have negligible effect on the process. The effects of copper and zinc ions on the sorption isotherms of leads are demonstrated in Figure 7.47.

The effect of temperature is more complicated in that metal sorption. Generally an increase in temperature within a certain range (e.g., 20°C–50°C)

TABLE 7.16
Biosorption Uptake of Zinc by Biomass

Biosorbent Material	q_{max} (mmol/g)	Operating Conditions pH	Temperature (°C)	Reference
Algae				
Cymodocea nodosa	0.71	5.5		159
Fungus				
Penicillium chrysogenum	0.09	NA		160
Rhizopus arrhizus	0.30	5.5		150
Bacterium				
Citrobacter sp.	0.12	NA		160
Streptoverticillium cinnamomeum	0.15	NA		160

TABLE 7.17
Biosorption Uptake of Nickel by Biomass

Biosorbent Material	q_{max} (mmol/g)	Operating Conditions pH	Temperature (°C)	Reference
Algae				
Sargassum fluitans	1.99	5		152
Ascophyllum nodosum	1.43	5		152
Fungus				
Polyporous versicolor	0.80	5	25	161
Rhizopus arrhizus	0.28	6		150

can enhance the soprtion. However, it may reduce the metal uptake when the temperature is above a certain value.

Higher temperature would increase the kinetic energy of the metal ions, which facilitates attachment of metal ions on the surface.[165–167] Table 7.18 shows the temperature effect on cadmium biosorption onto *Hypnea valentiae* biomass.[167] However, very high temperatures may cause distortion of some sites of the biosorbent. Hence, biosorption processes are normally operated at temperature range from 5°C to 35°C.

Biosorption kinetics is considered to have major advantages over other sorption technologies. The process is much faster than using activated carbons and metal oxides. Cationic heavy metal ion can be quickly removed as shown in Figures 7.48 and 7.49. An equilibrium time of 1 to 3 h is sufficient

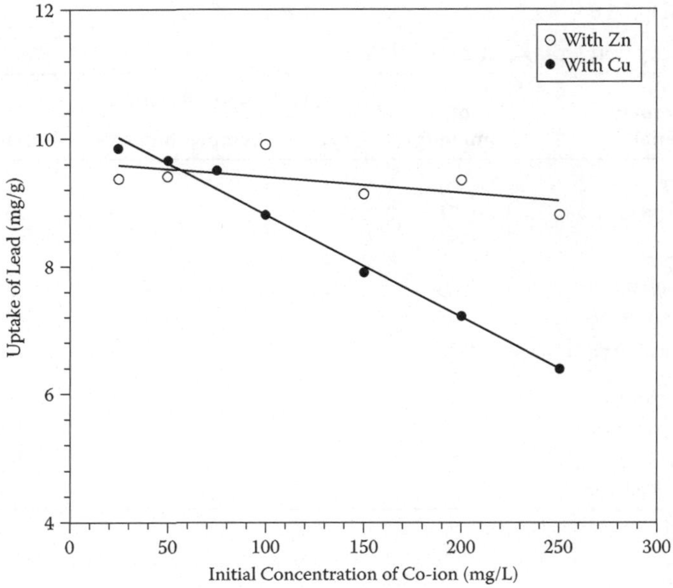

FIGURE 7.46
Effect of competitive metal ions on the biosorption of Pb(II) at 25°C in a binary solution. Initial Pb(II) concentration of 25 mg/L and algae dose of 2 g/L.[164]

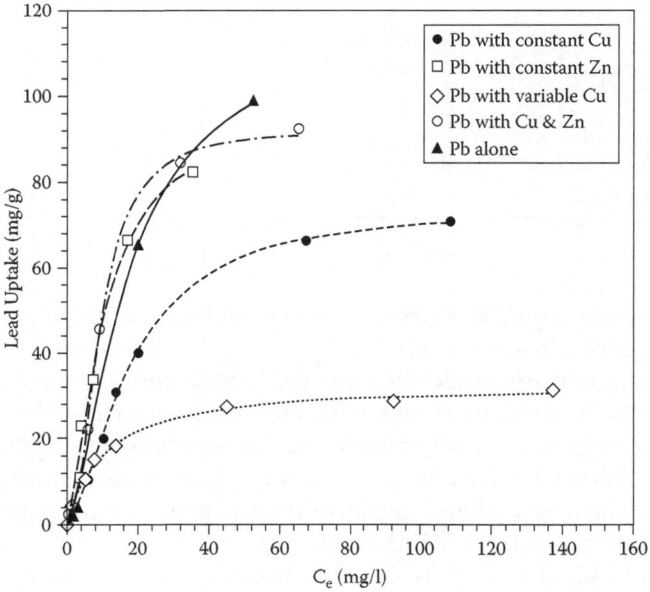

FIGURE 7.47
Effect of competitive metal ions on lead sorption isotherms.[164]

TABLE 7.18

Effect of Temperatures on Sorption Isotherms of Cadmium onto *Hypnea valentiae* Biomass

Temperature (°C)	q_{max} (mg/g)	b (L/mg)
30	15.8	0.053
40	18.8	0.049
50	20.9	0.046
60	28.7	0.012

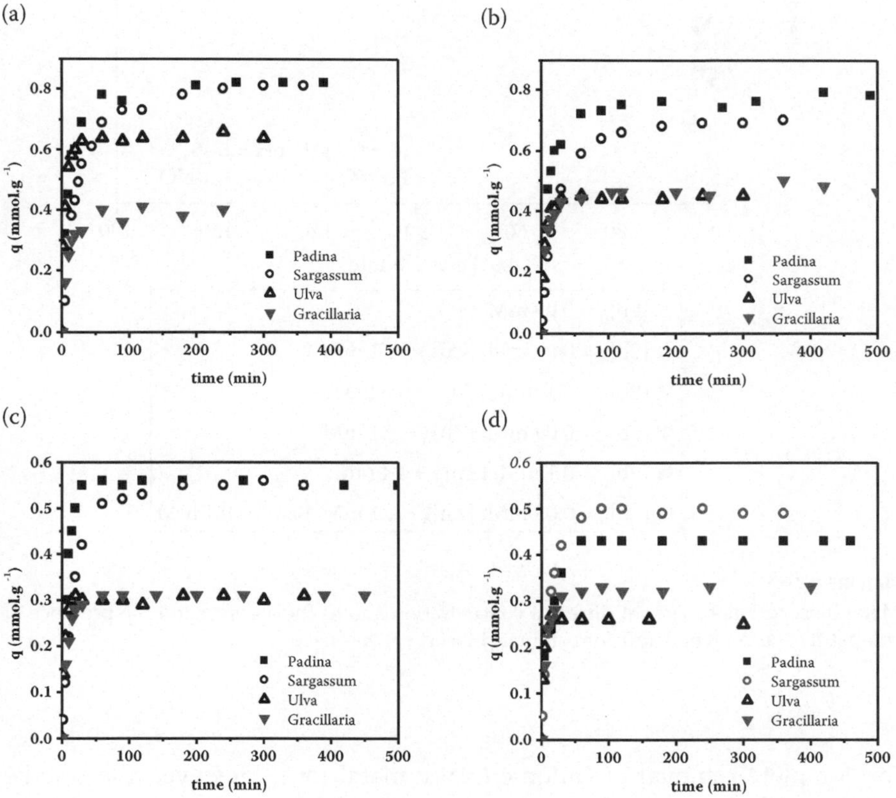

FIGURE 7.48
Kinetics of heavy metals biosorption by formaldehyde treated *Sargassum* sp. [MSW] = 1 g/L; $[Cu]_0$=1.0 mM, $[Ni]_0$ = 1.20 mM, $[Pb]_0$ = 1.20 mM.

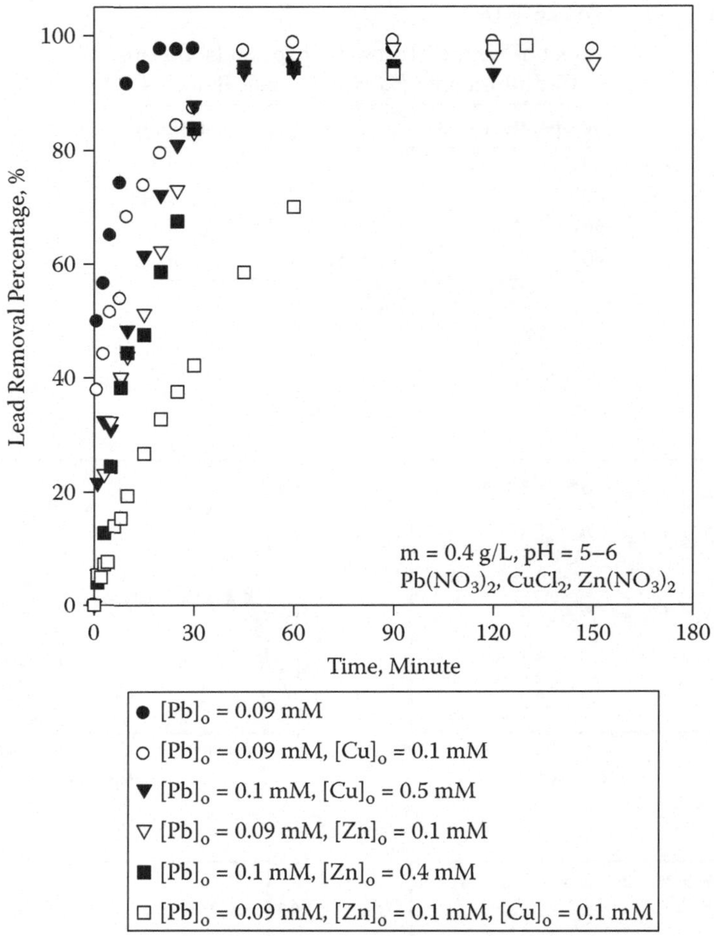

FIGURE 7.49
Metal biosorption kinetics at different concentration and in the absence and the presence of competitive heavy metal. Calcium alginate biosorbent was used.

for complete removal of cationic heavy metal ions. However, the uptake rate is much slower for the biosorption of anionic heavy metal ion as illustrated in Figure 7.50. Anionic heavy metal biosorption equilibrium may be established within 24 h. In addition, the presence of competitive metal ions can greatly delay biosorption as shown in Figure 7.49 (the case of Pb, Zn, and Cu).

Biosorption uptake of heavy metal ions initially occurs rapidly, which contributes 80% to 90% of the final removal.[9,140,141] A slower process then occurs, which contributes 10% to 20% of the final removal.

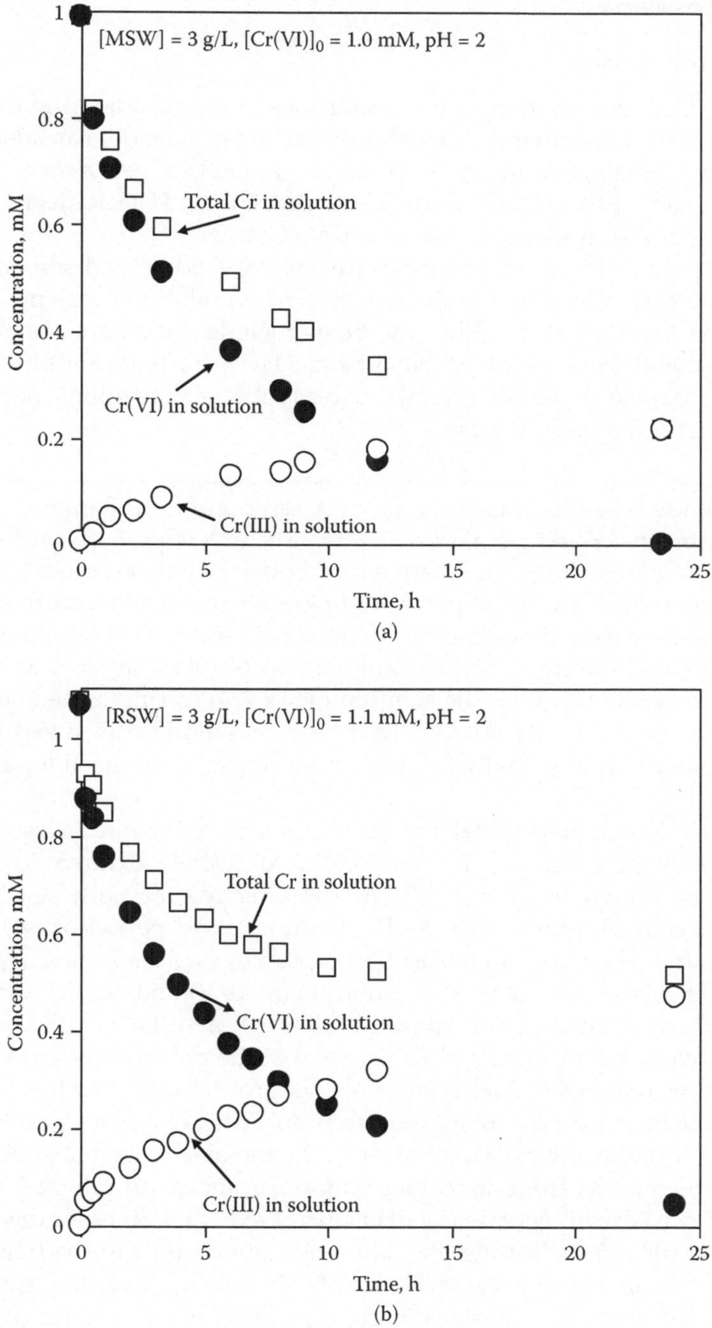

FIGURE 7.50
Kinetics of chromate biosorption onto raw and formaldehyde treated *Sargassum* sp. (RSW and MSW) at pH 2.[141]

7.4 Metal Oxide

Metal oxides can remove heavy metal ions through adsorption.[11–14,168–190] It has been well documented that aluminum and iron oxides can adsorb various heavy metals. The adsorption onto these metal oxides has been studied since the early 1970s.[171] The adsorption behaviors and mechanisms of both metal oxides are similar to those of activated carbons.

In the past 10 years, several innovative metal-oxide-based sorbents have been synthesized to meet the challenges of new pollution problems, such as arsenic in drinking water. These sorbents include the binary metal oxides and the metal oxide modified adsorbents. They are often multifunctional and can treat various contaminants such as cationic and anionic heavy metals and natural organic matters.

Case Study I: Fe-Mn Binary Oxide for Arsenic and Lead Removal

An innovative Fe–Mn binary oxide adsorbent was prepared by Zhang and co-workers (2007) for treatment of both arsenate and arsenite.[14] In the preparation of sorbent, potassium permanganate and iron(II) sulfate heptahydrate were first dissolved in deionized water. $FeSO_4$ solution was slowly added into the $KMnO_4$ solution; 5 M NaOH solution was simultaneously added to keep the solution pH of 7 to 8. The formed suspension was continuously stirred, aged, and washed by deionized water. The suspension was finally filtrated and dried, and was used for arsenic removal.

It was demonstrated that the Fe-Mn binary oxide had a very good adsorption performance. It was able to completely oxidize As(III) to As(V) as shown in Figure 7.51. It was effective for both As(V) and As(III) removal, particularly As(III) within a short period of operation (~12 hours). The maximal adsorption capacities (q_{max}) and the Langmuir constant related to heat of adsorption (b) of As(V) and As(III) were 0.93 and 1.59–1.77 mmol/g-sorbent, and 6.8×10^3 and 1.87×10^2 L/mmol, respectively. Figures 7.52 and 7.53 show that the pH effect is less significant in the removal of As(III) and As(V) in groundwater that has pH of 6 to 7.5. Furthermore, the ionic strength is not critical in the removal.

In addition to adsorption of arsenic, the metal oxide can also remove heavy metal ions from aqueous solution. As shown in Figure 7.54, the adsorption of lead increases as pH is increased; it becomes less sensitive to pH at pH > 5, indicating its wider application range in the treatment of heavy metal waste streams. Figure 7.55 shows the adsorption isotherm of lead. The maximum adsorption capacity (q_{max}) and Langmuir constant (b) are 54.6 mg/g and 0.053 L/mg, respectively. This demonstrates that the adsorption of the metal oxide is comparable to that of biosorbent and much better than activated carbons.

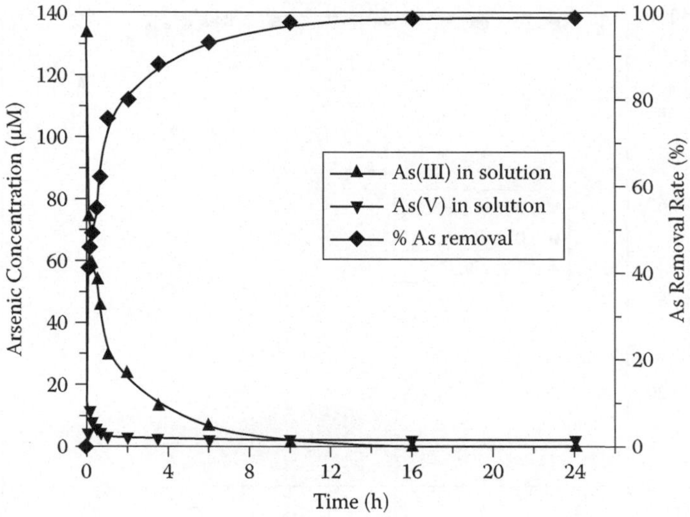

FIGURE 7.51
Arsenic adsorption history. Initial As(III) concentration and Fe-Mn binary oxide adsorbent content were 133 μMm and 200 mg/L, respectively. pH was controlled at 6.9.[14]

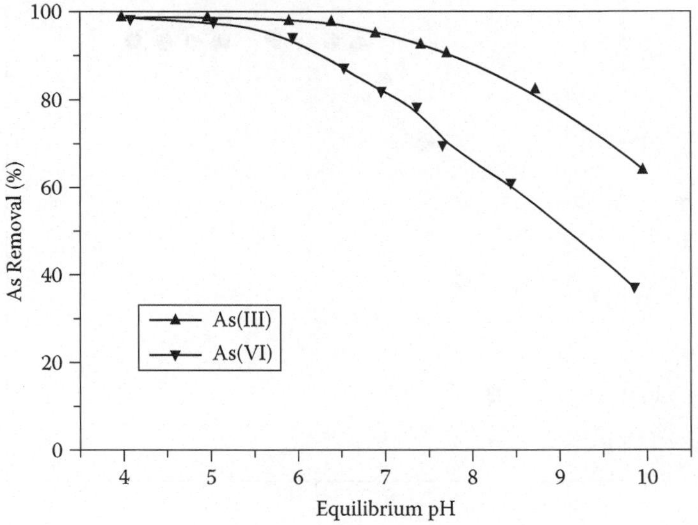

FIGURE 7.52
Effect of pH on arsenic adsorption onto Fe-Mn binary oxide. Initial As(III) concentration and adsorbent content were 133 μMm and 200 mg/L, respectively.[14]

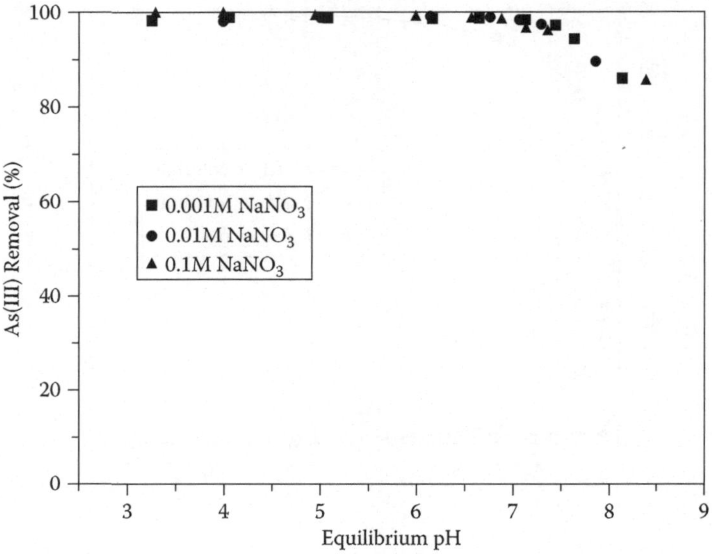

FIGURE 7.53
Effect of ionic strength on As(III) adsorption onto Fe-Mn binary oxide. Initial As(III) concentration and adsorbent content were 133 μMm and 200 mg/L, respectively.

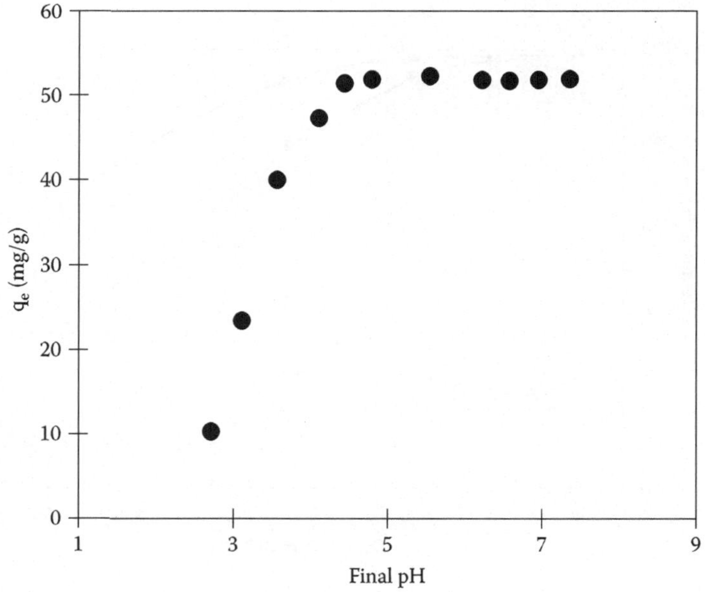

FIGURE 7.54
Effect of pH on lead adsorption onto Fe-Mn binary oxide.

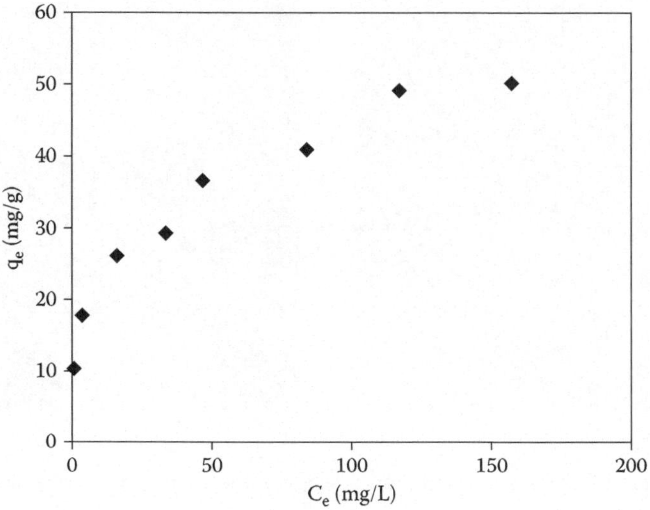

FIGURE 7.55
Adsorption isotherm of lead onto Fe-Mn binary oxide.

Case Study II: Zirconium-Based Nanoscale Sorbent for Arsenic Removal
A zirconium-based nanoscale sorbent was designed and fabricated for the adsorption of arsenic by Ma et al. (2011).[184,192] A mixture of $ZrOCl_2{\cdot}8H_2O$ and an acid was first prepared and mixed, leading to the formation of the white fine particles suspended in the solution. The new particles were harvested from the suspension by the centrifugation, which were washed with deionized water and dried. The resulted nanoscale sorbent is demonstrated in Figure 7.56. Most of them have particle sizes around 100 nm. The sorbent was placed in several solutions at different pH; the zirconium concentrations of leachates were measured. It was shown that the concentrations were below the detection limit of ICP-OES. This indicates that the nano particles would not cause harmful effects to human beings when the treated water is consumed.

The adsorption kinetics of anionic arsenic on the sorbent is of great importance for designing appropriate adsorption technologies. It can be found from Figure 7.57 that most of uptake of As(V) rapidly takes place in the first 6 h, followed by a relative slow process. The uptake of As(V) reaches equilibrium within 12 h. The result is much better than many available adsorbents for anionic pollutants removal as the sorbent size is within nanometers while others are micrometers.

The experimental result shown in Figure 7.58 demonstrates that the adsorption of As(V) on the sorbent is strongly pH dependent. The best As(V) uptake is in the pH range between 2.5 and 3.5. At an extremely acidic situation, the adsorption increases with an increase in pH, and

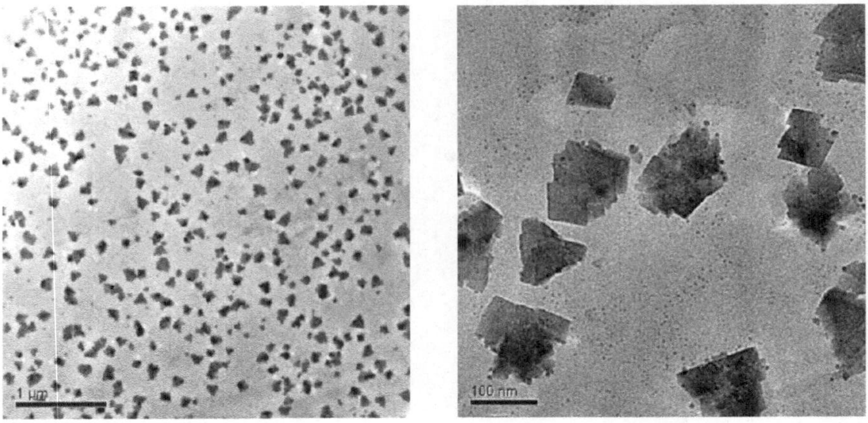

FIGURE 7.56
High-resolution transmission electron microscopy of the zirconium nanoparticle: (a) 4,000×; (b) 30,000×.[184]

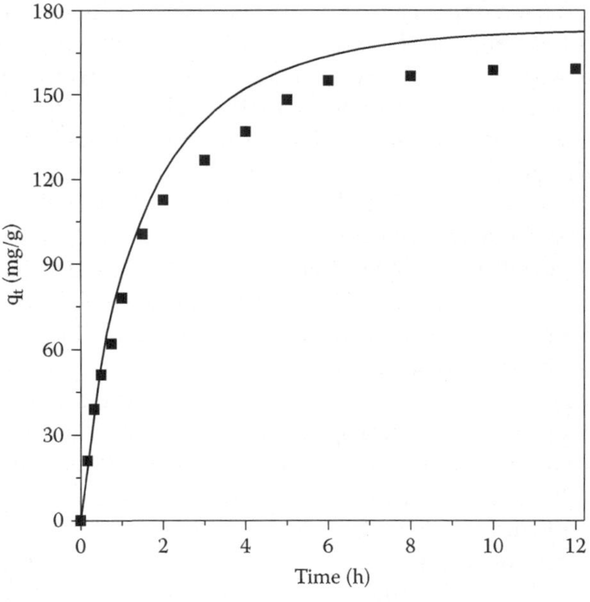

FIGURE 7.57
Adsorption kinetic of As(V) onto the sorbent (sorbent dose = 0.1 g/L, initial concentration = 20 mg/L, pH = 6.8, T = 20°C). Points and line represent experimental data and modeling fitting (by a surface diffusion control model).[184]

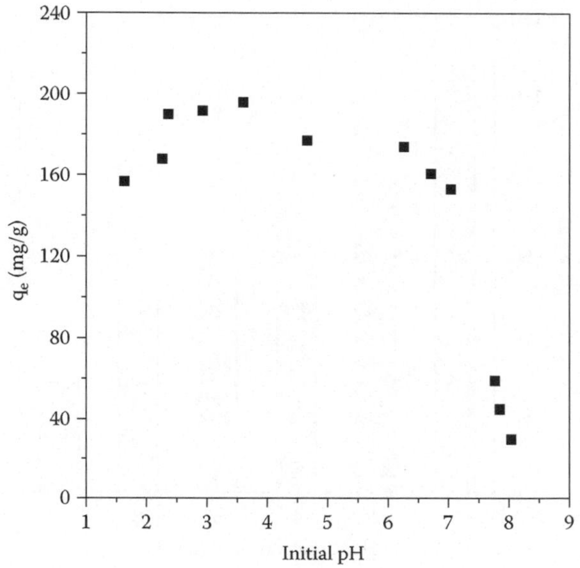

FIGURE 7.58
Effect of initial pH on As(V) adsorption (sorbent dose = 0.1 g/L, initial concentration = 20 mg/L, T = 20 ± 1°C).[184]

reaches a maximum uptake at pH around 3.0. The uptake of As(V) decreases with a further increasing pH.

The study on As(V) adsorption isotherm was conducted at pH 3.2 ± 0.1, the optimal pH for As(V) adsorption on the sorbent (Figure 7.58). It was found that the Langmuir isotherm gave the best fitting of the data. The maximum adsorption capacity (q_{max}) and Langmuir constant (b) are 256.4 mg/g and 1.564 L/mg, respectively. The q_{max} value is much larger than many other sorbents reported in the literature. The regression coefficient of 0.999 indicates that the arsenic uptake is due to the monolayer adsorption.

The presence of natural organic matters (e.g., humic acid, HA) and other anionic substances can reduce the adsorption of arsenic by the sorbent. As shown in Figure 7.59a, the adsorption of arsenate decreases by 27% and 74% in the presence of 2 and 10 mg/L HA, respectively. The valencies of arsenic species were studied by the voltammetry technique. Figure 7.59b shows that the arsenate and arsenite have characteristic peaks with voltage of 0.07 and 0.11 V, respectively. The solution sample after the adsorption has a peak with voltage of 0.07 V. These observations indicate no change in the valency of the arsenate during its adsorption onto the sorbent in the presence of the humic acid.

As shown in Figure 7.60, there is no significant influence on the As(V) adsorption capacity caused by the presence of fluoride or nitrate.

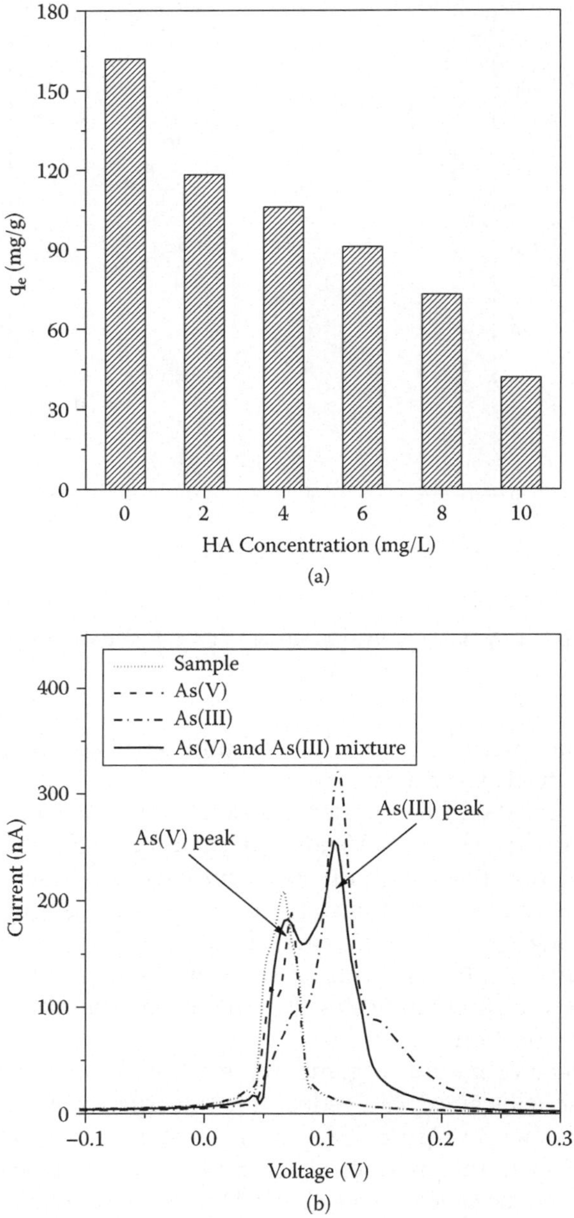

FIGURE 7.59
Effect of humic acid on adsorption of As(V): (a) the adsorption capacity; (b) arsenic valency. Experimental conditions: sorbent dose = 0.1 g/L, initial concentration = 20 mg/L, pH = 6.9 ± 0.1, T = 20°C.[184]

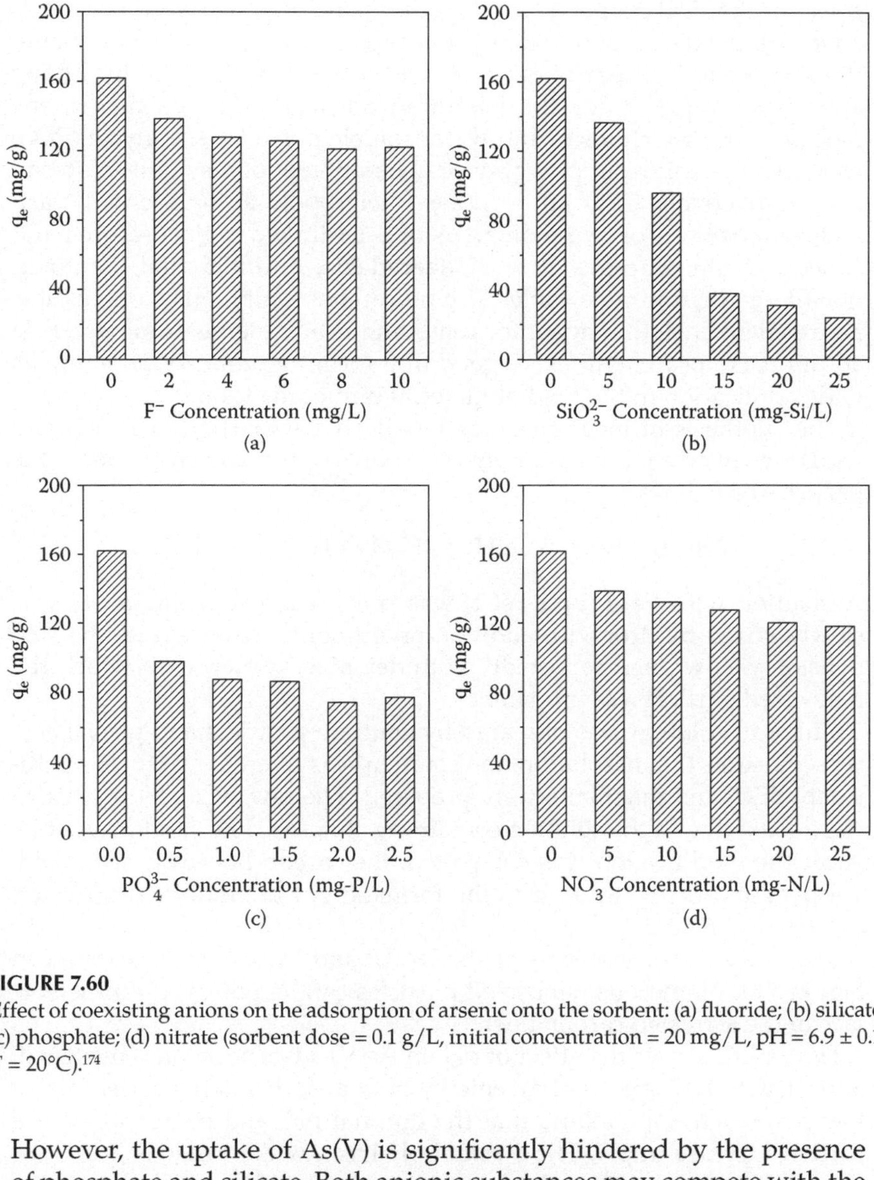

FIGURE 7.60
Effect of coexisting anions on the adsorption of arsenic onto the sorbent: (a) fluoride; (b) silicate; (c) phosphate; (d) nitrate (sorbent dose = 0.1 g/L, initial concentration = 20 mg/L, pH = 6.9 ± 0.1, T = 20°C).[174]

However, the uptake of As(V) is significantly hindered by the presence of phosphate and silicate. Both anionic substances may compete with the arsenic for the adsorption sites on the surface of sorbent.

Case Study III: Combination of Conventional Materials with Metal Oxides for Arsenic Removal

As a new generation of carbonaceous adsorbent, activated carbon fiber (ACF) has attracted considerable attention because of its rapid adsorption kinetics, high adsorption capacities for low-level contaminants, and a great mechanical property. On the other hand, magnetite iron particles have demonstrated good adsorption capacity for arsenic. As a result, the magnetite iron particles were used to modify the ACF by Zhang and co-workers[185,186] for the enhanced adsorption of arsenic. In the modification, the chitosan and water soluble polyethylene glycol (PEG) were used as a film support and for generation of pores on the chitosan film, respectively. Chitosan is a linear heteropolysaccharide with various compositions of two monomers of β-1,4-linked δ-glucosamine and *N*-acetyl-δ-glucosamine. It is biodegradable, antimicrobial, nontoxic, nonallergenic, and renewable. As chitosan has a high NH_2 content, it is an excellent chelating agent for removing heavy metals from water. In addition, chitosan manifests a good film-forming property, allowing its wide application in the field of material immobilization.

The synthesis of magnetite was based on the reaction of Fe(III) and Fe(II) ions in an aqueous ammonia solution to form nanosized magnetite particles as follows:

$$2FeCl_3 + FeCl_2 + 8NH_3 + 4H_2O \rightarrow Fe_3O_4 + 8NH_4Cl$$

A solution with $FeCl_2$ and $FeCl_3$ was prepared. The solution pH was slowly raised by dropwise addition of ammonia. After aging, the suspension was washed repeatedly with deionized water, with which the nanosized particle was prepared.

Chitosan solution was prepared by chitosan powders and glacial acetic acid. After the dissolution, the solution was filtered. Mixtures of chitosan, PEG, and magnetite were prepared. The raw ACF cloths (RACF) were soaked with the solution containing chitosan, PEG, and magnetite. After the modification was completed, the carbon fiber was separated, washed, and dried, leading to the formation of the modified activated carbon fiber (MACF).

The surface morphologies of the RACF and MACF are illustrated in Figure 7.61. Numerous nanosized particles can be observed on the surface of the activated carbon fiber.

Figure 7.62 shows the effect of pH on As(V) adsorption onto the RACF and MACF. The uptake of arsenic by both adsorbents increases from a low pH, reaches a maximum at the optimal pH, and decreases with a further increase in pH. The optimal pH has a wider range of 3 to 5 for the MACF, whereas it is rather narrow for the RACF at around pH 5. The maximum adsorption capacities (q_{max}) and Langmuir constants (b) of the MACF and RACF are 0.5 and 4.61 mg/g and 0.85 and 48.06 L/mg, respectively. Comparison of the adsorption performance shows that the modification of carbon by adding the magnetite sorbent significantly enhances the arsenic uptake capacity of the carbon filter. The MACF not only can adsorb organic matters but also can remove the arsenic contaminant from groundwater.

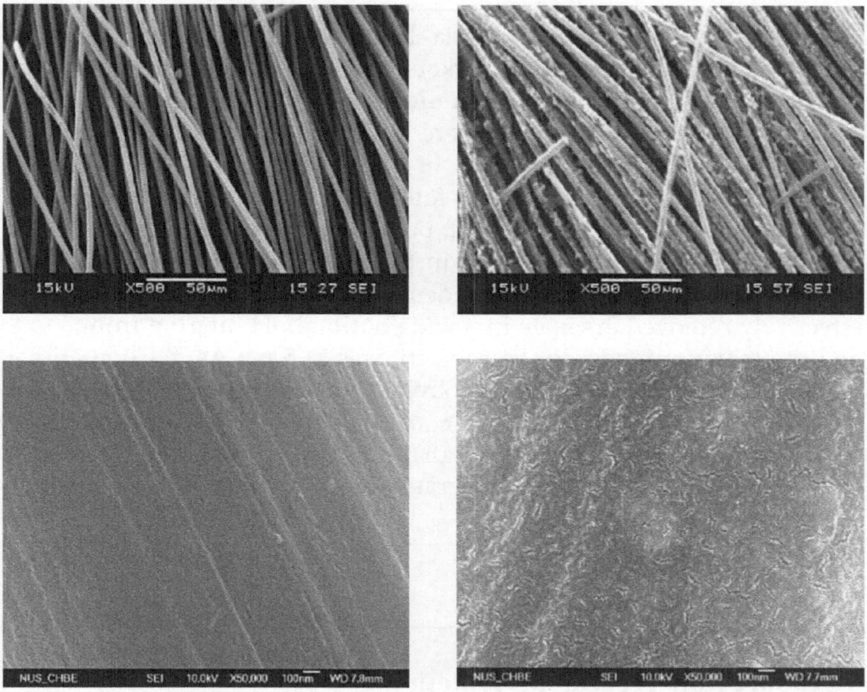

FIGURE 7.61
Surface morphologies of the RACF (left) and MACF (right).

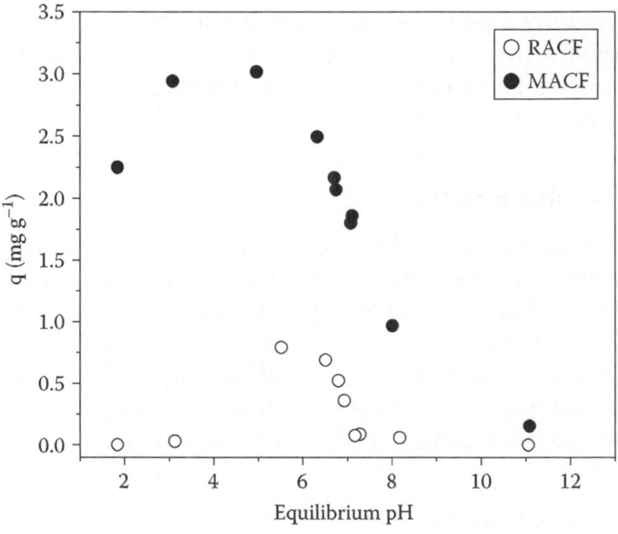

FIGURE 7.62
Effect of pH on arsenic adsorption. Experimental conditions: dosage = 0.6–0.7 g L^{-1}, adsorption time = 24 h.

Case Study IV: A PVDF/Zirconia blend flat sheet membrane was fabricated by Zheng and coworkers in order that it could adsorb arsenate and remove other contaminants such as microorganisms.[183] The hydrous zirconia particles were added into homogeneous casting solution consisted of a mixture of PVDF and PEG. The flat-sheet membrane was prepared. It was found that the addition of zirconia increased the hydrophilicity and porosity of the membrane, which led to a significant enhancement in the water flux of membrane. The batch adsorption experiments demonstrated that the membrane effectively removed arsenate in a wide optimal pH ranging from 3 to 8 and the maximum adsorption capacity was 21.5 mg-As/g. Competitive factors such as fluoride did not obviously affect the adsorption. The membrane demonstrated that it could effective remove arsenic and microorganisms (e.g., E. coli) through filtration. The membrane saturated by arsenate could be easily regenerated using an alkaline solution and subsequently be reused.

7.5 Adsorption Treatment System

Any sorption is a liquid–solid interaction process. Similar to the ion exchange process, a sorption process can be operated in batch, semicontinuous, or continuous modes. These can be through an upflow/downflow fixed-bed reactor, fluidized bed reactor, or continuous stirring treatment reactor. The downflow fixed-bed reactor is economic in operation and wideby used in ion-exchange and adsorption processes, but needs to carefully control the pressure loss along the column.

7.5.1 Fluidized Bed Reactor

Sorbent can be packed in a column and waste stream is pumped upflow as shown in Figure 7.63. When the flow rate reaches the fluidizing critical value, the sorbent-packed bed begins to expand and fluidize. This is a fluidized-bed operation. Generally, a fluidized-bed reactor causes less clogging than a fixed-bed reactor. The higher degree of mixing enables a faster sorption kinetics. The disadvantage is the bed expansion in the operation. Meanwhile, a more stable hydraulic condition should be maintained.

7.5.2 Stirred Tank Reactor

The operation of a stirred tank reactor can be batch or continuous. An amount of sorbent can be added at the inlet of the reactor or initially added into the reactor. The solid/liquid separation process can be accomplished either by

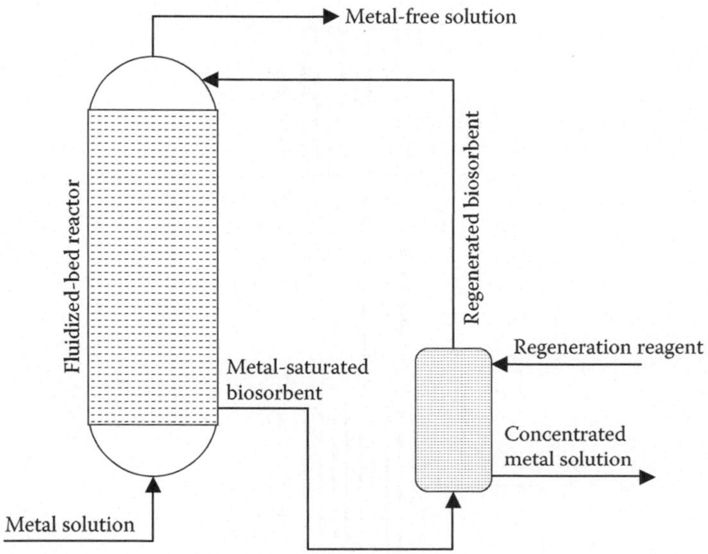

FIGURE 7.63
Fluidized bed reactor-based biosorption process.

gravity sedimentation in the same tank or by filtration by an additional unit. The operation of a stirred tank reactor is simpler than fixed and fluidized bed reactors; however, the operation cost is often higher. Figure 7.64 demonstrates a stirred tank reactor for metal sorption and recovery.

7.5.3 Fixed-Bed Reactor

Like ion exchange system, a sorption process can be implemented by operating a series of fixed beds.[78] Some of the beds are operated for metal uptake, while the other beds are for regeneration of used sorbents. A minimum requirement of column number is two. One is for biosorption for metal uptake, and the other one is for desorption for regeneration of metal-loaded sorbents. An auto-switch system is needed for the operation. Figure 7.65 demonstrates a fixed-bed reactor for metal sorption. The design of the reactor for metal removal can be performed by using computation programs; some of examples reported by Chen and coworkers are available in the literatures.[77,78]

If metal waste streams contain suspended solids, these solids can deposit on or between sorbent particles and increase pressure drop along the column. The uptake of metal can increase as high as 60% of sorbent weight, which reduces the porosity of the bed and decreases the flow rate. Too small particles of biosorbent are undesirable as they cause a higher pressure drop and increase operation cost. However, too-large particles can decrease the biosorption rate as the external mass transfer resistance is increased. Thus, selection of appropriate

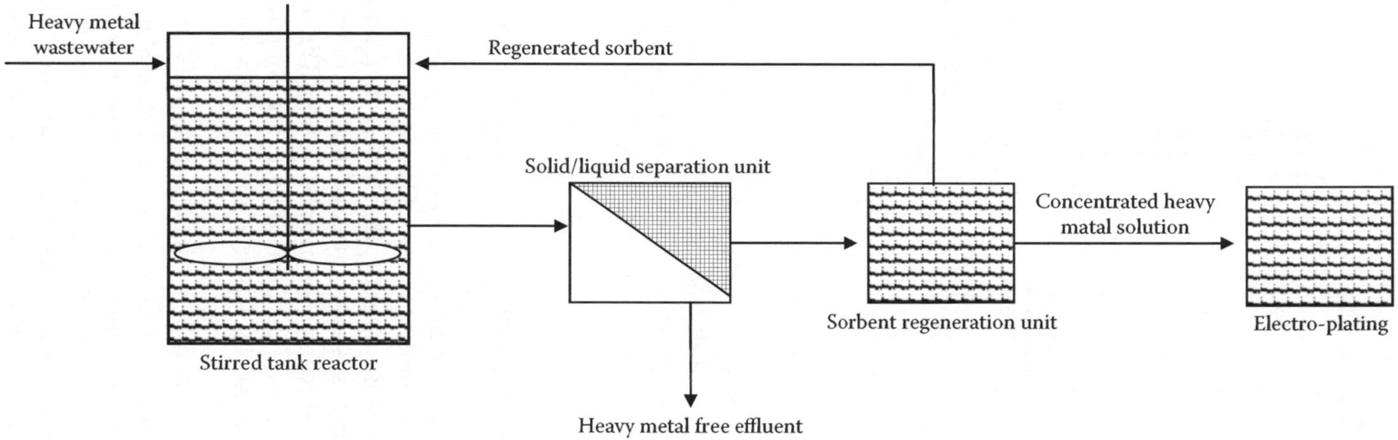

FIGURE 7.64
Stirred tank reactor-based sorption process.

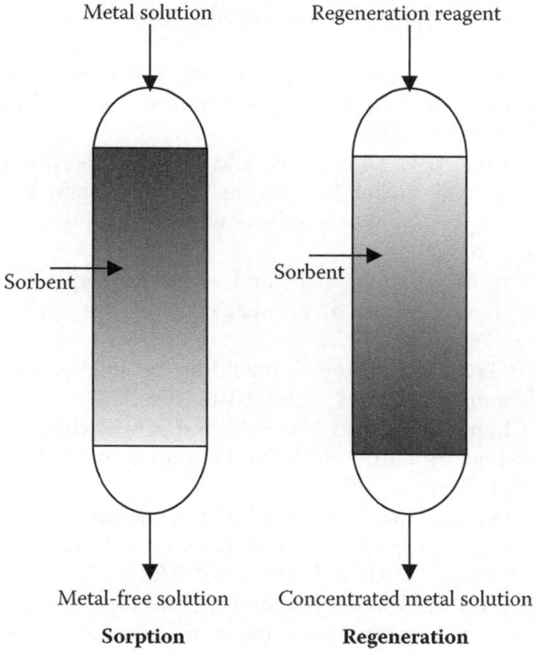

FIGURE 7.65
Fixed-bed-based biosorption process.

size is important for operation of sorption in a fixed-bed reactor. In application, a sorbent diameter of 2 mm is often adopted to meet this requirement.

References

1. Ho, Y. S., and El-Khaiary, M. I. Metal research trends in the environmental field. In *Heavy Metals in the Environment*. Chapter 1, 1–12, CRC, Boca Raton, FL, 2009.
2. Radovic, L. R., Moreno-Castilla, C., and Rivera-Utrilla, J. Carbon materials as adsorbents in aqueous solutions. *Chem. Phys. Carbon*, 27, 227–405, 2001.
3. Brierley, C. L. Bioremediation of metal-contaminated surface and groundwater. *Geomicrobiol. J.*, 8, 201–223,1990.
4. Davis, T. A.,Volesky, B., and Mucci, A. A review of the biochemistry of heavy metal biosorption by brown algae. *Water Res.*, 37(18), 4311–4330, 2003.
5. Kapoor, A., and Viraraghavan, T. Fungal biosorption—an alternative treatment option for heavy metal bearing wastewaters: A review. *Bioresour. Technol.*, 53(3), 195–206, 1995.
6. Yan, G. Y., and Viraraghavan, T. Heavy metal removal in a biosorption column by immobilized M-rouxii biomass. *Bioresour. Technol.*, 78(3), 243–249, 2001.

7. Gadd, G. M. Minireview: microbial metal transformations. *J. Microbiol.*, 39(2), 83–88, 2001.
8. Sheng, P. X., Tan, L. H., Chen, J. P., and Ting, Y. P. Biosorption performance of two brown marine algae for removal of chromium and cadmium. *J. Disper. Sci. Technol.*, 25(5), 679–686, 2004.
9. Sheng, P. X., Ting, Y. P., Chen, J. P., and Hong, L. Sorption of lead, copper, cadmium, zinc, and nickel by marine algal biomass: Characterization of biosorptive capacity and investigation of mechanisms. *J. Colloid Interf. Sci.*, 275(1), 131–141, 2004.
10. Mukhopadhyay, M., Noronha, S. B., and Suraishkumar, G. K. Kinetic modeling for the biosorption of copper by pretreated *Aspergillus niger* biomass. *Bioresour. Technol.*, 98(9), 1781–1787, 2007.
11. Evans, L. J. Chemistry of metal retention by soils—several processes are explained. *Environ. Sci. Technol.*, 23(9), 1046–1056, 1989.
12. Chen, J. P., Chong, K.H., and Wu, S.N. Surface Modification of a Granular Activated Carbon by Citric Acid for Enhancement of Copper Adsorption. *Carbon*, 41(10), 1979–1986, 2003
13. Dong, D. M., Liu, L., Hua, X. Y., and Lu, Y. Z. Comparison of lead, cadmium, copper and cobalt adsorption onto metal oxides and organic materials in natural surface coatings. *Microchem. J.*, 85(2), 270–275, 2007.
14. Zhang, G. S., Qu, J. H., Liu, H. J., Liu, R. P., and Li, G. T. Removal mechanism of As(III) by a novel Fe-Mn binary oxide adsorbent: Oxidation and sorption. *Environ. Sci. Technol.*, 41(13), 4613–4619, 2007.
15. Namasivayam, C., and Ranganathan, K. Effect of organic ligands on the removal of Pb(II), Ni(II) and Cd(II) by "waste" Fe(III)/Cr(III) hydroxide. *Water Res.*, 32 (3), 969–971, 1998.
16. Ouki, S. K., and Kavannagh, M. Performance of natural zeolites for the treatment of mixed metal-contaminated effluents. *Waste Manage Res.*, 15(4), 383–394, 1997.
17. Erdem, M., and Ozverdi, A. Lead adsorption from aqueous solution onto siderite. *Sep. Purif. Technol.*, 42(3), 259–264, 2005.
18. McKay, G., and Porter, J. F. Equilibrium parameters for the sorption of Cu, Cd and Zn ions onto peat. *J. Chem. Tech. Biotechnol.*, 69, 309–320, 1997.
19. Lister, S. K., and Line, M. A. Potential utilization of sewage sludge and paper mill waste for biosorption of metals from polluted waterways. *Bioresour. Technol.*, 79, 35–39, 2001.
20. Yang, L., Wu, S.N., and Chen, J.P. Modification of activated carbon by polyaniline for enhanced adsorption of aqueous arsenate. *Ind. Eng. Chem. Res.*, 46(7), 2133–2140, 2007.
21. Gupta, V. K., Mohan, D., and Sharma, S., Removal of lead from wastewater using bagasse fly ash—a sugar industry waste material. *Sep. Sci. Technol.*, 33(9), 1331–1343, 1998.
22. Karnitz, O., Gurgel, L. V. A., de Melo, J. C. P., Botaro, V. R., Melo, T. M. S., Gil, R. P. D. F., and Gil, L. F. Adsorption of heavy metal ion from aqueous single metal solution by chemically modified sugarcane bagasse. *Bioresour. Technol.*, 98(6), 1291–1297, 2007.
23. Farinella, N. V., Matos, G. D., and Arruda, M. A. Z. Grape bagasse as a potential biosorbent of metals in effluent treatments. *Bioresour. Technol.*, 98(10), 1940–1946, 2007.

24. Davila-Jimenez, M. M., Elizalde-Gonzalez, M. P., Geyer, W., Mattusch, J., and Wennrich, R. Adsorption of metal cations from aqueous solution onto a natural and a model biocomposite. *Colloid Surface A,* 219(1–3), 243–252, 2003.
25. Sun, G., and Shi, W. Sunflowers stalks as adsorbents for the removal of metal ions from wastewater. *Ind. Eng. Chem. Res.,* 37, 1324–1328, 1998.
26. Okieimen, F. E., Okundia, E. U., and Ogbeifun, D. E. Sorption of cadmium and lead ions on modified groundnut (Arachis-Hypogaea) husks. *J. Chem. Technol. Biot.,* 51(1), 97–103, 1991.
27. Lee, M. Y., Lee, S. H., Shin, H. J., Kajiuchi, T., and Yang, J. W. Characteristics of lead removal by crab shell particles. *Process Biochem.,* 33(7), 749–753, 1998.
28. Ajmal, M., Rao, R. A. K., and Siddiqui, B. A. Studies on removal and recovery of Cr(VI) from electroplating wastes. *Water Res.,* 30(6), 1478–1482, 1996.
29. Raji, C., and Anirudhan, T. S. Batch Cr(VI) removal by polyacrylamide-grafted sawdust: Kinetics and thermodynamics. *Water Res.,* 32(12), 3772–3780, 1998.
30. Khalid, N., Ahmad, S., Kiani, S. N., and Ahmed, J. Removal of mercury from aqueous solutions by adsorption to rice husks. *Sep. Sci. Technol.,* 34(16), 3139–3153, 1999.
31. Tarley, C. R. T., and Arruda, M. A. Z. Biosorption of heavy metals using rice milling by-products. Characterisation and application for removal of metals from aqueous effluents. *Chemosphere,* 54(7), 987–995, 2004.
32. Chuah, T. G., Jumasiah, A., Azni, I., Katayon, S., and Choong, S. Y. T. Rice husk as a potentially low-cost biosorbent for heavy metal and dye removal: An overview. *Desalination,* 175(3), 305–316, 2005.
33. Gaballah, I., and Kilbertus, G. Recovery of heavy metal ions through decontamination of synthetic solutions and industrial effluents using modified barks. *J. Geochem. Explor.,* 62(1–3), 241–286, 1998.
34. Reddy, B. R., Mirghaffari, N., and Gaballah, I. Removal and recycling of copper from aqueous solutions using treated Indian barks. *Resour. Conserv. Recy.,* 21(4), 227–245, 1997.
35. Al-Asheh, S., and Duvnjak, Z. Binary metal sorption by pine bark: Study of equilibria and mechanisms. *Sep. Sci. Technol.,* 33(9), 1303–1329, 1998.
36. Salim, R., Al-Subu, M., and Qashoa, S. Removal of lead from polluted water using decaying plant leaves. *J. Environ. Sci. Heal. A,* 29, 2087–2114, 1994.
37. Salim, R., Alsubu, M. M., and Sahrhage, E. Uptake of cadmium from water by beech leaves. *J. Environ. Sci. Health Part A: Environ. Sci. Eng. Toxic Hazard. Substance Cont.,* A27(3), 603–627, 1992.
38. Salim, R., Al-Subu, M., and Dawod, E. Efficiency of removal of cadmium from aqueous solutions by plant leaves and the effects of interaction of combinations of leaves on their removal efficiency. *J. Environ. Manage.,* 87(3), 521–532, 2008.
39. Sekar, M., Sakthi, V., and Rengaraj, S. Kinetics and equilibrium adsorption study of lead(II) onto activated carbon prepared from coconut shell. *J. Colloid Interf. Sci.,* 279(2), 307–313, 2004.
40. Lacour, S., Bollinger, J. C., Serpaud, B., Chantron, P., and Arcos, R. Removal of heavy metals in industrial wastewaters by ion-exchanger grafted textiles. *Anal. Chim. Acta,* 428(1), 121–132, 2001.
41. Zhou, D., Zhang, L. N., Zhou, J. P., and Guo, S. L. Cellulose/chitin beads for adsorption of heavy metals in aqueous solution. *Water Res.,* 38(11), 2643–2650, 2004.

42. Al-Asheh, S., Banat, F., and Al-Rousan, D. Beneficial reuse of chicken feathers in removal of heavy metals from wastewater. *J. Clean Prod.*, 11(3), 321–326, 2003.
43. Jin, X. C., *Sediment Pollution Chemistry*. 21–56, China Environmental Science Publisher, Beijing, 1992.
44. Weis, J. S., and Weis, P. Metal uptake, transport and release by wetland plants: Implications for phytoremediation and restoration. *Environ. Int.*, 30(5), 685–700, 2004.
45. Santona, L., Castaldi, P., and Melis, P. Evaluation of the interaction mechanisms between red muds and heavy metals. *J. Hazard. Mater.*, 136(2), 324–329, 2006.
46. Matos, G. D., and Arruda, M. A. Z. Vermicompost as natural adsorbent for removing metal ions from laboratory effluents. *Process Biochem.*, 39(1), 81–88, 2003.
47. Pavan, F. A., Lima, I. S., Lima, E. C., Airoldi, C., and Gushikem, Y. Use of Ponkan mandarin peels as biosorbent for toxic metals uptake from aqueous solutions. *J. Hazard. Mater.*, 137(1), 527–533, 2006.
48. Kadirvelu, K., Senthilkumar, P., Thamaraiselvi, K., and Subburam, V. Activated carbon prepared from biomass as adsorbent: Elimination of Ni(II) from aqueous solution. *Bioresour. Technol.*, 81(1), 87–90, 2002.
49. Bailey, S. E., Olin, T. J., Bricka, R. M., and Adrian, D. D. A review of potentially low-cost sorbents for heavy metals. *Water Res.*, 33(11), 2469–2479, 1999.
50. Babel, S., and Kurniawan, T. A. Low-cost adsorbents for heavy metals uptake from contaminated water: A review. *J. Hazard Mater.*, 97(1–3), 219–243, 2003.
51. Suhas, C., Carrott, P. J. M., and Carrott, M. M. L. R. Lignin—from natural adsorbent to activated carbon: A review. *Bioresour Technol.*, 98(12), 2301–2312, 2007.
52. Ishizaki, C., and Marti, I. Surface oxide structures on a commercial activated carbon. *Carbon*, 19(6), 409–412, 1981.
53. Corapcioglu, M. O., and Huang, C. P. The adsorption of heavy-metals onto hydrous activated carbon. *Water Res.*, 21(9), 1031–1044, 1987.
54. Ajmal, M., Khan, A. H., Ahmad, S., and Ahmad, A. Role of sawdust in the removal of copper(II) from industrial wastes. *Water Res.*, 32(10), 3085–3091, 1998.
55. Chen, J. P., and Lin, M. S. Equilibrium and kinetics of metal ion adsorption onto a commercial H-type granular activated carbon: Experimental and modeling studies. *Water Res.*, 35(10), 2385–2394, 2001.
56. Chen, J. P., and Lin, M. S. Surface charge and metal ion adsorption on an H-type activated carbon: Experimental observation and modeling simulation by the surface complex formation approach. *Carbon*, 39(10), 1491–1504, 2001.
57. Sigworth, E. A., and Smith, S. B. Adsorption of inorganic compounds by activated carbon. *J. Am. Water Works Ass.*, 64(6), 386–391, 1972.
58. Huang, C. P., and Wu, M. H. The removal of chromium(IV) from dilute aqueous solution by activated carbon. *Water Res.*, 11, 673–679, 1977.
59. Netzer, A., and Hughes, D. E. Adsorption of copper, lead and cobalt by activated carbon. *Water Res.*, 18(8), 927–933, 1984.
60. Chen, J. P.,Yiacoumi, S., and Blaydes, T. G. Equilibrium and kinetic studies of copper adsorption by activated carbon. *Sep. Technol.*, 6(2), 133–146, 1996.
61. Corapcioglu, M. O., and Huang, C. P. The surface-acidity and characterization of some commercial activated carbons. *Carbon*, 25(4), 569–578, 1987.
62. Reed, B. E., and Matsumoto, M. R. Modeling Cd adsorption in single and binary adsorbent (Pac) systems. *J. Environ. Eng.-ASCE*, 119(2), 332–348, 1993.

63. Chen, J. P., and Wu, S. Simultaneous adsorption of copper ions and humic acid onto an activated carbon. *J. Colloid Interf. Sci.*, 280(2), 334–342, 2004.
64. Chen, J. P., and Wu, S. N. Study on EDTA-chelated copper adsorption by granular activated carbon. *J. Chem. Technol. Biot.*, 75(9), 791–797, 2000.
65. Chen, J. P., and Wu, S. N. Acid/base-treated activated carbons: Characterization of functional groups and metal adsorptive properties. *Langmuir*, 20(6), 2233–2242, 2004.
66. Chen, J. P., and Lin, M. S. Comprehensive investigation of important factors governing metal-ion adsorption by an H-type granular activated carbon. *Sep. Sci. Technol.*, 35(13), 2063–2081, 2000.
67. Kneen, W. R., Rogers, J. W., and Simpson, P. *Chemistry: Facts, Patterns, and Principles*. Addison-Wesley, London, 1978.
68. Robinson, R. A., and Stokes, R. H. Electrolyte solutions: The measurement and interpretation of conductance, chemical potential and diffusion in solutions of simple electrolytes. Butterworth, London, 1959.
69. Martell, A. E., Smith, R. M., and Motekaitis, R. J. NIST critically selected stability constants of metal complexes database, version 3.0. Texas A&M University, College Station, TX, 1997.
70. Cookson, J. T. Adsorption mechanisms: The chemistry of organic adsorption on activated carbon. In *Carbon Adsorption Handbook*. Eds. Cheremisinoff, P.N. and Eller busch, E. 241–279. Ann Arbor Science, Ann Arbor, MI, 1978.
71. Boehm, H. P. Some aspects of the surface chemistry of carbon-blacks and other carbons. *Carbon*, 32(5), 759–769, 1994.
72. Huang, C. P., and Smith, E. H. Removal of Cd(II) from plating wastewater by an activated carbon process. In *Chemistry in Water Re-use*. Eds. Cooper, W. 355–412. Ann Arbor Science, Ann Arbor, MI, 1981.
73. Pesavento, M., Profumo, A., Alberti, G., and Conti, F. Adsorption of lead(II) and copper(II) on activated carbon by complexation with surface functional groups. *Anal. Chim. Acta*, 480(1), 171–180, 2003.
74. Reed, B. E., and Nonavinakere, S. K. Metal adsorption by activated carbon—effect of complexing ligands: Competing adsorbates, ionic-strength, and background electrolyte. *Sep. Sci. Technol.*, 27(14), 1985–2000, 1992.
75. Tan, T. C., and Teo, W. K. Combined effect of carbon dosage and initial adsorbate concentration on the adsorption-isotherm of heavy-metals on activated carbon. *Water Res.*, 21(10), 1183–1188, 1987.
76. Namasivayam, C., and Kadirvelu, K. Uptake of mercury(II) from wastewater by activated carbon from an unwanted agricultural solid by-product: Coirpith. *Carbon*, 37(1), 79–84, 1999.
77. Chen, J. P., and Wang, L. Characterization of metal adsorption kinetic properties in batch and fixed-bed reactors. *Chemosphere*, 54(3), 397–404, 2004.
78. Chen, J. P., Yoon, J. T., and Yiacoumi, S. Effects of chemical and physical properties of influent on copper sorption onto activated carbon fixed-bed columns. *Carbon*, 41(8), 1635–1644, 2003.
79. Kongsuwan, A., Patnukao, P., and Pavasant, P. Binary component sorption of Cu(II) and Pb(II) with activated carbon from Eucalyptus camaldulensis Dehn bark. *J. Ind. Eng. Chem.*, 15(4), 465–470, 2009.
80. Moreno-Pirajan, J. C., Garcia-Cuello, V. S., and Giraldo, L. The removal and kinetic study of Mn, Fe, Ni and Cu ions from wastewater onto activated carbon from coconut shells. *Adsorption*, 17(3), 505–514, 2011.

81. Moreno-Pirajan, J. C., and Giraldo, L. Activated carbon obtained by pyrolysis of potato peel for the removal of heavy metal copper (II) from aqueous solutions. *J. Anal. Appl. Pyrol.*, 90(1), 42–47, 2011.
82. Li, K. Q., Zheng, Z., and Li, Y. Characterization and lead adsorption properties of activated carbons prepared from cotton stalk by one-step H_3PO_4 activation. *J. Hazard. Mater.*, 181(1–3), 440–447, 2010.
83. Ahmed, Y. M., Al-Mamun, A., Muyibi, S. A., Al-Khatib, M. F. R., Jameel, A. T., and Alsaadi, M. A. Study of Pb adsorption by carbon nanofibers grown on powdered activated carbon. *J. Appl. Sci.*, 10(17), 1983–1986, 2010.
84. Kalavathy, H., Karthik, B., and Miranda, L. R. Removal and recovery of Ni and Zn from aqueous solution using activated carbon from *Hevea brasiliensis*: Batch and column studies. *Colloid Surface B*, 78(2), 291–302, 2010.
85. Lalhruaitluanga, H., Jayaram, K., Prasad, M. N. V., and Kumar, K. K. Lead(II) adsorption from aqueous solutions by raw and activated charcoals of *Melocanna baccifera* Roxburgh (bamboo): A comparative study. *J. Hazard. Mater.*, 175(1–3), 311–318, 2010.
86. Zhu, H. J., Jia, Y. F., Wu, X., and Wang, H. Removal of arsenic from water by supported nano zero-valent iron on activated carbon. *J. Hazard. Mater.*, 172(2–3), 1591–1596, 2009.
87. Ghosh, P. K. Hexavalent chromium[Cr(VI)] removal by acid modified waste activated carbons. *J. Hazard. Mater.*, 171(1–3), 116–122, 2009.
88. Zaini, M. A. A., Okayama, R., and Machida, M. Adsorption of aqueous metal ions on cattle-manure-compost based activated carbons. *J. Hazard. Mater.*, 170 (2–3), 1119–1124, 2009.
89. Ahmadpour, A., Zabihi, M., and Asl, A. H. Removal of mercury from water by carbonaceous sorbents derived from walnut shell. *J. Hazard. Mater.*, 167(1–3), 230–236, 2009.
90. Rao, M. M., Ramana, D. K., Seshaiah, K., Wang, M. C., and Chien, S. W. C. Removal of some metal ions by activated carbon prepared from *Phaseolus aureus* hulls. *J. Hazard. Mater.*, 166(2–3), 1006–1013, 2009.
91. Li, K. L., and Wang, X. H. Adsorptive removal of Pb(II) by activated carbon prepared from *Spartina alterniflora*: Equilibrium, kinetics and thermodynamics. *Bioresour. Technol.*, 100(11), 2810–2815, 2009.
92. Zhu, S. M.,Yang, N., and Zhang, D. Poly(*N,N*-dimethylaminoethyl methacrylate) modification of activated carbon for copper ions removal. *Mater. Chem. Phys.*, 113(2–3), 784–789, 2009.
93. Amuda, O. S., Adelowo, F. E., and Ologunde, M. O. Kinetics and equilibrium studies of adsorption of chromium(VI) ion from industrial wastewater using *Chrysophyllum albidum* (Sapotaceae) seed shells. *Colloid Surface B*, 68(2), 184–192, 2009.
94. Rao, M. M., Reddy, D. H. K. K., Venkateswarlu, P., and Seshaiah, K. Removal of mercury from aqueous solutions using activated carbon prepared from agricultural by-product/waste. *J. Environ. Manage.*, 90(1), 634–643, 2009.
95. Patnukao, P., Kongsuwan, A., and Pavasant, P. Batch studies of adsorption of copper and lead on activated carbon from *Eucalyptus camaldulensis* dehn. bark. *J. Environ. Sci. (China)*, 20(9), 1028–1034, 2008.
96. Imamoglu, M., and Tekir, O. Removal of copper(II) and lead(II) ions from aqueous solutions by adsorption on activated carbon from a new precursor hazelnut husks. *Desalination*, 228(1–3), 108–113, 2008.

97. Kula, I., Ugurlu, M., Karaoglu, H., and Celik, A. Adsorption of Cd(II) ions from aqueous solutions using activated carbon prepared from olive stone by $ZnCl_2$ activation. *Bioresour. Technol.*, 99(3), 492–501, 2008.
98. Genc-Fuhrman, H., Mikkelsen, P. S., and Ledin, A. Simultaneous removal of As, Cd, Cr, Cu, Ni and Zn from stormwater: Experimental comparison of 11 different sorbents. *Water Res.*, 41(3), 591–602, 2007.
99. Chang, C., and Ku, Y. The adsorption and desorption characteristics of EDTA-chelated copper-ion by activated carbon. *Sep. Sci. Technol.*, 30(6), 899–915, 1995.
100. Chang, C., and Ku, Y. The adsorption characteristics of EDTA-chelated copper ion by activated carbon in a column test. *Sep. Sci. Technol.*, 33(4), 483–501, 1998.
101. Huang, C. P., and Ostovic, F. B. Removal of cadmium(Ii) by activated carbon adsorption. *J. Environ. Eng. Div.-ASCE*, 104(5), 863–878, 1978.
102. Bhattacharyya, D., and Cheng, C. Y. R. Activated carbon adsorption of heavy-metal chelates from single and multicomponent systems. *Environ. Prog.*, 6(2), 110–118, 1987.
103. Lafrance, P., and Mazet, M. Adsorption of humic substances in the presence of sodium-salts. *J. Am. Water Works Ass.*, 81(4), 155–162, 1989.
104. Ferro-Garcia, M. A., Rivera-Utrilla, J., Bautista-Toledo, I., and Moreno-Castilla, C. Adsorption of humic substances on activated carbon from aqueous solutions and their effect on the removal of Cr(III) ions. *Langmuir*, 14(7), 1880–1886, 1998.
105. Herbelin, A., and Westall, J. FITEQL: A computer program for determination of chemical equilibrium constants from experimental data, Version 4.0. Department of Chemistry, Oregon State University, Corvallis, OR, 1999.
106. Parker, D. R., Zelazny, L. W., and Kinraide, T. B. Improvements to the program. *Geochem. Soil Sci. Soc. Am. J.*, 51(2), 488–491, 1987.
107. Lubal, P., Siroky, D., Fetsch, D., and Havel, J. The acidobasic and complexation properties of humic acids: Study of complexation of Czech humic acids with metal ions. *Talanta*, 47(2), 401–412, 1998.
108. Itabashi, H.,Yamazaki, D., Kawamoto, H., and Akaiwa, H. Evaluation of the complexing ability of humic acids using solvent extraction technique. *Anal. Sci.*, 17(6), 785–787, 2001.
109. Lorenzo, J. I., Nieto, O., and Beiras, R. Effect of humic acids on speciation and toxicity of copper to *Paracentrotus lividus* larvae in seawater. *Aquat. Toxicol.*, 58(1–2), 27–41, 2002.
110. Lee, E. M., and Koopal, L. K. Adsorption of cationic and anionic surfactants on metal oxide surfaces: Surface charge adjustment and competition effects. *J. Colloid Interf. Sci.*, 177(2), 478–489, 1996.
111. Budinova, T. K., Gergova, K. M., Petrov, N. V., and Minkova, V. N. Removal of metal-ions from aqueous-solution by activated carbons obtained from different raw-materials. *J Chem. Technol. Biot.*, 60(2), 177–182, 1994.
112. Marzal, P., Seco, A., Gabaldon, C., and Ferrer, J. Cadmium and zinc adsorption onto activated carbon: Influence of temperature, pH and metal/carbon ratio. *J Chem. Technol. Biot.*, 66(3), 279–285, 1996.
113. Leyva-Ramos, R., Rangel-Mendez, J. R., Mendoza Barron, J., Fuentes Rubio, L., and GuerreroCoronado, R. M. Adsorption of cadmium(II) from aqueous solution onto activated carbon. *Water Sci. Technol.*, 35(7), 205–211, 1997.
114. Huang, C. P., and Blankenship, D. W. The removal of mercury(Ii) from dilute aqueous-solution by activated carbon. *Water Res.*, 18(1), 37–46, 1984.

115. Figueiredo, J. L., Pereira, M. F. R., Freitas, M. M. A., and Orfao, J. J. M. Modification of the surface chemistry of activated carbons. *Carbon,* 37(9), 1379–1389, 1999.
116. Toles, C. A., Marshall, W. E., and Johns, M. M. Granular activated carbons from nutshells for the uptake of metals and organic compounds. *Carbon,* 35(9), 1407–1414, 1997.
117. Toles, C., Rimmer, S., and Hower, J. C. Production of activated carbons from a Washington lignite using phosphoric acid activation. *Carbon,* 34(11), 1419–1426, 1996.
118. Girgis, B. S., and Ishak, M. F. Activated carbon from cotton stalks by impregnation with phosphoric acid. *Mater. Lett.*, 39(2), 107–114, 1999.
119. Dandekar, A., Baker, R. T. K., and Vannice, M. A. Characterization of activated carbon, graphitized carbon fibers and synthetic diamond powder using TPD and drifts. *Carbon,* 36(12), 1821–1831, 1998.
120. Gomezserrano, V., Acedoramos, M., Lopezpeinado, A. J., and Valenzuelacalahorro, C. Oxidation of activated carbon by hydrogen-peroxide—study of surface functional-groups by FT-IR. *Fuel,* 73(3), 387–395, 1994.
121. Noh, J. S., and Schwarz, J. A. Effect of HNO3 treatment on the surface-acidity of activated carbons. *Carbon,* 28(5), 675–682, 1990.
122. Gomez-Serrano, V., Acedo-Ramos, M., Lopez-Peinado, A. J., and Valenzuela-Calahorro, C. Mass and surface changes of activated carbon treated with nitric acid. Thermal behavior of the samples. *Thermochim. Acta,* 291(1–2), 109–115, 1997.
123. Mazet, M., Farkhani, B., and Baudu, M. Influence of heat or chemical treatment of activated carbon onto the adsorption of organic-compounds. *Water Res.*, 28(7), 1609–1617, 1994.
124. Preston, D. R., Bitton, G., and Farrah, S. R. Enhancement of enterovirus infectivity in vitro by pretreating host-cell monolayers with the cationic polymer polyethyleneimine. *Appl. Environ. Microb.*, 56(1), 295–297, 1990.
125. Gajardo, R., Diez, J. M., Jofre, J., and Bosch, A. Adsorption-elution with negatively and positively-charged glass powder for the concentration of hepatitis-a virus from water. *J. Virol. Methods,* 31(2–3), 345–352, 1991.
126. Julien, F., Baudu, M., and Mazet, M. Relationship between chemical and physical surface properties of activated carbon. *Water Res.*, 32(11), 3414–3424, 1998.
127. delaPuente, G., Pis, J. J., Menendez, J. A., and Grange, P. Thermal stability of oxygenated functions in activated carbons. *J. Anal. Appl. Pyrol.*, 43(2), 125–138, 1997.
128. Shin, S., Jang, J., Yoon, S. H., and Mochida, I. A study on the effect of heat treatment on functional groups of pitch based activated carbon fiber using FTIR. *Carbon,* 35(12), 1739–1743, 1997.
129. Pastor-Villegas, J., Gomez-Serrano, V., Duran-Valle, C. J., and Higes-Rolando, F. J. Chemical study of extracted rockrose and of chars and activated carbons prepared at different temperatures. *J. Anal. Appl. Pyrol.*, 50(1), 1–16, 1999.
130. Menendez, J. A., Menendez, E. M., Iglesias, M. J., Garcia, A., and Pis, J. J. Modification of the surface chemistry of active carbons by means of microwave-induced treatments. *Carbon,* 37(7), 1115–1121, 1999.
131. Park, S. J., Park, B. J., and Ryu, S. K. Electrochemical treatment on activated carbon fibers for increasing the amount and rate of Cr(VI) adsorption. *Carbon,* 37(8), 1223–1226, 1999.
132. Toles, C. A., Marshall, W. E., and Johns, M. M. Surface functional groups on acid-activated nutshell carbons. *Carbon,* 37(8), 1207–1214, 1999.

133. Mostafa, M. R. Adsorption of mercury, lead and cadmium ions on modified activated carbons. *Adsorpt. Sci. Technol.*, 15(8), 551–557, 1997.
134. Ravindran, V., Stevens, M. R., Badriyha, B. N., and Pirbazari, M. Modeling the sorption of toxic metals on chelant-impregnated adsorbent. *AICHE J.*, 45(5), 1135–1146, 1999.
135. Lakov, L.,Vassileva, P., and Peshev, O. Sorption of heavy metals on pyrazolone-containing carbon sorbents. *Carbon*, 37(10), 1655–1657, 1999.
136. Monser, L., and Adhoum, N. Modified activated carbon for the removal of copper, zinc, chromium and cyanide from wastewater. *Sep. Purif. Technol.*, 26(2–3), 137–146, 2002.
137. Biniak, S., Pakula, M., Szymanski, G. S., and Swiatkowski, A. Effect of activated carbon surface oxygen- and/or nitrogen-containing groups on adsorption of copper(II) ions from aqueous solution. *Langmuir*, 15(18), 6117–6122, 1999.
138. Wang, S. B., and Lu, G. Q. Effects of acidic treatments on the pore and surface properties of Ni catalyst supported on activated carbon. *Carbon*, 36(3), 283–292, 1998.
139. Kratochvil, D., and Volesky, B. Biosorption of Cu from ferruginous wastewater by algal biomass. *Water Res.*, 32(9), 2760–2768, 1998.
140. Chen, J. P., and Yang, L. Chemical modification of *Sargassum* sp. for prevention of organic leaching and enhancement of uptake during metal biosorption. *Ind. Eng. Chem. Res.*, 44(26), 9931–9942, 2005.
141. Yang, L., and Chen, J. P. Biosorption of hexavalent chromium onto raw and chemically modified *Sargassum* sp. *Bioresour. Technol.*, 99(2), 297–307, 2008.
142. Chen, J. P., Hong, L. A., Wu, S. N., and Wang, L. Elucidation of interactions between metal ions and Ca alginate-based ion-exchange resin by spectroscopic analysis and modeling simulation. *Langmuir*, 18(24), 9413–9421, 2002.
143. Chen, J. P., and Yang, L., Study of a heavy metal biosorption onto raw and chemically modified *Sargassum* sp. via spectroscopic and modeling analysis. *Langmuir*, 22(21), 8906–8914, 2006.
144. Goldstone, M. E., and Lester, J. N. The balance of heavy-metals through sewage-treatment works. *Sci. Total Environ.*, 105, 259–266,1991.
145. Yu, Q. M., Matheickal, J. T., Yin, P. H., and Kaewsarn, P. Heavy metal uptake capacities of common marine macro algal biomass. *Water Res.*, 33(6), 1534–1537, 1999.
146. Jalali, R., Ghafourian, H., Asef, Y., Davarpanah, S. J., and Sepehr, S. Removal and recovery of lead using nonliving biomass of marine algae. *J. Hazard. Mater.*, 92(3), 253–262, 2002.
147. Volesky, B. Advances in biosorption of metals—selection of biomass types. *FEMS Microbiol. Rev.*, 14(4), 291–302, 1994.
148. Ozer, A., Ozer, D., and Ekiz, H. I. Application of Freundlich and Langmuir models to multistage purification process to remove heavy metal ions by using *Schizomeris leibleinii. Process Biochem.*, 34(9), 919–927, 1999.
149. Lo, W. H., Chua, H., Lam, K. H., and Bi, S. P. A comparative investigation on the biosorption of lead by filamentous fungal biomass. *Chemosphere*, 39(15), 2723–2736, 1999.
150. Fourest, E., Canal, C., and Roux, J. C. Improvement of heavy-metal biosorption by mycelial dead biomasses (Rhizopus-Arrhizus, Mucor-Miehei and Penicillium-Chrysogenum)—Ph control and cationic activation. *FEMS Microbiol. Rev.*, 14(4), 325–332, 1994.

151. Schmitt, D., Muller, A., Csogor, Z., Frimmel, F. H., and Posten, C. The adsorption kinetics of metal ions onto different microalgae and siliceous earth. *Water Res.*, 35(3), 779–785, 2001.
152. Davis, T. A.,Volesky, B., and Vieira, R. H. S. F. Sargassum seaweed as biosorbent for heavy metals. *Water Res.*, 34(17), 4270–4278, 2000.
153. Pagnanelli, F., Trifoni, M., Beolchini, F., Esposito, A., Toro, L., and Veglio, F. Equilibrium biosorption studies in single and multi-metal systems. *Process Biochem.*, 37(2), 115–124, 2001.
154. Esposito, A., Pagnanelli, F., Lodi, A., Solisio, C., and Veglio, F. Biosorption of heavy metals by *Sphaerotilus natans*: An equilibrium study at different pH and biomass concentrations. *Hydrometallurgy*, 60(2), 129–141, 2001.
155. Matheickal, J. T., Yu, Q. M., and Woodburn, G. M. Biosorption of cadmium(II) from aqueous solutions by pre-treated biomass of marine alga *Durvillaea potatorum*. *Water Res.*, 33(2), 335–342, 1999.
156. Rangsayatorn, N., Upatham, E. S., Kruatrachue, M., Pokethitiyook, P., and Lanza, G. R. Phytoremediation potential of *Spirulina (Arthrospira) platensis*: Biosorption and toxicity studies of cadmium. *Environ. Pollut.*, 119(1), 45–53, 2002.
157. Yin, P. H.,Yu, Q. M., Jin, B., and Ling, Z. Biosorption removal of cadmium from aqueous solution by using pretreated fungal biomass cultured from starch wastewater. *Water Res.*, 33(8), 1960–1963, 1999.
158. Arica, M. Y., Kacar, Y., and Genc, O. Entrapment of white-rot fungus *Trametes versicolor* in Ca-alginate beads: Preparation and biosorption kinetic analysis for cadmium removal from an aqueous solution. *Bioresour. Technol.*, 80(2), 121–129, 2001.
159. Sanchez, A., Ballester, A., Blazquez, M. L., Gonzalez, F., Munoz, J., and Hammaini, A. Biosorption of copper and zinc by *Cymodocea nodosa*. *FEMS Microbiol. Rev.*, 23(5), 527–536, 1999.
160. Puranik, P. R., Modak, J. M., and Paknikar, K. M. A comparative study of the mass transfer kinetics of metal biosorption by microbial biomass. *Hydrometallurgy*, 52(2), 189–197, 1999.
161. Dilek, F. B., Erbay, A., and Yetis, U. Ni(II) biosorption by polyporous versicolor. *Process Biochem.*, 37(7), 723–726, 2002.
162. Nourbakhsh, M. N., Kilicarslan, S., Ilhan, S., and Ozdag, H. Biosorption of Cr^{6+}, Pb^{2+} and Cu^{2+} ions in industrial waste water on *Bacillus* sp. *Chem. Eng. J.*, 85(2–3), 351–355, 2002.
163. Marques, P. A., Pinheiro, H. M., Teixeira, J. A., and Rosa, M. F. Removal efficiency of Cu^{2+}, Cd^{2+}, Pb^{2+} by waste brewery biomass: pH and cation association effects. *Desalination*, 124(1–3), 137–144, 1999.
164. Ei-Naas, M. H., Abu Ai-Rub, F., Ashour, I., and Al Marzouqi, M. Effect of competitive interference on the biosorption of lead(II) by *Chlorella vulgaris*. *Chem. Eng. Process*, 46(12), 1391–1399, 2007.
165. Nuhoglu, Y., and Oguz, E. Removal of copper(II) from aqueous solutions by biosorption on the cone biomass of *Thuja orientalis*. *Process Biochem.*, 38(11), 1627–1631, 2003.
166. Aravindhan, R., Maharshi, B., Sreeram, K. J., Rao, J. R., and Nair, B. U. Biosorption of cadmium metal ion from simulated wastewaters using Hypnea valentiae biomass: A kinetic and thermodynamic study. *Bioresour. Technol.*, 101(5), 1466–1470, 2010.

167. Sari, A., Uluozlu, O. D., and Tuzen, M. Equilibrium, thermodynamic and kinetic investigations on biosorption of arsenic from aqueous solution by algae (*Maugeotia genuflexa*) biomass. *Chem. Eng. J.*, 167(1), 155–161, 2011.
168. Smith, H. Edward. Uptake of heavy metals in batch systems by a recycled iron-bearing material. *Water Res.*, 30(10), 2424–2434, October 1996.
169. James, R. O., and MacNaughton, M. G. The adsorption of aqueous heavy metals on inorganic minerals. *Geochim. Cosmochim. Acta*, 41(11), 1549–1555, 1977.
170. Lai, C. H., and Chen, C. Y. Removal of metal ions and humic acid from water by iron-coated filter media. *Chemosphere*, 44(5), 1177–1184, 2001.
171. Davis, J. A., and Leckie, J. O. Surface ionization and complexation at the oxide/water interface II. Surface properties of amorphous iron oxyhydroxide and adsorption of metal ions. *J. Colloid Interf. Sci.*, 67(1), 90–107, 1978.
172. Erdemoğlu, M., and Sarıkaya, M. Effects of heavy metals and oxalate on the zeta potential of magnetite. *J. Colloid Interf. Sci.*, 300(2), 795–804, 2006.
173. Kikuchi, Y., Qian, Q., Machida, M., and Tatsumoto, H. Effect of ZnO loading to activated carbon on Pb(II) adsorption from aqueous solution. *Carbon*, 44(2), 195–202, 2006.
174. Han, R., Zou, W., Li, H., Li, Y., and Shi, J. Copper(II) and lead(II) removal from aqueous solution in fixed-bed columns by manganese oxide coated zeolite. *J. Hazard. Mater.*, 137(2), 934–942, 2006.
175. Lai, C. H., Lo, S. L., and Chiang, H. L. Adsorption/desorption properties of copper ions on the surface of iron-coated sand using BET and EDAX analyses. *Chemosphere*, 41(8), 1249–1255, 2000.
176. Recillas, S., Colón, J., Casals, E., González, E., Puntes, V., Sánchez, A., and Font, X. Chromium VI adsorption on cerium oxide nanoparticles and morphology changes during the process. *J. Hazard. Mater.*, 184(1–3), 425–431, 2010.
177. Kim, D.-H., Shin, M.-C., Choi, H.-D., Seo, C.-I., and Baek, K. Removal mechanisms of copper using steel-making slag: Adsorption and precipitation. *Desalination*, 223(1–3), 283–289, 2008.
178. Xu, Y., Boonfueng, T., Axe, L., Maeng, S., and Tyson, T. Surface complexation of Pb(II) on amorphous iron oxide and manganese oxide: Spectroscopic and time studies. *J. Colloid Interf. Sci.*, 299(1), 28–40, 2006.
179. Jenke, D. R., and Diebold, F. E. Electroprecipitation treatment of acid mine wastewater. *Water Res.*, 18(7), 855–859, 1984.
180. Chen, X., Lam, K. F., and Yeung, K. L. Selective removal of chromium from different aqueous systems using magnetic MCM-41 nanosorbents. *Chem. Eng. J.*, 172(2–3), 728–734, 2011.
181. Zhang, G., Qu, J., Liu, H., Liu, R., and Wu, R. Preparation and evaluation of a novel Fe–Mn binary oxide adsorbent for effective arsenite removal. *Water Res.*, 41(9), 1921–1928, 2007.
182. Ren, Z., Zhang, G., and Chen, J. P. Adsorptive removal of arsenic from water by an iron–zirconium binary oxide adsorbent. *J. Colloid Interf. Sci.*, 358(1), 230–237, 2011.
183. Zheng, Y. M., Zou, S. W., Nanayakkara, K. G. N., Matsuura, T., and Chen, J. P. Adsorptive removal of arsenic from aqueous solution by a PVDF/Zirconia blend flat-sheet membrane. *J. Membrane Sci.*, 374(1–2), 1–11, 2011.
184. Ma, Y., Zheng, Y. M., and Chen, J. P. A zirconium nanoparticle for significantly enhanced sorption of arsenate: Synthesis, characterization, measurement of performances, and mechanism study. *J. Colloid Interf. Sci.*, 354(2), 785–792, 2011.

185. Zhang, S. J., Li, X. Y., and Chen, J. P. An XPS study for mechanisms of arsenate adsorption onto a magnetite doped activated carbon fiber. *J. Colloid Interf. Sci.*, 343(1), 232–238, 2010.
186. Zhang, S. J., Li, X. Y., and Chen, J. P. Preparation and evaluation of a magnetite doped activated carbon fiber for enhanced arsenic removal. *Carbon*, 48, 60–67, 2010.
187. Zheng, Y. M., Lim, S. F., and Chen, J. P. Preparation and characterization of zirconium based magnetic sorbent for arsenate removal. *J. Colloid Interf. Sci.*, 338(1), 22–29, 2009.
188. Lim, S. F., Zheng, Y. M., and Chen, J. P. Removal of copper by calcium alginate encapsulated magnetic sorbent. *Chem. Eng. J.*, 152(2–3), 509–513, 2009.
189. Lim, S. F., Zheng, Y. M., and Chen, J. P. Organic arsenic adsorption onto a magnetic sorbent. *Langmuir*, 25(9), 4973–4978, 2009.
190. Lim, S. F., Zheng, Y. M., Zou, S. W., and Chen, J. P. Uptake of arsenate by an alginate-encapsulated magnetic sorbent: Process performance and characterization of adsorption chemistry. *J. Colloid Interf. Sci.*, 333(1), 33–39, 2009.
191. Lim, S.F., Zheng, Y. M., Zou, S. W., and Chen, J. P. Characterization of copper adsorption onto an alginate encapsulated magnetic sorbent by a combined FT-IR, XPS, and mathematical modeling study. *Environ. Sci. Technol.*, 42(7), 2551–2556, 2008.
192. Zheng, Y. M., Yu, L., and Chen, J. P. Removal of methylated arsenic using a nanostructured zirconia-based sorbent: Process performance and adsorption chemistry. *J. Colloid Interface Sci.*, 367(1), 362–369, 2012.

8

Calculation of Metal Ion Uptake in Environmental Systems

Metal ions in aqueous and solid phases are distributed according to physical, chemical, and biological factors and reactions. The distribution depends on three aspects: equilibrium, kinetics, and transfer. Equilibrium addresses the final state of all the species and is essentially controlled by chemical reactions present in all the species. Kinetics addresses the history or rate of changes, which depends on types of chemical reactions, properties of materials, and physical conditions (e.g., mixing and temperature). Transfer addresses the hydrodynamic effect on distribution of metal ions.

Sorption, ion exchange, and precipitation play essential roles in metal ion immobilization/mobilization. Precipitation can be easily described and modeled. However, sorption and ion exchange are more complicated. In this chapter, the sorption reaction and ion exchange are discussed. The principles that govern the reaction are summarized. Two mathematical models for determination of metal uptake kinetics are presented. Various empirical and theoretical tools are described together with several case studies.

8.1 Sorption Reaction

Not much work was done on interpretation of adsorption behavior before 1914. Even since, focus has been on adsorption of gaseous species and then organic substances in aqueous solution. Theories on adsorption of metal ions were studied since the early 1970s by a group of pioneering researchers.[1–17]

Sorption equilibrium modeling describes the relationship among various species in the sorption system, which include volume of solution, amount of sorbent, concentration of sorbates (before and after sorption), pH of solution, ionic strength, and competitive effects. Adsorption isotherm is the most important in the studies of sorption. This is due to the need to estimate the sorbent amount to be used, solution volume to be treated, and final concentration of sorbate, which are significant in the application of sorption technology. In the interaction of sorbent and heavy metal ions, solution chemistry, particularly pH and ionic strength, is essentially important, and it has been discussed in the previous chapters of this book.

In adsorption, contaminant is distributed in both the solution phase and solid phase. Concentration of contaminant in the solution phase (called concentration of contaminant in aqueous solution, C_e) is expressed in terms of mg/L, mole/L (M), mmole/L (mM), or ppm. Its concentration in solid phase (sorbent, q_e) is expressed in terms of mg/g or mmole/g and is calculated as follows:

$$q_e = \frac{V(C_o - C_e)}{m} \tag{8.1}$$

where V, C_0, and m are volume of aqueous solution, initial concentration of contaminant in aqueous solution, and mass of sorbent applied.

The relationship between concentration of contaminant in solution and sorbent (solid phase) at a fixed temperature under chemical conditions in a narrow range (e.g., pH) is called *adsorption isotherm.*

Metal ion adsorption equilibrium was initially studied with simplified adsorption systems containing only one sorbate at a fixed pH. The most common models used for sorption processes are the Langmuir and Freundlich isotherms.[12–16] They are still popular because of their simplicity of model parameters.

8.1.1 Langmuir Equation

The Langmuir equation was initially derived from kinetic studies. It was assumed that there were a number of adsorption sites evenly distributed on the surface of the adsorbent, on which adsorption of gas chemically or physically occurred.[12,13] An assumption of monolayer adsorption was also associated with the equation. For his work on adsorption chemistry, Langmuir was awarded the Nobel Prize in Chemistry in 1932.

The Langmuir equation remains an extensively used isotherm model for correlation of concentrations of sorbate in both solution and solid phases. It can be used for the determination of both the adsorption reaction constant (affinity) and maximum adsorption capacity.

The Langmuir adsorption model assumes that there are a fixed number of definite localized adsorption sites on a homogeneous surface, that each site can hold one adsorbate molecule (monolayer adsorption), and that there is no interaction between adsorbed molecules on neighboring sites. Adsorption occurs according to the following reaction:

$$S + X = SX \tag{8.2}$$

$$b = \frac{[SX]}{[S][X]} \tag{8.3}$$

$$[S]_t = [SX] + [S] \tag{8.4}$$

where $[S]_t$ is the total concentration of adsorbent site. b is the adsorption strength/adsorption reaction constant related to the heat of adsorption and indicates the affinity of the sorbate for the sorbent. [SX] and [S] are the concentrations of adsorbate in the solid and solution phases, respectively. Higher b values indicate that the sorbate is easier to be adsorbed onto the sorbent.

$$b = \frac{[SX]}{([S]_t - [SX])[X]} \tag{8.5}$$

Reorganizing Equation 8.5 will yield the following equation:

$$[SX] = \frac{b[S]_t[X]}{1 + b[X]} \tag{8.6}$$

Let

$$q_e = \frac{[SX]V}{m} \tag{8.7}$$

$$q_{\max} = \frac{[S]_t V}{m} \tag{8.8}$$

$$C_e = [X] \tag{8.9}$$

where $q_{\max}$ represents the saturated capacity of adsorption (i.e., the maximum adsorption capacity of the adsorbent). V and m are the volume of solution and the mass of adsorbent, respectively.

Therefore, we will have the following Langmuir equation:

$$q_e = \frac{q_{\max} bC_e}{1 + bC_e} \tag{8.10}$$

Note that the isotherm assumes monolayer surface sorption. However, molecular interaction is not negligible at higher surface coverage. Therefore, the Freundlich adsorption model should then be used.

The Langmuir isotherm curve is given in Figure 8.1. A change in $q_{\max}$ can lead to change in the adsorption capacity at higher equilibrium concentration in solution. Larger $q_{\max}$ will cause more adsorption, and lower $q_{\max}$ will lead to less adsorption. Change in b will change the adsorption; the obvious change is the shift of the curve toward, or backward from, the y-axis (q_e), as shown in Figure 8.1.

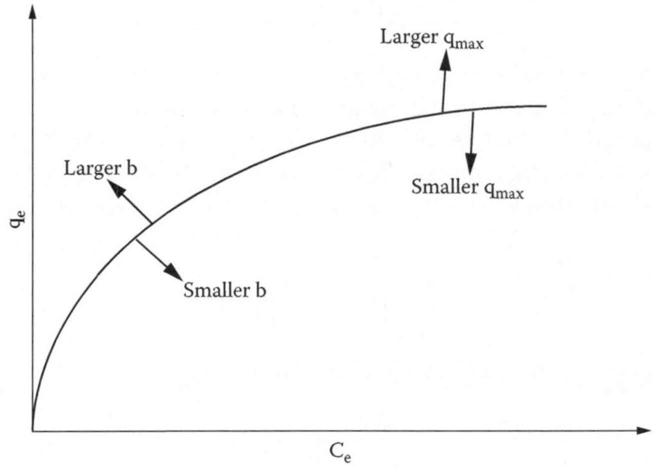

FIGURE 8.1
Illustration of a typical adsorption isotherm curve.

The two model parameters are q_{max} and b to be experimentally determined so that the Langmuir isotherm can be used. Equation 8.10 can be written as follows:

$$\frac{1}{q_e} = \frac{1}{q_{max}b} \times \frac{1}{C_e} + \frac{1}{q_{max}} \tag{8.11a}$$

$$\frac{C_e}{q_e} = \frac{1}{q_{max}b} + \frac{C_e}{q_{max}} \tag{8.11b}$$

A straight line can be plotted by the sorption isotherm data according to Equation 8.11a or 8.11b. Commercial software such as MS Excel can be used to calculate the intersect and slope of the line, by which the foregoing parameters can be determined. Figure 8.2a,b demonstrates the straight lines in the calculation. Note that the Langmuir constants from these two methods could be different from each other. The nonlinear regression approach may also be used.

EXAMPLE 8.1

An adsorption experiment was conducted for treatment of industrial wastewater. The data of the adsorption isotherm experiment are given in Table 8.1.

(a) Determine the parameters of Langmuir adsorption isotherm.
(b) Assume that the industrial wastewater with a volume of 100 m^3 and concentration of 10 mM will be treated by the adsorbent. The concentration of treated water is expected to be 0.1 mM.

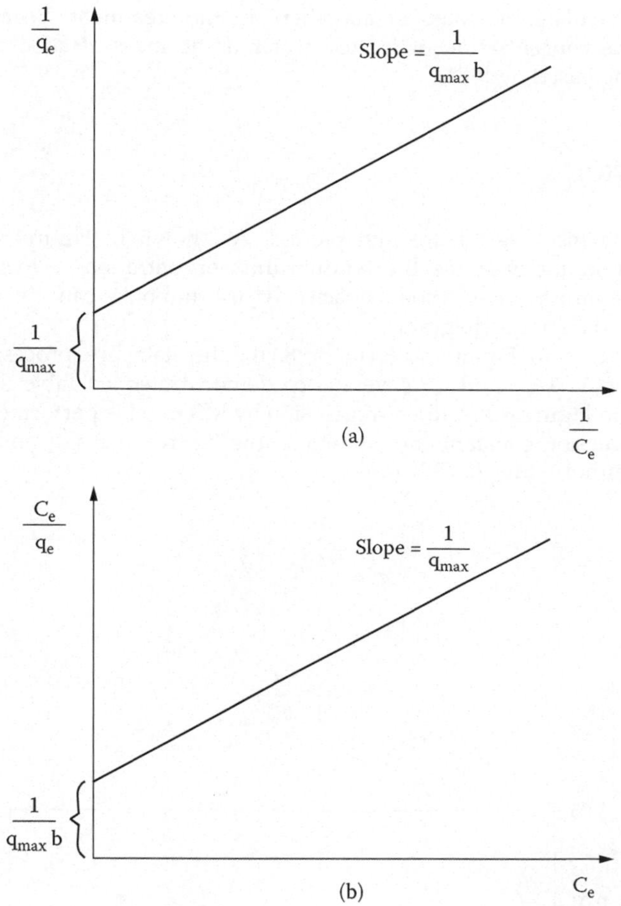

FIGURE 8.2
Determination of Langmuir constants: (a) versus $1/q_e$ versus $1/C_e$; (b) versus C_e/q_e versus C_e.

TABLE 8.1
Data of an Adsorption Isotherm Experiment

No.	C_e (mM)	q_e (mmol/g)
1.	0.177	0.040
2.	0.075	0.032
3.	0.143	0.038
4.	0.042	0.026
5.	0.020	0.017
6.	0.100	0.035
7.	0.012	0.012
8.	0.200	0.040
9.	0.007	0.008

Calculate the mass of adsorbent in the treatment. What is the concentration of treated water if the mass of adsorbent increased by 100%?

SOLUTION

(a)

The data in Table 8.1 are first plotted. As shown in Figure 8.3, the adsorption increases as the equilibrium concentration is increased. The maximum adsorption capacity of 0.4 mmol/g can be rightly estimated from the diagram.

According to Equations 8.11a or 8.11b, the data are processed as 1/qe versus $1/c_e$, and c_e/q_e versus c_e, demonstrated in Table 8.2 and shown in Figure 8.4. A linear regression by MS Excel is performed, and the regression constants are shown in the figure. The q_{max} and b are 0.0474 mmol/g and 28.78 L/mmol.

(b)

$$q_e = \frac{V(C_0 - C_e)}{m}$$

$$q_e = \frac{q_{max} b C_e}{1 + bC_e}$$

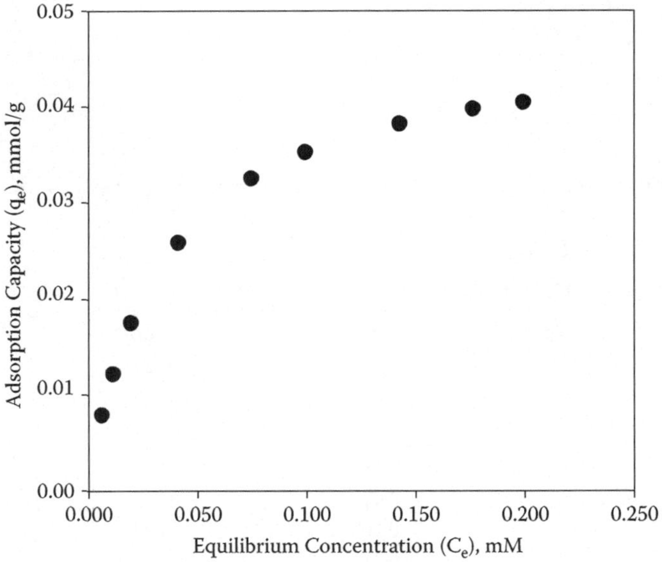

FIGURE 8.3
Adsorption isotherm.

TABLE 8.2

Determination of Langmuir Isotherm Constants

C_e (mM)	q_e (mmol/g)	$\frac{1}{C_e}$	$\frac{1}{q_e}$	$\frac{C_e}{q_e}$
0.177	0.040	5.663	25.249	4.458
0.075	0.032	13.333	30.871	2.315
0.143	0.038	6.972	26.208	3.759
0.042	0.026	24.06	38.733	1.61
0.020	0.017	50	57.747	1.155
0.100	0.035	10	28.427	2.843
0.012	0.012	83.99	82.662	0.984
0.200	0.040	5	24.762	4.953
0.007	0.008	146.78	128.694	0.877

$$\frac{V(C_0 - C_e)}{m} = \frac{q_{max} b C_e}{1 + bC_e}$$

$$m = \frac{V(C_0 - C_e)(1 + bC_e)}{q_{max} b C_e} \tag{8.12}$$

We have:

$V = 100\ m^3 = 100 \times 10^3$ L, $C_0 = 10$ mM, $C_e = 0.1$ mM, $q_{max} = 0.0474$ mmol/g, b = 28.78 L/mmol.

Thus,

$$m = \frac{100 \times 10^3 \times (10 - 0.1) \times (1 + 28.78 \times 0.1)}{0.0474 \times 28.78 \times 0.1} = 28{,}143 \text{ kg}$$

When the mass of sorbent is increased by 100% or it becomes 56,286 kg, Equation 8.12 is now reorganized as follows:

$$m q_{max} b C_e = V(C_0 - C_e)(1 + bC_e)$$

$$VbC_e^2 + (m q_{max} b + V - bC_0 V)C_e - VC_0 = 0$$

$$C_e = \frac{-(m q_{max} b + V - bC_0 V) + \sqrt{(m q_{max} b + V - bC_0 V)^2 + 4V^2 bC_0}}{2Vb} \tag{8.13}$$

Let

$$a_1 = m q_{max} b + V - bC_0 V$$

$$a_2 = 4V^2 bC_0$$

$$a_3 = Vb$$

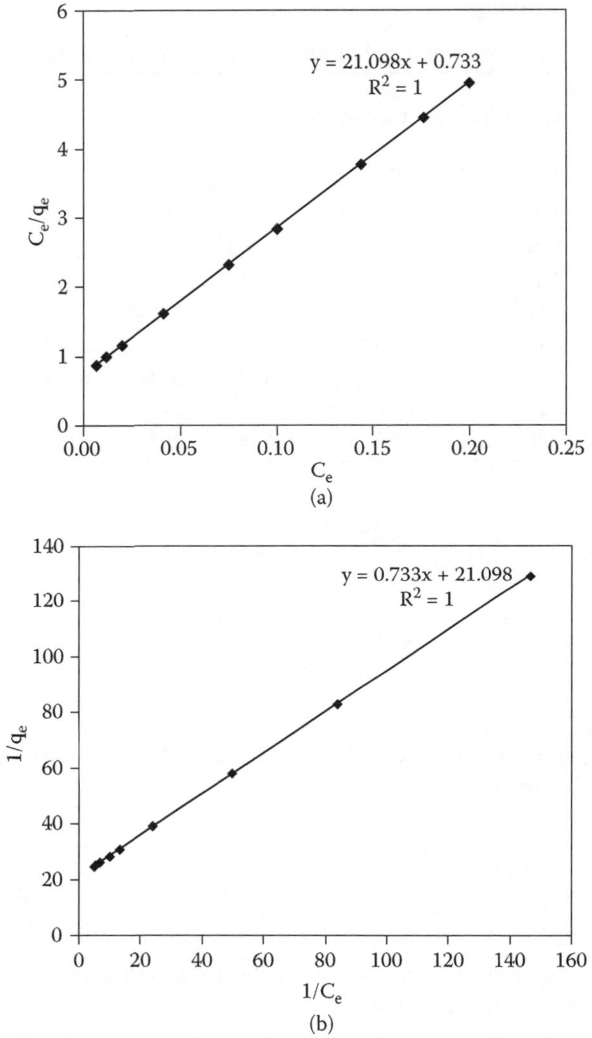

FIGURE 8.4
Determination of Langmuir constants: (a) C_e/q_e versus C_e, (b) $1/q_e$ versus $1/C_e$.

$$a_1 = 56286\times10^3\times0.0474\times28.78+100\times10^3$$
$$-28.78\times10\times100\times10^3 = 9.7105\times10^6$$

$$a_2 = 4(100\times10^3)^2\times28.78\times10 = 1.1512\times10^{13}$$

$$a_3 = 100\times10^3\times28.78 = 2.878\times10^8$$

$$C_e = \frac{-(9711893)+\sqrt{(9711893)^2+1.1512\times10^{17}}}{2\times2878000} = 0.02\text{mM}$$

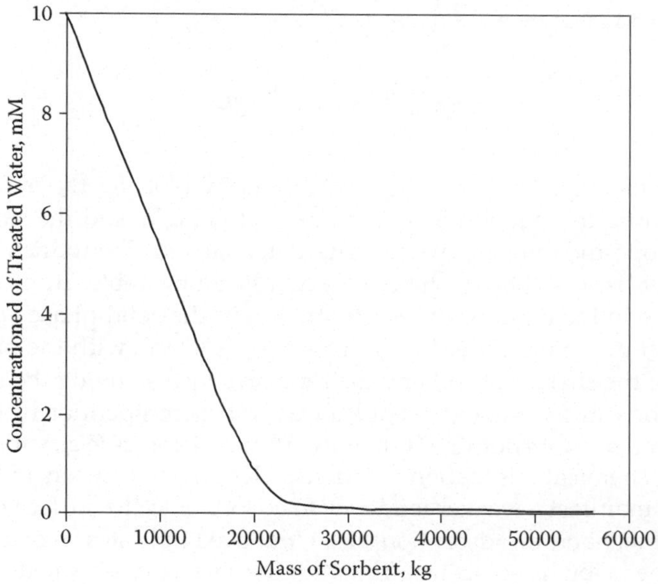

FIGURE 8.5
Effect of mass of sorbent on treatment of concentration of treated water.

Based on the Langmuir parameters, we can plot the concentration of treated water as a function of the mass of sorbent. As shown in Figure 8.5, a removal of 90% (1 mM) can be obtained when the mass of sorbent is around 20,000 kg. However, much more sorbent will have to be applied if 99% removal is required.

8.1.2 Freundlich Equation

Van Bemmelen in 1888 and Boedecker in 1895 proposed an equation to describe the adsorption isotherm.[11,14,15] This equation was the same as the Freundlich equation. This empirical equation was named as the Freundlich equation because Freundlich was the first person to report the great importance of the equation and popularize its use.[14,15]

The Freundlich equation is more empirical than the Langmuir equation; however, it has not been theoretically justified. It can be used in correlating metal ion adsorption data over a restricted range of concentrations. It has been accepted that the isotherm is based on multilayer adsorption. The equation is shown as follows:

$$q_e = K_F C_e^{1/n} \tag{8.14}$$

where K_F is the Freundlich adsorption constant related to the heat of adsorption, and $1/n$ is a constant known as the heterogeneity factor, which is related to surface heterogeneity.

Equation 8.14 can be written as

$$\log q_e = \log K_f + \frac{1}{n}\log C_e \quad (8.15)$$

The Freundlich constants can be determined by plotting the experimental data according to Equation 8.15. The intersect is $\log K_f$, and the slope is $1/n$. With the slope and intersect, we can then determine the Freundlich constants.

When $1/n$ is less than 1, the adsorption is unfavorable. In other words, the increase in the equilibrium concentration in the solid phase (q_e) is lower than that in the solution phase (C_e). This suggests that, with the progressive coverage of the surface of sorbent, the repulsive forces among the metal ions (e.g., electronic and Coulombic repulsion as well as competition by other substances) become more dominant than the attractive forces (e.g., van der Waals forces and chemical adsorption reactions). Hence, one can conclude that the metal ions manifest a weaker tendency to adsorb onto the surface of sorbent.

On the other hand, an adsorption with $1/n$ above 1 indicates favorable adsorption. The metal ion has a strong affinity for the surface of activated carbon.

8.1.3 Sips Model

The Sips model is an extension and combination of the Langmuir and Freundlich isotherms. The model suggests that equilibrium data follow the Freundlich curve at a lower initial solute concentration and follows the Langmuir pattern at higher solute concentrations. The equation is as follows:

$$q_e = \frac{q_{max} K_s C_e^{1/n}}{1 + K_s C_e} \quad (8.16)$$

where K_S is a constant related to the heat of adsorption.

8.1.4 Dubinin–Raduskevich Adsorption Model

The Dubinin–Raduskevich adsorption model is used to calculate the apparent free energy of adsorption as it is transferred to the surface of the solid from an infinite distance in the solution.

$$q_e = q_{max} e^{(-K_{DR}\varepsilon^2)} \quad (8.17)$$

$$\varepsilon = RT \ln\left(1 + \frac{1}{C_e}\right) \quad (8.18)$$

where K_{DR} is the constant related to the adsorption energy, ε is the Polanyi potential, R is the gas constant, and T is the temperature (K).

The adsorption model can be linearized by logarithmic transfer of both sides of Equation 8.17 as follows:

$$\ln q_e = \ln q_{max} - K_{DR}\varepsilon^2 \tag{8.19}$$

The model parameters, q_{max} and K_{DR} can be determined from the slope and intercept of the plot of $\ln q_e$ against ε^2.

The mean free energy of the adsorption reaction $\left(\overline{E_{DR}}\right)$ is defined as the free energy change when 1 mol of ion is transferred to the surface of the solid from an infinite distance in the solution. It can be determined by the following equation:[16]

$$\left(\overline{E_{DR}}\right) = \left(-2K_{DR}\right)^{-0.5} \tag{8.20}$$

The result can be used to identify whether chemisorption is involved. If the calculated mean free energy of adsorption reaction $\left(\overline{E_{DR}}\right)$ falls between 8 and 16 kJ mol^{-1}, it may indicate that chemisorption plays a significant role in the adsorption process.

8.1.5 Redlich–Peterson Model

The Redlich–Peterson model incorporates the advantageous significance of both the Freundlich and Langmuir models.[17,18] The model is reportedly able to represent the adsorption isotherm over a wide range of concentrations:

$$q_e = \frac{K_R C_e}{1 + \alpha_R C_e^{\beta}} \tag{8.21}$$

where K_R and α_R are the Redlich–Peterson isotherm constants and $\beta \leq 1$.

It approaches the linear isotherm at very low equilibrium concentration, the Freundlich model at high equilibrium concentration, and the Langmuir isotherm when $\beta = 1$. The linearization of the equation can not be used to determine the three-model parameters. Therefore, a nonlinearization program may have to be used in the estimation of parameters.

8.1.6 Toth Model

The Toth model is derived from potential theory and is applicable to heterogeneous adsorption:[19]

$$q_e = \frac{q_{max} K_T C_e}{\left(1 + (K_T C_e)^{n_T}\right)^{1/n_T}} \tag{8.22}$$

where K_T and n_T are the Toth isotherm parameters.

It assumes a quasi-Gaussian energy distribution. It is one of the empirical and popularly used equations since it exhibits the correct Henry-law-type behavior at both the low and high ends of concentration. However, these three-parameter models are less popular than the Langmuir or Freundlich models because of their complicated structures. There is no critical reason to use unnecessarily more complex models if simple two-parameter models can fit the data reasonably well.

8.1.7 Multicomponent Isotherms

The above models are for single-component adsorption. However, metal waste streams in reality contain a mixture of various metal ions. For example, there are always more than 10 metal ions in acid mine drainage and electroplating effluents. For practical applications, multicomponent adsorption models have to be used.

The frequently used multicomponent isotherms were documented by Tien.[20] They include the multicomponent Langmuir model, combinational Langmuir and Freundlich model, and Ideal Adsorbed Solution Theory (IAST). Note that these models are essentially valid for cases of adsorption of gaseous species. However, they may not be suitable for adsorption of metal ions, even though some successful cases have been reported in the literatures.

The multicomponent Langmuir model is generated from the corresponding single-component Langmuir model. Thus, by obtaining the Langmuir parameters, a_i and b_i, from the individual adsorbate adsorption, it is able to calculate the equilibrium concentration for multicomponent adsorption using Equation 8.23. This model may be used to describe isotherms of simultaneous adsorption of multiple divalent metal ions. Although its simple form is a great advantage, the extended Langmuir model may not predict the equilibrium concentrations of multiple metals accurately.

$$q_e^i = \frac{a_i C_e^i}{1 + \sum_{i=1}^{n} b_i C_e^i} \tag{8.23}$$

The combinational Langmuir and Freundlich model offers another solution for fitting experimental data. The equation is as follows:

$$q_e^i = \frac{a_i (C_e^i)^{1/n}}{1 + \sum_{i=1}^{n} b_i (C_e^i)^{1/n}} \tag{8.24}$$

The IAST developed by Radke and Prausnitz is capable of accurately predicting multicomponent adsorption equilibrium based on single-component adsorption.[21] The single-component adsorption isotherms can be chosen for convenience of calculation as well as to suit specific requirements of the system. For example, the Langmuir and Slips equations were nicely incorporated into the IAST for modeling of Cu-Zn, Cu-Cd, and Cd-Zn binary systems, and a better representation of experimental data was reported.[22]

8.1.8 Surface Complex Formation Model

The major advantage of the above models is their simplicity as there are several model parameters involved. Generally speaking, the models are "mathematical functions" and do not take solution chemistry into consideration. These models fail to describe accurate adsorption equilibrium under such chemical conditions as pH, ionic strength, and presence of competitive substances. For example, it has been shown in the previous chapter that metal ion sorption changes dramatically in a short pH range. If the parameters of an empirical equation (e.g., Langmuir equation) are based on experiments conducted at a certain pH, they cannot be used to calculate sorption equilibrium at a different pH. Moreover, they cannot provide a full mechanistic understanding of metal ion adsorption.

To overcome the aforementioned problems, several models have been developed for describing structurally the adsorption of heavy metal ions since the 1960s.[1–11] The theoretical models include the Gou–Chapman–Stern–Graham model; the ion-exchange model; the ion-solvent interaction model; the surface complex formation model; and the most recently developed model, with a combination of ion exchange and coordination reactions.[23–27] It has been shown that these models generally can describe not only the adsorption isotherm but also other important factors such as pH and presence of competitive ions. Based on the nature of the metal–ligand interaction, some models have been developed and used generally in the description of sorption processes.

The surface complex formation model (SCFM) (also called the surface complexation model) was developed in the early 1960s. It has been well documented that the model is suitable for modeling of metal sorption onto metal oxides and activated carbon.

The SCFM has the advantage of describing single- and multispecies metal adsorption as functions of various operational conditions such as pH, ionic strength, and competing factors. The model has proved to be successful in the simulation of sorption of metals onto metal oxides and activated carbon. However, it may not be suitable for metal sorption onto biosorbents, when ion exchange plays an important role. The SCFM essentially consists of a series of surface reactions, which consider reactions among hydrogen ions, metal ions, surface functional groups and other ionic species.

8.1.8.1 Model Description

According to the surface complex formation model, metal ion sorption depends on three interrelated processes: surface ionization, complex formation, and formation of an electrostatic double layer (EDL) adjacent to the surface of sorbent. Few models have been proposed for EDL formation. Different considerations of the electrostatic layers adjacent to solid surfaces lead to the formulations of the diffuse layer model, the Basic Stern model, the constant capacitance model, and the triple-layer model (TLM). The TLM seems more popular than others and thus is described as follows.

The schematic representation of the triple layer model is shown in Figure 8.6. Hydrogen ions are usually considered to be sorbed in the inner layer, and the other ions are regarded as being sorbed in the inner/outer layer. The other ions include free metal ions, free anions (e.g., chloride ions), and metal–ligand complexes that are, for example, metal hydroxide and metal chloride.

There were few theories to illustrate the location of sorption of ionic species. By comparing acid-base titration data with modeling results, Westall and Hohl concluded that there were no differences among various models.[27] On the other hand, Hayes and co-workers indicated that the location of adsorbed ions strongly depended on the relative bonding affinity of ions for the functional

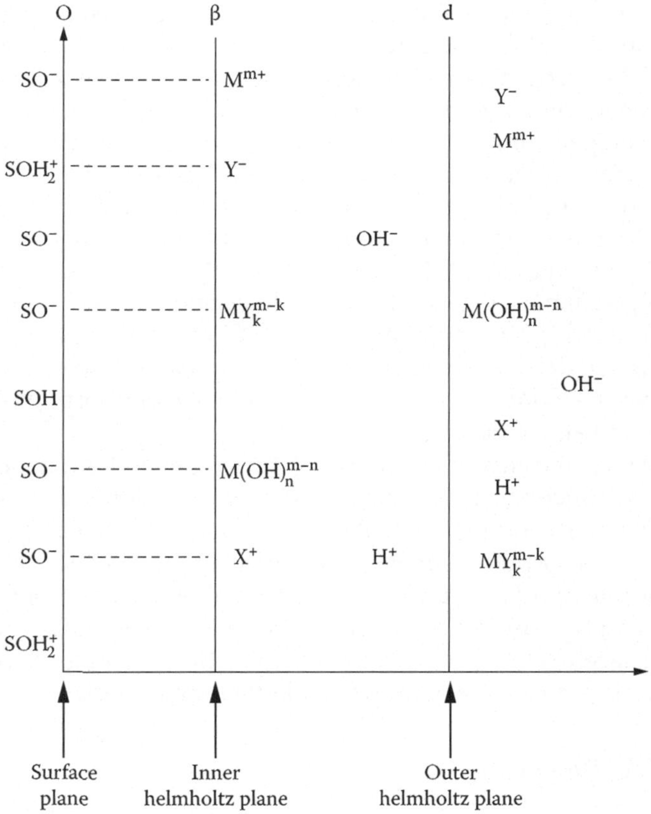

FIGURE 8.6
Schematic illustration of two-pK triple-layer surface complex formation model.

groups of the sorbents.[28,29] By investigating the effect of the ionic strength, one can distinguish the location of ions sorbed. The modeling results from Hayes and co-workers demonstrated that the ions with higher affinity for the functional group on the surfaces of sorbent are located (adsorbed) in the inner layer, whereas those with lower affinity were located/adsorbed in the outer layer. The triple-layer model is considered a better model as it can describe sorption when ions have different affinities for the functional groups of surfaces.

At the surface, functional groups for ion sorption are regarded as one averaged site or functional group, which is treated as an analogue to complexing ligands in any aqueous solution. Under different pH conditions, the functional group has different surface charges. The functional group is treated as a Lewis acid (normally symbolized as SOH). Because of the protonation and deprotonation reactions on the surface of sorbent, we have SO^- and SOH_2^+. In addition, a certain portion in the functional group goes through neither protonation nor deprotonation. As a result, we also have SOH.

On the other hand, multiple-site models have been proposed and used. For example, Reed and co-workers used three types of functional groups to represent the surface functional groups of activated carbon in their modeling, instead of one averaged functional group.[30–32] Their modeling results gave a good representation of their experimental data. Benjamin and Leckie studied competitive adsorption of copper, lead, and zinc ions onto amorphous iron hydroxide consisting of several groups of binding sites.[33,34] The model gave a good representation of the experimental data. Basically, the model was the same as the diffuse layer model, with the exception that multiple sites were involved in the calculations.

Better representations may be achieved if both multiple-site and triple-layer models are included in the calculations. The calculations may, however, become extremely complicated, and the number of parameters may be too large to be determined.

8.1.8.2 Surface Complex Formation Reactions

Metal ion sorption on the functional groups on the surfaces of the sorbents results in their removal from aqueous solutions. The functional groups are treated as analogues to complexing ligands in aqueous solution (e.g., Cu^{2+} complexed with EDTA).

Generally speaking, metal waste streams always have various inert ions such as sodium and potassium. Thus, an inert electrolyte (ionic strength) represented as XY should be included in the modeling simulation. In addition, variation of ionic strength in an experimental study may release information on the EDL as well as complexation of background electrolytes with surface functional groups on sorbent.

The general surface protolysis reactions, the electrolyte surface reactions, and the metal ion sorption reactions for a typical triple-layer two-pK SCFM are listed in Tables 8.3 and 8.4. The triple layer refers to the location of the ions in

TABLE 8.3

Mathematical Description of Interactions of Hydrogen and Electrolyte Surface Functional Groups by Two-Pk Triple-Layer Surface Complex Formation Model

Reactions	Equilibrium expressions[a]
Surface protolysis reaction $SOH + H^+ \Leftrightarrow SOH_2^+$ $SOH \Leftrightarrow SO^- + H^+$	$\frac{[SOH_2^+]}{[SOH]\{H^+\}} = K_{H1}\exp\left(-\frac{e\psi_o}{kT}\right)$ $\frac{[SO^-]\{H^+\}}{[SOH]} = K_{H2}\exp\left(\frac{e\psi_o}{kT}\right)$
Electrolyte, XY, surface reactions $SOH + X^+ \Leftrightarrow SO^-X^+ + H^+$ $SOH + H^+ + Y^- \Leftrightarrow SOH_2^+Y^-$	$\frac{[SO^-X^+]\{H^+\}}{[SOH]\{X^+\}} = K_X\exp\left(\frac{e(\psi_o - \psi_\beta)}{kT}\right)$ $\frac{[SOH_2^+Y^-]}{[SOH]\{H^+\}\{Y^-\}} = K_Y\exp\left(-\frac{e(\psi_o - \psi_\beta)}{kT}\right)$

[a] Expressions include electrostatic interactions; if electrostatic interactions are neglected, exponential terms equal to one.

TABLE 8.4

Mathematical Description of Interactions of Metal–Surface Functional Groups by Two-Pk Triple-Layer Surface Complex Formation Model

Reactions	Equilibrium Expressions[a]
Outer-sphere metal ion (M^{m+}) sorption[b]	
$nSOH + M^{m+} + lH_2O \Leftrightarrow (SO^-)_n M(OH)_l^{(m-l)+} + (n+l)H^+$	$\frac{\left[(SO^-)_n M(OH)_l^{(m-l)+}\right]\{H^+\}^{(n+l)}}{[SOH]^n\{M^{m+}\}} = K^n_{M(OH)_l} \times \exp\left(\frac{e(n\psi_o - (m-l)\psi_\beta)}{kT}\right)$
Inner-sphere metal ion (M^{m+}) sorption[a]	
$nSOH + M^{m+} + lH_2O \Leftrightarrow (SO)_n M(OH)_l^{(m-l-n)+} + (n+l)H^+$	$\frac{\left[(SO)_n M(OH)_l^{(m-l-n)+}\right]\{H^+\}^{(n+l)}}{[SOH]^n\{M^{m+}\}} = K^n_{M(OH)_l} \times \exp\left(\frac{e(n+l-m)\psi_o}{kT}\right)$

[a] Expressions include electrostatic interactions; if electrostatic interactions are neglected, exponential terms equal to one.

[b] n is the number of surface sites in mol/L, which reacts with one mol/L of M^{m+} in solution; $n = 1, 2, ..., N$. l defines the l-th hydrolysis product of M^{m+}; $l = 0, 1, 2, ..., L$; $l = 0$ corresponds to the simple ion.

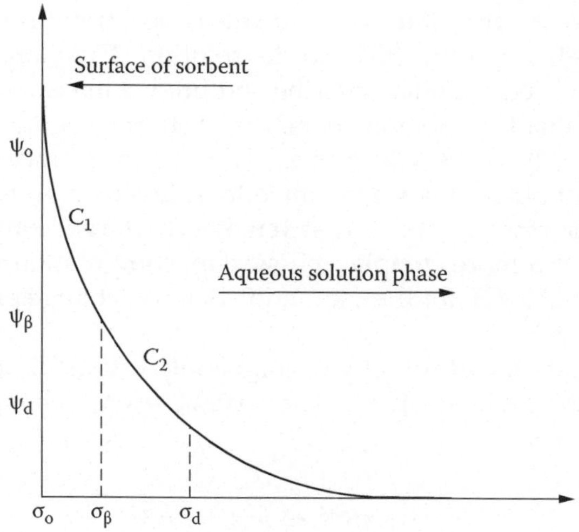

FIGURE 8.7
Surface potential and surface charge as a function of distance.

three different layers, that is, o, β, and d-layers, as shown in Figures 8.6 and 8.7. The two-pK refers to surface ionization corresponding to three types of surface species (SO^-, SOH, and SOH_2^+) due to the protonation and deprotonation reactions associated with two equilibrium constants (K_{H1} and K_{H2}).

Uptake of metal ions may result from the formation of surface–metal complexes $SO\text{–}M^{m+}$ or $SO^-M(OH)_n^{m-n}$, or a combination of both. In aqueous solution, such anionic ligands (Y^-) as chloride may form metal–ligand complexes (MY_k^{m-k}), which can then be adsorbed onto the surface functional groups of sorbent. Anionic (Y^-) (e.g., Cl^-) and cationic (X^+) (e.g., Na^+) substances (as ionic strength) may form surface complexes with the surface functional groups. In addition, complexation of metal ions by multidentate surface sites such as $(SO^-)_n\ M(OH)_l^{(m-1)+}$ may contribute to the binding of metal ions (not illustrated in Figure 8.6).

Except hydrogen ions, the ionic substances (including metal ions) are assumed to be adsorbed in the β-layer (see Figure 8.6). If the surface functional groups have stronger affinities for the ions, they could be bound in the inner layer as suggested by Hayes and co-workers.[28,29]

8.2 Ion Exchange

Ion exchange (IX) is an important water treatment process for the removal of unknown ionic substances from water.[35] Empirical equations such as the Langmuir isotherm may be used for the IX process. They are simple and easy to use; however, they fail to provide the concentrations of ions released from the ion exchange resin (IXR) to the solution. Therefore, the following approach can be used to better describe and predict ion exchange process.

It is assumed that the solution chemistry is simple enough that the target ions (to be removed) are the only species in the system. Competing effects due to the presence of other ions are assumed to be less important. Ion exchange reactions are the only reactions that remove the target ions. If the reader needs to consider a more complicated system, computation programs such as MINEQL and FITEQL for the estimation of ion exchange processes would need to be used.[36–39]

We assume that an IXR initially contains only cations B and the solution has the cations A (target ions).[37,40,41] The reversible ion exchange reaction can be represented by

$$mA^{n+} + n\overline{B^{m+}} \Leftrightarrow m\overline{A^{n+}} + nB^{m+} \tag{8.25}$$

where the stoichiometric coefficients m and n are the valencies of ions B and A, and the overbars denote the ion exchanger (solid phase).

Thus, we have

$$K_{A \to B} = \frac{\overline{\{A^{n+}\}}^m \{B^{m+}\}^n}{\{A^{n+}\}^m \overline{\{B^{m+}\}}^n} \tag{8.26a}$$

$$K_{A \to B} = \left(\frac{\overline{\{A^{n+}\}}}{\{A^{n+}\}}\right)^m \left(\frac{\{B^{m+}\}}{\overline{\{B^{m+}\}}}\right)^n \tag{8.26b}$$

where {} denotes the activity or concentration of species, and $K_{A \to B}$ is the equilibrium constant, otherwise referred to as the selectivity coefficient.

The thermodynamic constant $K_{A \to B}$ is a function of temperature and is determined by the standard free energy of the ion exchange reaction. Seasonal variations in temperature of up to 30°C are possible in practical ion exchange processes. Therefore, the effect of temperature cannot be ignored. Thermal regeneration induced by such temperature changes can be possible for some resins, and this causes redistribution of ions on the resins.

It is always preferable to obtain the constant through experimental measurement. However, if measurement is not allowed, one can use the following equation for rough estimation of constant:[37]

$$K_{A \to B} = \frac{K_A}{K_B} \tag{8.27}$$

where K_A and K_B can be found from Table 8.5.

TABLE 8.5

Selectivity Scale for Cations on 8% Cross-Linked Strong-Acid Resin and for Anions on Strong-Base Resins[37]

Cation	Selectivity	Anion	Selectivity
Li^+	1.0	HPO_4^{2-}	0.01
H^+	1.3	CO_3^{2-}	0.03
Na^+	2.0	OH^- (Type I)	0.06
UO^{2+}	2.5	F^-	0.1
NH_4^+	2.6	SO_4^{2-}	0.15
K^+	2.9	CH_3COO^-	0.2
Rb^+	3.2	HCO_3^-	0.4
Cs^+	3.3	OH^- (Type II)	0.65
Mg^{2+}	3.3	BrO_3^-	1.0
Zn^{2+}	3.5	Cl^-	1.0
Co^{2+}	3.7	CN^-	1.3
Cu^{2+}	3.8	NO_2^-	1.3
Cd^{2+}	3.9	HSO_4^-	1.6
Ni^{2+}	3.9	Br^-	3
Be^{2+}	4.0	NO_3^-	4
Mn^{2+}	4.1	I^-	8
Pb^{2+}	5.0	SO_4^{2-}	9.1
Ca^{2+}	5.2	SeO_4^{2-}	17
Sr^{2+}	6.5	CrO_4^{2-}	100
Ag^+	8.5		
Pb^{2+}	9.9		
Ba^{2+}	11.5		
Ra^{2+}	13.0		

The distribution of species A and B in both the solid and solution phases can be determined by the foregoing equations together with a mass balance equation. When the calculation is conducted, the species can be treated as those in solution reactions, with units of meq/L. If the ionic strength is not high, activity can be treated as concentration, which makes the calculation less complicated.

A method was developed to simplify the IX calculation that has been widely used in the water industry, which is described as follows.[37,38]

Let q and c represent concentration of species in the IX resin (solid phase) and solution phases; Equation 8.26b can then be written as

$$K_{A\to B} = \left(\frac{q_A}{c_A}\right)^m \left(\frac{c_B}{q_B}\right)^n \tag{8.28}$$

Let Q be the total exchange capacity of the IX resin (eq/L) and C be the total ionic concentration in the solution (eq/L), and also let X and $\overline{X}$ represent the percentage of species in solution and solid phases. We then have:

$$q_A = \overline{X_A} \times Q \quad (8.29a)$$

$$q_B = \overline{X_B} \times Q \quad (8.29b)$$

$$c_A = X_A \times C \quad (8.30a)$$

$$c_B = X_B \times C \quad (8.30b)$$

$$\overline{X_A} + \overline{X_B} = 1 \quad (8.31a)$$

$$X_A + X_B = 1 \quad (8.32b)$$

Thus, Equation 8.28 can be rewritten as

$$K_{A \to B} = \left(\frac{\overline{X_A}}{X_A} \right)^m \left(\frac{X_B}{\overline{X_B}} \right)^n \left(\frac{Q}{C} \right)^{m-n} \quad (8.33)$$

Combining Equations 8.29a, 8.29b, 8.30a and 8.30b yields:

$$K_{A \to B} = \left(\frac{\overline{X_A}}{X_A} \right)^m \left(\frac{1 - X_A}{1 - \overline{X_A}} \right)^n \left(\frac{Q}{C} \right)^{m-n} \quad (8.34)$$

$$\frac{\overline{X_A}^m}{(1 - \overline{X_A})^n} = K_{A \to B} \frac{X_A^m}{(1 - X_A)^n} \left(\frac{Q}{C} \right)^{m-n} \quad (8.35)$$

The foregoing reaction and equations can be further simplified for some special cases. For the monovalent ion exchange process, $n = 1$ and $m = 1$:

$$A^+ + \overline{B^+} \Leftrightarrow \overline{A^+} + B^+ \quad (8.36)$$

Equation 8.35 can then be simplified, and yields:

$$\overline{X_A} = \frac{K_{A \to B} X_A}{1 + (K_{A \to B} - 1) X_A} \quad (8.37)$$

Figure 8.8 illustrates the distribution of species A in both resin and solution with different selectivity values in monovalent ion exchange process. As shown, more species A are present in the solid phase when the $K_{A\rightarrow B}$ value is higher than 1. When $K_{A\rightarrow B}$ is equal to unity, 50% of A is in the solid, while the remaining 50% is in the solution.

In the exchange of calcium or magnesium ions A for sodium or hydrogen ions B in a water treatment process, $n = 2$, $m = 1$, then

$$mA^{n+} + n\overline{B^{m+}} \Leftrightarrow m\overline{A^{n+}} + nB^{m+}$$

$$K_{A\rightarrow B} = \left(\frac{\overline{X_A}}{X_A}\right)^m \left(\frac{1-X_A}{1-\overline{X_A}}\right)^n \left(\frac{Q}{C}\right)^{m-n}$$

$$A^{2+} + 2\overline{B^{+}} \Leftrightarrow \overline{A^{2+}} + 2B^{+} \tag{8.38}$$

$$K_{A\rightarrow B} = \left(\frac{c_B}{q_B}\right)^2 \frac{q_A}{c_A} \tag{8.39}$$

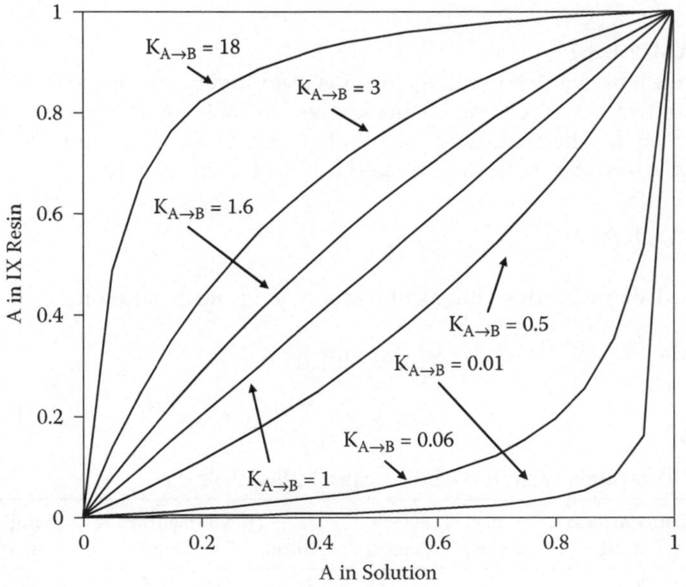

FIGURE 8.8
Distribution of species A in both resin and solution with different selectivity values in the monovalent ion exchange process.

Or

$$K_{A\to B} = \left(\frac{\overline{X_A}}{X_A}\right)\left(\frac{1-X_A}{1-\overline{X_A}}\right)^2 \frac{C}{Q} \tag{8.40}$$

An alternative parameter to determine the preference of an ion exchanger for a particular ionic species is the separation factor (α_B^A), which is defined as

$$\alpha_B^A = \frac{\overline{X_A} / X_A}{\overline{X_B} / X_B} \tag{8.41}$$

Note that the equilibrium constant ($K_{A\to B}$) of the IX reaction is the same as the separation factor (α_B^A) when both *m* and *n* are equal to unity.

The separation factor ultimately reflects the likelihood of exchange occurring between particular ions. It provides information on the degree of interaction between the ions and the ion-exchange resin. If $\alpha_B^A = 1$, it indicates that the resin shows no preference for A over B. If it is above unity, A interacts more strongly with the resin than B.

The separation factor relies on the design of the active sites on the resins and the solution chemistry. Mechanisms contributing to chemical selectivity reportedly include the hard-soft acid-base principle, hydrophilicity/hydrophobicity and waste properties (e.g. pH).

EXAMPLE 8.2

A strong-base anion exchange resin is used to remove nitrate ions from groundwater with characteristics as given in Table 8.6. The resin capacity is 2.5 eq/L. The selectivity for nitrate is 4.0. Determine the maximum volume of water that can be treated if 10 L of resins is used.

SOLUTION

1. Determine the equivalent fraction of nitrate in solution:

$$X_{NO_3^-} = 2.0/5.0 = 0.4$$

TABLE 8.6

List of the Components in the Groundwater in Example 8.2

Cation	Concentration, mg/L	mg/meq	meq/L	Anion	Concentration, mg/L	mg/meq	meq/L
Ca^{2+}	30	20.04	1.5	Cl^-	106.5	35.5	3.0
Mg^{2+}	12.15	12.15	1.0	SO_4^{2-}	0	48.03	0
Na^+	57.5	23.00	2.5	NO_3^-	100	50	2.0
Total		—	5.0	Total			5.0

2. Determine the equivalent fraction of nitrate on the resin:

$$NO_3^- + \overline{R-X^-} \Leftrightarrow \overline{R-NO_3^-} + X^-$$

$$K_{NO_3^- \to X^-} = 4$$

$$\text{Given } \overline{X_A} = \frac{K_{A\to B} X_A}{1+(K_{A\to B}-1)X_A},$$

We have:

$$\overline{X_{NO_3^-}} = \frac{4\times 0.4}{1+(4-1)\times 0.4} = 72.7\%$$

This means 72.7% of resin sites are occupied by nitrate ions.

3. Compute the maximum useful capacity Y:

$$Y = 2.5 \text{ eq/L} \times 72.7\% = 1.83 \text{ eq/L} = 1830 \text{ meq/L}$$

4. Compute the volume of water (V) that can be treated when 1 L of resin is used:

$$V = (1830 \text{ meq/L of resin })/(2.0 \text{ meq/L of water})$$

$$= 915 \text{ L of water per liter of resin}$$

5. Thus, the total volume of water that can be treated when 10 L of resins is used = 915 × 10 = 9150 L

EXAMPLE 8.3

A strong-acid cation exchanger is used to remove cadmium ions from a waste acidic solution. Cadmium ions (A) are exchanged with hydrogen ions (B) in the exchange. The selectivity scale values of A and B can be found in Table 8.5. The total exchange capacity of the cation exchange (Q) is 1.9 eq/L. At the total ionic concentration in the solution (C) of 0.1 eq/L, determine the equilibrium equivalents of cadmium ions in the exchanger when the cadmium concentration is 2.25 g/L.

SOLUTION

$$A^{2+} + 2\overline{B^+} \Leftrightarrow \overline{A^{2+}} + 2B^+$$

$$K_{A\to B} = \frac{c_B^2 q_A}{c_A q_B^2}$$

From Table 8.5, we have $K_A = 3.9$ and $K_B = 1.3$; thus:

$$K_{A\to B} = \frac{K_A}{K_B} = \frac{3.9}{1.3} = 3$$

The copper concentration in the solution at equilibrium (C_A) = 2.25/112.4 = 0.02 M = 0.04 N:

$$Q = q_A + q_B$$

$$C_B = C - C_A = 0.1 - 0.04$$

$$= 0.06 \text{ N}$$

$$q_B = Q - q_A$$

$$= 1.9 - q_A$$

Noted that C_A, C_B, q_A, and q_B have units of eq/L or N.

Thus, we have:

$$3 = \frac{(0.1-0.04)^2 \times q_A}{0.04 \times (1.9-q_A)^2}$$

Solving the foregoing equation gives:

q_A= 0.84 N

8.3 Hybrid Model

The mechanisms that cause heavy metal removal by some natural or synthesized materials may be more complicated than those for "pure" metal oxide, activated carbon, and ion exchange resin. Two distinctly different mechanisms may be involved.

8.3.1 Metal Uptake by Biosorbent

Metal removal by biosorbents originally from the seawater environment is a typical case that involves ion exchange and surface complex formation. The biosorbents contain substantial amounts of alkaline earth metal ions (i.e., calcium (Ca) and magnesium (Mg)), which are chemically bound onto the functional groups of the biosorbents. When heavy metal ions are brought into contact with the biosorbents, both calcium and magnesium ions exchange with heavy metal ions, which are ion exchange reactions. In addition, there may be some functional groups not associated with the

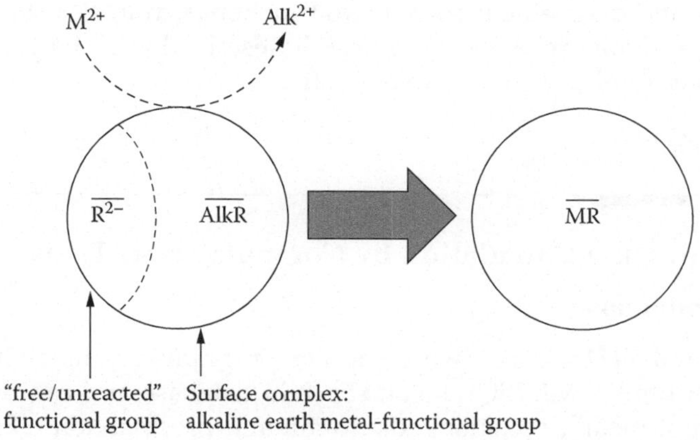

FIGURE 8.9
Basic mechanism for heavy metal biosorption.

calcium and magnesium ions. They may play a role in the metal uptake due to the formation of metal–ligand complexes, which can be described by the surface complex formation model. Chen and co-workers developed a hybrid IX–SCF model that was successful in describing and predicting metal uptake by seaweeds and calcium alginate.[24–26] The theoretical description now follows.

Generally speaking, the biosorbents are polymers containing a series of functional groups. For example, alginate is a polysaccharide-like polymer, and contains three different functional groups: –COO⁻ (carboxylate), -C-O-C- (ether), and –OH (alcohol). It can be assumed that there is one generalized type of functional group (represented as R^{2-}) in the metal ion removal process, as shown in Figure 8.9. Ion exchange is found to be the main avenue through which metal ions are stripped away.

When a biosorbent is immersed in an acid, alkaline earth metal ions such as calcium ions can be replaced by hydrogen ions. Similarly, in an aqueous solution of metal salts, an exchange between the metal ions (M^{2+}) and alkaline earth metal ions Alk^{2+} inside of sorbent takes place. In addition, there may be some "free/unreacted" functional groups in the virgin biosorbent, leading to the formation of an ion pair.

$$\overline{R^{2-}} + M^{2+} = \overline{RM}$$

8.3.2 Metal Uptake by Composite Sorbent

When the sorbent contains not only biomaterials but also other adsorptive substances such as metal oxides, additional reactions must be included. For example, the model for the composite sorbent with alginate, alkaline earth metal ions, and magnetite involved two functional groups, surface complex formation, and ion exchange reactions.[26] It described well the pH effect, the sorption isotherm, and competitive sorption.

8.4 Equilibrium Calculation by Computational Tools

8.4.1 Introduction

MINEQL and FITEQL are two computer programs for modeling chemical equilibrium.[41,42] MINEQL has its built-in database and can be used for calculation of metal speciation, adsorption and precipitation.[41] FITEQL, on the other hand, does not have a database, and the information on chemical reactions and their equilibrium constants must be provided.[27] The advantage of FITEQL is that it can optimize model parameters such as equilibrium constants. KINEQL, probably the only calculation model for metal adsorption kinetics that includes SCFM also uses the same tableau approach as MINEQL and FITEQL.[5]

MINEQL has a few versions, some of which are still being used, including MINEQL+[39]. The source code (written in Fortran) has since been used as a base for many chemical equilibrium and kinetics models.

MINEQL essentially originated from a computation program called REDEQL developed by James Morgan and Francois Morel at MIT in 1972. The REDEQL became the prototype for MINEQL, which was subsequently developed by John Westall and Francois Morel in 1975. MINEQL, written in FORTRAN code, used a generic tableau approach to describe equilibria and mass balance in aqueous systems, which has since been used in various textbooks on aqueous chemistry. Since 1975, MINEQL was further developed to various versions such as WATEQ (by USGS in the early 1980s), and MINTEQ (by USEPA in the late 1980s). A company called *Environmental Research Software* has further developed MINEQL since the early 1990s. MINEQL+ 4.6 is the most recent Windows-based version of MINEQL and provides a user-friendly interface.

Two key terms have to be defined before the formulation of the mathematical model. They are the *components* and *species*, which are used in the equilibrium model, MINEQL.[5,42] According to the definitions, the *components* mathematically form the independent basis set, on which every *species* can be defined, and upon which the mass balance equations are based. Chemically, the *components* are a set of chemical entities that enable every *species* to be represented as the product of a reaction involving only these *components*. In addition, no *component* can be represented as the product of a reaction involving only the other *components*. *Species* represent log-linear combinations of the *components*; a *species* is the product of a chemical reaction involving the *components* as reactants. The concepts of component and species can be illustrated in the following examples. Appendix A provides more detailed description of MINEQL.

8.4.2 Mathematical Description

Assuming a system in which there are M_x aqueous species with concentrations x_i, M_y adsorbed/complexated and ion-exchanged species with concentrations y_i, and M_p precipitated species with activities of unity, one can express the concentrations or activities of these three types of species in terms of N_a aqueous components with concentrations c_j, N_s sorbent components with concentrations s_j, and two electrostatic components C_0 and C_β. The equilibrium relationship on the concentrations of the three species is described as follows:[5,42]

$$x_i = K_i^x \prod_{k=1}^{N_a} c_k^{a_{ik}^x} \qquad i = 1,2,\ldots,M_x, \tag{8.42a}$$

$$y_i = K_i^y \left(\prod_{k=1}^{N_a} c_k^{a_{ik}^y} \right) \left(\prod_{k=1}^{N_s} s_k^{b_{ik}^y} \right) c_o^{a_{io}^y} c_\beta^{a_{i\beta}^y} \qquad i = 1,2,\ldots,M_y, \tag{8.42b}$$

$$1 = K_i^p \prod_{k=1}^{N_a} c_k^{a_{ik}^p}, \qquad i = 1,2,\ldots,M_p, \tag{8.42c}$$

where

K_i^x = modified equilibrium constant of the i-th aqueous species,

a_{ik}^x = stoichiometric coefficient of the k-th aqueous component in the i-th aqueous species,

K_i^y = modified equilibrium constant of the i-th sorbed species,

a_{ik}^y = stoichiometric coefficient of the k-th aqueous or electrostatic component in the i-th sorbed species,

b_{ik}^y = stoichiometric coefficient of the k-th sorbent component in the i-th sorbed species,

K_i^p = modified equilibrium constant of the i-th precipitated species,

a_{ik}^p = stoichiometric coefficient of the k-th aqueous component in the i-th precipitated species.

The equilibrium constants in Equations 8.42a–8.42c include activity corrections. Because of the EDL adjacent to the sorbent surface, electrostatic components c_0 and c_β are included in Equation 8.42b. The EDL consists of the charged surface and an excess of counterions over co-ions that are diffused in the solution as shown in Figures 8.6 and 8.7. The electrostatic components are energy terms related to the electrical potentials ψ_o and ψ_β at o and β-layers of the EDL as follows:

$$c_o = \exp\left(-\frac{e\psi_o}{kT}\right), \tag{8.43a}$$

$$c_\beta = \exp\left(-\frac{e\psi_\beta}{kT}\right), \tag{8.43b}$$

where *e*, *k*, and *T* are the electron charge, Boltzmann constant, and absolute temperature, respectively. The electrical potentials, ψ_o, ψ_β, and ψ_d in the three layers of the EDL are related to the corresponding surface charge densities, σ_o, σ_β, and σ_d, through the charge/potential relationships defined from the specific model chosen for the EDL. For the triple-layer model illustrated in Figures 8.6 and 8.7, these relationships are:[5]

$$C_1(\psi_o - \psi_\beta) = \sigma_o, \tag{8.44a}$$

$$C_2(\psi_\beta - \psi_d) = \sigma_\beta, \tag{8.44b}$$

$$\sigma_d = -\sqrt{8\varepsilon RTI}\sinh\left(\frac{e\psi_d}{2kT}\right), \tag{8.44c}$$

where C_1 and C_2 are the capacitances of the EDL, ε is the permittivity of the medium, *R* is the gas constant, and *I* is the ionic strength.

From the electroneutrality condition, we have

$$\sigma_o + \sigma_\beta + \sigma_d = 0 \tag{8.45}$$

The surface charge densities, σ_o and σ_β, can be obtained from the sorbed species concentrations as

$$\sigma_o = \sum_{i=1}^{M_y} a_{io}^y y_i, \tag{8.46a}$$

$$\sigma_\beta = \sum_{i=1}^{M_y} a_{i\beta}^y y_i, \tag{8.46b}$$

The above equations may be solved to give the total sorbed concentration of the j-th aqueous component, S_j, given by

$$S_j = \sum_{i=1}^{M_y} a_{ij}^y y_i, \qquad j = 1,2,\ldots,N_a. \tag{8.47}$$

Note that Equations 8.42a–8.42c are essential to the modeling of metal and the sorption. However, the terms due to the EDL may not be necessairly used in order to avoid complicities in the modeling.

8.4.3 Determination of Model Parameters

As the equilibrium constants of models are unknown, they must be determined mathematically with experimental data as an input. Generally speaking, the simulation is conducted with assumed chemical reactions, of which results are compared with experimental data. The optimization of parameters is achieved when a criterion is met. FITEQL (e.g., Version 4.0) uses a nonlinear least-squares optimization approach for the parameter searching/optimization. Other optimization programs can also be used, including Sequential Quadratic Programming (SQP) and PEST (a commercial program by Scientific Software Group).

When the MINEQL code is used for equilibrium or kinetics modeling, the SQP optimization routine is recommended for the optimization of model parameters.[5,43,44] The objective function is defined as follows:

$$\mathrm{SHAT} = \left(\frac{\sum_{j=1}^{NP} \left(\frac{y_{j,\mathrm{measured}} - y_{j,\mathrm{calculated}}}{y_{j,\mathrm{measured}}} \right)^2}{\mathrm{NP} - \mathrm{NPAR} - 1} \right)^{0.5} \tag{8.48}$$

where $y_{j,\mathrm{measured}}$ and $y_{j,\mathrm{calculated}}$ correspond to the experimental data and calculated results, respectively; NP is the number of experimental data; and NPAR is the number of parameters to be found.

The unknown model parameters are of two types: equilibrium constants and total concentrations of components. In order to obtain these parameters, two experiments normally have to be performed and used as input files for determining the parameters. They are potentiometric titration experiments of sorbents and equilibrium experiments of metal ion sorption (adsorption isotherm or pH effect).

The pH versus the concentration of hydrogen or hydroxide ions added obtained from acid-base titration experiments provides basic characteristics of the sorbent. For the SCFM, the equilibrium constants of surface protolysis reactions (K_{H1} and K_{H2}) and of electrolyte surface reactions (K_X and K_Y), the physical constants of EDL (the capacitances C_1 and C_2), and the total concentration of sorbent (mole/gram sorbent) can be obtained. For the hybrid IX-SCF model, the total concentration of functional group and the equilibrium constants of the aforementioned reactions can be obtained.

Once these parameters are obtained, equilibrium constants of metal ion sorption reactions can be found based on sorption equilibrium experimental data and the model parameters (e.g., K_{H1}, and K_{H2}) obtained from the modeling study of titration data. The search of all these parameters can be carried out by comparing experimental data with model predictions and by using optimization techniques.

When the parameters are obtained, the speciation of ions at equilibrium is well defined. The modes can then be used to predict sorption behaviors (e.g., competitive sorption). Several case studies that illustrate the application of SCFM and the hybrid IX-SCF model follow.

8.5 Case Studies of Adsorption Equilibrium

8.5.1 Heavy Metal Adsorption onto Activated Carbon

An experimental study of heavy sorption by an H-type activated carbon was carried out by Chen and Lin.[44] The modeling simulation for the sorption of three heavy metal ions is conducted by the two-pK triple-layer model with surface reactions given in Table 8.7. Note that *components* are listed on the left-hand side of the reactions, whereas *species* are listed on the right-hand side of the reactions.

8.5.1.1 Surface Charge Properties

Modeling of surface charge versus pH is carried out based on surface reactions (1) to (4) of Table 8.7. The titration data are used as an input in the searching of model parameters.

MINEQL (in FORTRAN code) with the additional subroutine SQP is used for searching constants of reactions, capacitances, and concentration of functional groups.[5] The minimization of the difference between the experimental values of surface charge and the calculated values is set as the objective function (Equation 8.48). The values of these parameters are found and listed in Table 8.8.

Figure 8.10a,b compares modeling results with the experimental data of the surface change densities of activated carbon as a function of pH in different ionic strengths and types of insert ionic species. The SCF model generally gives a better representation of the pH-dependent surface charge density

TABLE 8.7

Surface Reactions in Heavy Metal Sorption

No.	Surface Reactions
1.	$SOH + H^+ + \exp(-y_o) \overset{K_{H1}}{\Leftrightarrow} SOH_2^+$
2.	$SOH - H^+ - \exp(-y_o) \overset{K_{H2}}{\Leftrightarrow} SO^-$
3.	$SOH + Y^- + H^+ + \exp(-y_o) - \exp(-y_\beta) \overset{K_y}{\Leftrightarrow} SOH_2^+Y^-$
4.	$SOH + Na^+ - H^+ - \exp(-y_o) + \exp(-y_\beta) \overset{K_{Na}}{\Leftrightarrow} SO^-Na^+$
5.	$SOH + M^{2+} - H^+ - \exp(-y_o) + 2\exp(-y_\beta) \overset{K_M}{\Leftrightarrow} SO^-M^{2+}$
6.	$SOH + M^{2+} - 2H^+ - \exp(-y_o) + \exp(-y_\beta) \overset{K_{MOH}}{\Leftrightarrow} SO^-MOH^+$
7.	$SOH + M^{2+} - H^+ + Cl^- - \exp(-y_o) + \exp(-y_\beta) \overset{K_{MCl}}{\Leftrightarrow} SO^-MCl^+$

Note: $y_o = e\psi_o / kT$ referred to o-layer and $y_\beta = e\psi_\beta / kT$ to β-layer; M = Cu, Zn, or Co; Y = Cl, NO_3, or ClO_4.

TABLE 8.8

Model Parameters of Surface Protolysis Reaction and Electrolyte Surface Reactions

Parameter	Values
C_1	31.11 mF/cm^2
C_2	23.14 mF/cm^2
K_{H1}	$10^{3.75}$ M^{-1}
K_{H2}	$10^{-10.74}$ M
K_X	$10^{2.94}$ M^{-2} (X=Cl); $10^{3.79}$ M^{-2} (X=NO_3); $10^{4.00}$ M^{-2} (X=ClO_4)
K_{Na}	$10^{-9.36}$
Adsorbent concentration	10 g/L

as well as the effect of ionic strength. In addition, the model can accurately predict a pH_{pzc} of 7.16 (Figure 8.10a).

Inert ionic species form ionic strength. Different background electrolyte as ionic strength may affect differently the surface charge of activated carbon. When $NaNO_3$ and $NaClO_4$ are used as electrolyte in potentiometric titration, the surface charge density behaves differently as shown in Figures 8.10b and 8.10c.

In the modeling of surface charge density in $NaNO_3$ and $NaClO_4$, or Surface Reactions (1), (2), and (4) of Table 8.7 remain unchanged, while

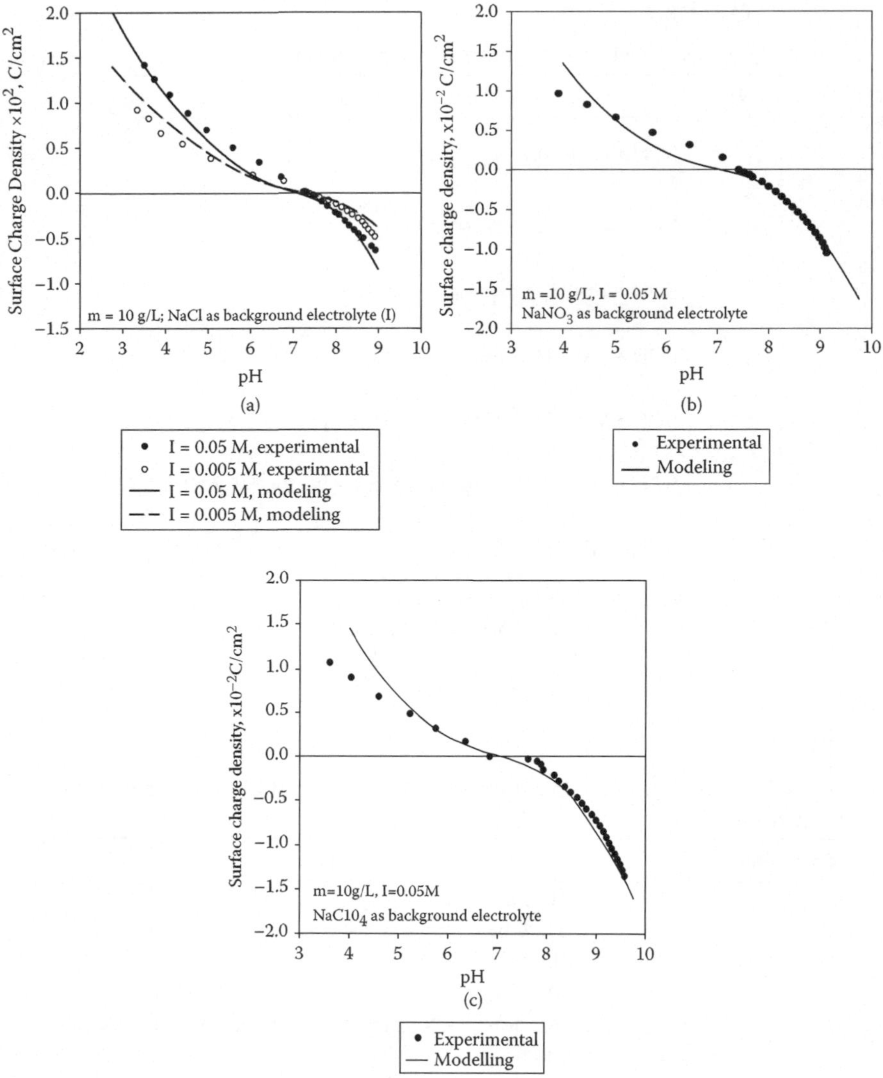

FIGURE 8.10
Illustration of surface charge on activated carbon as a function of pH: modeling results versus experimental data.

Reaction (3) is changed according to the change in the background electrolyte. The constant for Reaction (3) as $10^{3.79}$ or $10^{4.00}$ M^{-2} is found when $NaNO_3$ or $NaClO_4$ is used as background electrolyte, respectively. As shown in Figure 8.10b, the model describes well the effect of pH on the surface charge density. In addition, the values from the modeling in $NaNO_3$ and $NaClO_4$ were almost the same.

The sensitivity analysis of these parameters of C_1, C_2, K_{H1}, K_{H2}, K_{Na}, and K_{Cl} is performed, with the results shown in Figure 8.11. The analysis would provide us ideas regarding changes in the surface charge density when the model parameters are varied.

Figure 8.11a demonstrates that the change in surface charge density is negligible in pHs 5 to 8 when C_1 is increased from 31 to 45 mF/cm^2 or C_2 from 23 to 35 mF/cm^2. At pH < 5, higher surface charge density increases as C_2 is increased. More importantly, the estimated pH$_{pzc}$ values always remain the same, no matter how C_1 and C_2 are varied.

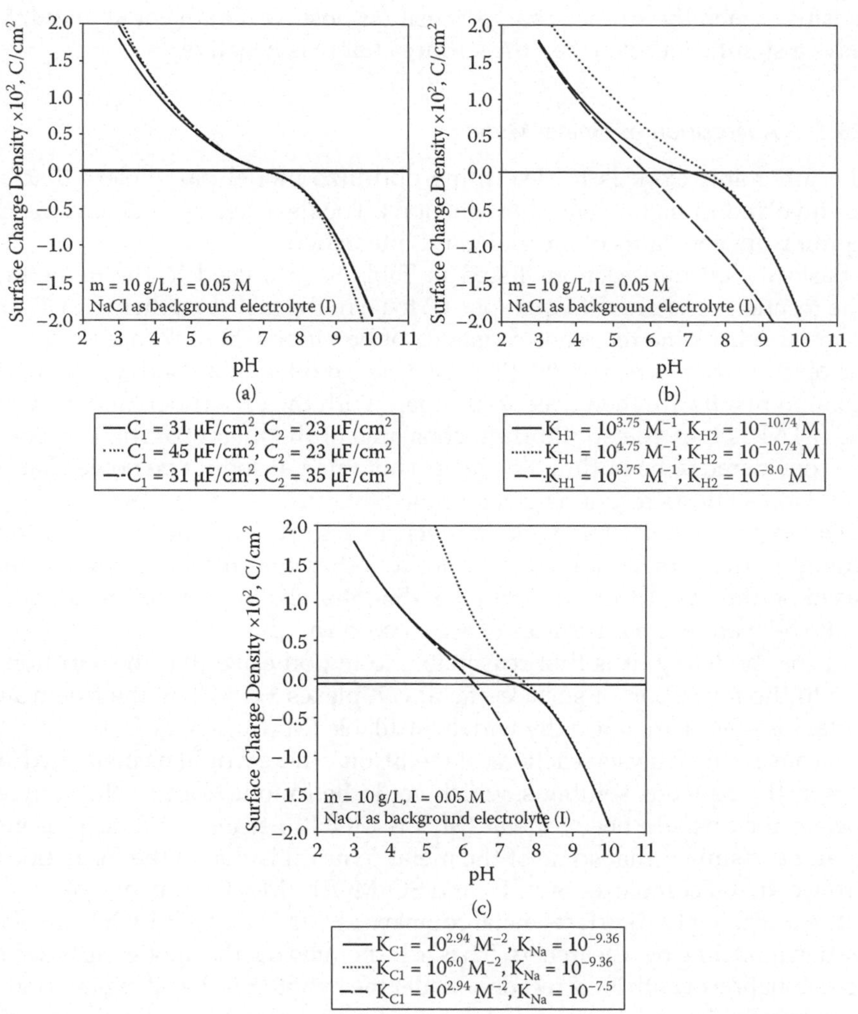

FIGURE 8.11
Sensitivity of model parameter on simulation of surface charge. (a) C_1 and C_2; (b) K_{H1} and K_{H2}; (c) K_{Cl} and K_{Na}.

Both K_{H1} and K_{H2} play important roles in the surface charge density, as shown in Figure 8.11b. Unlike the effect of C_1 and C_2, variation of K_{H1} and K_{H2} can change the location of pH_{pzc}. As shown in Figure 8.10b, the effect of K_{H1} is important at $pH < pH_{pzc}$, while K_{H2} played a key role at $pH > pH_{pzc}$. When K_{H1} is increased or K_{H2} is decreased while the other parameter (K_{H1} or K_{H2}) is kept the same, the pH_{pzc} value increases or decreases. The pH_{pzc} shifts to a lower value when K_{H2} is increased or K_{H1} is decreased.

As illustrated in Figure 8.11c, both K_{Cl} and K_{Na} have the similar trends as K_{H1} and K_{H2} (shown in Figure 8.10b). pH_{pzc} shifts to a larger value (from left to right) as K_{Cl} is increased or K_{Na} is decreased (not shown). K_{Cl} becomes more sensitive when the surface charge density is positive. On the other hand, K_{Na} plays a significant role when the surface charge is negative.

8.5.1.2 Adsorption of Heavy Metals

The foregoing work is to determine/optimize model parameters that are not involved in metal–ligand interactions. The next step is to determine the equilibrium constants of metal–ligand interactions.

Basically, all the reactions listed in Table 8.7 are used in the modelling. The reaction constants of Equations (1) to (4) in the table together with C_1 and C_2 are used. As the reactions constants of Equations (5) to (7) are unknown, the optimization program (SQP) is used. The modeling is conducted; the calculation results are then used to compare with the experimental data. Once the SHAT is met, the equilibrium constants of the metal sorption reactions are considered to be optimized and are obtained. It should be noted that the metal ion solution reactions given in the literature[10] are included.

The experimental observations given in Figure 8.12 illustrate that ionic strength affects metal ion adsorption. It is therefore reasonable to assume, based on the EDL theory described earlier, that the metal ions are adsorbed in the β-layer near the surfaces of activated carbon.

In the modeling, it is first reasonable to hypothesize that the sorption is due to the formation of surface–metal complexes SO^-M^{2+} as the free heavy metal ions (M^{2+}) are normally widely available (M=Cu, Zn, or Co).

Because of hydrolysis reactions of metal ions, there would be metal hydroxides in the aqueous solutions, which could be in ionic forms. The simplest species for divalent heavy metal (M^{2+}) is MOH^+. Because of their presence, it can be assumed that some of the metal removal is due to the formation of surface–metal complexes SO^-M^{2+} and SO^-MOH^+ (M = Cu, Zn, or Co).

The foregoing two surface–metal complexes SO^-M^{2+} and SO^-MOH^+ are often used in modeling by researchers. It has to be pointed out that modeling based on a combination of both is correct only under the hypothesis that all ligands (other than hydroxide) do not form metal–ligand complexes. Normally, these metal salts are metal nitrate and metal perchlorate salts (e.g., $Cu(NO_3)_2$ and $Cu(ClO_4)_2$).

Normally, there are complexing anions/ligands (anionic ligands) (other than hydroxide) in aqueous solution that can form soluble metal–ligand complexes.

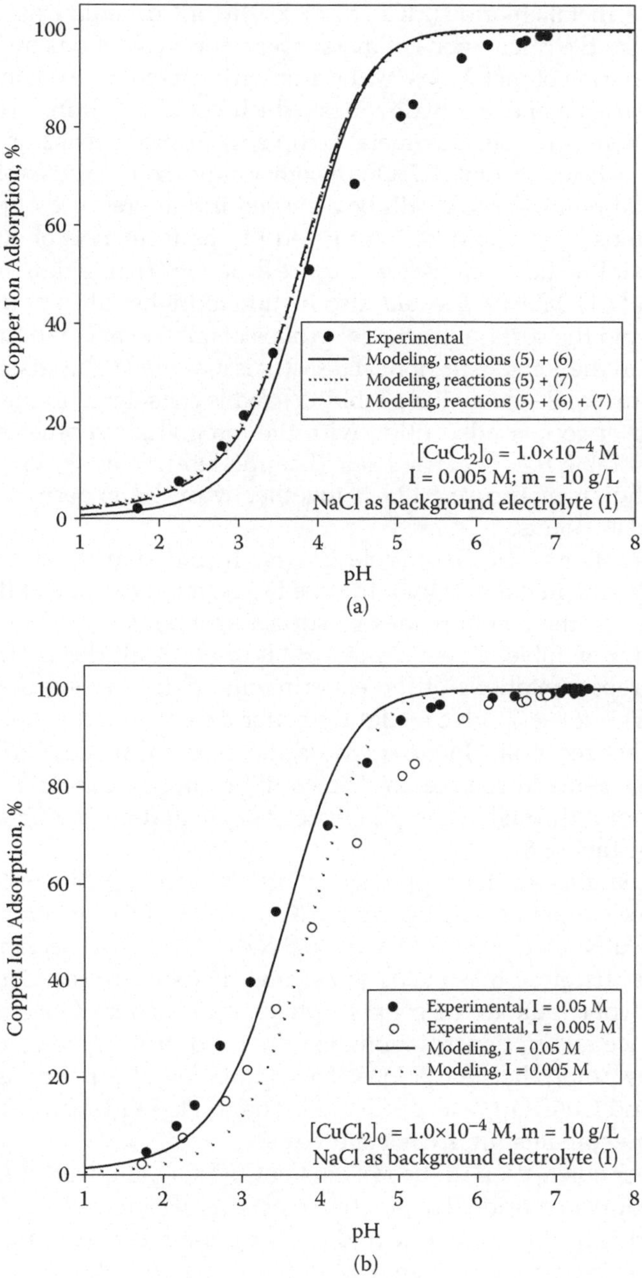

FIGURE 8.12
Modeling of effects of pH and ionic strength on metal sorption. (a) pH; (b) pH and ionic strength.

For example, metal chloride (MCl^+, M = Cu, Zn, or Co) could exist in pHs 1 to 5 according to the calculations of metal speciation in solutions by MINEQL.[39] The complexation of metal ions by these anions can reduce the concentrations of free metal ions and the hydroxides, which could result in a reduction in the production of the surface–metal complexes formed according to the surface reactions listed in Table 8.7. On the other hand, some studies showed that greater metal removal could still be achieved in the presence of complexing anions/ligands. This could be interpreted by the formation of surface complexes containing these anions (e.g., Surface Reaction (7) in Table 8.7) and others (SO^-M^{2+} and SO^-MOH^+). It could also be due to higher affinity between the metal ions and the surface functional groups of adsorbents than that between the metal and the complexing ions. The MCl^+ is assumed to be adsorbed as SO^-MCl^+ (see Surface Reaction (7) of Table 8.7) and is considered in the modeling.

Modeling of copper adsorption with the foregoing hypothesis (reactions (1) to (7) of Table 8.7) is carried out. The modeling results represented by lines are shown in Figure 8.12a–b together with the experimental results represented as points.

Surface reactions (1) to (4) of Table 8.7 were included in all the calculations. Additionally, the metal was hypothesized to be removed due to three mechanisms, that is, the combinations of surface reactions (5) + (7), (5) + (6), or (5) + (6) + (7) of Table 8.7. As shown in Figure 8.13, all these combinations give good representations of the experimental data. Among them, surface reactions (5) + (6) + (7) give a slightly better description for the adsorption than the other reactions. In other words, metal removal is due to the formation of surface–metal complexes. The equilibrium constants for these three reactions found through a computer program, including MINEQL and SQP, are listed in Table 8.8.

The calculations on the copper speciation by MINEQL show that copper exists as free copper and its hydroxide (Cu^{2+} and $CuOH^+$) when the solution contained $Cu(NO_3)_2$ or $Cu(ClO_4)_2$ with $NaNO_3$ or $NaClO_4$ as background electrolyte. It is therefore reasonable to assume that surface reactions (1) to (6) of Table 8.7 contribute to the copper uptake. As a matter of fact, the validity of the surface adsorption reactions for the adsorption process can be better tested by doing the simulations. The simulation of copper adsorption in $Cu(NO_3)_2$ and $Cu(ClO_4)_2$ was conducted. The modeling results nicely match the experimental data (figure not shown).

The copper adsorption in the systems of $Cu(NO_3)_2$ and $Cu(ClO_4)_2$ is higher than with copper chloride. The solution reactions of copper chloride ($CuCl_n^{2-n}$, n = 1, 2, and 3) could compete with the adsorption reactions for copper ions, essentially reducing copper removal. Additionally, we should note that the equilibrium constants for nitrate or perchlorate adsorption reactions are $10^{3.79}$ M^{-2} and $10^{4.00}$ M^{-2} for NO_3^- and ClO_4^-, respectively, which are larger than for the chloride ($10^{2.94}$ M^{-2}) (shown in Table 8.8). This results in more chloride ions being available in the solution phase, which could increase the solution complexation of chloride with more copper ions.

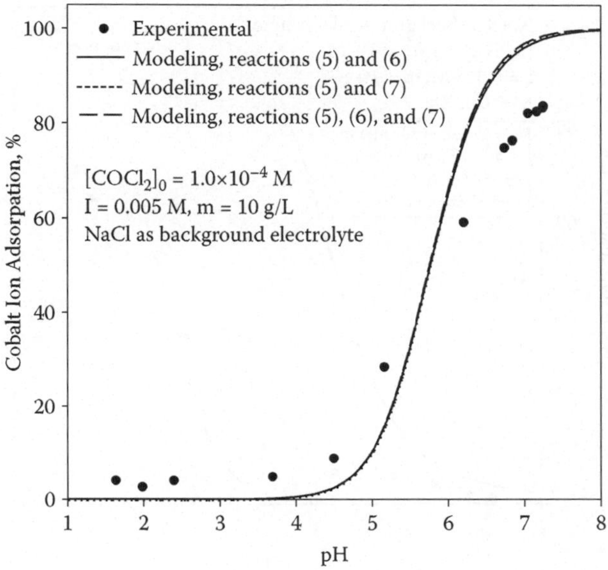

FIGURE 8.13
Modeling of effects of pH on zinc and cobalt sorption.

The 2-pK-TL-SCFM was also used to elucidate the initial copper concentration effect on the removal. The model successfully represents the experimental observations (figure not shown) with *a single set of model parameters*. The equilibrium constants are independent of pH, ionic strength, and other factors. Some studies in the literature have reported good modeling results; however, the model parameters are usually specific (e.g., dependent on ionic strength or concentration).

Several adsorption reactions given in Table 8.7 were used for modeling of cobalt adsorption. As shown in Figure 8.13, combinations of reactions listed in the figure seem to have no different in the representation of experimental data.

8.5.1.3 Multiple-Species Metal Ion Adsorption

Multiple-metal-ion adsorption can be predicted based on the reactions and their constants obtained from the single-metal-ion adsorption modeling discussed previously for the cases of copper, zinc, and cobalt. As illustrated in Figures 8.14 and 8.15, the model describes well the competition among the various metal ions, in agreement with the observed experimental data. It has been pointed out that all the prediction is based on *one single set of model parameters*. The model can predict both bi- and triple-species metal adsorption systems shown in Figures 8.14 and 8.15, respectively. One can see from here that a complicated multiple-metal-ion adsorption process can be elucidated on the basis of a combination of a few simple single-metal-ion

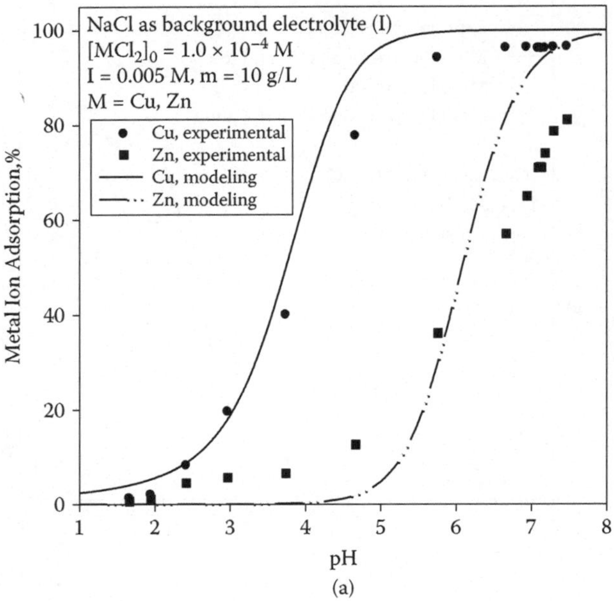

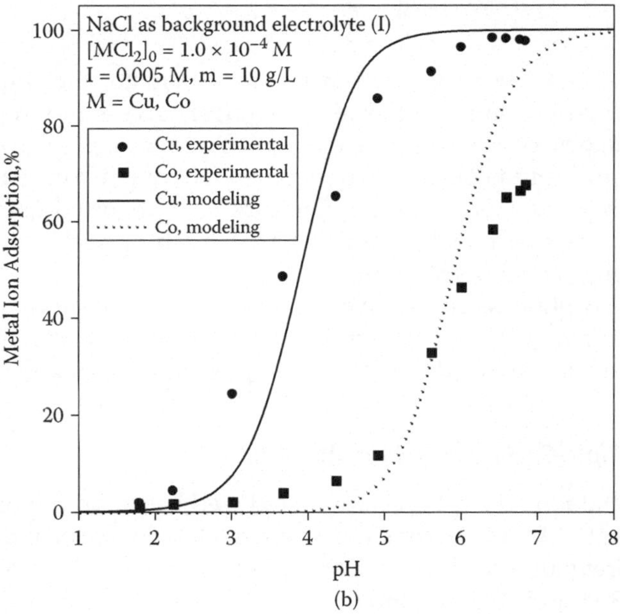

FIGURE 8.14
Prediction of adsorption equilibrium of multiple-metal-ions.

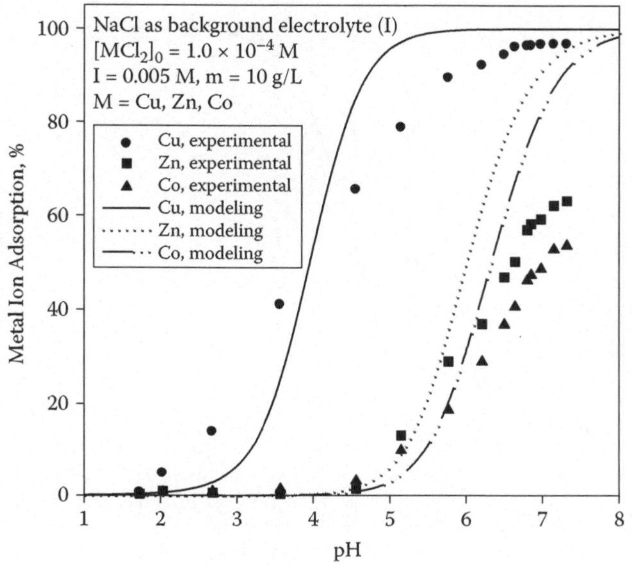

FIGURE 8.15
Prediction of adsorption equilibrium of multiple-metal-ions.

adsorption reactions. This is the major advantage of the 2-pK-TL-SCFM over other empirical models (e.g., Langmuir and Freundlich equations).

Based on surface reactions (5) to (7) in Table 8.6, the sensitivity analysis of the constants is carried out with the results depicted in Figure 8.16. With a decrease in K_{Cu}, the adsorption curve shifts from left to right, indicating less metal ion removal at low pH. Interestingly, all the adsorption curves with various K_{Cu} values are parallel to each other (figures not shown). When K_{Cu} is decreased below 10^{-3}, no further difference in the removal could be observed (not shown).

The slope of the adsorption curve increases as K_{CuOH} is increased. No further change is observed when K_{CuOH} is below the limit of $10^{-8.0}$ M (not shown). Changing K_{CuCl} can alter the removal in the lower pH area. As K_{CuCl} is increased from $10^{-2.4}$ to $10^{-2.0}$ M^{-1}, the removal increases from 2% to 15% at pH 1.0 as illustrated in Figure 8.16. Meanwhile, the slope decreases when K_{CuCl} is increased.

With the results from the foregoing sensitivity analysis of various parameters and experimental observations, it would be much easier to identify key adsorption reactions. Suitable adsorption reactions, therefore, can be selected and used in modeling studies.

8.5.2 Heavy Metal Sorption onto a Calcium-Alginate Encapsulated Magnetic Sorbent

A calcium-alginate encapsulated magnetic sorbent was a compositive developed by Lim and co-workers[25,45–47] for the removal of heavy metal ions from

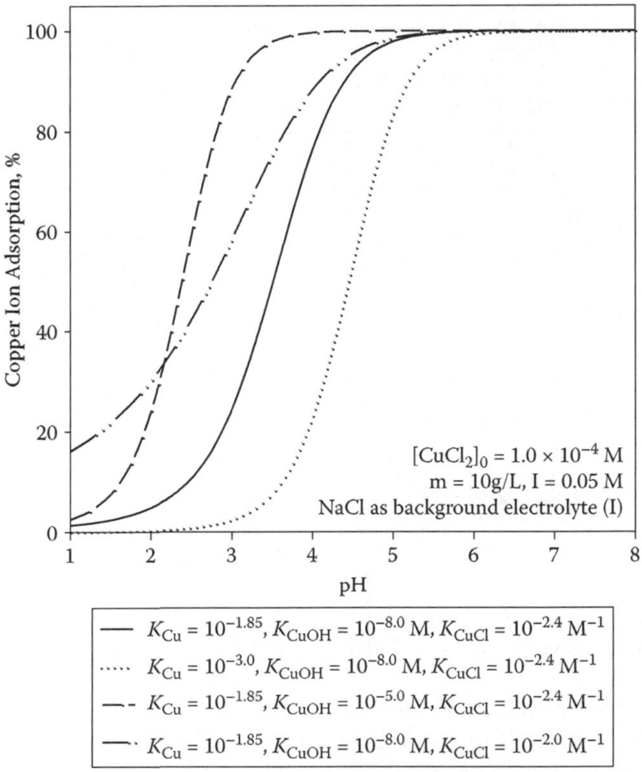

FIGURE 8.16
Sensitivity of model parameter on simulation of metal adsorption.

aqueous solutions. It was demonstrated that it could effectively remove copper and arsenic from various waste solutions. The magnetic sorbent was composed of polysaccharide-like polymer (alginate) and iron oxide (magnetite), and calcium ions.

The SCFMs and ion exchange models have their own application ranges. They are obviously not suitable for modeling of heavy metals onto the sorbent. For example, Schiewer and Volesky showed that an ion exchange model gave a nice description of metal biosorption onto a protonated seaweed.[48] However, it is obviously not suitable as the magnetic sorbent is not protonated and also contained calcium and magnetite. The SCFM clearly cannot be used, because the sorbent has an ion exchange property. Therefore, neither SCFM nor IX model can be used to predict the sorption process.

It was observed that a larger amount of calcium ions was released during the heavy metal ion removal.[25] In addition, metal surface complexes may be formed due to the presence of magnetite. As such, it was hypothesized that the metal uptake was due to ion exchange and metal complex formation/coordination.

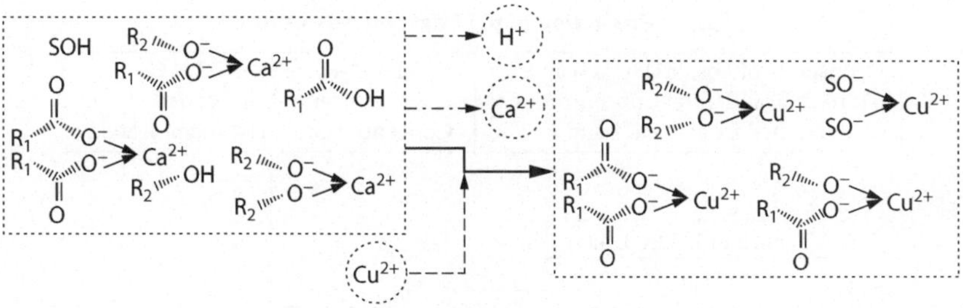

FIGURE 8.17
Schematic diagram of copper uptake by magnetic sorbent.

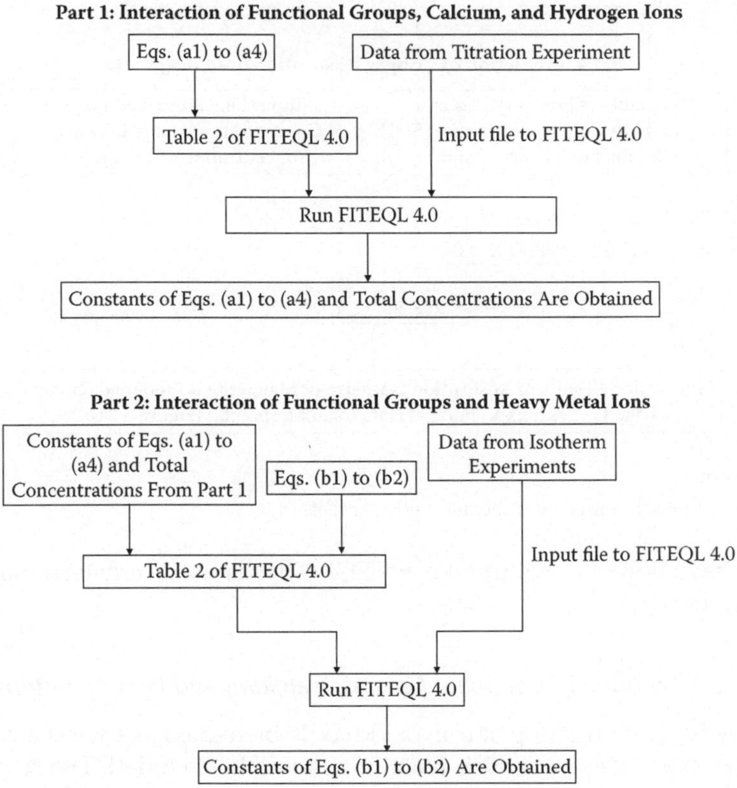

FIGURE 8.18
Searching reaction constants and total concentrations.

In the modeling study, it was assumed to have one generalized functional group of R^{2-} due to the presence of alginate and one generalized functional group of SOH due to the presence of iron oxide. Figure 8.17 provides a general schematic diagram of metal sorption onto the sorbent. Figures 8.18 and 8.19 show the procedures in the searching of model parameters as well

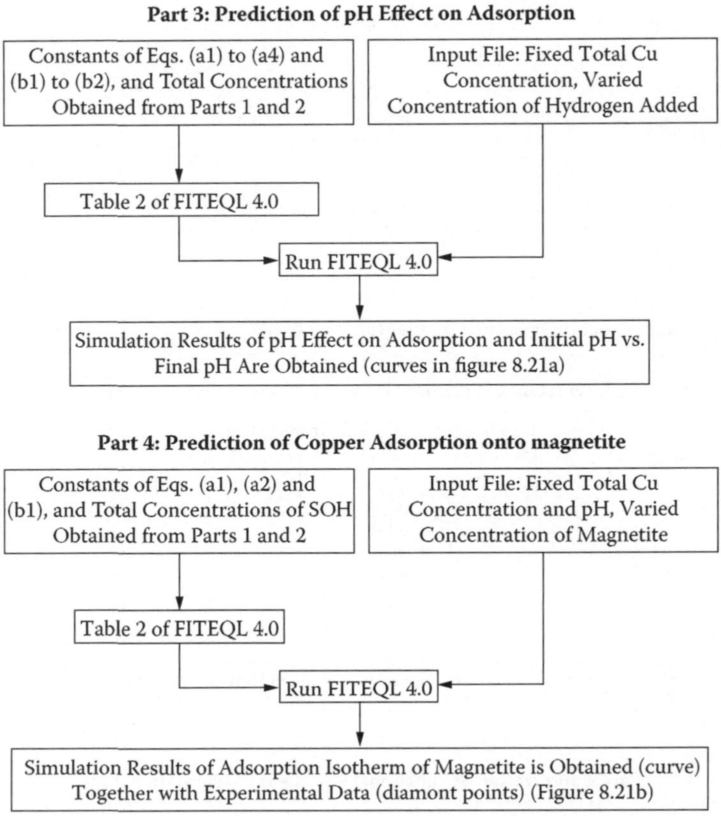

FIGURE 8.19
Prediction of metal uptake onto the magnetic sorbent.

as the prediction of metal uptake. FITEQL 4.0 was used in determination of model parameter.[7]

8.5.2.1 Interaction of Functional Groups, Calcium, and Hydrogen Ions

Since the functional group of iron oxide could be treated as a weak acid,[1,9] the protonation and deprotonation reactions should be included. The respective reactions included in the modeling are shown in Equations (a1) and (a2) in Table 8.10. As the alginate was used for the encapsulation of magnetic iron oxide, the calcium (for ion exchange) and alginate (for complexion) must be included in the modeling. As such, Equations (a3) and (a4) listed in Table 8.10 are included in the calculation.

The components (FITEQL) involved in the interaction of magnetic sorbent with hydrogen ions and heavy metal ions are calcium ions, alginate (represented by R^{2-}), iron oxide (represented by SOH), and copper ions. The interaction of calcium alginate and hydrogen ions (e.g., acid-titration) is essentially controlled by ion exchange and basic coordination reactions. A reorganization of Equations (a3) and (a4) of Table 8.9 would lead to an ion exchange reaction between calcium and hydrogen ions as follows:

$$2H^{+} + \overline{R^{2-}} = \overline{H_2R}$$

$$Ca^{2-} - \overline{R^{2-}} = \overline{CaR}$$

$$2H^{+} + \overline{CaR} = Ca^{2+} + \overline{H_2R}$$

It has to be pointed out that the selection of R, H, and Ca as the components are necessary in order that the equilibrium constants can be optimized and obtained. If one takes CaR, H, and Ca as components for the modeling, it is mathematically unacceptable due to the definition of component and species in FITEQL. Note that CaR (if taken as a component) would include Ca, which is against the rule of MINEQL and FITEQL.

The equilibrium reaction constants of Equations (a1) to (a4) and the total content of functional groups (R and SOH) can be determined by FITEQL 4.0 (7). The modeling of titration is conducted with the data from acid-base titration experiments as an input file, in order that the equilibrium constants of reactions (a1) to (a4) and total concentrations of components (R and SOH in Table 8.9) can be obtained. As shown in Figure 8.20, the titration curve at an ionic strength of 0.1 M is virtually the same as that at an ionic strength of 0.001 M. Therefore, the equilibrium reaction constants and the contents of Ca, R, and SOH for both cases are treated as identical. The simulation process can be found in Part 1 of Figure 8.18. All the model parameters determined by the modeling simulation are given in Table 8.9.

As shown in Figure 8.20, the model (Reactions (a1) to (a4) of Table 8.10) provides an excellent representation of the titration data. The equilibrium constants for Equations (a1) and (a2) are quite close to those of iron oxides reported by Westall and Hohl,[27] which clearly indicates no significant change in the physicochemical properties of metal oxide during the fabrication of the magnetic sorbent. The constants for Equations (a3) and (a4) are slightly higher than those previously reported[24], which may result from the difference in the preparation of sorbent. The calcium alginate sorbent was prepared by the direct formation in the calcium chlorine solution,[24] while the

TABLE 8.9

Model Parameters of Metal Surface Reactions

	Heavy metal ions (M^{2+})		
Parameter	**Cu**	**Zn**	**Co**
K_M	$10^{-1.85}$	$10^{-4.16}$	$10^{-4.31}$
K_{MOH}	$10^{-8.00}$ M	$10^{-16.38}$ M	$10^{-14.82}$ M
K_{MCl}	$10^{-2.40}$ M^{-1}	—	—

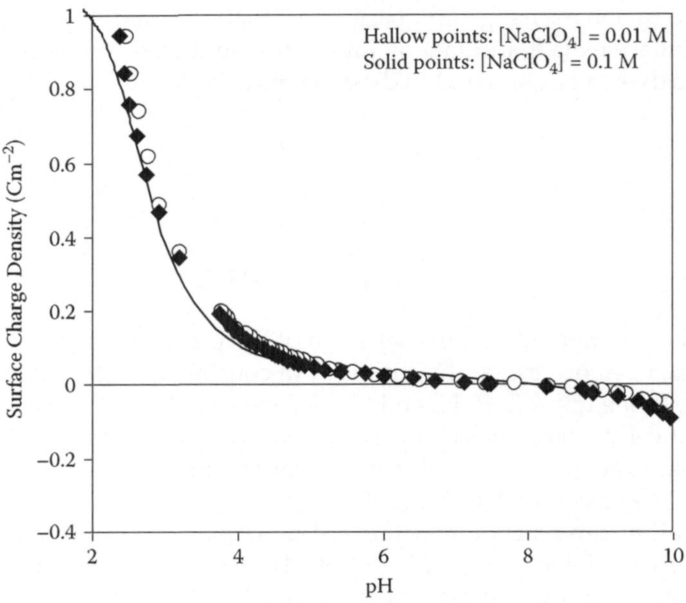

FIGURE 8.20
Surface charge as a function of pH (1 $g{\cdot}L^{-1}$) under different ionic strength. Points represent experimental data, while line represents modeling results.[25]

TABLE 8.10
List of Reactions in Modeling of Metal Uptake[25]

Interaction between functional groups and hydrogen ions		
Reactions	**Equilibrium constant**	
$SOH = SO^- + H^+$	$10^{-8.988}$ M	(a1)
$SOH + H^+ = SOH_2^+$	$10^{6.289}$ M^{-1}	(a2)
$R^{2-} + 2H^+ = H_2R$	$10^{9.391}$ M^{-2}	(a3)
$R^{2-} + Ca^{2+} = CaR$	$10^{6.065}$ M^{-1}	(a4)

Interaction between functional groups and heavy metal ion				
	[$NaClO_4$]			
Reactions	**0**	**0.005 M**	**0.05 M**	
$SOH + Cu^{2+} = SOCu^+ + H^+$	$10^{0.045}$	$10^{0.711}$	$10^{0.022}$	(b1)
$R^{2-} + Cu^{2+} = CuR$	$10^{7.346}$ M^{-1}	$10^{6.663}$ M^{-1}	$10^{6.198}$ M^{-1}	(b2)

Total content of functional groups and calcium ion		
$T_{SOH} = 1.277 \times 10^{-4}$ $mol{\cdot}g^{-1}$	$T_R = 9.184 \times 10^{-4}$ $mol{\cdot}g^{-1}$	$T_{Ca} = 8.352 \times 10^{-4}$ $mol{\cdot}g^{-1}$

magnetic sorbent was formed by the direct formation in the calcium chlorine solution, followed by freeze-drying (e.g., solidification in the formation of sorbent). The total SOH content of 1.277×10^{-4} mol·g^{-1} and the total R content of 9.184×10^{-4} mol·g^{-1} indicate the limitation of the maximum metal uptake capacity, which is no more than the sum of both, namely, 1.046×10^{-3} mol·g^{-1}.

8.5.2.2 Interaction of Functional Groups and Heavy Metal Ions

When heavy metal ions are present in the solution, it is assumed that additional reactions occur between the metal ions and the functional groups/Ca^{2+} according to Equations (b1) and (b2) of Table 8.10 and Figure 8.17. It is hypothesized that an adsorption reaction occurs between the heavy metal ions and the functional groups of magnetite (iron oxide) (Equation (b1)) and the alginate (Equation (b2)), respectively. In addition, the heavy metal would be exchanged with the calcium ions originally existing in the magnetic sorbent, which reaction can be easily described by reorganizing Equations (a4) and (b2) as follows:

$$Cu^{2+} + \overline{R^{2}} = \overline{CuR}$$

$$Ca^{2+} + \overline{R^{2-}} = \overline{CaR}$$

$$\overline{\qquad\qquad\qquad\qquad\qquad}$$

$$Cu^{2} + \overline{CaR} = Ca^{2+} + \overline{CuR}$$

Based on all the equations in Table 8.10 together with the data from the adsorption isothermal experiment, the equilibrium constants of Equations (b1) and (b2) can be obtained by using FITEQL 4.0. The procedure in simulation can be found in Part 2 of Figure 8.18.

Figure 8.21b shows that the model represents well the adsorption isothermal data. Table 8.10 shows that K_{CaR} is lower than K_{CuR}, which suggests that copper has a higher affinity for the functional group (R) than calcium. The difference in the affinity serves as the "driving force" for the heavy metal sorption onto the sorbent, which is further supported by the findings in the literature. For example, Haug reported that the affinity of alginic acid to metal ions followed the order: $Pb^{2+} > Cu^{2+} > Cd^{2+} > Ba^{2+} > Sr^{2+} > Ca^{2+} > Co^{2+} > Ni^{2+} > Mn^{2+} > Mg^{2+}$.[49] In addition, Table 8.10 shows that K_{CuR} is higher at lower ionic strength. The negative effect of ionic strength on the adsorption likely due to the competition caused by the sodium ions (ionic strength).

8.5.2.3 Prediction of pH Effect

The parameters discussed earlier are obtained by the optimization technique in FITEQL 4.0. From a mathematical standpoint, it would be reasonable to have more than one set of parameters that can represent the experimental data (e.g., titration and adsorption isotherm) well. Modeling simulation may inaccurately predict metal uptake processes under different operational conditions.

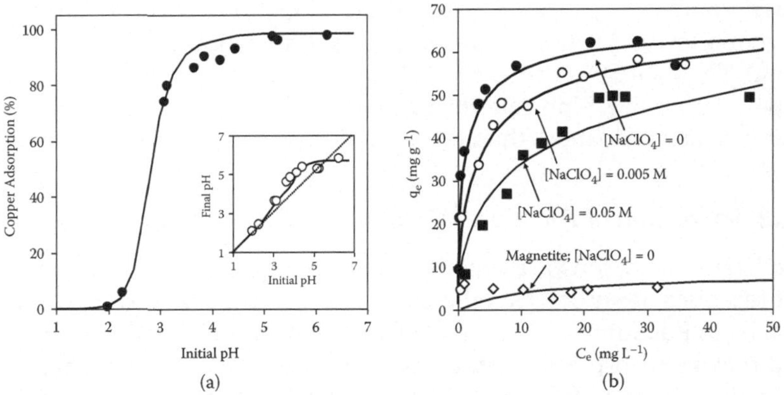

FIGURE 8.21
Copper removal of magnetic sorbent: (a) pH effect ($[Cu]_0 = 1 \times 10^{-4}$ M, $[NaClO_4] = 0.005$ M, m = 0.5 g L^{-1}); (b) sorption isotherm at different ionic strengths ($m = 0.5$ g·L^{-1}, pH = 5). Points represent experimental data, while lines represent modeling results.[25]

In order to test the validity of the model, the pH effect the metal removal is simulated by using the parameters in Table 8.10. The experimental conditions, such as initial metal concentration and sorbent dosage, are used as an input to the model. The details of the simulation can be found in Part 3 of Figure 8.19.

As shown in Figure 8.21, the model successfully describes the pH effect on copper uptake by the magnetic sorbent. More importantly, the model can nicely simulate the final pH in the adsorption as a function of initial pH. The prediction of final pH based on initial pH, dosage, and initial metal concentration, to the best knowledge of the authors, seldom appears in the literature. The description of the adsorption isotherm of magnetite is also conducted by Equations (a1), (a2), and (b1) of Table 8.10. The modeling result matches the experimental data well as illustrated in Figure 8.21b.

8.6 Modeling of Adsorption Kinetics

The diffusion model can be employed to simulate the metal uptake kinetics for sorbents and ion exchange resins.[5,20,50,51] This model works with an assumption of the "two-step mass transport mechanism": the adsorbate first transfers through the external liquid film from the bulk solution and subsequently diffuses inside the sorbent, before finally being adsorbed by functional groups. Depending on the physical properties of the sorbent, there are two major assumptions: surface diffusion and pore diffusion.

Empirical models such as pseudo 1st- and 2nd-order equations have been widely used in evaluation of kinetics data. Appendix B provides general description of these models together with some parameters. Note that these models may provide nice fitting of experimental data. However they fail to provide fundamental understanding of adsorption processes.

8.6.1 Surface Diffusion Control Model

The mathematical equations and corresponding initial and boundary conditions are expressed as follows:[20] The model can be derived according to Appendix C.

$$D_S \frac{\partial}{\partial r}\left(r^2 \frac{\partial q}{\partial r}\right) = r^2 \frac{\partial q}{\partial t}, \quad 0 \le r \le a_p, t \ge 0 \tag{8.49}$$

$$q = 0 \quad 0 \le r \le a_p, t < 0 \tag{8.50}$$

$$\frac{\partial q}{\partial r} = 0 \quad r = 0 \tag{8.51}$$

$$D_S \rho_p \frac{\partial q}{\partial r} = k_f (C - C_s) \tag{8.52}$$

where C and q are the concentration of the sorbate in bulk and in solid phases, respectively; C_s is the aqueous phase concentration at the particle surface, in equilibrium with the corresponding concentration in the solid phase q^*; D_S is the surface diffusivity within the particle; ρ_p is the particle density; r is the radius distance measured from the center of particle; a_p is the particle radius; k_f is the external mass transfer coefficient; and t is the time.

The *modeling* can be conducted with the Langmuir or Frendlich adsorption isotherm constants (q_{max} and b or k_f and 1/n), and physical parameters such as particle radius and density. Two key parameters of surface diffusivity and external mass transfer coefficient must be available for the modeling, which can be obtained from the literature or from a theoretical estimation.

The diffusivity D_s and the external mass transfer coefficient k_f are important in the removal rate. It is not recommended to obtain the two parameters simultaneously by fitting the model results with the kinetic experimental data. The external mass transfer coefficient may first be estimated by the following equation:[50]

$$\frac{k_f d}{D} = 2 + 0.44\left(\frac{\rho d U_L}{\mu}\right)^{0.5}\left(\frac{\mu}{\rho D}\right)^{0.38} \tag{8.53}$$

where D is the ion diffusivity in the bulk solution (m^2/s), U_L is the superficial liquid velocity, d is the particle diameter (m), ρ is the liquid density (kg/m^3), and μ is the liquid viscosity (kg/m×s).

The typical value of bivalence metal ion diffusivity in the solutions is about 1×10^{-9} m^2/s. U_L can be calculated directly from the stirring speed. The influence of the density difference between the liquid and the solids on the U_L may be ignored.[51]

Typically, the D_s and k_f values are 10^{-10} to 10^{-11} m^2/s and 10^{-5} to 10^{-3} m/s, respectively. The value of D_s depends on the properties of the sorbent. The k_f value not only depends on the properties of the sorbent, but also affects the operational conditions. A higher mixing strength can cause an increase in the external mass transfer, leading to a higher value of k_f. On the other hand, the k_f value in the slow mixed/stirred solution is low.

The modeling of adsorption kinetics can be illustrated by the work done by Chen et al.[51] In the study, a biosorbent was used for lead removal. The normal stirring speed in the experiments of metal biosorption is assumed to be 180 rpm. It is assumed that the liquid in the cylinder tank moves at an angular speed of 180 rpm around the symmetric axle of the tank. The averaged thread speed along the radius of the tank can therefore be regarded as U_L. The U_L of 0.565 m/s can be estimated. Therefore, the calculated value of k_f from the foregoing equation is 2.41×10^{-4} m/s.

D_s can be found by fitting the modeling results with the experimental data once the k_f value is found. Alternatively, the D_s value can be estimated by a method in the literature.[50] The estimated D_s value ($D_{s,\,theory}$) is related to the bulk solutions (D) and the particle porosity ($\in$) as follows:

$$D_{s,theory} = D\left[\frac{\in}{2-\in}\right]^2 \tag{8.54}$$

The typical value of D is 1×10^{-9} m^2/s. $\in$ can be estimated by replacing the actual fractional void volume with the weight fraction of the imbibed solvent. For example, the value of $\in$ measured is 0.48. Thus, the $D_{s,theory}$ calculated is 9.97×10^{-11} m^2/s. A comparison of the $D_{s,theory}$ with the D_s available in the literature shows that they are quite close to each other.[51,52]

As shown in Figure 7.57 of Chapter 7 and Figures 8.22 and 8.23, the surface diffusion control model describes well the adsorption kinetics of the calcium alginate bead (biosorbent). Figure 8.22 demonstrates that the external mass transfer coefficient increases by nearly 10 times when the solution stirring is changed from slow stirring to normal stirring. The diffusivity, however, still remains unchanged.

The existence of insert inorganic ions may affect the adsorption kinetics. As shown in Figure 8.23, the D_s for this case increases from 2.95×10^{-11} to 3.0×10^{-10} m^2/s because of the presence of the sodium ion. The diffusion inside the particles is enhanced. The k_f value in the presence of the sodium ion is nearly changed.

8.6.2 Pore Diffusion Control Model

The control equation and its corresponding initial and boundary conditions are as follows:

$$\left[\varepsilon_p + \rho_p f'(c)\right]\frac{\partial c}{\partial t} = \frac{D_p}{r^2}\frac{\partial}{\partial}\left(r^2\frac{\partial c}{\partial r}\right) \tag{8.55}$$

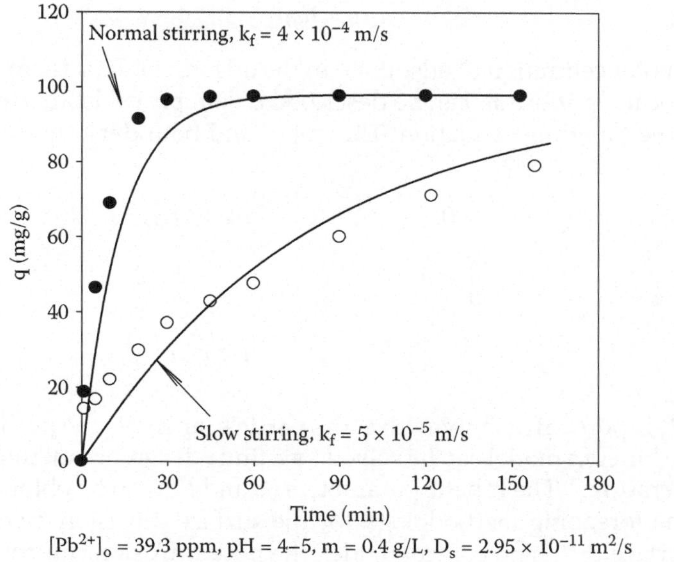

FIGURE 8.22
Effect of stirring speed on adsorption kinetics. Calcium alginate bead was used.

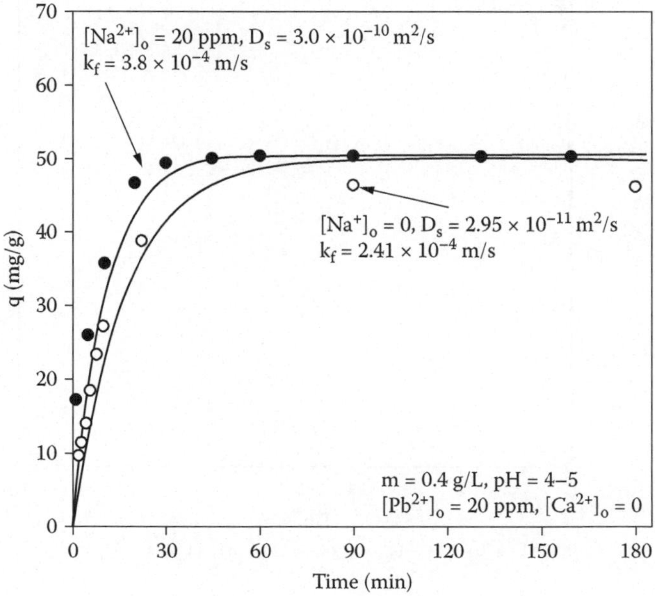

FIGURE 8.23
Effect of sodium concentration on adsorption kinetics. Calcium alginate bead was used.

$$q = f(c) \tag{8.56}$$

where c is concentration of adsorbate in the adsorbent. It is assumed that the instant local equilibrium can be described by em pirical isotherm equations such as the Langmuir equation. The initial and boundary conditions are as follows:

$$c = 0, \qquad 0 \le r \le r_p, \ t=0 \tag{8.57}$$

$$\frac{\partial c}{\partial r} = 0 \qquad r = 0, \ t > 0 \tag{8.58}$$

$$\frac{\partial c}{\partial r} = \frac{k_f}{D_p}(C - C_s) \qquad r = r_p, \ t > 0 \tag{8.59}$$

where D_p is pore diffusivity within the particle, and ε_p is the porosity of particle. The kinetics model can be solved by a finite difference method available in the literature.[20] The kinetic parameters (k_f and D_p) can be obtained according to the foregoing methodology for the surface diffusion model. Similar to the surface diffusion control model, the pore diffusion control model can describe well the adsorption of heavy metal ion onto porous sorbents such as activated carbon illustrated in Figure 8.24.[52] The k_f values for r_p of 0.034 and 0.018 cm are 1×10^{-4} and 6×10^{-4} m/s, respectively.

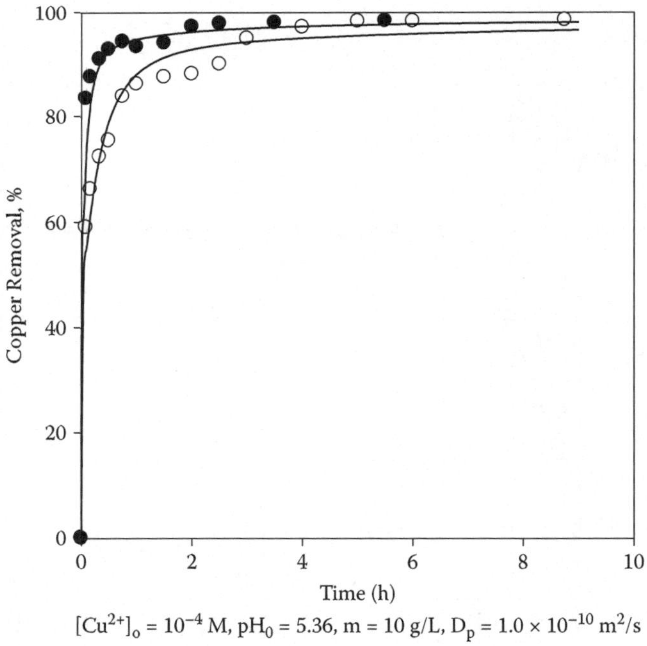

FIGURE 8.24
Kinetics of copper adsorption onto an activated carbon.[52] pH not controlled, hollow points (r_p = 0.034 cm), solid points (r_p = 0.018 cm), curves representing modeling results.

Appendix A: Introduction of MINEQL Modeling

As defined in Section 8.4.1, concentrations (activities) of all the chemicals involved in any system (liquid, gas and solid) can be determined by a set of components and species. The relationship between components and species given in Figure A-1 can be defined as follows assuming that there are a total of *M* components selected.

$$[Species_i] = function\ of\ [Component_1], \ldots, [Component_M] \quad \text{(a-1)}$$

For example, water has the following reaction:

$$H_2O = H^+ + OH^- \qquad K_w = 10^{-14} \quad \text{(a-2)}$$

If we select H^+ as a component, the concentration of OH^- can then be calculated as:

$$H_2O - H^+ = OH^- \quad \text{(a-3)}$$

$$[OH^-] = K_w[H^+]^{-1} \quad \text{(a-4)}$$

The calculation of concentrations of components and species is based on three principles:

1. Mass balance
2. Charge balance (Electroneutrality)
3. Chemical reactions

The mass balance is referred to the fact that the total concentration of an element is contributed by the concentrations of component and species that

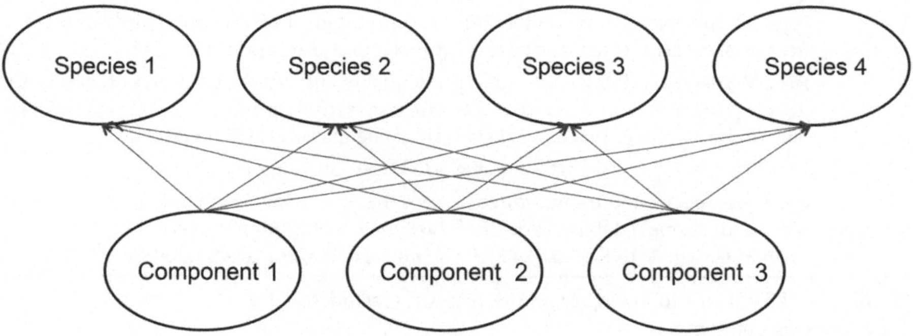

FIGURE A-1
Illustration of formation of species from components. Note that any species is formed by more than one components.

have the element. For example, copper can be selected as a component in the system that has copper chloride ($CuCl_2$). Assume that we have the reactions below:

$$Cu^{2+} + H_2O = Cu(OH)^+ + H^+ \quad \text{(a-5a)}$$

$$Cu^{2+} + H_2O - H^+ = Cu(OH)^+ \quad \text{(a-5b)}$$

$$Cu^{2+} + Cl^- = CuCl^+ \quad \text{(a-6)}$$

$$Cu^{2+} + 2H_2O = Cu(OH)_2(s) + 2H^+ \quad \text{(a-7a)}$$

$$Cu^{2+} + 2H_2O - 2H^+ = Cu(OH)_2(s) \quad \text{(a-7b)}$$

TABLE A-1

Definitions of Species in MINEQL[6]

Type	Definitions
I	**Type I Species** is essentially the same as Component, which is selected from a component list (database, e.g., given in Table A-2) to represent reactants in all chemical reactions that lead to the formation of species.
II	**Type II Species** is complex, which is due to the chemical reactions of components. Solution reactions, adsorption reactions and ion exchange reactions of components lead to the formation of Type II Species. The solution reactions are provided in commercial versions of MINEQL (e.g., MINEQL 4.5/4.6), while adsorption and ion exchange reactions must be creased by users.
III	**Type III Species** is a fixed species or fixed entities. Any Type III Species has a fixed activity, which may be fixed pH, solid phase present in infinite supply, and gas present at a fixed partial pressure (e.g. CO_2 in open system).
VI	**Type VI Species** is a mineral that initially present in a finite quantity and is subject to complete dissolution during the approach to equilibrium.
V	**Type V Species** is defined as a solid phase that is initially not present, but may form during reaction (equilibrium). If the actual ion activity product (Q in Chapter 3) is above the solubility product constant (K_{sp}), the minerals will precipitate.
VI	**Type VI Species** represents those excluded from calculation.
VII	**Type VII Species** represents added species that cannot be found in the thermodynamic database. Note that this type is not used in the recent versions of MINEQL (e.g., MINEQL 4.5/4.6); it can be created and placed together with Type I.

In MINEQL 4.5/4.6, the following types of species are included:
Type I Species: components
Type II Species: aqueous species with inclusive of adsorbed species
Type III Species: fixed entities
Type V Species: dissolved solids
Type VI Species: species not included

The total concentration of copper $[Cu]_T$ is contributed by Component of $[Cu^{2+}]$, and species of $[CuOH^+]$, $[CuCl^+]$, and $[Cu(OH)_2]$. As such, $[Cu]_T$ can be determined by:

$$[Cu]_T = [Cu^{2+}] + [CuOH^+] + [CuCl^+] + [Cu(OH)_2] \quad \text{(a-8)}$$

The charge balance or electroneutrality indicates that all the negatively charged components and species must be in balance with all the positively charged components and species. If the mass balance for hydrogen ion is involved, the charge balance becomes unnecessary.

The chemical reactions are referred to all the reactions with involvement of components. For example, the chemical reactions in copper chloride system are given in Equations (a-2) and (a-5) to (a-7).

Seven types of species may be used in the operation of MINEQL.[(5-10)] The definitions are given in Table A-1.[(6)] Note that the later version of MINEQL (e.g. MINEQL+) is slightly different to the traditional versions.[39] For example, Type 7 Species is no longer used.

MINEQL can be used for calculation of solution situation (single point) by providing total concentrations of components, by which the concentrations of species can be obtained. For example, pH and concentrations of copper

TABLE A-2

List of Components Typically used in MINEQL[6]

ID Number	Name	ID Number	Name	ID Number	Name	ID Number	Name
001	e^-	150	Ca^{2+}	460	Mg^{2+}	800	Sr^{2+}
002	H_2O	160	Cd^{2+}	470	Mn^{2+}	870	Tl^+
020	Ag^+	180	CL^-	471	Mn^{3+}	871	$Tl(OH)_3$
030	Al^+	210	Cr^{2+}	490	NH_4^+	890	U^{3+}
060	H_3AsO_3	211	$Cr(OH)_2^+$	491	NO_2^-	891	U^{4+}
061	H_3AsO_4	212	CrO_4^{2-}	492	NO_3^-	892	UO_2^+
090	H_3BO_3	220	Cs^+	500	Na^+	893	UO_2^{2+}
100	Ba^{2+}	230	Cu^+	540	Ni^{2+}	900	V^{2+}
130	Br^-	231	CU^{2+}	580	PO_4^{3-}	901	V^{3+}
140	CO_3^{2-}	270	F^-	600	Pb^{2+}	902	VO^{2+}
141	Fulvate	280	Fe^{2+}	680	Rb^+	903	VO_2^+
142	Humate	281	Fe^{3+}	730	HS^-	950	Zn^{2+}
143	Acetate	330	H^+	731	S	990	SOH1
144	Tartrate	360	Hg_2^{2+}	732	SO_4^{2-}	991	SOH2
145	Glycine	361	$Hg(OH)_2$	760	HSe^-	992	XPSIO
146	Salicylate	380	I^-	761	$HSeO_3^-$	993	XPSIB
147	Glutamate	410	K^+	762	SeO_4^{2-}	994	XPSID
148	Phathalate	440	Li^+	770	H_4SiO_4	995	SOHB

species (e.g. Cu^{2+}, $CuCl^+$ and $CuOH^+$) can be obtained if one provides the total concentration of copper chlorine for the modeling by the program.

MINEQL can also be used to obtain the change of species' concentrations as a function of variable such as pH. This can be performed by using the "Titration" mode in the program. For example, effect of pH on metal adsorption can be obtained by changing pH of solution.[39]

Appendix B: Empirical Kinetics Model

B-1 Pseudo First and Second-order Equations

The pseudo first-order equation is a simple kinetic model that describes the process of adsorption. It was first suggested by Lagergren for the adsorption in a solid/liquid system.[3] It has been used since for metal ion adsorption kinetics by many authors. Its linear form is as follows:

$$ln(q_e - q_t) = lnq_e - k_1 t \tag{b-1}$$

where k_1 is the rate constant of pseudo first-order adsorption, q_e is the amount of adsorbed species at equilibrium, and q_t is the amount of adsorbed species at time t. Note that q_e is dependent upon the operational conditions such as pH, initial concentration, temperature, and ionic strength.

The pseudo first-order kinetic equation can be applied only for the rapid initial stage and is not appropriate for the entire process.

The pseudo second-order kinetic equation is expressed as follows.

$$\frac{t}{q_t} = \frac{1}{k_2 q_e^2} + \frac{t}{q_e} \tag{b-2}$$

where k_2 is the rate constant of pseudo second-order adsorption.

The initial rate constant (H) is defined as:

$$H = k_2 q_e^2 \tag{b-3}$$

The pseudo second-order model can be applied for the entire adsorption process. A good agreement with second-order kinetic equation has been observed with metal ion adsorption reported in literature. The rate-limiting step may result from chemical adsorption. The model may also suggest that two active sites participate in uptake of contaminant(s).

Either pseudo first-order model or second-order model reportedly well describes adsorption kinetics. Sometimes, both can provide best-fitting for adsorption data. Tables B-1 and B2 provide some of kinetics parameters of the above equations.[53–80]

TABLE B-1

List of Constants of First-Order Reaction Kinetics

Adsorbent	Adsorbate	Dosage (g/L)	pH	Initial Concentration (mg/L)	T (K)	k_1	References
H_3PO_4-activated rubber wood sawdust	Cu(II)	5	6	10	303	3.36×10^{-2} min^{-1}	[53]
	Cu(II)	5	6	20	303	1.82×10^{-2} min^{-1}	
	Cu(II)	5	6	30	303	1.81×10^{-2} min^{-1}	
	Cu(II)	5	6	40	303	2.62×10^{-2} min^{-1}	
Activated carbon derived from acrylonitrile-divinylbenzene copolymer	Cr(VI)	0.6	2	30	295	3.60×10^{-3} h^{-1}	[54]
	Cr(VI)	0.6	2	30	308	2.74×10^{-3} h^{-1}	
	Cr(VI)	0.6	2	30	318	3.24×10^{-3} h^{-1}	
	Cr(VI)	0.6	2	30	328	6.52×10^{-3} h^{-1}	
	Cr(VI)	0.6	2	30	338	7.35×10^{-3} h^{-1}	
Palygorskite clay	Pb(II)	7.5	6	500	303	0.308 h^{-1}	[55]
Chitosan functionalized with 2[-bis-(pyridylmethyl)aminomethyl]-4-methyl-6-formylphenol	Cu(II)	2	5.5	100	298	1.8×10^{-2} min^{-1}	[56]
	Cd(II)	2	2	100	298	4.2×10^{-2} min^{-1}	
	Ni(II)	2	3	100	298	7.1×10^{-3} min^{-1}	
Activated carbon prepared from coconut shell	Pb(II)	1	4.5	10	303	2.85×10^{-2} min^{-1}	[57]
	Pb(II)	1	4.5	20	303	5.10×10^{-2} min^{-1}	
	Pb(II)	1	4.5	30	303	1.60×10^{-2} min^{-1}	
	Pb(II)	1	4.5	40	303	1.99×10^{-2} min^{-1}	
	Pb(II)	1	4.5	50	303	2.85×10^{-2} min^{-1}	

(Continued)

TABLE B-1 (*Continued*)

List of Constants of First-Order Reaction Kinetics

Adsorbent	Adsorbate	Dosage (g/L)	pH	Initial Concentration (mg/L)	T (K)	k_1	References
Leersia hexandra Swartz biomass	Cr(VI)	5	2	10	298	8.56×10^{-2} min^{-1}	[58]
	Cr(III)	5	5	10	298	4.02×10^{-2} min^{-1}	
Oxidized carbon	Cr(VI)	2.5	4	30	298	0.326 h^{-1}	[59]
	Cr(VI)	2.5	4	40	298	0.320 h^{-1}	
	Cr(VI)	2.5	4	50	298	0.307 h^{-1}	
Untreated carbons	Cr(VI)	2.5	4	30	298	0.215 h^{-1}	
	Cr(VI)	2.5	4	40	298	0.229 h^{-1}	
	Cr(VI)	2.5	4	50	298	0.219 h^{-1}	
Activated alumina	Pb(II)	5	5	10	303	7.16×10^{-2} min^{-1}	[60]
	Pb(II)	5	5	25	303	7.55×10^{-2} min^{-1}	
	Pb(II)	5	5	50	303	7.57×10^{-2} min^{-1}	
	Cd(II)	5	5	10	303	5.57×10^{-2} min^{-1}	
	Cd(II)	5	5	25	303	5.20×10^{-2} min^{-1}	
	Cd(II)	5	5	50	303	5.36×10^{-2} min^{-1}	
Kaolinite	Ni(II)	2	5.7	50	303	2.64×10^{-2} min^{-1}	[61]
Montmorillonite	Ni(II)	2	5.7	50	303	3.08×10^{-2} min^{-1}	[61]
Poly(hydroxy zirconium) modified kaolinite	Ni(II)	2	5.7	50	303	2.42×10^{-2} min^{-1}	[61]
TBA-kaolinite	Ni(II)	2	5.7	50	303	2.69×10^{-2} min^{-1}	[61]
TBA-montmorillonite	Ni(II)	2	5.7	50	303	3.06×10^{-2} min^{-1}	[61]

Fertilizer industry waste material	Cr(VI)	4	2	50	303	4.21×10^{-2} min^{-1}	[62]
	Cr(VI)	4	2	100	303	4.95×10^{-2} min^{-1}	
Montmorillonite	Cd(II)	2	7	50	303	1.9×10^{-2} min^{-1}	[63]
	Co(II)	2	7	50	303	2.7×10^{-2} min^{-1}	
	Cu(II)	2	7	50	303	2.8×10^{-2} min^{-1}	
	Pb(II)	2	7	50	303	2.8×10^{-2} min^{-1}	
	Ni(II)	2	7	50	303	3.1×10^{-2} min^{-1}	
Acid-activated montmorillonite	Cd(II)	2	7	50	303	2.4×10^{-2} min^{-1}	
	Co(II)	2	7	50	303	2.6×10^{-2} min^{-1}	
	Cu(II)	2	7	50	303	3.1×10^{-2} min^{-1}	
	Pb(II)	2	7	50	303	4.0×10^{-2} min^{-1}	
	Ni(II)	2	7	50	303	3.2×10^{-2} min^{-1}	
Untreated alga	Ni(II)	0.7	5	100	298	3.92×10^{-2} min^{-1}	[64]
	Ni(II)	0.7	5	200	298	4.05×10^{-2} min^{-1}	
Acid-treated alga	Ni(II)	0.7	5	100	298	3.39×10^{-2} min^{-1}	[64]
	Ni(II)	0.7	5	200	298	3.62×10^{-2} min^{-1}	
Silica–dithizone	Hg(II)	2	6	1003	298	2.55×10^{-2} min^{-1}	
	Hg(II)	2	6	1003	308	2.94×10^{-2} min^{-1}	[65]
	Hg(II)	2	6	1003	318	2.60×10^{-2} min^{-1}	
	Hg(II)	2	6	1003	323	3.00×10^{-2} min^{-1}	
Aminated mesoporous alumina	Cu(II)	3	5.8	10	298	0.6554 h^{-1}	[66]
	Cu(II)	3	5.8	15	298	0.4624 h^{-1}	
	Cu(II)	3	5.8	20	298	0.2865 h^{-1}	
Protonated mesoporous alumina	Cu(II)	2	5.8	10	298	0.4586 h^{-1}	
	Cu(II)	2	5.8	15	298	0.2429 h^{-1}	
	Cu(II)	2	5.8	20	298	0.2015 h^{-1}	

(Continued)

TABLE B-1 (*Continued*)

List of Constants of First-Order Reaction Kinetics

Adsorbent	Adsorbate	Dosage (g/L)	pH	Initial Concentration (mg/L)	T (K)	k_1	References
Hydrogel Fd1-1	Cu(II)	2	3	300	298	$0.17\ min^{-1}$	[67]
Hydrogel Fd1-10	Cu(II)	2	3	300	298	$3.8 \times 10^{-2}\ min^{-1}$	
Hydrogel Fd2-10	Cu(II)	2	3	300	298	$3.2 \times 10^{-2}\ min^{-1}$	
Carboxylate-functionalized cation exchanger Fd2-10	Cr(III)	2	7	10	303	$9.9 \times 10^{-3}\ min^{-1}$	[68]
	Cr(III)	2	7	25	303	$7.0 \times 10^{-3}\ min^{-1}$	
	Cr(III)	2	7	50	303	$7.8 \times 10^{-3}\ min^{-1}$	
	Cr(III)	2	7	100	303	$9.2 \times 10^{-3}\ min^{-1}$	
	Cr(III)	2	7	50	303	$7.8 \times 10^{-3}\ min^{-1}$	
	Cr(III)	2	7	50	313	$12.6 \times 10^{-3}\ min^{-1}$	
	Cr(III)	2	7	50	323	$13.3 \times 10^{-3}\ min^{-1}$	
	Cr(III)	2	7	50	333	$15.2 \times 10^{-3}\ min^{-1}$	
Carboxylate-functionalized IXR	Ur(VI)	1	3	100	293	$2.6 \times 10^{-2}\ min^{-1}$	[69]
Activated carbon	Ur(VI)	1	3	150	293	$1.88 \times 10^{-2}\ min^{-1}$	
	Ur(VI)	1	3	200	293	$1.75 \times 10^{-2}\ min^{-1}$	[69]
	Ur(VI)	1	3	300	293	$0.99 \times 10^{-2}\ min^{-1}$	
Calcium hydroxyapatite	Zn(II)	10	6	-	303	$1.06 \times 10^{-2}\ min^{-1}$	[70]
	Zn(II)	10	6	-	303	$1.91 \times 10^{-2}\ min^{-1}$	

Wine processing waste sludge	Cr(III)	10	4	40	303	0.98 min^{-1}	[71]
	Cr(III)	10	4	70	303	1.46 min^{-1}	
	Cr(III)	10	4	130	303	1.55 min^{-1}	
	Cr(III)	10	4	160	303	1.80 min^{-1}	
Wine processing waste sludge Chitosan	Cr(III)	10	4	100	293	0.98 min^{-1}	[71]
	Cr(III)	10	4	100	303	1.46 min^{-1}	
	Cr(III)	10	4	100	313	1.55 min^{-1}	
	Cr(III)	10	4	100	323	1.8 min^{-1}	
Activated carbon prepared by Polygonum orientale Linn	Pb(II)	0.6	5	40	298	7.88×10^{-2} min^{-1}	[72]
	Pb(II)	0.6	5	60	298	9.81×10^{-2} min^{-1}	
	Pb(II)	0.6	5	80	298	0.1214	
Hazelnut shell activated carbon	Cu(II)	3	6	25	293	0.105 min^{-1}	[73]
	Cu(II)	3	6	50	293	4.28×10^{-2} min^{-1}	
	Cu(II)	3	6	100	293	3.81×10^{-2} min^{-1}	
	Cu(II)	3	6	200	293	5.98×10^{-2} min^{-1}	
	Cu(II)	0.5	6	50	293	3.46×10^{-2} min^{-1}	
	Cu(II)	1.5	6	50	293	3.37×10^{-2} min^{-1}	
	Cu(II)	2.0	6	50	293	3.91×10^{-2} min^{-1}	
NiO nanoparticle	Cd(II)	0.5	6	100	303	6.74×10^{-2} min^{-1}	[74]
	Cu(II)	0.5	6	100	303	7.46×10^{-2} min^{-1}	
Zinc oxide nanoparticles	Zn(II)	0.5	5.5	100	303	6.51×10^{-2} min^{-1}	[75]
	Cd(II)	0.5	5.5	100	303	7.78×10^{-2} min^{-1}	
	Hg(II)	0.5	5.5	100	303	6.72×10^{-2} min^{-1}	
Zirconia nanoparticle	As(III)	0.2	8-9	37.5	298	0.13 h^{-1}	[76]

TABLE B-2

List of Constants of First-Order Reaction Kinetics

Adsorbent	Adsorbate	Dosage (g/L)	pH	Initial Concentration (mg/L)	T (K)	k_2	H	References
H_3PO_4-activated rubber wood sawdust	Cu(II)	5	6	10	303	7.32×10^{-1} g/mg min^{-1}	2.89 mg/g min^{-1}	[53]
	Cu(II)	5	6	20	303	1.31×10^{-1}g/mg min^{-1}	1.84 mg/g min^{-1}	
	Cu(II)	5	6	30	303	9.37×10^{-2} g/mg min^{-1}	2.05 mg/g min^{-1}	
	Cu(II)	5	6	40	303	8.46×10^{-2} g/mg min^{-1}	2.79 mg/g min^{-1}	
Activated carbon derived from acrylonitrile-divinylbenzene copolymer	Cr(VI)	0.6	2	30	295	1.81×10^{-3} g/mg min^{-1}	4.31 mg/g min^{-1}	
	Cr(VI)	0.6	2	30	308	2.12×10^{-3} g/mg min^{-1}	4.88 mg/g min^{-1}	
	Cr(VI)	0.6	2	30	318	3.87×10^{-3} g/mg min^{-1}	8.79 mg/g min^{-1}	[54]
	Cr(VI)	0.6	2	30	328	4.97×10^{-3} g/mg min^{-1}	11.82 mg/g min^{-1}	
	Cr(VI)	0.6	2	30	338	6.50×10^{-3} g/mg min^{-1}	15.45 mg/g min^{-1}	
Palygorskite clay	Pb(II)	7.5	6	500	303	2.92×10^{-2} g/mg h^{-1}		[55]
Chitosan functionalized with 2[-bis-(pyridylmethyl) aminomethyl] -4-methyl-6-formylphenol	Cu(II)	2	5.5	100	298	1.7×10^{-3} g/mg min^{-1}	4.95 mg/g min^{-1}	[56]
	Cd(II)	2	2	100	298	8.7×10^{-3} g/mg min^{-1}	3.16 mg/g min^{-1}	
	Ni (II)	2	3	100	298	1.3×10^{-3} g/mg min^{-1}	0.24 mg/g min^{-1}	
Leersia hexandra Swartz biomass	Cr(VI)	5	2	10	298	2.77×10^{-1}g/mg min^{-1}	0.61 mg/g min^{-1}	[58]
	Cr(III)	5	5	10	298	2.0×10^{-1} g/mg min^{-1}	1.15 mg/g min^{-1}	
Oxidized carbons	Cr(VI)	2.5	4	20	298	4.17×10^{-1} g/mg h^{-1}	13.62 mg/g h^{-1}	[59]
	Cr(VI)	2.5	4	30	298	9.4×10^{-2} g/mg h^{-1}	5.20 mg/g h^{-1}	
	Cr(VI)	2.5	4	40	298	2.3×10^{-2} g/mg h^{-1}	2.49 mg/g h^{-1}	

Un-oxidized carbons	Cr(VI)	2.5	4	20	298	8.75×10^{-1} g/mg h^{-1}	16.31 mg/g h^{-1}	[59]
	Cr(VI)	2.5	4	30	298	3.31×10^{-1} g/mg h^{-1}	9.88 mg/g h^{-1}	
	Cr(VI)	2.5	4	40	298	1.27×10^{-1} g/mg h^{-1}	5.28 mg/g h^{-1}	
Activated alumina	Pb(II)	5	5	10	303	1.23×10^{-1} g/mg min^{-1}	0.496 mg/g min^{-1}	[60]
	Pb(II)	5	5	25	303	4.89×10^{-2} g/mg min^{-1}	1.15 mg/g min^{-1}	
	Pb(II)	5	5	50	303	2.34×10^{-2} g/mg min^{-1}	2.04 mg/g min^{-1}	
	Cd(II)	5	5	10	303	1.96×10^{-1}g/mg min^{-1}	0.35 mg/g min^{-1}	
	Cd(II)	5	5	25	303	7.73×10^{-2} g/mg min^{-1}	0.81 mg/g min^{-1}	
	Cd(II)	5	5	50	303	3.88×10^{-1} g/mg min^{-1}	1.53 mg/g min^{-1}	
Fertilizer industry waste material	Cr(VI)	4	2	50	303	4.21×10^{-2} g/mg min^{-1}		[62]
	Cr(VI)	4	2	100	303	4.95×10^{-1} g/mg min^{-1}		
Montmorillonite	Cd(II)	2	7	50	303	3.1×10^{-2} g/mg min^{-1}		[63]
	Co(II)	2	7	50	303	4.7×10^{-2} g/mg min^{-1}		
	Cu(II)	2	7	50	303	1.54×10^{-1} g/mg min^{-1}		
	Pb(II)	2	7	50	303	8.4×10^{-2} g/mg min^{-1}		
	Ni(II)	2	7	50	303	5.3×10^{-2} g/mg min^{-1}		
Acid-activated Montmorillonite	Cd(II)	2	7	50	303	1.11×10^{-1} g/mg min^{-1}		[63]
	Co(II)	2	7	50	303	5.4×10^{-1} g/mg min^{-1}		
	Cu(II)	2	7	50	303	1.59×10^{-1} g/mg min^{-1}		
	Pb(II)	2	7	50	303	1.13×10^{-1} g/mg min^{-1}		
	Ni(II)	2	7	50	303	5.50×10^{-2} g/mg min^{-1}		
Untreated alga	Ni(II)	0.7	5	100	298	7.15×10^{-4} g/mg min^{-1}		[64]
	Ni(II)	0.7	5	200	298	3.91×10^{-4} g/mg min^{-1}		
Acid-treated alga	Ni(II)	0.7	5	100	298	8.27×10^{-4} g/mg min^{-1}		
	Ni(II)	0.7	5	200	298	4.15×10^{-4} g/mg min^{-1}		

(Continued)

TABLE B-2 (*Continued*)

List of Constants of First-Order Reaction Kinetics

Adsorbent	Adsorbate	Dosage (g/L)	pH	Initial Concentration (mg/L)	T (K)	k_2	H	References
Aminated mesoporous alumina	Cu (II)	3	5.8	10	298	5.59×10^{-2} g/mg min^{-1}	2.43 mg/g min^{-1}	[66]
	Cu (II)	3	5.8	15	298	4.50×10^{-2} g/mg min^{-1}	2.83 mg/g min^{-1}	
	Cu (II)	3	5.8	20	298	3.64×10^{-2} g/mg min^{-1}	1.90 mg/g min^{-1}	
Protonated mesoporous alumina	Cu (II)	2	5.8	10	298	2.06×10^{-2} g/mg min^{-1}	2.07 mg/g min^{-1}	[66]
	Cu (II)	2	5.8	15	298	9.50×10^{-3} g/mg min^{-1}	2.01 mg/g min^{-1}	
	Cu (II)	2	5.8	20	298	5.59×10^{-2} g/mg min^{-1}	2.43mg/g min^{-1}	
Activated carbon	Ur(VI)	1	3	100	293	4.09×10^{-3} g/mg min^{-1}	2.66 mg/g min^{-1}	[69]
	Ur(VI)	1	3	150	293	1.64×10^{-3} g/mg min^{-1}	1.08 mg/g min^{-1}	
	Ur(VI)	1	3	200	293	1.21×10^{-3} g/mg min^{-1}	0.54 mg/g min^{-1}	
	Ur(VI)	1	3	300	293	1.56×10^{-4} g/mg min^{-1}	0.11 mg/g min^{-1}	
Activated carbon prepared from Polygonum orientale Linn	Pb(II)	0.6	5	40	298	7.22×10^{-2} g/mg min^{-1}	243.9 mg/g min^{-1}	[72]
	Pb(II)	0.6	5	60	298	4.12×10^{-2} g/mg min^{-1}	322.6 mg/g min^{-1}	
	Pb(II)	0.6	5	80	298	3.31×10^{-2} g/mg min^{-1}	344.8 mg/g min^{-1}	
Hazelnut shell activated carbon	Cu(II)	3	6	25	293	1.21×10^{-2} g/mg min^{-1}	0.964 mg/g min^{-1}	[73]
	Cu(II)	3	6	50	293	1.8×10^{-3} g/mg min^{-1}	0.493 mg/g min^{-1}	
	Cu(II)	3	6	100	293	2.0×10^{-4} g/mg min^{-1}	0.219 mg/g min^{-1}	
	Cu(II)	3	6	200	293	8.0×10^{-5} g/mg min^{-1}	0.117 mg/g min^{-1}	
	Cu(II)	0.5	6	50	293	3.0×10^{-5} g/mg min^{-1}	0.076 mg/g min^{-1}	
	Cu(II)	1.5	6	50	293	6.0×10^{-5} g/mg min^{-1}	0.108 mg/g min^{-1}	
	Cu(II)	2.0	6	50	293	3.1×10^{-4} g/mg min^{-1}	0.176 mg/g min^{-1}	
NiO nanoparticle	Cd(II)	0.5	6	100	303	1.78×10^{-3} g/mg min^{-1}		[74]

	Cu(II)	0.5	6	100	303	7.13×10^{-3} g/mg min^{-1}		
Zinc oxide nanoparticles	Zn(II)	0.5	5.5	100	303	1.87×10^{-3} g/mg min^{-1}		[75]
	Cd(II)	0.5	5.5	100	303	2.69×10^{-3} g/mg min^{-1}		
	Hg(II)	0.5	5.5	100	303	2.53×10^{-3} g/mg min^{-1}		
Zirconia nanoparticle	As(III)	0.2	8-9	37.5	298	0.07 g/mmol h^{-1}		[76]
Herbaceous peat	Cu (II)	1	5.5	19.05	294	2.8 g/mg min^{-1}	9.58 mg/g min^{-1}	[77]
	Cu (II)	1	5.5	28.575	294	1.22 g/mg min^{-1}	9.09 mg/g min^{-1}	
	Cu (II)	1	5.5	38.1	294	0.47 g/mg min^{-1}	5.88 mg/g min^{-1}	
Chitosan	Cu(II)	1		59.52	303	6.93×10^{-4} g/mg min^{-1}		[78]
	Cu(II)	1		89.6	303	4.08×10^{-4} g/mg min^{-1}		
	Cu(II)	1		124.8	303	3.31×10^{-4} g/mg min^{-1}		
	Cu(II)	1		186.88	303	2.27×10^{-4} g/mg min^{-1}		
Carbon nanotubes	Pb(II)	0.5	5	10	298	9.20×10^{-3} g/mg min^{-1}		[79]
	Pb(II)	0.5	5	20	298	1.16×10^{-2} g/mg min^{-1}		
	Pb(II)	0.5	5	30	298	5.30×10^{-3} g/mg min^{-1}		
Sphagnum moss peat	Cu(II)	4	5	20	298	9.64×10^{-2} g/mg min^{-1}	3.19 mg/g min^{-1}	[80]
	Cu(II)	4	5	50	298	1.68×10^{-2} g/mg min^{-1}	1.32 mg/g min^{-1}	
	Cu(II)	4	5	100	298	1.13×10^{-2} g/mg min^{-1}	1.73 mg/g min^{-1}	
	Cu(II)	4	5	200	298	8.91×10^{-3} g/mg min^{-1}	1.81 mg/g min^{-1}	
	Ni(II)	4	5	10	298	1.75×10^{-2} g/mg min^{-1}	0.804 mg/g min^{-1}	
	Ni(II)	4	5	25	298	4.27×10^{-2} g/mg min^{-1}	0.643 mg/g min^{-1}	
	Ni(II)	4	5	50	298	2.64×10^{-2} g/mg min^{-1}	1.01 mg/g min^{-1}	
	Ni(II)	4	5	100	298	1.59×10^{-2} g/mg min^{-1}	0.905 mg/g min^{-1}	
	Pb(II)	4	5	35	298	9.19×10^{-1} g/mg min^{-1}	17.3 mg/g min^{-1}	
	Pb(II)	4	5	50	298	3.35×10^{-1} g/mg min^{-1}	12.7 mg/g min^{-1}	
	Pb(II)	4	5	100	298	6.98×10^{-2} g/mg min^{-1}	10.6 mg/g min^{-1}	
	Pb(II)	4	5	210	298	1.79×10^{-2} g/mg min^{-1}	10.9 mg/g min^{-1}	

B-2 Double-exponential Model

Michard et al. proposed a diffusion controlled mechanism for two-step adsorption kinetics.[81] During the first step, a rapid metal uptake takes place, which involves external and internal diffusion. Subsequently, a slow step prevails: the intraparticle diffusion controls the adsorption rate and finally the metal uptake reaches its equilibrium.

Different interpretations were proposed.[82,83] In the first phase, the sorbate molecules bind with the active sites on the adsorbent surface where the sorbate-sorbate interaction is negligible, leading to the formation of a monolayer. As this monolayer approaches the saturation, the rearrangement may start with a further increase in the sorbate molecules, which constitutes the second step.

The following double-exponential expression is used to describe the two-step mechanism.[84]

$$q = q_e - \frac{D_1}{m} e^{-K_{D_1} t} - \frac{D_2}{m} e^{-K_{D_2} t} \tag{b-4}$$

where m is the mass of adsorbent, D_1 and D_2 are adsorption rate parameters (mmol L^{-1}) of the rapid and the slow step, respectively, and K_{D_1} and K_{D_2} are parameters (min^{-1}) controlling the mechanism.

If the exponential term corresponding to the rapid process is assumed to be negligible on the overall kinetics (K_{D_1} is much greater than K_{D_2}), Equation (b-3) can be simplified, which allow the determination of D_2 and K_{D_2} as follows:

$$q = q_e - \frac{D_2}{m} e^{-K_{D_2} t} \tag{b-5}$$

Wilczak and Keinath correlated the two-step kinetics of the adsorption of lead and copper onto activated carbon using this model.[84] It is noted that it can also describe a process where the adsorbent offers two different types of adsorption sites.

B-3 Simplified Intraparticle Diffusion Model

In most cases, adsorption rate in batch is controlled by intraparticle diffusion of adsorbates. A simplified intraparticle diffusion model was proposed by Weber and Morris.[85] The fractional approach to equilibrium changes according to a function of $(Dt/r_p^2)^{1/2}$, where r_p is the particle radius, and D is the diffusivity of solutes within the particle. The rate of intraparticle diffusion can be simply expressed as following:

$$q = k_{id}\, t^{1/2} + a \tag{b-6}$$

where k_{id} is the rate constant for intraparticle diffusion. The intercept of the plot, a, reflects the boundary layer effect. The larger the intercept, the greater is the contribution of the surface sorption in the rate-limiting step.[85,86]

Fitting the experimental data in an intraparticle diffusion plot is a most commonly used technique for identifying the mechanism in the adsorption process. If the plot of q versus $t^{1/2}$ gives a straight line, the adsorption process is suggested to be controlled by intraparticle diffusion only. If the data exhibit multi-linear plots, then two or more steps influence the adsorption process. A good linearity indicates that the intraparticle diffusion is the rate-controlling step.

Appendix C: Surface Diffusion Model

Major assumptions in the kinetics modeling

1. The sorbent is spherical.
2. Surface diffusion is predominant in adsorption kinetics, and is independent of concentration.

Mass balance of species to be adsorbed on a spherical sorbent

$$\text{Mass in-Mass out} = \text{Accumulation} \tag{c-1}$$

The general formula of Fick's first law is expressed as follows:

$$F = -D\frac{\partial c}{\partial x} \tag{c-2}$$

The mass balance can be established based on Figure C-1 shown above, leading to:

$$\begin{aligned} &-4\pi(r+\Delta r)^2 \Delta t D_s \rho_a \left.\frac{\partial q(r+\Delta r,t)}{\partial r}\right|_{r+\Delta r} - \left(-4\pi r^2 \Delta t D_s \rho_a \left.\frac{\partial q(r,t)}{\partial r}\right|_r\right) \\ &= -[4\pi r^2 \Delta r \rho_a q(r,t+\Delta t) - 4\pi r^2 \Delta r \rho_a q(r,t)] \end{aligned} \tag{c-3}$$

where

D_s = surface diffusion coefficient

q = surface concentration (adsorption capacity), $q\ (r, t)$

r = radius distance measured from the center

t = time

ρ_a = apparent density of sorbent

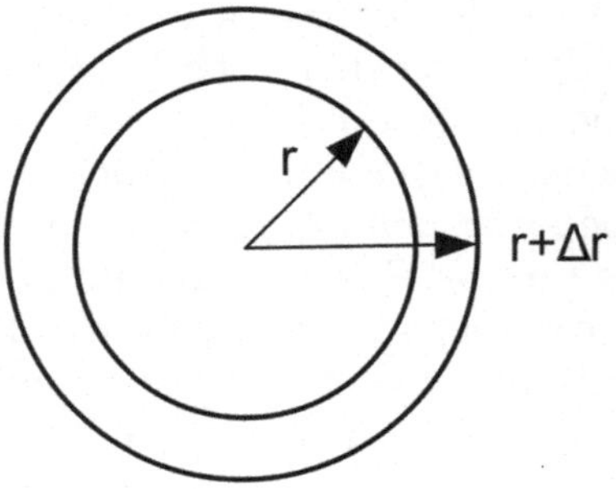

FIGURE C-1
Illustration of spherical sorbent for modeling of adsorption kinetics.

Notice that Δr is negative, as we have: $r - (r+\Delta r) = -\Delta r$

Equation (c-3) is divided by $4\pi\rho_a\Delta r\Delta t$, which yields to:

$$\frac{(r+\Delta r)^2 D_s \left.\frac{\partial q(r+\Delta r,t)}{\partial r}\right|_{r+\Delta r} - r^2 D_s \left.\frac{\partial q(r,t)}{\partial r}\right|_{r}}{\Delta r} = r^2 \frac{q(r,t+\Delta t)-q(r,t)}{\Delta t} \quad \text{(c-4)}$$

Let Δt, $\Delta r \to 0$

$$D_s \frac{\partial}{\partial r}\left(r^2 \frac{\partial q}{\partial r}\right) = r^2 \frac{\partial q}{\partial t} \quad \text{(c-5)}$$

or

$$\frac{\partial q}{\partial t} = \frac{D_s}{r^2}\frac{\partial}{\partial r}\left(r^2 \frac{\partial q}{\partial r}\right) \quad \text{(c-6)}$$

Equation (c-6) is the homogeneous solid diffusion equation accounting for mass transfer within the particle (2nd order, linear).

The initial condition is as follows:

$$q(t=0, 0 \le r \le r_p) = 0 \quad \text{(c-7)}$$

Boundary conditions

$$\frac{\partial}{\partial r} q(r=0, t \ge 0) = 0 \quad \text{(c-8)}$$

$$\frac{r_p^2}{\rho_a}(C - c_s) = \frac{\partial}{\partial t}\int_0^{a_p} q r^2\, dr, \qquad t \ge 0, r = r_p \quad \text{(c-9)}$$

where:
k_f = external mass transfer coefficient or film transfer coefficient
C = concentration of adsorbate in solution
c_s = concentration of adsorbate at external surface of particle
r_p = particle radius

The second boundary condition states that the rate of mass transfer through the film equals the rate of change in average surface concentration and is derived from the following mass balance:

$$4\pi r_p^2 k_f (C - c_s) = \frac{\partial M}{\partial t} = \frac{\partial}{\partial t}\int_0^{a_p} 4\pi r^2 \rho_a q \, dr \qquad \text{(c-10)}$$

Mass balance on batch reactor

$$\frac{dC}{dt} = \frac{3Mk_f}{V\varepsilon r_p \rho_a}(C - c_s) \qquad \text{(c-11)}$$

where
ε = porosity of batch adsorber
V = solution volume
M = weight of adsorbent in reactor

It states that the change in bulk solution concentration C with time equals the amount of adsorbed.

Among the various models, the diffusion-controlled models are often used for activated carbon and biopolymers. Most of the diffusion-controlled models discussed in the literature use a simplified expression (e.g., the Freundlich and Langmuir models) to describe the local equilibrium relationship.[20]

References

1. Stumm, W., Sigg, L., and Sulzberger, B. *Chemistry of the Solid-Water Interface: Processes at the Mineral-Water and Particle-Water Interface in Natural Systems.* John Wiley & Sons, New York, 1992.
2. Morel, F., and Hering, J. G. *Principles and Applications of Aquatic Chemistry.* John Wiley & Sons, New York, 1983.
3. Snoeylink, V. L., and Jenkins, D. *Water Chemistry.* John Wiley & Sons, New York, 1980.
4. Benjamin, M. M. *Water Chemistry.* McGraw-Hill, New York, 2001.
5. Yiacoumi, S., and Tien, C. *Kinetics of Metal Adsorption from Aqueous Solutions.* Springer-Verlag, New York, 1993.

6. Brown, D. S., and Allison, J. D. MINTEQA1, An Equilibrium Metal Speciation Model: User's Manual, EPA/600/3-87/012, United States Environmental Protection Agency, Athens, Georgia, Oct 1987.
7. Herbelin, A. L., and Westall, J. C. FITEQL 4.0: A computer program for determination of chemical equilibrium constants from experimental data; Report 99-01. Department of Chemistry, Oregon State University, Corvallis, 1999.
8. Werner, S., and Morgan, J. J. An introduction emphasizing chemical equilibria in natural waters.In *Aquatic Chemistry*, 2nd ed. John Wiley & Sons, New York, 1981.
9. Dzombak, D. A., and Morel, F. M. M. *Surface Complexation Modeling: Hydrous Ferric Oxide*. Wiley-Interscience, New York, 1990.
10. Stumm, W., and Morgan, J. J. Chemical equilibria and rates in natural waters. In *Aquatic Chemistry*, 3rd ed. John Wiley & Sons, New York, 1996.
11. Browski, A. D. Adsorption—from theory to practice. *Adv. Colloid Interface Sci.*, 93, 135–224, 2001.
12. Langmuir, I. The adsorption of gases on plane surfaces of glass, mica and platinum. *J. Am. Chem. Soc.*, 40, 1361–1403, 1918.
13. Langmuir, I. The constitution and fundamental properties of solids and liquids: Part I. Solids. *J. Am. Chem. Soc.*, 38, 2219–2576, 1916.
14. Freundlich, H. Over the adsorption in solution. *J. Phys. Chem.*, 57, 385–470, 1906.
15. Freundlich, H. *Colloid and Capillary Chemistry*. Methuen, London 1926.
16. Afzal, M., Hasany, S. M., Ahmad, H., and Mahmood, F. Adsorption studies of cerium on lead dioxide from aqueous solution. *J. Radioanal. Nucl. Chem.*, 170, 309–319, 1993.
17. Redilich, O., Peterson, D. L., and Peterson, D. L. A useful adsorption isotherm. *J. Phys. Chem.*, 63, 1024, 1959.
18. Allen, S. J., Mckay, G., and Porter, J. F. Adsorption isotherm models for basic dye adsorption by peat in single and binary component systems. *J. Colloid Interface Sci.*, 280, 332–333, 2004.
19. Toth, J. State equations of the solid-gas interface layers. *Acta. Chim. Hung*, 69, 311–328, 1971.
20. Tien, C. *Adsorption Calculations and Modeling*. Butterworth-Heinemann, Boston, MA, 1994.
21. Radke, C. J., and Prausnitz, J. M. Thermodynamics of multi-solute adsorption from dilute liquid solutions. *J. Am. Inst. Chem. Engrs.*, 18, 761–768, 1972.
22. Ko, D. C., Cheung, C. W., Choy, K. K., Porter, J. F., and McKay, G. Sorption equilibria of metal ions on bone char. *Chemosphere*, 54, 273–281, 2004.
23. Schiewer, S., and Volesky, B. Modeling of the proton-metal ion exchange in biosorption. *Environ. Sci. Technol.*, 29, 3049–3058, 1995.
24. Chen, J. P., Hong, L., Wu, S., and Wang, L. Elucidation of interactions between metal ions and Ca Alginate-based ion-exchange resin by spectroscopic analysis and modeling simulation. *Langmuir.*, 18, 9413–9421, 2002.
25. Lim, S. F., Zheng, Y. M., Zou, S. W., and Chen, J. P. Characterization of copper adsorption onto an alginate encapsulated magnetic sorbent by a combined FT-IR, XPS, and mathematical modeling study. *Environ. Sci. Technol.*, 42, 2551–2556, 2008.
26. Chen, J. P., and Yang, L. Study of a heavy metal biosorption onto raw and chemically modified *Sargassum* sp. via spectroscopic and modeling analysis. *Langmuir.*, 22, 8906–8914, 2006.

27. Westall, J., and Hohl, H. A comparison of electrostatic models for the oxide/solution interface. *Adv. Colloid Interface Sci.*, 12, 265–294, 1980.
28. Hayes, K. F., and Leckie, J. O. Modeling ionic strength effects on cation adsorption at hydrous oxide/solution interfaces. *J. Colloid Interface Sci.*, 115, 564–572, 1987.
29. Hayes, K. F., Papelis, C., and Leckie, J. O. Modeling ionic strength effects on anion adsorption at hydrous oxide/solution interfaces. *J. Colloid Interface Sci.*, 125, 717–726, 1987.
30. Reed, B. E., and Matsumoto, M. R. Modeling heavy metal adsorption in a multi-adsorbent (powdered activated carbon) system. National Conference on Environmental Engineering, American Society of Civil Engineers, 420–427, 1990.
31. Reed, B. E., and Matsumoto, M. R. Modeling Cd adsorption in single and binary adsorbent (PAC) systems. *J. Environ. Eng.*, 119, 332–348, 1993.
32. Reed, B. E., Jensen, J. N., and Matsumoto, M. R., Acid-base characteristics of powdered-activated-carbon surfaces. *J. Environ. Eng.*, 119, 585–590.
33. Benjamin, M. M., and Leckie, J. O. Multiple-site adsorption of Cd, Cu, Zn, and Pb on amorphous iron oxyhydroxide. *J. Colloid Interface Sci.*, 79, 209–221, 1981.
34. Benjamin, M. M., and Leckie, J. O. Competitive adsorption of Cd, Cu, Zn, and Pb on amorphous iron oxyhydroxide. *J. Colloid Interface Sci.*, 83, 410–419, 1981.
35. Athanasiadis, K., and Helmreich, B. Influence of chemical conditioning on the ion exchange capacity and on kinetic of zinc uptake by clinoptilolite. *Water Res.* 39, 1527–1532, 2005.
36. Helfferich, F., *Ion exchange*, McGraw-Hill, New York, 1962.
37. Seader J. D., and Henley, E. J. *Separation Process Principles*. John Wiley & Sons, New York, 1998.
38. Chen, J. P., Yang, L., Ng, W. J., and Wang, L. K. P. Ion exchange process, advanced physicochemical treatment processes. *Handbook of Environmental Engineering Series*, 4(8), 257–288, 2006.
39. Schecher, W. D., and McAvoy, D. C. *MINEQL+ Chemical Equilibrium Modeling System*. Version 4.5 for Windows. Environmental Research Software. Hallowell, ME, 2001.
40. Clifford, D. A. *Ion Exchange and Inorganic Adsorption, in Water Quality and Treatment*, 5th ed., D. Letterman (Ed.). McGraw-Hill, New York, 1999.
41. Anderson, R. E. Ion-exchange separation. In *Handbook of Separation Techniques for Chemical Engineers*, 3rd ed., P. A. Schweitzer (Ed.). McGraw-Hill, New York, 1996.
42. Westall, J. C., Zachary, J. L., and Morel, F. M. M. MINEQL: A Computer Program for the Calculation of Chemical Equilibrium Composition of Aqueous Systems. Technical Note No. 18, Massachusetts Institute of Technology, Cambridge, MA, 1976.
43. Biegler, T. Improved infeasible path optimization for sequential modular simulators: 1. The interface. *Comput. Chem. Eng.*, 9, 245–256, 1985.
44. Chen, J. P., and Lin, M. S. Surface charge and multiple metal ions adsorption on an H-type activated carbon: Experimental observation and modeling simulation by the surface complex formation approach. *Carbon*, 39, 1491–1504, 2001.
45. Lim, S. F., Zheng, Y. M., and Chen, J. P. Removal of copper by calcium alginate encapsulated magnetic sorbent. *Chem. Eng. J.*, 152, 509–513, 2009.
46. Lim, S. F., Zheng, Y. M., and Chen, J. P. Organic arsenic adsorption onto a magnetic sorbent. *Langmuir*, 25, 4973–4978, 2009.

47. Lim, S. F., Zheng, Y. M., Zou, S. W., and Chen, J. P. Uptake of arsenate by an alginate-encapsulated magnetic sorbent: Process performance and characterization of adsorption chemistry. *J. Colloid Interface Sci.*, 333, 33–39, 2009.
48. Schiewer, S., and Volesky, B. Modeling of proton-metal ion exchange in biosorption. *Environ. Sci. Technol.*, 29, 3049–3058, 1995.
49. Haug, A. The affinity of some divalent metals to different types of alginates. *Acta Chem. Scand.*, 15, 1794–1795, 1961.
50. Slater, M. J. *The Principle of Ion Exchange Technology*. Butterworth-Heinemann, Boston, MA, 1991.
51. Chen, J. P., Wang, L., and Zou, S. W. Determination of lead biosorption properties by experimental and modeling simulation study. *Chem. Eng. J.*, 131, 209–215, 2007.
52. Chen, J. P., and Wang, L. Characterization of metal adsorption kinetic properties in batch and fixed-bed reactors. *Chemosphere*, 54, 397–404, 2004.
53. Kalavathy, M., Karthikeyan, T., Rajgopal, S., Miranda, L. Kinetic and isotherm studies of Cu (II) adsorption onto H3PO4-activated rubber wood sawdust. *J. Colloid Interface Sci.*, 292, 354–362, 2005.
54. Duranoğlu, D., Trochimczuk, A. W., Beker, U. Kinetics and thermodynamics of hexavalent chromium adsorption onto activated carbon derived from acrylonitrile-divinylbenzene copolymer. *Chem. Eng. J.*, 187, 193–202, 2012.
55. Chen, H., Wang, A. Kinetic and isothermal studies of lead ion adsorption onto palygorskite clay. *J. Colloid Interface Sci.*, 307, 309–316, 2007.
56. Justi, K., Fávere, V., Laranjeira, M., Neves, A., Peralta, R. Kinetics and equilibrium adsorption of Cu (II), Cd (II), and Ni (II) ions by chitosan functionalized with 2 [-bis-(pyridylmethyl) aminomethyl]-4-methyl-6-formylphenol. *J. Colloid Interface Sci.*, 291, 369–374, 2005.
57. Sekar, M., Sakthi, V., Rengaraj, S. Kinetics and equilibrium adsorption study of lead (II) onto activated carbon prepared from coconut shell. *J. Colloid Interface Sci.*, 279, 307–313, 2004.
58. Li, J., Lin, Q., Zhang, X., Yan, Y. Kinetic parameters and mechanisms of the batch biosorption of Cr (VI) and Cr (III) onto Leersia hexandra Swartz biomass. *J. Colloid Interface Sci.*, 333, 71–77, 2009.
59. AL-Othman, Z. A., Ali, R., Naushad, M. Hexavalent chromium removal from aqueous medium by activated carbon prepared from peanut shell: Adsorption kinetics, equilibrium and thermodynamic studies. *Chem. Eng. J.*, 184, 238–247, 2012.
60. Naiya, T., Bhattacharya, A., Das, S. Adsorption of Cd (II) and Pb (II) from aqueous solutions on activated alumina. *J. Colloid Interface Sci.*, 333, 14–26, 2009.
61. Gupta, S., Bhattacharyya, K. Adsorption of Ni (II) on clays. *J. Colloid Interface Sci.*, 295, 21–32, 2006.
62. Gupta, V., Rastogi, A., Nayak, A. Adsorption studies on the removal of hexavalent chromium from aqueous solution using a low cost fertilizer industry waste material. *J. Colloid Interface Sci.*, 342, 135–141, 2010.
63. Bhattacharyya, K., Gupta, S. Adsorptive accumulation of Cd (II), Co (II), Cu (II), Pb (II), and Ni (II) from water on montmorillonite: Influence of acid activation. *J. Colloid Interface Sci.*, 310, 411–424, 2007.
64. Gupta, V., Rastogi, A., Nayak, A. Biosorption of nickel onto treated alga (Oedogonium hatei): Application of isotherm and kinetic models. *J. Colloid Interface Sci.*, 342, 533–539, 2009.

65. Cestari, A., Vieira, E., Lopes, E., da Silva, R. Kinetics and equilibrium parameters of Hg (II) adsorption on silica-dithizone. *J. Colloid Interface Sci.*, 272, 271–276, 2004.
66. Rengaraj, S., Kim, Y., Joo, C., Yi, J. Removal of copper from aqueous solution by aminated and protonated mesoporous aluminas: kinetics and equilibrium. *J. Colloid Interface Sci.*, 273, 14–21, 2004.
67. Orozco-Guareño, E., Santiago-Gutiérrez, F., Morán-Quiroz, J. L., Hernandez-Olmos, S. L., Soto, V., Cruzc, W., Manríquezd, R., Gomez-Salazar, S. Removal of Cu(II) ions from Aqueous Streams Using Poly(Acrylic Acid-co-Acrylamide) Hydrogels. *J. Colloid Interface Sci.*, 349, 583–593, 2010.
68. Anirudhan, T., Radhakrishnan, P. Chromium (III) removal from water and wastewater using a carboxylate-functionalized cation exchanger prepared from a lignocellulosic residue. *J. Colloid Interface Sci.*, 316, 268–276, 2007.
69. Mellah, A., Chegrouche, S., Barkat, M. The removal of uranium (VI) from aqueous solutions onto activated carbon: Kinetic and thermodynamic investigations. *J. Colloid Interface Sci.*, 296, 434–441, 2006.
70. Sheha, R. Sorption behavior of Zn (II) ions on synthesized hydroxyapatites. *J. Colloid Interface Sci.*, 310, 18–26, 2007.
71. Li, Y., Liu, C., Chiou, C. Adsorption of Cr (III) from wastewater by wine processing waste sludge. *J. Colloid Interface Sci.*, 273, 95–101, 2004.
72. Wang, L., Zhang, J., Zhao, R., Li, Y., Li, C., Zhang, C. L. Adsorption of Pb(II) on activated carbon prepared from Polygonum orientale Linn.: Kinetics, isotherms, pH, and ionic strength studies. *Bioresour. Technol.*, 101, 5808–5814, 2010.
73. Demirbas, E., Dizge, N., Sulak, M. T. Kobya, M. Adsorption kinetics and equilibrium of copper from aqueous solutions using hazelnut shell activated carbon. *Chem. Eng. J.*, 148, 480–487, 2009.
74. Sheela, T., Nayaka, Y. A. Kinetics and thermodynamics of cadmium and lead ions adsorption on NiO nanoparticles *Chem. Eng. J.*, 191, 123–131, 2012.
75. Sheela, T., Nayaka, Y. A. Viswanatha, R. Kinetics and thermodynamics studies on the adsorption of Zn(II), Cd(II) and Hg(II) from aqueous solution using zinc oxide nanoparticles. *Powder Technol.*, 217, 163–170, 2012.
76. Zheng, Y. M., Yu, L., Wu, D., Chen, J. P. Removal of arsenite from aqueous solution by a zirconia nanoparticle. *Chem. Eng. J.*, 188, 15–22, 2012.
77. Gündoan, R., Acemiolu, B., Alma, M. Copper (II) adsorption from aqueous solution by herbaceous peat. *J. Colloid Interface Sci.*, 269, 303–309, 2004.
78. Wu, F., Tseng, R., Juang, R., Kinetic modeling of liquid-phase adsorption of reactive dyes and metal ions on chitosan. *Water Res.*, 35, 613–618, 2001.
79. Li, Y., Di, Z., Ding, J., Wu, D., Luan, Z., Zhu, Y.Q., Adsorption thermodynamic, kinetic and desorption studies of Pb^{2+} on carbon nanotubes. *Water Res.*, 39, 605–609, 2005.
80. Ho, Y.S., McKay, G., (2000) The kinetics of sorption of divalent metal ions onto sphagnum moss peat. *Water Res.*, 34, 735–742, 2000.
81. Michard, P., Guibal, E., Vincent, T. and Le Cloirec, P. Sorption and desorption of uranyl ions by silica gel: pH, particle size and porosity effects, Microporous Mater. 5, 309–24, 1996.
82. Varshney, K. G., Khan, A. A., Gupta, U. and Maheshwari, S. M. Kinetics of adsorption of phosphamidon on antimony(V) phosphate cation-exchange-evaluation of the order of reaction and some physical parameters, Physicochem. *Eng. Aspects*. 113, 19–23, 1996.

83. Sarkar, D. and Chattoraj, D. K. Activation parameters for kinetics of protein adsorption at silica/water interface, *J. Colloid Interf. Sci.*, 157, 219–226, 1993.
84. Wilczak, A. and Keinath, T. M. Kinetics of sorption and desorption of copper (II) and lead (II) on activated carbon, *Water Environ. Res.*, 65, 238–244, 1993.
85. Weber, W. J. and Morris, J. C. Kinetics of adsorption on carbon from solution, *J. Sanit. Eng. Div.*, ASCE 89 (SA2), 31–39, 1963.
86. Lo, S. L. and Leckie, J. O. Internal mass transport limited adsorption of cadmium on amorphous aluminum oxide, *J. Ch. Inst. Eng.*, 12, 579–589, 1989.

Index

D

E

N

O

P